De Gruyter Studies in Mathematics 38

Vassili N. Kolokoltsov

Markov Processes, Semigroups and Generators

De Gruyter

Mathematical Subject Classification 2010: Primary: 60-01, 60-02; Secondary: 60G51, 60G52, 60H10, 60H30, 60G50, 60J25, 60K35, 60F17, 81Q05, 81S40.

ISBN 978-3-11-025010-7
e-ISBN 978-3-11-025011-4
ISSN 0179-0986

Library of Congress Cataloging-in-Publication Data

Kolokol'tsov, V. N. (Vasilii Nikitich)
Markov processes, semigroups, and generators / by Vassili N. Kolokoltsov.
p. cm. – (De Gruyter studies in mathematics ; 38)
Includes bibliographical references and index.
ISBN 978-3-11-025010-7 (alk. paper)
1. Markov processes. 2. Semigroups. 3. Group theory – Generators. I. Title.
QA274.7.K65 2011
519.2'33–dc22
2010050783

Bibliographic information published by the Deutsche Nationalbibliothek

The Deutsche Nationalbibliothek lists this publication in the Deutsche Nationalbibliografie; detailed bibliographic data are available in the Internet at http://dnb.d-nb.de.

Typesetting: Da-TeX Gerd Blumenstein, Leipzig, www.da-tex.de
Printing and binding: Hubert & Co. GmbH & Co. KG, Göttingen
∞ Printed on acid-free paper

Printed in Germany

www.degruyter.com

To the memory of my parents,
Nikita and Svetlana Kolokoltsov,
who helped me much to become a scientist

Preface

Markov processes represent a universal model for a large variety of real life random evolutions. The wide flow of new ideas, tools, methods and applications constantly pours into the ever-growing stream of research on Markov processes that rapidly spreads over new fields of natural and social sciences, creating new streamlined logical paths to its turbulent boundary. Even if a given process is not Markov, it can be often inserted into a larger Markov one (Markovianization procedure) by including the key historic parameters into the state space.

Markov processes are described probabilistically by the distributions on their trajectories (often specified by stochastic differential equations) and analytically by the Markov semigroups that specify the evolution of averages and arise from the solutions to a certain class of integro-differential (or pseudo-differential) equations, which is distinguished by the preservation of the positivity property (probabilities are positive). Thus the whole development stands on two legs: stochastic analysis (with tools such as the martingale problem, stochastic differential equations, convergence of measures on Skorohod spaces), and functional analysis (weighted Sobolev spaces, pseudo-differential operators, operator semigroups, methods of Hilbert and Fock spaces, Fourier analysis).

The aim of the monograph is to give a concise (but systematic and self-contained) exposition of the essentials of Markov processes (highly nontrivial, but conceptually excitingly rich and beautiful), together with recent achievements in their constructions and analysis, stressing specially the interplay between probabilistic and analytic tools. The main point is in the construction and analysis of Markov processes from the 'physical picture' – a formal pre-generator that specifies the corresponding evolutionary equation (here the analysis really meets probability) paying particular attention to the universal models (analytically – general positivity-preserving evolutions), which go above standard cases (e.g. diffusions and jump-type processes).

The introductory Part I is an enlarged version of the one-semester course on Brownian motion and its applications given by the author to the final year mathematics and statistics students of Warwick University. In this course, Browninan motion was studied not only as the simplest continuous random evolution, but as a basic continuous component of complex processes with jumps. Part I contains mostly well-known material, though written and organized with a point of view that anticipates further developments. In some places it provides more general formulations than usual (as with the duality theorem in Section 1.9 or with the Holtzmark distributions in Section 1.5) and new examples (as in Section 2.11).

Part II is based mainly on the author's research. To facilitate the exposition, each chapter of Part II is composed in such a way that it can be read almost independently of others, and it ends with a section containing comments on bibliography and related topics. The main results concern:

(i) various constructions and basic continuity properties of Markov processes, including processes stopped or killed at the boundary (as well as related boundary points classification and sensitivity analysis),

(ii) in particular, heat kernel estimates for stable-like processes,

(iii) limiting processes for position-dependent continuous time random walks (obtained by a random time change from the Markov processes) and related fractional (in time) dynamics,

(iv) the rigorous Feynman path-integral representation for the solutions of the basic equations of quantum mechanics, via jump-type Markov processes.

We also touch upon the theory of stochastic monotonicity, stochastic scattering, stochastic quasi-classical (also called small diffusion) asymptotics, and stochastic control. An important development of the methods discussed here is given by the theory of nonlinear Markov processes (including processes on manifolds) presented in the author's monograph [196]. They are briefly introduced at the end of Chapter 5.

It is worth pointing out the directions of research closely related to the main topic of this book, but not touched here. These are Dirichlet forms, which can be used for constructing Markov processes instead of generators, Mallivin calculus, which is a powerful tool for proving various regularity properties for transition probabilities, log-Sobolev inequalities, designed to systematically analyze the behavior of the processes for large times, and processes on manifolds. There exists an extensive literature on each of these subjects.

The book is meant to become a textbook and a monograph simultaneously, taking more features of the latter as the exposition advances. I include some exercises, their weight being much more sound at the beginning. The exercises are supplied with detailed hints and are meant to be doable with the tools discussed in the book. The exposition is reasonably self-contained, with pre-requisites being just the standard math culture (basic analysis and linear algebra, metric spaces, Hilbert and Banach spaces, Lebesgue integration, elementary probability). We shall start slowly from the prerequisites in probability and stochastic processes, omitting proofs if they are well presented in university text books and not very instructive for our purposes, but stressing ideas and technique that are specially relevant. Streamlined logical paths are followed to the main ideas and tools for the most important models, by-passing wherever possible heavy technicalities (say, by working with Lévy processes instead of general semi-martingales, or with left-continuous processes instead of predictable ones).

A methodological aspect of the presentation consists in often showing various perspectives for key topics and giving several proofs of main results. For example, we begin the analysis of random processes with several constructions of the Brownian motion: 1) via binary subdivisions anticipating the later given Itô approach to constructing Markov evolutions, 2) via tightness of random-walk approximations, anticipating the later given LLN for non-homogeneous random walks, 3) via Hilbert-space methods leading to Wiener chaos that is crucial for various developments, for instance for Malliavin calculus and Feynman path integration, 4) via the Kolmogorov continuity theorem. Similarly, we give two constructions of the Poisson process, several constructions of basic stochastic integrals, several approaches to proving functional CLTs (via tightness of random walks, Skorohod embedding and the analysis of generators). Further on various probabilistic and analytic constructions of the main classes of Markov semigroups are given. Every effort was made to introduce all basic notions in the most clear and transparent way, supplying intuition, developing examples and stressing details and pitfalls that are crucial to grasp its full meaning in the general context of stochastic analysis. Whenever possible, we opt for results with the simplest meaningful formulation and quick direct proof.

As teaching and learning material, the book can be used on various levels and with different objectives. For example, short courses on an introduction to Brownian motion, Lévy and Markov processes, or on probabilistic methods for PDE, can be based on Chapters 2, 3 and 4 respectively, with chosen topics from other parts. Let us stress only that the celebrated Itô's lemma is not included in the monograph (it actually became a common place in the textbooks). More advanced courses with various flavors can be built on part II, devoted, say, to continuous-time random walks, to probabilistic methods for boundary value problems or for the Feynman path integral.

Finally, let me express my gratitude to Professor Niels Jacob from the University of Wales, Swansea, an Editor of the De Gryuter Studies in Mathematics Series, who encouraged me to write this book. I am also most grateful to Professor Nick Bingham from Imperial College, London, for reading the manuscript carefully and making lots of comments that helped me to improve the overall quality immensely.

Coventry, November 2010 V. N. Kolokoltsov

Notations

Numbers and sets

- $a \vee b = \max(a, b)$, $a \wedge b = \min(a, b)$
- $\mathbb{N}$ and $\mathbb{Z}$ are the sets of natural and integer numbers
- $\mathbb{C}^d$ and $\mathbb{R}^d$ are the complex and real d-dimensional spaces, $|x|$ or $\|x\|$ for a vector $x \in \mathbb{R}^d$ denotes its Euclidean norm, (x, y) or xy denotes the scalar product of the vectors $x, y \in \mathbb{R}^d$
- $\mathrm{Re}\, a$ and $\mathrm{Im}\, a$ are the real and imaginary part of a complex number a
- $B_r(x)$ (resp. B_r) is the ball of radius r centered at x (resp. at the origin)
- S^d is the d-dimensional unit sphere in $\mathbb{R}^{d+1}$
- $\mathbb{R}_+$ (resp. $\bar{\mathbb{R}}_+$) is the the set of positive (resp. non-negative) numbers
- $\bar{\Omega}$ and $\partial\Omega$ are the closure and the boundary respectively of the subset Ω in a metric space
- Ω^S is the set of all mappings $S \to \Omega$
- $[x]$ is the integer part of a real number x

Functions

- $B(S)$ (resp. $C(S)$ or $C_b(S)$) for a complete metric space (S, ρ) (usually $S = \mathbb{R}^d$, $\rho(x, y) = \|x - y\|$) is the Banach space of bounded Borel measurable (resp. bounded continuous) functions on S equipped with the sup-norm

$$\|f\| = \sup_{x \in S} |f(x)|$$

- $C_c(S) \subset C(S)$ consists of functions with a compact support
- $C_{\mathrm{Lip}}(S) \subset C(S)$ consists of Lipschitz continuous functions f, i.e. $|f(x) - f(y)| \leq \kappa\rho(x, y)$ with a constant κ; $C_{\mathrm{Lip}}(S)$ is a Banach space under the norm

$$\|f\|_{\mathrm{Lip}} = \sup_x |f(x)| + \sup_{x \neq y} |f(x) - f(y)|/|x - y|$$

- $C_\infty(S) \subset C(S)$ consists of f such that $\lim_{x \to \infty} f(x) = 0$, i.e. $\forall \epsilon\, \exists$ a compact set $K : \sup_{x \notin K} |f(x)| < \epsilon$ (it is a closed subspace of $C(S)$ if S is locally compact)

- $C^k(\mathbb{R}^d)$ or $C_b^k(\mathbb{R}^d)$ (sometimes shortly C^k) is the Banach space of k times continuously differentiable functions with bounded derivatives on $\mathbb{R}^d$ with the norm being the sum of the sup-norms of the function itself and all its partial derivative up to and including order k
- $C_{\text{Lip}}^k(\mathbb{R}^d)$ is the subspace of $C^k(\mathbb{R}^d)$ with all derivative up to and including order k being Lipschitz continuous; it is a Banach space equipped with the norm

$$\|f\|_{C_{\text{Lip}}^k} = \|f\|_{C^{k-1}} + \|f^{(k)}\|_{\text{Lip}}$$

- $C_c^k(\mathbb{R}^d) = C_c(\mathbb{R}^d) \cap C^k(\mathbb{R}^d)$
- $\nabla f = (\nabla_1 f, \dots, \nabla_d f) = (\frac{\partial f}{\partial x_1}, \dots, \frac{\partial f}{\partial x_d})$, $f \in C^1(\mathbb{R}^d)$
- $L^p(\Omega, \mathcal{F}, \mu)$, $p \geq 1$, is the usual Banach space of (the equivalence classes of) measurable functions f on the measure space Ω such that

$$\|f\|_p = \left(\int |f|^p(x)\mu(dx) \right)^{1/p} < \infty$$

- $L^\infty(\Omega, \mathcal{F}, \mathbf{P})$ is the Banach space of (the equivalence classes of) measurable functions f on the measure space Ω with a finite sup-norm

$$\|f\| = \text{ess sup}_{x \in \Omega} |f(x)|$$

- $S(\mathbb{R}^d) = \{f \in C^\infty(\mathbb{R}^d) : \forall k, l \in \mathbb{N}, |x|^k \nabla^l f \in C_\infty(\mathbb{R}^d)\}$ is the Schwartz space of rapidly deceasing functions
- $(f, g) = \int f(x)g(x)\,dx$ denotes the scalar product for functions f, g on $\mathbb{R}^d$ or on a general measure space
- $\mathbf{1}_M$ is the indicator function of a set M (equals one or zero according to whether its argument is in M or otherwise)
- sgn is the sign function taking values $+1, 0, -1$ for positive, vanishing and negative values of the argument respectively
- $f = O(g)$ means $|f| \leq Cg$ for some constant C
- $f = o(g)_{n \to \infty} \iff \lim_{n \to \infty}(f/g) = 0$

Measures

- $\mathcal{M}(S)$ (resp. $\mathcal{P}(S)$) is the set of finite (positive) Borel measures (resp. probability measures) on a metric space S
- $\mathcal{M}^{\text{signed}}(S)$ defines the Banach space of finite signed Borel measures on a metric space S
- $|\nu|$ for a signed measure ν is its (positive) total variation measure
- $(f, \mu) = \int_S f(x)\mu(dx)$ for $f \in C(S)$, $\mu \in \mathcal{M}(S)$

Matrices and linear operators

- A^T is the transpose to a matrix A
- $\mathrm{Ker}(A)$, $\mathrm{Sp}(A)$, $\mathrm{tr}(A)$ are the kernel, spectrum and trace of the operator A
- $\|A\|_B$ is the norm of the operator A in a Banach space B
- $\|A\|_{B\to C}$ is the norm of the operator A as a mapping between Banach spaces B and C
- $C([0,t],B)$ is the Banach space of continuous functions on $[0,t]$ with values in the Banach space B equipped with the sup-norm $\|f\| = \sup_{s\in[0,t]} \|f(s)\|$

Probability

- $\mathbf{E}$ and $\mathbf{P}$ define the expectation and probability, $\mathbf{E}_x, \mathbf{P}_x$ for $x \in S$ (respectively $\mathbf{E}_\mu, \mathbf{P}_\mu$ for $\mu \in \mathcal{P}(S)$) are the expectation and probability with respect to an S-valued process started at x (respectively with the initial distribution μ)

Standard abbreviations

a.s.	almost sure
i.i.d.	independent identically distributed
l.h.s.	left-hand side
r.h.s.	right-hand side
r.v.	random variable
BM	Brownian motion
CLT	central limit theorem
CTRW	continuous time random walk
LLN	law of large numbers
ODE	ordinary differential equation
OU	Ornstein–Uhlenbeck (process)
PDE	partial differential equation
PDO	partial differential operator
ΨDE	pseudo differential equation
ΨDO	pseudo differential operator
SDE	stochastic differential equation

Contents

Part I

Introduction to stochastic analysis

Chapter 1

Tools from probability and analysis

This chapter is meant to supply the preliminary material needed for reading the book. Though we do give most of the proofs (sometimes sketchy), some fundamental facts are only formulated. The criterion used for the omission of the proofs was two-fold. On the one hand, these proofs are not deeply connected with (nor very instructive for the understanding of) the main body of this text, and are non-trivial, so that their proper exposition would be time and space consuming; and on the other hand, they are quite standard by now and are widely represented in university textbooks. To set the ground for probability, we recall the notion of a measure space, but we do assume readers to be acquainted with the definition and basic properties of integrals with respect to an abstract measure including dominated and monotone convergence theorems. In the next three sections we collect the basic facts from standard probability texts, see e.g. Applebaum [19], Jacod and Protter [146], Shiryaev [293] and Kallenberg [154], so that references are not given to each formulated result separately. Afterwards we introduce infinitely divisible and stable distributions. Then we recall the basic topologies used routinely in stochastic analysis. And finally we introduce the analytic tools (fractional derivatives, pseudo-differential operators and semigroups) used in what follows. Readers with a sound background in probability and/or analysis may wish to skip some or all sections of this introductory chapter.

1.1 Essentials of measure and probability

A collection $\mathcal{F}$ of subsets of a given set S is called a *σ-algebra* if

(i) $S \in \mathcal{F}$;

(ii) $A \in \mathcal{F} \Rightarrow S \setminus A \in \mathcal{F}$;

(iii) (σ-additivity) $\bigcup_{n=1}^{\infty} A_n \in \mathcal{F}$ whenever $A_n \in \mathcal{F}$ for any $n \in \mathbb{N}$.

The pair $(S, \mathcal{F})$ is called a *measurable space.*

A *measure* on $(S, \mathcal{F})$ is a mapping $\mu : \mathcal{F} \mapsto [0, \infty]$ such that $\mu(\emptyset) = 0$ and *σ-additivity* holds:

$$\mu\Big(\bigcup_{n=1}^{\infty} A_n\Big) = \sum_{n=1}^{\infty} \mu(A_n)$$

for any sequence A_n of mutually disjoint sets in $\mathcal{F}$. The triple $(S, \mathcal{F}, \mu)$ is called a *measure space.* A measure μ is called *finite* if its *total mass* $\mu(S)$ is finite, *σ-finite* if there exists a sequence A_n, $n \in \mathbb{N}$, of subsets of $\mathcal{F}$ such that $S = \bigcup_{n=1}^{\infty} A_n$ and $\mu(A_n) < \infty$ for all n.

A measure space $(\Omega, \mathcal{F}, \mu)$ is called a *probability space* whenever $\mu(\Omega) = 1$. In this case μ is called a *probability measure* and the subsets from $\mathcal{F}$ are called *events*.

An extension of the notion of a measure that does not assume positivity is sometimes useful as well. Namely a *signed measure* (respectively a *complex measure*) of *finite variation* can be defined as a set function ϕ on a measurable space $(S, \mathcal{F})$ that is given by the integral

$$\phi(A) = \int_S f(x)\mu(dx), \quad A \in \mathcal{F}, \tag{1.1}$$

where μ is a (positive) finite measure on $(S, \mathcal{F})$ and f is a real (respectively complex-valued) function on S integrable with respect to μ.[1] The *total variation norm* of ϕ is defined as

$$\|\phi\| = \int |f(x)|\mu(dx).$$

The set of signed (respectively complex) measures of finite variation on $(S, \mathcal{F})$ is easily seen to be a real (respectively complex) Banach space when equipped with this norm. The *total variation measure* of ϕ is defined as the measure

$$|\phi|(dx) = |f(x)|\mu(dx),$$

so that

$$\phi(dx) = \sigma(x)|\phi|(dx)$$

with σ taking only three values $0, 1, -1$. As for usual measures, the following extension is sometimes useful. A set function ϕ is called a *σ-finite* signed (or complex) measure if it has a representation (1.1) with a σ-finite measure μ and a bounded measurable real (respectively complex) function f.

For a metric space S, e.g. a subset of $\mathbb{R}^d$, the smallest σ-algebra $\mathcal{B}(S)$ containing all its open subsets is called the *Borel σ-algebra* of S. Its elements are called *Borel sets* and any measure on $(S, \mathcal{B}(S))$ is called a *Borel measure*. The simplest example of a Borel measure is given by Lebesgue measure on $\mathbb{R}^d$. A Borel measure is called a *Radon measure* if it is finite on any compact set. One can also define a *signed or complex Radon measure* as a set function that becomes a signed or complex measure of finite variation when reduced to any compact set.

Throughout this book our processes will live in Euclidean spaces $\mathbb{R}^d$. However, the distribution of a $\mathbb{R}^d$-valued process is a distribution on a certain space of trajectories of such a process, and the latter space is often specified as a rather nontrivial infinite-dimensional metric space (Skorohod space). Hence the necessity to work with measures on general metric spaces, even when analyzing finite-dimensional processes.

---

[1] Alternatively signed measures can be defined axiomatically, in which case this representation follows from the so-called Hahn decomposition theorem.

For a collection Γ of the subsets of a set Ω the σ-algebra $\sigma(\Gamma)$ *generated by* Γ is the minimal σ-algebra containing all sets from Γ.

An important method of constructing measures is via the products. Namely, for a finite or a countable family of measure spaces $(S_i, \mathcal{F}_i, \mu_i)$, $i = 1, 2, \dots$, the *product measure space* $(S, \mathcal{F}, \mu)$ is defined, where $S = S_1 \times S_2 \times \cdots$, $\mathcal{F} = \mathcal{F}_1 \otimes \mathcal{F}_2 \otimes \cdots$ – the σ-algebra generated by the sets $A_1 \times \cdots \times A_n$, $A_i \in \mathcal{F}_i$, $n \in \mathbb{N}$, and $\mu = \mu_1 \times \mu_2 \times \cdots$ is the *product measure* uniquely specified by the prescription $\mu(A_1 \times \cdots \times A_n) = \mu_1(A_1) \cdots \mu_n(A_n)$.

For a measure space $(S, \mathcal{F}, \mu)$ a subset of S is called *negligible* or a *null set* if it is a subset of a $N \in \mathcal{F}$ with $\mu(N) = 0$. The σ-algebra $\bar{\mathcal{F}}$ of the subsets of S of the form $A \cup B$, with $A \subset \mathcal{F}$, B negligible and the measure $\bar{\mu}$ on it defined on these sets as $\bar{\mu}(A \cup B) = \mu(A)$, are called respectively the *completion* of $\mathcal{F}$ and μ (with respect to μ). In particular, for $S \subset \mathbb{R}^d$ the completion of $\mathcal{B}(S)$ with respect to Lebesgue measure is called the σ-algebra of *Lebesgue measurable sets* in S.

For a probability space $(\Omega, \mathcal{F}, \mu)$ one says that some property depending on $\omega \in \Omega$ holds *almost surely* (briefly *a.s.*) or *with probability* 1 if there exists a negligible set $N \in \mathcal{F}$ such that this property holds for all $\omega \in \Omega \setminus N$.

A handy tool of probability theory is given by the following famous result called the *Borel–Cantelli lemma.*

Theorem 1.1.1. *If a sequence of events A_n, $n \in \mathbb{N}$, on a probability space $(\Omega, \mathcal{F}, \mathbf{P})$ is such that $\sum_n \mathbf{P}(A_n) < \infty$, then a.s. only a finite number of A_n can occur.*

Proof. Let $B = \{\omega \in \Omega : \text{infinite number } A_n \text{ occur}\}$. Then

$$B = \bigcap_n \Big(\bigcup_{k \geq n} A_k \Big)$$

and

$$\mathbf{P}(B) \leq \mathbf{P}\Big(\bigcup_{k \geq n} A_k \Big) \leq \sum_{k \geq n} \mathbf{P}(A_k) \to 0,$$

as $n \to \infty$. Hence $\mathbf{P}(B) = 0$. □

If $(S_i, \mathcal{F}_i)$, $i = 1, 2$, are measurable spaces, a mapping $f : S_1 \to S_2$ is called *$(\mathcal{F}_1, \mathcal{F}_2)$-measurable* if $f^{-1}(A) \in \mathcal{F}_1$ whenever $A \in \mathcal{F}_2$. Two measurable spaces are called *Borel isomorphic*, or just *isomorphic*, if there exists a bijection $f : S \to T$ such that both f and f^{-1} are measurable. A measurable space S is called a *Borel space* if it is isomorphic to a Borel subset of $[0, 1]$. A deep result of measure theory states that a complete metric space is a Borel space. This result is very convenient, as it allows one to establish certain general facts by proving them only for the real line (see below the randomization lemma). In our book we shall use only Borel spaces and measures.

If S_1, S_2 are metric spaces equipped with their Borel σ-algebras, such a mapping is said to be *Borel measurable* or briefly *Borel*. Speaking about measurable mapping with values in $\mathbb{R}^d$ one usually means that $\mathbb{R}^d$ is equipped with its Borel σ-algebra.

For a probability space $(\Omega, \mathcal{F}, \mathbf{P})$ the measurable mappings $X : \Omega \to \mathbb{R}^d$ are called *random variables* (briefly r.v.), or sometimes random vectors in case $d > 1$. More generally, for a metric space S the Borel measurable mappings $\Omega \to S$ are called *S-valued random variables* or *random elements on S*. The σ-algebra $\sigma(X)$ *generated by a r.v.* X is the smallest σ-algebra containing the sets $\{X \subset B\}$ for all Borel sets B.

The *law* (or the *distribution*) of a random variable is the Borel probability measure p_X on S defined as $p_X = \mathbf{P} \circ X^{-1}$. In other words,

$$p_X(A) = \mathbf{P}(X^{-1}(A)) = \mathbf{P}(\omega \in \Omega : X(\omega) \in A) = \mathbf{P}(X \in A).$$

For example, if X takes only finite number of values, then the law p_X is a sum of Dirac δ-measures.

Clearly, if μ is a probability measure on $\mathbb{R}^d$, then the identical mapping in $\mathbb{R}^d$ defines a $\mathbb{R}^d$-valued random vector with the law μ defined on the probability space $(\mathbb{R}^d, \mathcal{B}, \mu)$. It turns out that for a family of laws depending measurably on a parameter one can specify a family of random variables defined on a single probability space and depending measurably on this parameter. This is shown in the following *randomization lemma*:

Lemma 1.1.1. *Let $\mu(x, dz)$ be a family of probability measures on a Borel space Z depending measurably on a parameter x from another measurable space X (such a family is called a probability kernel from X to Z). Then there exists a measurable function $f : X \times [0,1] \to Z$ such that if θ is uniformly distributed on* $[0,1]$, *then $f(X, \theta)$ has distribution $\mu(x, .)$ for every $x \in X$.*

Proof. Since Z is a Borel space, it is sufficient to prove the statement for $Z = [0,1]$. In this case f can be defined by the explicit formula, the *probability integral transformation*, that represents a standard method (widely used in practical simulations), for obtaining a random variable from a given one-dimensional distribution:

$$f(s,t) = \sup\{x \in [0,1] : \mu(s, [0,x]) < t\}.$$

Clearly this mapping depends measurably on s. Moreover, the events $\{f(s,t) \le y\}$ and $\{t \le \mu(s, [0,y])\}$ coincide. Hence, for a uniform θ

$$\mathbf{P}(f(s,\theta) \le x) = \mathbf{P}(t \le \mu(s, [0,y])) = \mu(s, [0,x]). \qquad \square$$

Two r.v. X and Y are called *identically distributed* if they have the same probability law. For a real (i.e. one-dimensional) r.v. X its *distribution function* is defined by

$F_X(x) = p_X((-\infty, x])$. A real r.v. X has a *continuous distribution* with a *probability density function* f if $p_X(A) = \int_A f(x)dx$ for all Borel sets A.

For an $\mathbb{R}^d$-valued r.v. X on a probability space $(\Omega, \mathcal{F}, \mu)$ and a Borel measurable function $f : \mathbb{R}^d \mapsto \mathbb{R}^m$ the *expectation* $\mathbf{E}$ of $f(X)$ is defined as

$$\mathbf{E}f(X) = \mathbf{E}(f(X)) = \int_\Omega f(X(\omega))\mu(d\omega) = \int_{\mathbb{R}^d} f(x)p_X(dx). \tag{1.2}$$

X is called *integrable* if $\mathbf{E}(|X|) < \infty$.

Exercise 1.1.1. Convince yourself that the two integral expressions in (1.2) really coincide. Hint: first choose f to be an indicator, then use linearity and approximation.

For two $\mathbb{R}^d$-valued r.v. $X = (X_1, \dots, X_d)$ and $Y = (Y_1, \dots, Y_d)$ the $d \times d$-matrix with the entries $\mathbf{E}[(X_i - \mathbf{E}(X_i))(Y_j - \mathbf{E}(Y_j))]$ is called the *covariance* of X and Y and is denoted $\mathrm{Cov}(X, Y)$. In case $d = 1$ and $X = Y$ the number $\mathrm{Cov}(X, Y)$ is called the *variance* of X and is denoted by $\mathrm{Var}(X)$ and sometimes also by σ_X^2. Expectation and variance supply two basic numeric characteristics of a random variable.

The random variables X and Y are called *uncorrelated* whenever $\mathrm{Cov}(X, Y) = 0$. Random variables $X_1, \dots, X_n$ are called *independent* whenever

$$\mathbf{P}(X_1 \in A_1, X_2 \in A_2, \dots, X_n \in A_n) = \mathbf{P}(X_1 \in A_1)\mathbf{P}(X_2 \in A_2)\cdots\mathbf{P}(X_n \in A_n)$$

for all Borel A_j. Clearly in this case $\mathrm{Cov}(X_i, X_j) = 0$ for all $i \neq j$ (i.e. independent variables are uncorrelated) and

$$\mathrm{Var}(X_1 + \cdots + X_n) = \mathrm{Var}(X_1) + \cdots + \mathrm{Var}(X_n). \tag{1.3}$$

As an easy consequence of the definition of the expectation constitute the following inequalities whose importance to the probability analysis is difficult to overestimate.

Theorem 1.1.2. *Markov's inequality: If X is a non-negative random variable, then for any $\epsilon > 0$*

$$\mathbf{P}(X \geq \epsilon) \leq \frac{\mathbf{E}X}{\epsilon}.$$

Chebyshev's inequality: For any $\epsilon > 0$ and a random variable Y

$$\mathbf{P}(|Y - \mathbf{E}Y| > \epsilon) \leq \frac{\mathrm{Var}(Y)}{\epsilon^2}.$$

Jensen's inequality: If g is a convex (respectively concave) function, then

$$g(\mathbf{E}(X)) \leq \mathbf{E}(g(X))$$

(respectively vice versa) whenever X and $g(X)$ are both integrable.

Proof. Evident inequalities

$$\mathbf{E}X \ge \mathbf{E}(X\mathbf{1}_{X\ge\epsilon}) \ge \epsilon\mathbf{E}\mathbf{1}_{X\ge\epsilon} = \epsilon\mathbf{P}(X \ge \epsilon)$$

imply Markov's one. Applying Markov's inequality with $X = |Y - \mathbf{E}Y|^2$ yields Chebyshev's one. Finally, if g is convex, then for any x_0 there exists a $\lambda(x_0)$ such that $g(x) \ge g(x_0) + (x - x_0)\lambda(x_0)$ for all x. Choosing $x_0 = \mathbf{E}X$ and $x = X$ yields

$$g(X) \ge g(\mathbf{E}X) + (X - \mathbf{E}X)\lambda(\mathbf{E}X).$$

Passing to the expectations leads to Jensen's inequality. Concave g are analyzed similarly. □

Exercise 1.1.2. Let X, Y be a random variable and a d-dimensional random vector respectively on a probability space. Show that for a continuous $g : \mathbb{R}^d \to \mathbb{R}$

$$\mathbf{E}(Xg(Y)) = \int_{\mathbb{R}^d} g(y)\nu(dy),$$

where ν is the signed measure $\nu(B) = \mathbf{E}(X\mathbf{1}_B(Y))$. Hint: start with indicator functions g.

A more complicated inequality that we are going to mention here is the following *Kolmogorov's inequality* that states that for the sums $S_m = \xi_1 + \xi_2 + \cdots + \xi_m$ of independent zero mean random variables $\xi_1, \xi_2, \dots$ one has

$$\mathbf{P}(\max_{1\le m\le n} |S_m| > \epsilon) \le \frac{\mathbf{E}|S_n|^2}{\epsilon^2}. \tag{1.4}$$

We shall not prove it here, but we shall establish later on its far-reaching extension: the Doob maximum inequality (notice only that both proofs, as well as other modification like Ottaviani's maximal inequality from Theorem 2.6.2, are based on the same idea of stopping at a point where the maximum is achieved). Doob's maximum inequality implies directly the following more general form of Kolmogorov's inequality (under the same assumptions as in (1.4)):

$$\mathbf{P}(\max_{1\le m\le n} |S_m| > \epsilon) \le \frac{\mathbf{E}|S_n|^p}{\epsilon^p} \tag{1.5}$$

for any $p \in [0, 1]$. In order to appreciate the beauty of this estimate it is worth noting that they give precisely the same estimate for $\max|S_m|$ as one would get for S_n itself via the rough Markov–Chebyshev inequality.

Let us recall now the four basic notions of the convergence of random variables. Let X and X_n, $n \in \mathbb{N}$, be S-valued random variables, where (S, ρ) is a metric space with the distance ρ. One says that X_n converges to X

1. *almost surely* or *with probability* 1 if $\lim_{n\to\infty} X_n(\omega) = X(\omega)$ almost surely;
2. *in probability* if for any $\epsilon > 0$ $\lim_{n\to\infty} \mathbf{P}(\rho(X_n, X) > \epsilon) = 0$;
3. *in distribution* if p_{X_n} *weakly converges* to p_X, i.e. if

$$\lim_{n\to\infty} \int_{\mathbb{R}^d} f(x) p_{X_n}(dx) = \int_{\mathbb{R}^d} f(x) p_X(dx)$$

for all bounded continuous functions f.

If S is $\mathbb{R}^d$, or a Banach space, X is said to converge *in* L^p $(1 \le p < \infty)$ if $\lim_{n\to\infty} \mathbf{E}(|X_n - X|^p) = 0$.

To visualize these notions, let us start with two examples.

1. Consider the following sequence of indicator functions $\{X_n\}$ on $[0,1]$: $\mathbf{1}_{[0,1]}$, $\mathbf{1}_{[0,1/2]}$, $\mathbf{1}_{[1/2,1]}$, $\mathbf{1}_{[0,1/3]}$, $\mathbf{1}_{[1/3,2/3]}$, $\mathbf{1}_{[2/3,1]}$, $\mathbf{1}_{[0,1/4]}$, $\mathbf{1}_{[1/4,2/4]}$, etc. Then $X_n \to 0$ as $n \to \infty$ in probability and in all L^p, $p \ge 1$, but not a.s. In fact $\limsup X_n(x) = 1$ and $\liminf X_n(x) = 0$ for each x so that $X_n(x) \to X(x)$ nowhere.
2. Choosing $X_n = X'$ for all n with X' distributed like X but independent of it, shows that $X_n \to X$ in distribution does not imply in general $X_n - X \to 0$.

The following statement gives instructive criteria for convergence in probability and a.s. and establish the link between them.

Proposition 1.1.1. 1. *$X_n \to X$ in probability if and only if*

$$\lim_{n\to\infty} \mathbf{E}\left(\frac{\rho(X_n, X)}{1+\rho(X_n, X)}\right) = 0 \iff \lim_{n\to\infty} \mathbf{E}(1 \wedge \rho(X_n, X)) = 0. \tag{1.6}$$

2. *$X_n \to X$ a.s. if and only if*

$$\lim_{m\to\infty} \mathbf{P}(\sup_{n\ge m} \rho(X_n, X) > \epsilon) = 0 \tag{1.7}$$

for all $\epsilon > 0$.

3. *Almost sure convergence implies convergence in probability.*
4. *Any sequence converging in probability has a subsequence converging a.s.*

Proof. 1. Convergence in probability follows from (1.6), because by Chebyshev's inequality

$$\mathbf{P}(\rho(X_n, X) > \epsilon) = \mathbf{P}(1 \wedge \rho(X_n, X) > \epsilon) \le \frac{1}{\epsilon}\mathbf{E}(1 \wedge \rho(X_n, X))$$

for $\epsilon \in (0,1)$. The converse statement follows from the inequalities

$$\mathbf{E}\left(\frac{\rho(X_n, X)}{1+\rho(X_n, X)}\right) \le \mathbf{E}(1 \wedge \rho(X_n, X)) \le \epsilon + \mathbf{P}(\rho(X_n, X) > \epsilon).$$

2. The event $X_n \to X$ is the complement of the event

$$B = \bigcup_{r \in \mathbb{Q}} B_r, \quad B_r = \bigcap_{m \in \mathbb{Q}} \{\sup_{n \ge m} |X_n - X| > 1/r\},$$

i.e., a.s. convergence is equivalent to $\mathbf{P}(B) = 0$ and hence to $\mathbf{P}(B_r) = 0$ for all r.

3. This is an obvious consequence of either of statements 1 or 2.

4. If X_n converge in probability, using statement 2, we can choose a subsequence X_k such that

$$\mathbf{E} \sum_k (1 \wedge \rho(X_k, X)) = \sum_k \mathbf{E}(1 \wedge \rho(X_k, X)) < \infty,$$

implying that $\sum_k (1 \wedge \rho(X_k, X)) < \infty$ a.s. and hence $\rho(X_k, X) \to 0$ a.s. □

Proposition 1.1.2. *L^p-convergence $\Rightarrow$ convergence in probability $\Rightarrow$ weak convergence. Finally, weak convergence to a constant implies convergence in probability.*

Proof. The first implication follows from Chebyshev's inequality.

For the second one assume S is $\mathbb{R}^d$. Decompose the integral $\int |f(X_n(\omega)) - f(X(\omega))| \mathbf{P}(d\omega)$ into the sum $I_1 + I_2 + I_3$ of three terms over the sets $\{|X_n - X| > \delta\}$, $\{|X_n - X| \le \delta, |X| > K\}$ and $\{|X_n - X| \le \delta, |X| \le K\}$. First choose K such that $\mathbf{P}(|X| > K + 1) < \epsilon$. Next, by the uniform integrability of f on the ball of radius K (here we use its compactness), choose δ such that $|f(x) - f(y)| < \epsilon$ for $|x - y| < \delta$. By the convergence in probability, choose N such that $\mathbf{P}(|X_n - X| > \delta) < \epsilon$ for $n > N$. Then

$$I_1 + I_2 + I_3 \le 3\epsilon \|f\| + \epsilon.$$

For general metric spaces S a proof can be obtained from statements 1 and 4 of Proposition 1.1.1.

Finally, the last statement follows from (1.6). □

Exercise 1.1.3. If probability measures p_n on $\mathbb{R}^d$ converge weakly to a measure p as $n \to \infty$, then the sequence $p_n(A)$ converges to $p(A)$ for any open or closed set A such that $p(\partial A) = 0$ (where ∂A is the boundary of A).

A family H of $L^1(\Omega, \mathcal{F}, \mu)$ is called *uniformly integrable* if

$$\lim_{c \to \infty} \sup_{X \in H} \mathbf{E}(|X| \mathbf{1}_{|X| > c}) = 0.$$

Proposition 1.1.3. *If either* (i) $\sup_{X \in H} \mathbf{E}(|X|^p) < \infty$ *for a* $p > 1$, *or* (ii) *there exists an integrable r.v. Y s.t. $|X| \le Y$ for all $X \in H$, then H is uniformly integrable.*

Proof. Follows from the inequalities.

$$\text{(i)} \qquad \mathbf{E}(|X|\mathbf{1}_{|X|>c}) < \frac{1}{c^{p-1}}\mathbf{E}(|X|^p\mathbf{1}_{|X|>c}) < \frac{1}{c^{p-1}}\mathbf{E}(|X|^p),$$

$$\text{(ii)} \qquad \mathbf{E}(|X|\mathbf{1}_{|X|>c}) < \mathbf{E}(Y\mathbf{1}_{Y>c}).$$

□

Proposition 1.1.4. *If $X_n \to X$ a.s. and $\{X_n\}$ is uniformly integrable, then $X_n \to X$ in L^1.*

Proof. Decompose the integral $\int |X_n - X|p(d\omega)$ into the sum of the three integrals over the domains $\{|X_n - X| > \epsilon\}$, $\{|X_n - X| \leq \epsilon, |X| \leq c\}$ and $\{|X_n - X| \leq \epsilon, |X| > c\}$. These can be made small respectively because $X_n \to X$ in probability (as it holds a.s.), by dominated convergence and by uniform integrability. □

Two famous theorems of integration theory, the dominated and monotone convergence theorems, give easy-to-use criteria for a.s. convergence to imply convergence in L_1.

The following famous result allows one to transfer weak convergence to a.s. convergence by an appropriate coupling.

Theorem 1.1.3 (Skorohod coupling). *Let $\xi, \xi_1, \xi_2, \dots$ be a sequence of random variables with values in a separable metric space S such that $\xi_n \to \xi$ weakly as $n \to \infty$. Then there exists a probability space with some S-valued random variables $\eta, \eta_1, \eta_2, \dots$ distributed as $\xi, \xi_1, \xi_2, \dots$ respectively and such that $\eta_n \to \eta$ a.s. as $n \to \infty$.*

The following celebrated convergence result is one of the oldest in probability theory.

Theorem 1.1.4 (Weak law of large numbers). *If $\xi_1, \xi_2, \dots$ is a collection of i.i.d. random variables with $\mathbf{E}\xi_j = m$ and $\operatorname{Var}\xi_j < \infty$, then the means $(\xi_1 + \dots + \xi_n)/n$ converge to m in probability and in L^2.*

Proof. By (1.3)

$$\operatorname{Var}\left(\frac{\xi_1 + \dots + \xi_n}{n} - m\right) = \operatorname{Var}\frac{(\xi_1 - m) + \dots + (\xi_n - m)}{n} = \frac{\operatorname{Var}\xi_1}{n},$$

implying convergence in L_2. Hence by Chebyshev's inequality

$$\mathbf{P}\left(\left|\frac{\xi_1 + \dots + \xi_n}{n} - m\right| > \epsilon\right) \leq \frac{\operatorname{Var}\xi_1}{n\epsilon^2},$$

implying convergence in probability. □

Using the stronger Kolmogorov's inequality allows one to get the following improvement.

Theorem 1.1.5 (Strong law of large numbers). *Let S_n denote the sums $\xi_1 + \xi_2 + \dots + \xi_n$ for a sequence $\xi_1, \xi_2, \dots$ of independent zero-mean random variables such that $\mathbf{E}|\xi_j|^2 = \sigma^2 < \infty$ for all j. Then the means S_n/n converge to 0 a.s.*

Proof. By (1.7) we have to show that

$$\lim_{m\to\infty} \mathbf{P}\left(\sup_{n\ge m}\left|\frac{S_n}{n}\right| > \epsilon\right) = 0.$$

Denote by A_k the events

$$A_k = \left\{\max_{2^{k-1}\le n<2^k}\left|\frac{S_n}{n}\right| > \epsilon\right\}.$$

Then by (1.4)

$$\mathbf{P}(A_k) \le \mathbf{P}(\max_{2^{k-1}\le n<2^k}|S_n| > \epsilon 2^{k-1}) \le 2^{-k}\frac{4\sigma^2}{\epsilon^2}.$$

Hence the sum $\sum_k \mathbf{P}(A_k)$ converges. Consequently

$$\mathbf{P}\left(\sup_{n\ge 2^{m-1}}\left|\frac{S_n}{n}\right| > \epsilon\right) \le \sum_{k=m}^{\infty}\mathbf{P}(A_k) \to 0$$

as $m \to \infty$ for any ϵ. □

Remark 1. Using (1.5) instead of (1.4) allows us to prove the above theorem under a weaker assumption that $\mathbf{E}|\xi_j|^p = \omega < \infty$ for some $p > 1$. It is instructive to see where this proof breaks down in case $p = 1$. By more involved arguments one can still prove the strong LLN if only $\mathbf{E}|\xi_j| < \infty$, but assuming that ξ_j are i.i.d.

The following result is routinely used in stochastic analysis to check a validity of a certain property for elements of $\sigma(\Gamma)$, where Γ is a collection of subsets closed under intersection. According to the theorem it is sufficient to check that the validity of this property is preserved under set subtraction and countable unions.

Theorem 1.1.6 (Monotone class theorem). *Let $\mathcal{S}$ be a collection of subsets of a set Ω s.t.*

(i) $\Omega \in \mathcal{S}$,

(ii) $A, B \in \mathcal{S} \Rightarrow A \setminus B \in \mathcal{S}$,

(iii) $A_1 \subset A_2 \subset \dots \subset A_n \in \mathcal{S} \Rightarrow \bigcup_n A_n \in \mathcal{S}$.

If a collection of subsets Γ belongs to S and is closed under pairwise intersection, then $\sigma(\Gamma) \in \mathcal{S}$.

Exercise 1.1.4. For $S \subset \mathbb{R}^d$ the *universal* σ-field $\mathcal{U}(S)$ is defined as the intersection of the completions of $\mathcal{B}(S)$ with respect to all probability measures on S. The $(\mathcal{U}(S), \mathcal{B}(S))$-measurable functions are called *universally measurable.* Show that a real valued function f is universally measurable if and only if for every probability measure μ on S there exists a Borel measurable function g_μ such that $\mu\{x : f(x) \neq g_\mu(x)\} = 0$. Hint for "only if" part: show that

$$f(x) = \inf\{r \in \mathbb{Q} : x \in U(r)\}, \quad \text{where } U(r) = \{x \in S : f(x) \leq r\}.$$

Since $U(r)$ belong to the completion of the Borel σ-algebra with respect to μ there exist $B(r)$, $r \in \mathbb{Q}$, such that

$$\mu\Big(\bigcup_{r\in\mathbb{Q}} (B(r)\Delta U(r))\Big) = 0.$$

Define

$$g_\mu(x) = \inf\{r \in \mathbb{Q} : x \in B(r)\}.$$

1.2 Characteristic functions

As we already mentioned, expectation and variance supply two basic numeric characteristics of a random variable. Some additional information on its behavior can be obtained from higher moments. A complete analytical description of a random variable is given by the characteristic function, which we recall briefly in this section.

If p is a probability measure on $\mathbb{R}^d$ its *characteristic function* is the function $\phi_p(y) = \int e^{i(y,x)} p(dx)$. For a $\mathbb{R}^d$-valued r.v. X its *characteristic function* is defined as the characteristic function $\phi_X = \phi_{p_X}$ of its law p_X, i.e.

$$\phi_X(y) = \mathbf{E}e^{i(y,X)} = \int_{\mathbb{R}^d} e^{i(y,x)} p_X(dx).$$

Any characteristic function is a continuous function, which clearly follows from the inequalities

$$|\phi_X(y+h) - \phi_X(y)| \leq \mathbf{E}|e^{ihX} - 1| \leq \max_{|x|\leq a} |e^{ihx} - 1| + 2\mathbf{P}(|X| > a). \tag{1.8}$$

Theorem 1.2.1 (Riemann–Lebesgue lemma). *If a probability measure p has a density, then ϕ_p belongs to $C_\infty(\mathbb{R}^d)$. In other words, the inverse Fourier transform*

$$f \mapsto F^{-1}f(y) = (2\pi)^{-d/2} \int e^{i(y,x)} f(x)dx$$

is a bounded linear operator $L^1(\mathbb{R}^d) \to C_\infty(\mathbb{R}^d)$.

Sketch of the proof. Reduce to the case, when f is a continuously differentiable function with a compact support. For this case use integration by parts.

For a vector $m \in \mathbb{R}^d$ and a positive definite $d \times d$-matrix A, a r.v. X is called *Gaussian* (or has *Gaussian distribution*) *with mean* m *and covariance* A, denoted by $N(m, A)$, whenever its characteristic function is

$$\phi_{N(m,A)}(y) = \exp\left\{i(m, y) - \frac{1}{2}(y, Ay)\right\}.$$

It is easy to deduce that $m = \mathbf{E}(X)$ and $A_{ij} = \mathbf{E}((X_i - m_i)(X_j - m_j))$ and that if A is non-degenerate, $N(m, A)$ random variables have distribution with the pdf

$$f(x) = \frac{1}{(2\pi)^{d/2}\sqrt{\det(A)}} \exp\left\{-\frac{1}{2}(x - m, A^{-1}(x - m))\right\}.$$

It is useful to observe that if X_1 and X_2 are independent $\mathbb{R}^d$-valued random variables with laws μ_1, μ_2 and characteristic functions ϕ_1 and ϕ_2, then $X_1 + X_2$ has the characteristic function $\phi_1\phi_2$ and the law given by the *convolution* $\mu_1 \star \mu_2$ defined by

$$(\mu_1 \star \mu_2)(A) = \int_{\mathbb{R}^d} \mu_1(A - x)\mu_2(dx).$$

Similarly, for independent random variables $X_1, \dots, X_n$ with the laws $\mu_1, \dots, \mu_n$ and characteristic functions $\phi_1, \dots, \phi_n$, the sum $X_1 + \cdots + X_n$ has the characteristic function $\phi_1 \cdots \phi_n$ and the law $\mu_1 \star \cdots \star \mu_n$. In particular, if $X_1, \dots, X_n$ are independent identically distributed (common abbreviation *i.i.d.*) random variables, then the sum $X_1 + \cdots + X_n$ has the characteristic function ϕ_1^n and the law $\mu_1 \star \cdots \star \mu_1$.

The next exercise anticipates the discussion of weak compactness or measures given at the end of this chapter.

Exercise 1.2.1. Show that if probability distributions p_n on $\mathbb{R}^d$, $n \in \mathbb{N}$, converge weakly to a probability distribution p, then

(i) the family p_n is *tight*, i.e.

$$\forall \epsilon > 0 \, \exists K > 0 : \forall n, \; p_n(|x| > K) < \epsilon;$$

(ii) their characteristic functions ϕ_n converge uniformly on compact sets.

Hint: for (ii) use tightness and representation (1.8) to show that the family ϕ_n is equicontinuous, i.e.

$$\forall \epsilon \, \exists \delta : |\phi_n(y + h) - \phi(y)| < \epsilon \quad \forall h < \delta, n \in \mathbb{N},$$

which implies uniform convergence.

Theorem 1.2.2 (Glivenko's theorem). *If ϕ_n, $n \in \mathbb{N}$, and ϕ are the characteristic functions of probability distributions p_n and p on $\mathbb{R}^d$, then $\lim_{n\to\infty} \phi_n(y) = \phi(y)$ for each $y \in \mathbb{R}^d$ if and only if p_n converge to p weakly.*

Theorem 1.2.3 (Lévy's theorem). *If ϕ_n, $n \in \mathbb{N}$, is a sequence of characteristic functions of probability distributions on $\mathbb{R}^d$ and $\lim_{n\to\infty} \phi_n(y) = \phi(y)$ for each $y \in \mathbb{R}^d$ for some function ϕ, which is continuous at the origin, then ϕ is itself a characteristic function (and so the corresponding distributions converge weekly as above).*

The following exercise suggests using Lévy's theorem to prove a particular case of the fundamental Prohorov criterion for tightness.

Exercise 1.2.2. Show that if a family of probability measures p_α on $\mathbb{R}^d$ is tight, then it is *relatively weakly compact*, i.e. any sequence of this family has a weakly convergent subsequence. Hint: tight $\Rightarrow$ family of characteristic functions is equicontinuous (by (1.8)), and hence is relatively compact in the topology of uniform convergence on compact sets. Finally use Lévy's theorem.

Exercise 1.2.3. (i) Show that a finite linear combination of $\mathbb{R}^d$-valued Gaussian random variables is again a Gaussian r.v.

(ii) Show that if a sequence of $\mathbb{R}^d$-valued Gaussian random variables converges in distribution to a random variable, then the limiting random variable is again Gaussian.

(iii) Show that if (X, Y) is a $\mathbb{R}^2$-valued Gaussian random variables, then X and Y are uncorrelated if and only if they are independent.

Theorem 1.2.4 (Bochner's criterion). *A function $\phi : \mathbb{R}^d \mapsto \mathbb{C}$ is a characteristic function of a probability distribution if and only if it satisfies the following three properties:*

(i) $\phi(0) = 1$;

(ii) *ϕ is continuous at the origin;*

(iii) *ϕ is positive definite, which means that*

$$\sum_{j,k=1}^{d} c_j \bar{c}_k \phi(y_j - y_k) \geq 0$$

for all real $y_1, \dots, y_d$ and all complex $c_1, \dots, c_d$.

Remark 2. To prove the "only if" part of Bochner's theorem is easy. In fact:

$$\sum_{j,k=1}^{d} c_j \bar{c}_k \phi_X(y_j - y_k) = \int_{\mathbb{R}^d} \sum_{j,k=1}^{d} c_j \bar{c}_k e^{i(y_j - y_k, x)} p_X(dx)$$
$$= \int_{\mathbb{R}^d} \Big(\sum_{j=1}^{d} c_j e^{i(y_j,x)}\Big)^2 p_X(dx) \geq 0.$$

1.3 Conditioning

Formally speaking, probability can be considered as a part of measure theory. What actually makes it special and fills it with new intuitive and practical content is *conditioning*. On the one hand, conditioning is a method for updating our perception of the probability of an event based on the information received (conditioning on an event). On the other hand, it is a method for characterizing random variables from their coarse description that neglects certain irrelevant details, like increasing the scale of an atlas or an image (conditioning with respect to a partition or subalgebra).

Assume a finite *partition* $\mathcal{A} = \{A_i\}$ of our probability space $(\Omega, \mathcal{F}, \mathbf{P})$ is given, i.e. it is decomposed into the union of non-intersecting measurable subsets $A_1, \ldots, A_m$. Assume that for certain purposes we do not need to distinguish the points belonging to the same element of the partition. In other words, we would like to reduce our original probability space to the simpler one $(\Omega, \mathcal{F}_{\mathcal{A}}, \mathbf{P})$, where $\mathcal{F}_{\mathcal{A}}$ is the finite σ-algebra generated by the partition $\mathcal{A}$ (that consists of all unions of the elements of this partition). Now, if we have a random variable X on $(\Omega, \mathcal{F}, \mathbf{P})$, how should we reasonably project it on the reduced probability space $(\Omega, \mathcal{F}_{\mathcal{A}}, \mathbf{P})$? Clearly such a projection $\tilde{X}$ should be measurable with respect to $\mathcal{F}_{\mathcal{A}}$, meaning that it should be constant on each A_i. Moreover, we want the averages of X and $\tilde{X}$ to coincide on each A_i. This implies that the value of $\tilde{X}$ on A_i should equal the average value of X on A_i. The random variable $\tilde{X}$, obtained in this way, is denoted by $\mathbf{E}(X|\mathcal{F}_{\mathcal{A}})$ and is called the *conditional expectation* of X given the σ-algebra $\mathcal{F}_{\mathcal{A}}$ (or equivalently, given the partition $\mathcal{A}$). Hence, by definition,

$$\mathbf{E}(X|\mathcal{F}_{\mathcal{A}})(\omega) = \int_{A_i} X(\omega)\mathbf{P}(d\omega)/\mathbf{P}(A_i), \quad \omega \in A_i, \tag{1.9}$$

for all $i = 1, \ldots, m$. Equivalently, $\mathbf{E}(X|\mathcal{F}_{\mathcal{A}})(\omega)$ is defined as a random variable on $(\Omega, \mathcal{F}_{\mathcal{A}}, \mathbf{P})$ such that

$$\int_A \mathbf{E}(X|\mathcal{F}_{\mathcal{A}})(\omega)\mathbf{P}(d\omega) = \int_A X(\omega)\mathbf{P}(d\omega)$$

for any $A \in \mathcal{F}_{\mathcal{A}}$.

This definition can be straightforwardly extended to arbitrary subalgebras. Namely, for a random variable X on a probability space $(\Omega, \mathcal{F}, \mathbf{P})$ and a σ-subalgebra $\mathcal{G}$ of $\mathcal{F}$, the *conditional expectation* of X given $\mathcal{G}$ is defined as a random variable $\mathbf{E}(X|\mathcal{G})$ on the probability space $(\Omega, \mathcal{G}, \mathbf{P})$ such that

$$\int_A \mathbf{E}(X|\mathcal{G})(\omega)\mathbf{P}(d\omega) = \int_A X(\omega)\mathbf{P}(d\omega)$$

for any $A \in \mathcal{G}$. Clearly, if such a random variable exists it is uniquely defined (up to the natural equivalence of random variables in $(\Omega, \mathcal{G}, \mathbf{P})$), because the difference of any two random variables with the required property has vanishing integrals over any measurable set in $(\Omega, \mathcal{G}, \mathbf{P})$, and hence this difference vanishes a.s. However, the existence of conditional expectation is not so obvious for infinite subalgebras. In fact, the defining equation (1.9) does not make sense in case $\mathbf{P}(A_i) = 0$. Hence in the general case, another approach to the construction of conditional expectation is needed, which we now describe.

For a given measure space $(S, \mathcal{F}, \mu)$, a measure ν on $(S, \mathcal{F})$ is called *absolutely continuous* with respect to μ if $\nu(A) = 0$ whenever $A \in \mathcal{F}$ and $\mu(A) = 0$. Two measures are called *equivalent* if they are mutually absolutely continuous.

Theorem 1.3.1 (Radon–Nikodym theorem). *If μ is σ-finite and ν is finite and absolutely continuous with respect to μ, then there exists a unique (up to almost sure equality) non-negative measurable function g on S such that for all $A \in \mathcal{F}$*

$$\nu(A) = \int_A g(x)\mu(dx).$$

This g is called the Radon–Nikodym derivative of ν with respect to μ and is often denoted $d\nu/d\mu$.

Let X be an integrable random variable on a probability space $(\Omega, \mathcal{F}, \mathbf{P})$ and let $\mathcal{G}$ be a sub-σ-algebra of $\mathcal{F}$. If $X \geq 0$ everywhere, the formula

$$Q_X(A) = \mathbf{E}(X\mathbf{1}_A) = \int_A X(\omega)\mathbf{P}(d\omega)$$

for $A \in \mathcal{G}$ defines a measure Q_X on $(\Omega, \mathcal{G})$ that is obviously absolutely continuous with respect to $\mathbf{P}$. The r.v. $dQ_X/d\mathbf{P}$ on $(\Omega, \mathcal{G}, \mathbf{P})$ is called the *conditional expectation of X with respect to $\mathcal{G}$*, and is usually denoted $\mathbf{E}(X|\mathcal{G})$. If X is not supposed to be positive one defines the *conditional expectation* as $\mathbf{E}(X|\mathcal{G}) = \mathbf{E}(X^+|\mathcal{G}) - \mathbf{E}(X^-|\mathcal{G})$. Clearly this new definition complies with the previous one, as so defined $Y = \mathbf{E}(X|\mathcal{G})$ is a r.v. on $(\Omega, \mathcal{G}, \mathbf{P})$ satisfying

$$\int_A Y(\omega)\mathbf{P}(d\omega) = \int_A X(\omega)\mathbf{P}(d\omega) \tag{1.10}$$

for all $A \in \mathcal{G}$ or, equivalently,

$$\mathbf{E}(YZ) = \mathbf{E}(XZ) \tag{1.11}$$

for any bounded $\mathcal{G}$-measurable Z.

If $X = (X_1, \dots, X_d) \in \mathbb{R}^d$, then

$$\mathbf{E}(X|\mathcal{G}) = (\mathbf{E}(X_1|\mathcal{G}), \dots, \mathbf{E}(X_n|\mathcal{G})).$$

The following result collects the basic properties of the conditional expectation.

Theorem 1.3.2. (i) $\mathbf{E}(\mathbf{E}(X|\mathcal{G})) = \mathbf{E}(X)$;

(ii) *if Y is $\mathcal{G}$-measurable, then* $\mathbf{E}(XY|\mathcal{G}) = Y\mathbf{E}(X|\mathcal{G})$ *a.s.;*

(iii) *if Y is $\mathcal{G}$-measurable and X is independent of $\mathcal{G}$, then a.s.*

$$\mathbf{E}(XY|\mathcal{G}) = Y\mathbf{E}(X),$$

and more generally

$$\mathbf{E}(f(X,Y)|\mathcal{G}) = G_f(Y) \tag{1.12}$$

a.s. for any bounded Borel function f, where $G_f(y) = \mathbf{E}(f(X,y))$ *a.s.;*

(iv) *if $\mathcal{H}$ is a sub-σ-algebra of $\mathcal{G}$ then* $\mathbf{E}(\mathbf{E}(X|\mathcal{G})|\mathcal{H}) = \mathbf{E}(X|\mathcal{H})$ *a.s. (this property is called the chain rule for conditioning);*

(v) *the mapping* $X \mapsto \mathbf{E}(X|\mathcal{G})$ *is an orthogonal projection* $L^2(\Omega, \mathcal{F}, \mathbf{P}) \to L^2(\Omega, \mathcal{G}, \mathbf{P})$*;*

(vi) $X_1 \leq X_2 \Rightarrow \mathbf{E}(X_1|\mathcal{G}) \leq \mathbf{E}(X_2|\mathcal{G})$ *a.s.;*

(vii) *the mapping* $X \mapsto \mathbf{E}(X|\mathcal{G})$ *is a linear contraction* $L^1(\Omega, \mathcal{F}, \mathbf{P}) \to L^1(\Omega, \mathcal{G}, \mathbf{P})$.

Exercise 1.3.1. Prove the above theorem. Hint: (ii) consider first the case with Y being an indicator function of a $\mathcal{G}$-measurable set; (v) assume $X = Y + Z$ with Y from $L^2(\Omega, \mathcal{G}, \mathbf{P})$ and Z from its orthogonal complement and show that $Y = \mathbf{E}(X|\mathcal{G})$. (vi) Follows from an obvious remark that $X \geq 0 \Rightarrow \mathbf{E}(X|\mathcal{G}) \geq 0$.

Remark 3. Property (v) above can be used to give an alternative construction of conditional expectation by-passing the Radon–Nikodym theorem.

If Z is a r.v. on $(\Omega, \mathcal{F}, \mathbf{P})$ one calls $\mathbf{E}(X|\sigma(Z))$ the *conditional expectation of X with respect to Z* and denotes it briefly by $\mathbf{E}(X|Z)$.

The measurability of $\mathbf{E}(X|Z)$ with respect to $\sigma(Z)$ implies that $\mathbf{E}(X|Z)$ is a constant on any Z-level set $\{\omega : Z(\omega) = z\}$. One denotes this constant by $\mathbf{E}(X|Z = z)$ and calls it the *conditional expectation of X given $Z = z$*. From statement (iv) of

Theorem 1.3.2 it follows that

$$\mathbf{E}(X) = \int \mathbf{E}(X|Z)(\omega)\mathbf{P}(d\omega) = \int \mathbf{E}(X|Z=z)p_Z(dz) \tag{1.13}$$

(the second equality is obtained by applying (1.2) to $f(Z(\omega)) = \mathbf{E}(X|Z)(\omega)$).

Let X and Z be $\mathbb{R}^d$ and respectively $\mathbb{R}^m$-valued r.v. on $(\Omega, \mathcal{F}, \mathbf{P})$, and let $\mathcal{G}$ be a sub-σ-algebra of $\mathcal{F}$. The *conditional probability of X given $\mathcal{G}$* and *X given $Z = z$* respectively are defined as

$$\begin{aligned}
\mathbf{P}_{X|\mathcal{G}}(B;\omega) &\equiv \mathbf{P}(X \subset B|\mathcal{G})(\omega) = \mathbf{E}(\mathbf{1}_B(X)|\mathcal{G})(\omega), \quad \omega \in \Omega;\\
\mathbf{P}_{X|Z=z}(B) &\equiv \mathbf{P}(X \subset B|Z=z) = \mathbf{E}(\mathbf{1}_B(X)|Z=z),
\end{aligned}$$

for Borel sets B, or equivalently through the equations

$$\begin{aligned}
\mathbf{E}(f(X)|\mathcal{G})(\omega) &= \int_{\mathbb{R}^d} f(x)\mathbf{P}_{X|\mathcal{G}}(dx;\omega),\\
\mathbf{E}(f(X)|Z=z) &= \int_{\mathbb{R}^d} f(x)\mathbf{P}_{X|Z=z}(dx)
\end{aligned}$$

for bounded Borel functions f. Of course $\mathbf{P}_{X|Z=z}(B)$ is just the common value of $\mathbf{P}_{X|Z}(B;\omega)$ on the set $\{\omega : Z(\omega) = z\}$.

It is possible to show (though this is not obvious) that, for any $\mathbb{R}^d$-r.v. X, the *regular conditional probability of X given $\mathcal{G}$* exists, i.e. such a version of conditional probability that $\mathbf{P}_{X|\mathcal{G}}(B,\omega)$ is a probability measure on $\mathbb{R}^d$ as a function of B for each ω (notice that from the above discussion the required additivity of conditional expectations hold a.s. only so that they may fail to define a probability even a.s.) and is $\mathcal{G}$-measurable as a function of ω. Hence one can define *conditional r.v.* $X_{\mathcal{G}}(\omega)$, $X_Z(\omega)$ and $X_{Z=z}$ as r.v. with the corresponding conditional distributions.

Proposition 1.3.1. *For a Borel function h*

$$\mathbf{E}h(X,Z) = \int h(x,z)\mathbf{P}_{X|Z=z}(dx)p_Z(dz) \tag{1.14}$$

whenever the l.h.s. is well defined.

Proof. It is enough to show this for the functions of the form $h(X,Z) = f(X)\mathbf{1}_{Z\in C}$ for a measurable C. And from (1.13) it follows that

$$\begin{aligned}
\mathbf{E}f(X)\mathbf{1}_{Z\in C} &= \int_{Z\in C} \mathbf{E}(f(X)|Z)(\omega)\mathbf{P}(d\omega) = \int_C \mathbf{E}(f(X)|Z=z)p_Z(dz)\\
&= \int_C \int_{\mathbb{R}^d} f(x)\mathbf{P}_{X|Z=z}(dx)p_Z(dz).
\end{aligned}$$

□

For instance, if X, Z are discrete r.v. with joint probability $\mathbf{P}(X = i, Z = j) = p_{ij}$, then the conditional probabilities $p(X = i|Z = j)$ are given by the usual formula $p_{ij}/\mathbf{P}(Z = j)$.

On the other hand, if X, Z are r.v. with a joint probability density function $f_{X,Z}(x,z)$, then the conditional r.v. $X_{Z=z}$ has a probability density function

$$f_{X_{Z=z}}(x) = f_{X,Z}(x,z)/f_Z(z)$$

whenever f_Z does not vanish. In order to see this, one has to compare (1.14) with the equation

$$\begin{aligned} \mathbf{E}h(X,Z) &= \int h(x,z)f(x,z)\,dxdz \\ &= \int h(x,z)\frac{f(x,z)}{f_Z(z)}\,dxf_Z(z)\,dz. \end{aligned}$$

Theorem 1.3.3. *Let X be a integrable variable on $(\Omega, \mathcal{F}, \mathbf{P})$ and let $\mathcal{G}_n$ be either*

(i) *an increasing sequence of sub-σ-algebras of $\mathcal{F}$ with $\mathcal{G}$ being the minimal σ-algebra containing all $\mathcal{G}_n$, or*

(ii) *a decreasing sequence of sub-σ-algebras of $\mathcal{F}$ with $\mathcal{G} = \bigcap_{n=1}^{\infty} \mathcal{G}_n$.*

Then a.s. and in L^1

$$\mathbf{E}(X|\mathcal{G}) = \lim_{n\to\infty} \mathbf{E}(X|\mathcal{G}_n). \tag{1.15}$$

Furthermore, if $X_n \to X$ a.s. and $|X_n| < Y$ for all n, where Y is an integrable random variable, then a.s. and in L^1

$$\mathbf{E}(X|\mathcal{G}) = \lim_{n\to\infty} \mathbf{E}(X_n|\mathcal{G}_n). \tag{1.16}$$

Proof. We shall sketch the proof of the convergence in L^1 (a.s. convergence is a bit more involved, and we shall neither prove, nor use it), say, for increasing sequences. Any r.v. of the form $\mathbf{1}_B$ with $B \in \mathcal{G}$ can be approximated in L^2 by a $\mathcal{G}_n$-measurable r.v. ξ_n. Hence the same holds for any r.v. from $L^2(\Omega, \mathcal{F}, \mathbf{P})$. As $\mathbf{E}(X|\mathcal{G}_n)$ is the best approximation (L^2-projection) for $\mathbf{E}(X|\mathcal{G})$ one obtains (1.15) for $X \in L^2(\Omega, \mathcal{F}, \mathbf{P})$, and hence for $X \in L^1(\Omega, \mathcal{F}, \mathbf{P})$ by density arguments. Next,

$$\mathbf{E}(X_n|\mathcal{G}_n) - \mathbf{E}(X|\mathcal{G}) = \mathbf{E}(X_n - X|\mathcal{G}_n) + (\mathbf{E}(X|\mathcal{G}_n) - \mathbf{E}(X|\mathcal{G})).$$

Since $|X_n| < Y$ and $X_n \to X$ a.s. one concludes that $X_n \to X$ in L^1 by dominated convergence. Hence as $n \to \infty$

$$\mathbf{E}(\mathbf{E}|X_n - X||\mathcal{G}_n) = \mathbf{E}|X_n - X| \to 0. \qquad \square$$

Theorem 1.3.4. *If $X \in L^1(\Omega, \mathcal{F}, \mathbf{P})$, the family of r.v. $\mathbf{E}(X|\mathcal{G})$, as $\mathcal{G}$ runs through all sub-σ-algebra of $\mathcal{F}$, is uniformly integrable.*

Proof.

$$\mathbf{1}_{|\mathbf{E}(X|\mathcal{G})|>c}\mathbf{E}(X|\mathcal{G}) = \mathbf{E}(X\mathbf{1}_{|\mathbf{E}(X|\mathcal{G})|>c}|\mathcal{G}),$$

because $\{|\mathbf{E}(X|\mathcal{G})| > c\} \in \mathcal{G}$. Hence

$$\begin{aligned}\mathbf{E}(\mathbf{1}_{|\mathbf{E}(X|\mathcal{G})|>c}\mathbf{E}(X|\mathcal{G})) &\le \mathbf{E}(\mathbf{1}_{|\mathbf{E}(X|\mathcal{G})|>c}|X|) \\ &\le \mathbf{E}(|X|\mathbf{1}_{|X|>d}) + d\mathbf{P}(|\mathbf{E}(X|\mathcal{G})| > c) \\ &\le \mathbf{E}(|X|\mathbf{1}_{|X|>d}) + \frac{d}{c}\mathbf{E}(|X|),\end{aligned}$$

where in the last inequality Markov's inequality was used. First choose d to make the first term small, then c to make the second one small. □

One says that two sigma algebra $\mathcal{G}_1$, $\mathcal{G}_2$ coincide on a set $A \in \mathcal{G}_1 \cap \mathcal{G}_2$ whenever $A \cap \mathcal{G}_1 = A \cap \mathcal{G}_2$.

Theorem 1.3.5 (Locality of conditional expectation). *Let the σ-algebras $\mathcal{G}_1, \mathcal{G}_2 \in \mathcal{F}$ and the random variables $X_1, X_2 \in L^1(\Omega, \mathcal{F}, \mathbf{P})$ be such that $\mathcal{G}_1 = \mathcal{G}_2$ and $X_1 = X_2$ on a set $A \in \mathcal{G}_1 \cap \mathcal{G}_1$. Then $\mathbf{E}(X_1|\mathcal{G}_1) = \mathbf{E}(X_2|\mathcal{G}_2)$ a.s. on A.*

Proof. The sets $\mathbf{1}_A\mathbf{E}(X_1|\mathcal{G}_1)$ and $\mathbf{1}_A\mathbf{E}(X_2|\mathcal{G}_2)$ are both $\mathcal{G}_1 \cap \mathcal{G}_2$-measurable, and for any $B \subset A$ such that $B \in \mathcal{G}_1$ (and hence $B \in \mathcal{G}_2$)

$$\int_B \mathbf{E}(X_1|\mathcal{G}_1)\mathbf{P}(d\omega) = \int_B X_1\mathbf{P}(d\omega) = \int_B X_2\mathbf{P}(d\omega) = \int_B \mathbf{E}(X_2|\mathcal{G}_2)\mathbf{P}(d\omega). \quad \square$$

1.4 Infinitely divisible and stable distributions

Infinitely divisible laws studied here occupy an honored place in probability, because of their extreme modeling power in the variety of situations. They form the cornerstone for the theory of Lévy processes discussed later.

A probability measure μ on $\mathbb{R}^d$ with a characteristic function ϕ_μ is called *infinitely divisible* if, for all $n \in \mathbb{N}$, there exists a probability measure ν such that $\mu = \nu \star \cdots \star \nu$ (n times) or equivalently $\phi_\mu(y) = f^n(y)$ with f being a characteristic function of a probability measure.

A random variable X is called *infinitely divisible* whenever its law p_X is infinitely divisible. This is equivalent to the existence, for any n, of i.i.d. random variable Y_j, $j = 1, \dots, n$, such that $Y_1 + \cdots + Y_n$ has the law p_X.

For example, any Gaussian distribution is clearly infinitely divisible.

Another key example is given by a *Poisson random variable with mean (or parameter) $c > 0$*, which is a random variable N with the non-negative integers as range and law

$$\mathbf{P}(N = n) = \frac{c^n}{n!}e^{-c}.$$

One easily checks that $\mathbf{E}(N) = \mathrm{Var}(N) = c$ and that the characteristic function of N is $\phi_N(y) = \exp\{c(e^{iy} - 1)\}$. This implies that N is infinitely divisible.

Of importance is the following generalization. Let $Z(n)$, $n \in \mathbb{N}$, be a sequence of $\mathbb{R}^d$-valued i.i.d. random variables with the common law μ_Z. The random variable $X = Z(1)+\cdots+Z(N)$ is called a *compound Poisson random variable*. It represents a random walk (each step specified by a random variable distributed like $Z(1)$) with a random (Poisson) number of steps. Let us check that

$$\phi_X(y) = \exp\left\{\int_{\mathbb{R}^d} (e^{i(y,x)} - 1)c\mu_Z(dx)\right\}. \tag{1.17}$$

In fact,

$$\begin{aligned}
\phi_X(y) &= \sum_{n=0}^{\infty} \mathbf{E}(\exp\{i(y, Z(1) + \cdots + Z(N))\}|N = n)\mathbf{P}(N = n) \\
&= \sum_{n=0}^{\infty} \mathbf{E}(\exp\{i(y, Z(1) + \cdots + Z(n))\})\frac{c^n}{n!}e^{-c} \\
&= \sum_{n=0}^{\infty} \phi_Z^n(y)\frac{c^n}{n!}e^{-c} = \exp\{c(\phi_Z(y) - 1)\}.
\end{aligned}$$

A Borel measure ν on $\mathbb{R}^d$ is called a *Lévy measure* if $\nu(\{0\}) = 0$ and

$$\int_{\mathbb{R}^d} \min(1, x^2)\nu(dx) < \infty.$$

The major role played by these measures in the theory of infinite divisibility is revealed by the following fundamental result.

Theorem 1.4.1 (The Lévy–Khintchine formula). *For any $b \in \mathbb{R}^d$, a positive definite $d \times d$-matrix G and a Lévy measure ν the function*

$$\phi(u) = \exp\left\{i(b, u) - \frac{1}{2}(u, Gu) + \int_{\mathbb{R}^d} [e^{i(u,y)} - 1 - i(u, y)\mathbf{1}_{B_1}(y)]\nu(dy)\right\} \tag{1.18}$$

is a characteristic function of an infinitely divisible measure, where B_a denotes a ball of radius a in $\mathbb{R}^d$. Conversely, any infinite divisible distribution has a characteristic function of form (1.18).

Proof. We shall prove only the simpler first part. For the converse statement see e.g. [20], [302] and references therein. If any function of form (1.18) is a characteristic function, then it is infinitely divisible (as its roots have the same form). To show the latter we introduce the approximations

$$\phi_n(u) = \exp\left\{i\left(b - \int_{B_1 \setminus B_{1/n}} y\nu(dy), u\right) - \frac{1}{2}(u, Gu) + \int_{\mathbb{R}^d \setminus B_{1/n}} (e^{i(u,y)} - 1)\nu(dy)\right\}.$$

Each ϕ_n is a characteristic function (of the convolution of a normal distribution and an independent compound Poisson) and $\phi_n(u) \to \phi(u)$ for any u. By the Lévy theorem in order to conclude that ϕ is a characteristic function, one needs to show that ϕ is continuous at zero. This is easy (check it!). □

The function η appearing under the exponent in the representation $\phi(u) = e^{\eta(u)}$ of form (1.18) is called the *characteristic exponent* or *Lévy exponent* or *Lévy symbol* of ϕ (or of its distribution). The vector b in (1.18) is called the *drift vector* and G is called the matrix of *diffusion coefficients*.

Theorem 1.4.2. *Any infinitely divisible probability measure μ is a weak limit of a sequence of compound Poisson distributions.*

Proof. Let ϕ be a characteristic function of μ so that $\phi^{1/n}$ is the characteristic function of its "convolution root" μ_n. Define

$$\phi_n(u) = \exp\{n[\phi^{1/n}(u) - 1]\} = \exp\left\{\int_{\mathbb{R}^d} (e^{i(u,y)} - 1)n\mu_n(dy)\right\}.$$

Each ϕ_n is a ch.f. of a compound Poisson process and

$$\phi_n = \exp\{n(e^{(1/n)\ln\phi(u)} - 1)\} \to \phi(u), \quad n \to \infty.$$

The proof completes by Glivenko's theorem. □

An important class of infinitely divisible distributions constitute the so-called stable laws. A probability law in $\mathbb{R}^d$, its characteristic function ϕ and a random variable X with this law are called *stable (respectively strictly stable)* if for any integer n there exist a positive constant c_n and a real constant γ_n (resp. if additionally $\gamma_n = 0$) such that

$$\phi(y) = [\psi(y/c_n)\exp\{i\gamma_n y\}]^n.$$

In other words, the sum of any number of i.i.d. copies of X is distributed like X up to a shift and scaling. Obviously, it implies that ϕ is infinitely divisible and therefore $\log\phi$ can be presented in the Lévy–Khintchine form with appropriate b, A, ν.

Theorem 1.4.3. *If ϕ is stable, then there exists an $\alpha \in (0,2]$, called the index of stability such that:*

(i) *if $\alpha = 2$, then $\nu = 0$ in the representation* (1.18), *i.e. the distribution is normal;*

(ii) *if $\alpha \in (0,2)$, then in the representation* (1.18) *the matrix G vanishes and the radial part of the Lévy measure ν has the form $|\xi|^{-(1+\alpha)}$, i.e.*

$$\log\phi_\alpha(y) = i(b,y) + \int_0^\infty \int_{S^{d-1}} \left(e^{i(y,\xi)} - 1 - \frac{i(y,\xi)}{1+\xi^2}\right) \frac{d|\xi|}{|\xi|^{1+\alpha}}\mu(ds), \quad (1.19)$$

where ξ is presented by its magnitude $|\xi|$ and the unit vector $s = \xi/|\xi| \in S^{d-1}$ in the direction ξ, and μ is some (finite) measure in S^{d-1}.

The classical proof can be found e.g. in Feller [111] or Samorodnitski and Taqqu [288].

The integration in $|\xi|$ in (1.19) can be carried out explicitly, as the following result shows.

Theorem 1.4.4. *The stable exponent* (1.19) *can be written in the form*

$$\log \phi_\alpha(y) = i(\tilde{b}, y) - \int_{S^{d-1}} |(y,s)|^\alpha \Big(1 - i \operatorname{sgn}((y,s)) \tan \frac{\pi\alpha}{2}\Big) \tilde{\mu}(ds), \quad \alpha \neq 1, \tag{1.20}$$

$$\log \phi_\alpha(y) = i(\tilde{b}, y) - \int_{S^{d-1}} |(y,s)| \Big(1 + i \frac{2}{\pi} \operatorname{sgn}((y,s)) \log |(y,s)|\Big) \tilde{\mu}(ds), \quad \alpha = 1, \tag{1.21}$$

where

$$\tilde{b} = b + a_\alpha \int_{S^{d-1}} s\mu(ds), \quad \tilde{\mu} = \begin{cases} \sigma_\alpha \cos(\pi\alpha/2)\mu, & \alpha \neq 1 \\ \pi\mu/2, & \alpha = 1 \end{cases} \tag{1.22}$$

with some constants a_α and σ_α specified below. The measure $\tilde{\mu}$ on S^{d-1} is called sometimes the spectral measure of a stable law.

Proof. For $\alpha \in (0,1)$ and a real p

$$\int_0^\infty (e^{irp} - 1) \frac{dr}{r^{1+\alpha}} = -\frac{\Gamma(1-\alpha)}{\alpha} e^{-i\pi\alpha \operatorname{sgn} p/2} |p|^\alpha, \tag{1.23}$$

where $\operatorname{sgn} p$ is of course the sign of p. In fact, one presents the integral on the r.h.s. of (1.23) as the limit as $\epsilon \to 0_+$ of

$$\int_0^\infty (e^{-(\epsilon - ip)r} - 1) \frac{dr}{r^{1+\alpha}}. \tag{1.24}$$

Let, say, $p > 0$. Then

$$\epsilon - ip = (\epsilon^2 + p^2)^{1/2} e^{-i\theta}$$

with $\tan \theta = p/\epsilon$. So by the Cauchy theorem one can rotate the contour of integration in (1.24) through the angle θ. Changing the variable r to $s = e^{-i\theta} r$ in the integral thus obtained yields for (1.24) the expression

$$e^{-i\theta\alpha} \int_0^\infty (e^{-(\epsilon^2+p^2)^{1/2}s} - 1) \frac{ds}{s^{1+\alpha}},$$

which equals (by integration by parts)

$$-e^{-i\theta\alpha} \frac{(\epsilon^2 + p^2)^{1/2}}{\alpha} \int_0^\infty e^{-(\epsilon^2+p^2)^{1/2}s} s^{-\alpha} \, ds = -e^{-i\theta\alpha} \frac{(\epsilon^2 + p^2)^{\alpha/2}}{\alpha} \Gamma(1-\alpha).$$

Passing to the limit $\epsilon \to 0_+$ (and hence $\theta \to \pi/2$) yields (1.23). In case $p < 0$ one would have to rotate the contour of integration in (1.24) in the opposite direction.

Next, for $\alpha \in (1,2)$ and $p > 0$ integration by parts gives

$$\int_0^\infty \frac{e^{irp} - 1 - irp}{r^{1+\alpha}} dr = \frac{ip}{\alpha} \int_0^\infty (e^{ipr} - 1) \frac{dr}{r^\alpha},$$

and then by (1.23)

$$\int_0^\infty \frac{e^{irp} - 1 - irp}{r^{1+\alpha}} dr = \frac{\Gamma(\alpha - 1)}{\alpha} e^{-i\pi\alpha/2} p^\alpha. \tag{1.25}$$

Note that the real parts of both (1.23) and (1.23) are positive. From (1.23), (1.25) it follows that for $\alpha \in (0,2), \alpha \neq 1$,

$$\int_0^\infty \left(e^{irp} - 1 - \frac{irp}{1+r^2} \right) \frac{dr}{r^{1+\alpha}} = i a_\alpha p - \sigma_\alpha e^{-i\pi\alpha/2} p^\alpha \tag{1.26}$$

with

$$\sigma_\alpha = \alpha^{-1}\Gamma(1-\alpha), \qquad a_\alpha = -\int_0^\infty \frac{dr}{(1+r^2)r}, \qquad \alpha \in (0,1), \tag{1.27}$$

$$\sigma_\alpha = -\alpha^{-1}\Gamma(\alpha-1), \quad a_\alpha = \int_0^\infty \frac{r^{2-\alpha} dr}{1+r^2}, \qquad \alpha \in (1,2). \tag{1.28}$$

The case of $\alpha = 1$ is a bit more involved. In order to deal with it observe that

$$\begin{aligned}\int_0^\infty \frac{e^{irp} - 1 - ip\sin r}{r^2} dr &= -\int_0^\infty \frac{1 - \cos rp}{r^2} dr + i \int_0^\infty \frac{\sin rp - p \sin r}{r^2} dr \\ &= -\frac{1}{2}\pi p - ip \log p.\end{aligned}$$

In fact, the real part of this integral is evaluated using a standard fact that $f(r) = (1 - \cos r)/(\pi r^2)$ is a probability density (with the characteristic function $\psi(z)$ that equals to $1 - |z|$ for $|z| \le 1$ and vanishes for $|z| \ge 1$), and the imaginary part can be presented in the form

$$\begin{aligned}\lim_{\epsilon \to 0} \left[\int_\epsilon^\infty \frac{\sin pr}{r^2} dr - p \int_\epsilon^\infty \frac{\sin r}{r^2} dr \right] - p \lim_{\epsilon \to 0} \int_\epsilon^{p\epsilon} \frac{\sin r}{r^2} dr \\ = -p \lim_{\epsilon \to 0} \int_1^p \frac{\sin \epsilon y}{\epsilon y^2} dy = -p \int_1^p \frac{dy}{y},\end{aligned}$$

which implies the required formula. Therefore, for $\alpha = 1$

$$\int_0^\infty \left(e^{irp} - 1 - \frac{irp}{1+r^2} \right) \frac{dr}{r^{1+\alpha}} = i a_1 p - \frac{1}{2}\pi p - ip \log p \tag{1.29}$$

with

$$a_1 = \int_0^\infty \frac{\sin r - r}{(1+r^2)r^2}\, dr. \tag{1.30}$$

Formulae (1.26)–(1.30) yield (1.20) and (1.21). □

Exercise 1.4.1. Check that $\tilde{\mu}$ is continuous in (1.22), i.e. that

$$\lim_{\alpha\to 1} \sigma_\alpha \cos\left(\frac{\pi\alpha}{2}\right) = \frac{\pi}{2}. \tag{1.31}$$

Hint: if $\alpha < 1$, then

$$\lim_{\alpha\to 1} \frac{\Gamma(1-\alpha)}{\alpha} \cos\left(\frac{\pi\alpha}{2}\right) = \lim_{\alpha\to 1} \frac{\Gamma(2-\alpha)}{\alpha(1-\alpha)} \cos\left(\frac{\pi\alpha}{2}\right) = \lim_{\alpha\to 1} \frac{1}{1-\alpha} \cos\left(\frac{\pi\alpha}{2}\right).$$

For instance, if $d = 1$, S^0 consists of two points. Denoting their $\tilde{\mu}$-measures by μ_1, μ_{-1} one obtains for $\alpha \neq 1$ that

$$\log \phi_\alpha(y) = i\tilde{b}y - |y|^\alpha \left[(\mu_1 + \mu_{-1}) - i \operatorname{sgn} y(\mu_1 - \mu_{-1}) \tan \frac{\pi\alpha}{2}\right].$$

This can be written also in the form

$$\log \phi_\alpha(y) = i\tilde{b}y - \sigma|y|^\alpha \exp\left\{i\frac{\pi}{2}\gamma \operatorname{sgn} y\right\} \tag{1.32}$$

with some $\sigma > 0$ and a real γ such that $|\gamma| \leq \alpha$, if $\alpha \in (0,1)$, and $|\gamma| \leq 2-\alpha$, if $\alpha \in (1,2)$.

If the spectral measure $\tilde{\mu}$ is symmetric, i.e. $\tilde{\mu}(-\Omega) = \tilde{\mu}(\Omega)$ for any $\Omega \subset S^{d-1}$, then $\tilde{b} = b$ and formulas (1.25), (1.26) both give the following simple expression:

$$\log \phi_\alpha(y) = i(b,y) - \int_{S^{d-1}} |(y,s)|^\alpha \tilde{\mu}(ds). \tag{1.33}$$

In particular, if the measure $\tilde{\mu}$ is the Lebesgue measure ds on the sphere, then

$$\log \phi_\alpha(y) = i(b,y) - \sigma|y|^\alpha, \tag{1.34}$$

with the constant

$$\sigma = \int_{S^{d-1}} |\cos\theta|^\alpha ds = 2\pi^{(d-1)/2} \frac{\Gamma((\alpha+1)/2)}{\Gamma((\alpha+d)/2)} \tag{1.35}$$

(where $\theta \in [0,\pi]$ denotes the angle between a point on the sphere and its north pole, directed along y), called the scale of a stable distribution.

Exercise 1.4.2. Check the second equation in (1.35).

One sees readily that the characteristic function $\phi_\alpha(y)$ with $\log \phi_\alpha(y)$ from (1.20) and (1.21) and (1.33) with vanishing $\tilde{b}$ enjoy the property that $\phi_\alpha^n(y) = \phi_\alpha(n^{1/\alpha}y)$. Therefore all stable distributions with index $\alpha \neq 1$ and symmetric distributions with $\alpha = 1$ can be made strictly stable, if centered appropriately.

1.5 Stable laws as the Holtzmark distributions

Possibly the first appearance of non-Gaussian stable laws in physics was due to Holtzmark [134], who showed that the distribution of the gravitation force (acting on any given object), caused by the infinite collection of stars distributed uniformly in $\mathbb{R}^3$ (see below for the precise meaning of this) is given by the $3/2$-stable symmetric distribution in $\mathbb{R}^3$. This distribution is now called the *Holtzmark distribution* and is widely used in astrophysics and plasma physics. We shall sketch a deduction of this distribution in a more general context than usual, showing in particular that, choosing an appropriate power decay of a potential force, any stable law can be obtained in this way, that is as a distribution of the force caused by the infinite collection of points in $\mathbb{R}^d$, distributed uniformly in sectors.

Suppose the force between a particle placed at a point $x \in \mathbb{R}^d$ and a fixed object at the origin is given by

$$F(x) = \gamma x|x|^{-m-1}, \tag{1.36}$$

where γ is a real constant and m is a positive constant. In the classical example of the gravitational or Coulomb forces (the Holtzmark case) $d = 3, m = 2$ and γ depends on the physical parameters of the particles (mass, charge, etc). Suppose now that the position x of a particle is random and is uniformly distributed in the ball B_R of the radius R in $\mathbb{R}^d$. Then the characteristic function of the force between this particle and the origin is

$$\phi_1(p) = |B_R|^{-1} \int_{B_R} e^{i(p,F(x))} dx = 1 + |B_R|^{-1} \int_{B_R} (e^{i(p,F(x))} - 1) dx,$$

where $|B_R|$ denote the volume of B_R. If there are N independent uniformly distributed particles in B_R, then the characteristic function of the force induced by all these particles is clearly

$$\phi_N(p) = \left[1 + |B_R|^{-1} \int_{B_R} (e^{i(p,F(x))} - 1) dx\right]^N.$$

Assume now that the number of particles N is proportional to the volume with a certain fixed density $\lambda > 0$, that is $N = \lambda|B_R|$. We are interested in the limit of the corresponding distribution as $R \to \infty$ (the constant density equation $N = \lambda|B_R|$ makes precise the idea of 'uniform distribution in $\mathbb{R}^d$' mentioned above). The resulting limiting distribution of stars in $\mathbb{R}^d$ is called a *Poisson point process with intensity* λ and will be studied in more detail in Chapter 3.

Thus we are looking for the limit

$$\phi(p) = \lim_{R\to\infty} \left[1 + |B_R|^{-1} \int_{B_R} (e^{i(p,F(x))} - 1) dx\right]^{\lambda|B_R|}.$$

By the property of the exponential function this limit exists and equals

$$\phi(p) = \exp\{-\lambda\omega(p)\}, \tag{1.37}$$

whenever the limit

$$\omega(p) = \lim_{R\to\infty} \int_{B_R} (1 - e^{i(p,F(x))})dx$$

exists. Using (1.36) this rewrites as

$$\omega(p) = \lim_{R\to\infty} \int_{B_R} (1 - \exp\{i\gamma(p,x)|x|^{-m-1}\}),$$

or (noting that the imaginary part vanishes by symmetry) as

$$\omega(p) = \lim_{R\to\infty} \int_{B_R} (1 - \cos\{\gamma(p,x)|x|^{-m-1}\}).$$

Choosing spherical coordinates (r, θ, ϕ) in $\mathbb{R}^d$, $r > 0, \theta \in [0, \pi], \phi \in S^{d-2}$ with the main axis $\{\theta = 0\}$ given by the direction p and using

$$dx = r^{d-1}\, dr \sin^{d-2}\theta \, d\theta \, d\phi$$

yields

$$\omega(p) = |S^{d-2}| \lim_{R\to\infty} \int_0^R r^{d-1} dr \int_0^\pi \sin^{d-2}\theta (1 - \cos\{\gamma|p|r^{-m}\cos\theta\}) d\theta, \tag{1.38}$$

where $|S^{d-2}|$ is the surface area of the $(d-2)$-dimensional unit sphere S^{d-2}. In order to see when this improper Riemann integral exists, let us observe that for large r (and any fixed $|p|$) we have

$$1 - \cos\{\gamma|p|r^{-m}\cos\theta\} \sim \gamma^2|p|^2 r^{-2m}\cos^2\theta,$$

so that the integrability condition reads as $d - 1 - 2m < -1$, or just

$$\alpha = d/m < 2. \tag{1.39}$$

Assuming this holds, make the change of variable r to $z = |p|r^{-m}$ in (1.38). Then

$$r = (z/|p|)^{-1/m}, \quad dr = -\frac{1}{m}|p|^{1/m} z^{-1-1/m}\, dz,$$

so that

$$\omega(p) = c(\gamma)|p|^\alpha, \tag{1.40}$$

where $\alpha = d/m$ and

$$c(\gamma) = \frac{1}{m}|S^{d-2}| \int_0^\infty z^{-\alpha-1}\,dz \int_0^\pi \sin^{d-2}\theta(1-\cos\{\gamma z\cos\theta\})\,d\theta. \qquad (1.41)$$

Putting together (1.37) and (1.40) we see that the characteristic function $\phi(p)$ of the limiting distribution coincides with the characteristic function of α-stable symmetric distribution in $\mathbb{R}^d$, and condition (1.39) coincides with the general bound for the index of stability.

Remark 4. Since $\phi(p)$ given by (1.37) and (1.40) is obtained as the limit of characteristic functions the arguments above yield another proof of the fact that the functions of the form $\exp\{-c|p|^\alpha\}$ with $c > 0$ and $\alpha \in (0,2)$ are infinitely divisible characteristic functions.

Exercise 1.5.1. Extend the above construction to obtain nonsymmetric stable laws by assuming that the density of particles in the cone generated by a surface element ds is proportional to $\mu(ds)$ with some measure μ in S^{d-1}.

1.6 Unimodality of probability laws

We shall sketch here the theory of unimodality of probability laws. More details and historical comments can be found e.g. in Dharmadhikari and Joag-Dev [99] and in the appendix to [179]. Loosely speaking, *unimodality* means that the law is bell-shaped, that is the probability is peaked at some place and then its concentration decreases when moving away from this peak (mode) in any direction. Despite its clear practical importance, this property is not often discussed in textbooks, as such a discussion requires nontrivial excursions to geometrical topics connected with tomography. We shall undertake this excursion for completeness aiming eventually at proving unimodality of symmetric stable laws, which will be used in Chapter 7 for obtaining effective two-sided estimates for the heat kernels of stable and stable-like processes.[2]

In one dimension the corresponding rigorous definition is obvious. Namely, a probability law (or a finite measure) on the real line with a continuous density function f is called *unimodal* with the *mode (or vertex)* $a \in \mathbb{R}$ if f is non-decreasing on $(-\infty, a)$ and non-increasing on (a, ∞). More generally, if a law is given by its distribution function F, it is *unimodal* with the mode $a \in \mathbb{R}$, if $F(x)$ is convex (possibly not strictly) on $(-\infty, a)$ and concave on (a, ∞).

The simplest example represents all normal laws. Other natural examples are stable laws that we shall discuss later.

In several dimensions a proper definition is not so straightforward. In fact, there exist several reasonable definitions of unimodality for finite-dimensional distributions.

[2] Readers not interested in this development may skip this section.

A measure on $\mathbb{R}^d$ with a density f is called *convex unimodal*, if the function f has convex sets of upper values, i.e. the sets $\{x : f(x) \geq c\}$ are convex for all c.

One of the disadvantages of this definition is the fact that the class of convex unimodal measures is not closed under convex linear combinations. The following two more general concepts improve the situation, at least for the symmetric case with which we shall deal here.

The uniform probability law μ_A supported by a compact symmetric convex $A \subset \mathbb{R}^d$ is called an *elementary unimodal symmetric distribution* in $\mathbb{R}^d$. A centrally symmetric finite measure on $\mathbb{R}^d$ is called *central convex unimodal* (CCU), if it is a weak limit of a sequence of the finite linear combinations of elementary symmetric unimodal measures.

A measure μ on $\mathbb{R}^d$ is called *monotone unimodal*, if for any $y \in \mathbb{R}^d$ and any centrally symmetric convex body $M \subset \mathbb{R}^d$, the function $\mu(M + ty)$ is non increasing for $t > 0$.

Exercise 1.6.1. (i) If a convex unimodal measure is centrally symmetric, then it is CCU and monotone unimodal.

(ii) Normal laws are unimodal in all three senses defined above.

The classes of CCU and monotone unimodal measures clearly are closed under linear combinations with positive coefficients and CCU measures are preserved when passing to the weak limit. Notice also that since any convex set can be approximated by a sequence of convex sets with nonempty interiors, the definition of CCU will not change, if one considers there only the compact convex sets with nonempty interiors. Such convex sets will be called *convex bodies*.

To obtain the main properties of unimodal measures one needs some elements of the Brunn–Minkowski theory of mixed volumes, which we now recall.

Theorem 1.6.1 (Brunn–Minkowski inequality). *For any non-empty compact sets $A, B \subset \mathbb{R}^d$*

$$V^{1/d}(A + B) \geq V^{1/d}(A) + V^{1/d}(B), \tag{1.42}$$

or more generally

$$V^{1/d}(t_1 A + t_2 B) \geq t_1 V^{1/d}(A) + t_2 V^{1/d}(B) \tag{1.43}$$

for any positive t_1, t_2, where V denotes the standard volume in $\mathbb{R}^d$ and the sum of two sets is defined as usual by $A + B = \{a + b : a \in A, b \in B\}$.

Proof. This classical result has a long history and can be proved by different methods, see e.g. Schneider [291]. We sketch here a beautiful elementary proof taken from Burago and Zalgaller [70]. Namely, let us say that a compact set A in $\mathbb{R}^d$ is *elementary*, if it is the union of the finite number $l(A)$ of non-degenerate cuboids with sides parallel to the coordinate axes and such that their interiors do not intersect. Each compact set A can be approximated by a sequence of elementary sets A_i so that $V(A_i) \to V(A)$. Therefore, it suffices to prove (1.42) for elementary sets only. Consider first the case when each of A, B consists of only one cuboid with edges $a_i > 0$ and $b_i > 0$ respectively. Then (1.42) takes the form

$$\prod_{i=1}^{d} (a_i + b_i)^{1/d} \geq \prod_{i=1}^{d} a_i^{1/d} + \prod_{i=1}^{d} b_i^{1/d},$$

which follows from the inequality

$$\Big(\prod_{i=1}^{d} \frac{a_i}{a_i + b_i}\Big)^{1/d} + \Big(\prod_{i=1}^{d} \frac{b_i}{a_i + b_i}\Big)^{1/d} \leq \frac{1}{d}\sum_{i=1}^{d} \frac{a_i}{a_i + b_i} + \frac{1}{d}\sum_{i=1}^{d} \frac{b_i}{a_i + b_i} = 1.$$

For general non-empty elementary sets A, B the proof can be carried out by induction over $l(A) + l(B)$. Assume that (1.42) is true when $l(A) + l(B) \leq k - 1$. Suppose that $l(A) \geq 2$. Clearly there exists a hyperplane P which is orthogonal to one of the coordinate axes and which splits A into elementary sets A', A'' such that $l(A') < l(A)$ and $l(A'') < l(A)$. Then $V(A') = \lambda V(A)$ with some $\lambda \in (0, 1)$. Since parallel translations do not change volumes, one can choose the origin of coordinates on the plane P and then shift the set B so that the same hyperplane P splits B into sets B', B'' with $V(B') = \lambda V(B)$. Plainly $l(B') \leq l(B), l(B'') \leq l(B)$. The pairs of sets A', B', and A'', B'' each lies in its own half-space with respect to P and in each pair there are no more than $k - 1$ cuboids. Hence

$$\begin{aligned} V(A + B) &\geq V(A' + B') + V(A'' + B'') \\ &\geq [V^{1/d}(A') + V^{1/d}(B')]^d + [V^{1/d}(A'') + V^{1/d}(B'')]^d \\ &= \lambda[V^{1/d}(A) + V^{1/d}(B)]^d + (1 - \lambda)[V^{1/d}(A) + V^{1/d}(B)]^d \\ &= [V^{1/d}(A) + V^{1/d}(B)]^d, \end{aligned}$$

which completes the proof of (1.42). Inequality (1.43) is a direct consequence of (1.42). □

Consider now a convex body A in $\mathbb{R}^d$. Let S be a k-dimensional subspace, $k < d$. The k-dimensional *X-ray* of A (or, in other terminology, the *section function*, or the *k-plane Radon transform* of A) *parallel to* S is the function of $x \in S^{\perp}$ defined by the formula $X_S A(x) = V_k(A \cap (S + x))$, where V_k denotes the k-dimensional volume.

Corollary 1. *For any convex body A and any k-dimensional subspace S the function $(X_S A)^{1/k}$ is concave on its support. If A is also symmetric, then $X_S A$ has a maximum at the origin.*

Proof. In fact if $A_0 = A \cap (x+S)$ and $A_1 = A \cap (y+S)$, then

$$(X_S A((1-t)x+ty))^{1/k} \geq V_k^{1/k}((1-t)A_0+tA)) \geq (1-t)V_k^{1/k}(A_0)+tV_k^{1/k}(A_1),$$

the first and second inequality following respectively from the convexity of A and (1.43). □

Corollary 2 (Fary–Redei lemma). *Let B_1 and B_2 be two centrally symmetric convex bodies. Then the function $(\mathbf{1}_{B_1} \star \mathbf{1}_{B_2})^{1/d}$ is concave on its support (where $\star$ denotes the convolution).*

Proof. Let us consider two d-dimensional planes L_1 and L_2 in $\mathbb{R}^{2d}$ intersecting only at the origin and having angle $\phi \leq \pi/2$ between them. Let M_1 and M_2 denote the bodies which are equal to B_1 and B_2 respectively, but lie on the planes L_1 and L_2 respectively. Then the measure with density

$$\mathbf{1}_{B_1} \star \mathbf{1}_{B_2}(x) = V_d((x - B_1) \cap B_2) = V_d((x - B_1) \cap B_2) \tag{1.44}$$

can be considered as the limit as $\phi \to 0$ of the measures μ_ϕ whose densities f_ϕ are convolutions of the indicators of M_1 and M_2 in $\mathbb{R}^{2d}$. Notice that though $\mathbf{1}_{M_i}$, $i = 1, 2$, do not have densities with respect to the volume in $\mathbb{R}^{2d}$, their convolution does. In fact, if $\phi = \pi/2$, then

$$\mu_{\pi/2}(A_1 \times A_2) = V_d(A_1 \cap M_1) \times V_d(A_2 \cap M_2)$$

for any $A_i \subset L_i, i = 1, 2$, implying that $f_{\pi/2} = \mathbf{1}_{M_1+M_2}$. Generally

$$f_\phi = \mathbf{1}_{M_\phi}(\sin\phi)^{-d} = \mathbf{1}_{M_1+M_2}(\sin\phi)^{-d},$$

as the linear transformation of $\mathbb{R}^{2d}$ which is the identity on L_1 and which makes L_2 perpendicular to L_1, has the determinant $(\sin\phi)^d$. Hence

$$\begin{aligned}(\mathbf{1}_{B_1} \star \mathbf{1}_{B_2})^{1/d}(x) &= \lim_{\phi\to 0} \frac{1}{\sin\phi} V_d^{1/d}((x + L_1^\perp) \cap M_\phi) \\ &= \lim_{\phi\to 0} (\sin\phi)^{-1}(X_{L_1^\perp} M_\phi(x))^{1/d},\end{aligned}$$

and the concavity of $(\mathbf{1}_{B_1} \star \mathbf{1}_{B_2})^{1/d}$ follows from Corollary 1. □

As a direct corollary of the Brunn–Minkovski theory developed above, one obtains the following basic properties of finite-dimensional unimodality.

Theorem 1.6.2. (i) *The class* CCU *is closed with respect to convolution* $\star$.

(ii) *All* CCU *measures are monotone unimodal.*

(iii) *Let a* CCU *measure* μ *has a continuous density* f. *Then for any unit vector* v *the function* $f(tv)$ *is non-increasing on* $\{t \geq 0\}$, *and moreover, for any* $m < d$ *and any* m*-dimensional subspace* S *the integral of* f *over the plane* $tv + S$ *is a non-increasing function on* $\{t \geq 0\}$. *(In other words, the Radon transform of* f *is non-increasing, as is the Radon transform of the restriction of* f *to any subspace.)*

Remark 5. Surprisingly enough the converse of the statement (ii) does not hold.

Proof. (i) It is enough to prove that the function (1.44) is convex unimodal for arbitrary centrally symmetric convex bodies B_1 and B_2. But this follows directly from the Fary–Redei lemma.

(ii) Let us prove it here only for CCU measures with densities (and we shall use it only in this case). Notice first that if $\mu = \mu_A$ with some compact convex A, and if M is compact, the required statement about the function $\mu(M + ty)$ follows directly from the Fary–Redei lemma and (1.44). For a non-compact set M, the statement is obtained by a trivial limiting procedure. For a general absolutely continuous μ it is again obtained by a limiting procedure.

(iii) It is a consequence of (ii). For instance, to prove that $f(tv)$ is non-increasing one supposes that $f(t_1 v) > f(t_2 v)$ with some $t_1 > t_2$ and then uses statement (ii) with a set M being the ball B_ϵ of sufficiently small radius ϵ to get a contradiction. □

We shall discuss now the unimodality of stable laws. This property is important for the study of the asymptotic behavior of the stable distributions (and more generally stable-like processes introduced later on) as it allows one to describe the behaviour of stable densities between the regions of "large" and "small" distances.

It was proven by Yamazato [320] (see also Zolotarev [327]) that all stable distributions on the real line are unimodal. We shall neither use nor prove this rather nontrivial fact. Instead we shall concentrate on the unimodality of finite-dimensional symmetric laws obtained by Kanter.

Theorem 1.6.3. *All symmetric stable laws are unimodal.*

Proof. This will be given in three steps.

Step 1. *Reduction to the case of finite Lévy measure.* Recall that the density of a general symmetric stable law with the index of stability $\alpha \in (0, 2)$ (we shall not consider the case of $\alpha = 2$ which is the well-known Gaussian distribution) is given by the Fourier transform

$$S(x, \alpha, \mu) = \frac{1}{(2\pi)^d} \int \psi_\alpha(p) e^{ipx} \, dp$$

of the characteristic function ψ_α, which is given either by (1.19) with vanishing A and a symmetric measure μ on S^{d-1}. For any $\epsilon > 0$ consider the finite Lévy measure

$$\nu_\epsilon(d|\xi|, ds) = \begin{cases} |\xi|^{-1-\alpha} d|\xi|\mu(ds), & |\xi| \geq \epsilon \\ \epsilon^{-1-\alpha} d|\xi|\mu(ds), & |\xi| \leq \epsilon \end{cases} \tag{1.45}$$

and the corresponding infinitely divisible distribution with the characteristic function ψ_α^ϵ defined by the formula

$$\log \psi_\alpha^\epsilon(y) = \int_0^\infty \int_{S^{d-1}} \left(e^{i(y,\xi)} - 1 - \frac{i(y,\xi)}{1+\xi^2} \right) \nu_\epsilon(d|\xi|, ds). \tag{1.46}$$

Let P_ϵ denote the corresponding probability distribution. One sees that $\psi_\alpha^\epsilon \to \psi_\alpha$ as $\epsilon \to 0$ uniformly for y from any compact set, because

$$\begin{aligned} |\log \psi_\alpha^\epsilon(y) - \log \psi_\alpha(y)| &\leq \int_0^\epsilon \int_{S^{d-1}} \left| e^{i(y,\xi)} - 1 - \frac{i(y,\xi)}{1+\xi^2} \right| |\xi|^{-1-\alpha} d|\xi|\mu(ds) \\ &= O(1)|y|^2 \int_0^\epsilon |\xi|^{1-\alpha} d|\xi| = O(1)|y|^2|\epsilon|^{2-\alpha}. \end{aligned}$$

The convergence of characteristic functions (uniform on compacts) implies the weak convergence of the corresponding distributions. Therefore, it is enough to prove the unimodality of the distribution P_ϵ for any ϵ.

Step 2. *Reduction to the unimodality of the Lévy measure.* We claim now that in order to prove the unimodality of P_ϵ it is suffice to prove the unimodality of the Lévy measure (1.45). In fact, since this measure is finite and symmetric, formula (1.46) can be rewritten in the form

$$\log \psi_\alpha^\epsilon(y) = \int_0^\infty \int_{S^{d-1}} e^{i(y,\xi)} \nu_\epsilon(d|\xi|, ds) - C_\epsilon$$

with some constant C_ϵ, so that up to a positive multiplier

$$\psi_\alpha^\epsilon(y) = \exp\left\{ \int_0^\infty \int_{S^{d-1}} e^{i(y,\xi)} \nu_\epsilon(d|\xi|, ds) \right\}.$$

Expanding exp in the Taylor series shows that $\psi_\alpha^\epsilon(y)$ is the Fourier transform of a limit (uniform on compacts) of the finite linear combinations of ν_ϵ and its convolutions with itself. Therefore our assertion follows from Theorem 1.6.2.

Step 3. *Completion.* It remains to prove that the Lévy measure (1.45) is unimodal. To this end, notice that any measure μ on S^{d-1} can be approximated weakly by a sequences of discrete measures concentrated on countably many points. Hence, by linearity, it is enough to prove the unimodality of measure (1.45) in the case of $\mu(ds)$ concentrated in one point only. But in this case measure (1.45) is one-dimensional and the statement is obvious, which completes the proof of the theorem. □

The same arguments prove the following fact.

Proposition 1.6.1. *If the Lévy measure* ν *of an infinitely divisible distribution* F *in* $\mathbb{R}^d$ *(with polar coordinates* $|\xi|, s = \xi/|\xi|$*) has the form*

$$\nu(d\xi) = f(|\xi|)\, d|\xi|\, \mu(ds)$$

with any finite (centrally) symmetric measure μ *on* S^{d-1} *and any non-increasing function* f*, then* F *is symmetric unimodal.*

1.7 Compactness for function spaces and measures

The properties of separability, metrizability, compactness and completeness for a topological space S are crucial for the analysis of S-valued random processes. Here we shall recall the relevant notions for the basic function spaces and measures highlighting the main ideas with examples while omitting lengthy proofs.

Recall that a topological (e.g. metric) space is called *separable* if it contains a countable dense subset. It is useful to have in mind that separability is a topological property, unlike, say, completeness, that depends on the choice of the distance (for example, an open interval and the line $\mathbb{R}$ are homeomorphic, but the usual distance is complete for the line and not complete for the interval). The following standard examples show that separability does not go without saying.

Examples. 1. The Banach spaces l_∞ of bounded sequences of real (or complex) numbers $a = (a_1, a_2, \dots)$ equipped with the sup norm $\|a\| = \sup_i |a_i|$ is not separable, because its subset of sequences with values in $\{0, 1\}$ is not countable, but the distance between any two such (not coinciding) sequences is one.

2. The Banach spaces $C(\mathbb{R}^d)$, $L_\infty(\mathbb{R}^d)$, $\mathcal{M}^{\text{signed}}(\mathbb{R}^d)$ are not separable, because they contain a subspace isomorphic to l_∞.

3. The Banach spaces $C_\infty(\mathbb{R}^d)$, $L_p(\mathbb{R}^d)$, $p \in [1, \infty)$, are separable, which follows from the Stone–Weierstrass theorem.

Turning to compactness, recall that a subset of a metric space is called *relatively compact* if its closure is compact. Let us start with the space $C([0, T], S)$ of continuous functions on an interval with values in a metric space S equipped with its usual sup-norm. A useful characteristic of a S-valued function x on $[0, t]$ is the *modulus of continuity*

$$w(x, t, h) = \sup\{\rho(x(s), x(r)) : r - h \le s \le r \le t\}, \quad h > 0.$$

As one easily sees a measurable function x is continuous on $[0, t]$ if and only if $\lim_{h\to 0} w(x, t, h) = 0$. Moreover, the following well-known characterization of compactness is available for continuous functions (see e.g. [323]).

Theorem 1.7.1 (Arzelà–Ascoli theorem). *Let S be a complete metric spaces and $T > 0$. A set H in $C([0,T], S)$ is relatively compact in the sup-norm topology if and only if the sets $\pi_t(H)$ are relatively compact in S for t from a dense subset of $[0,T]$ and*

$$\lim_{h\to 0} \sup_{f\in H} w(x,T,h) = 0.$$

Here π_t denotes the evaluation map: $\pi_t f = f(t)$.

As we shall see, the main classes of stochastic processes, martingales and regular enough Markov process, have modifications with *càdlàg* paths meaning that they are right-continuous and have left limits (the word càdlàg is a French acronym). This is a quite remarkable fact taking into account the general Kolmogorov result on the existence of processes on the space of all (even nonmeasurable) paths. If (S,ρ) is a metric space, the set of S-valued càdlàg functions on a finite interval $[0,T]$, $T \in \mathbb{R}_+$ or on the half-line $\mathbb{R}_+$ is usually denoted by $D = D([0,T], S)$ or $D = D(\mathbb{R}_+, S)$, and is called the *Skorohod path space*. We shall often write $D([0,T], S)$ for both these cases, meaning that T can be finite or infinite.

Proposition 1.7.1. *If $x \in D([0,T], S)$, then for any $\delta > 0$ there can exist only finitely many jumps of x on $[0,T]$ of a size exceeding δ.*

Proof. Otherwise the jumps exceeding δ would have an accumulation point on $[0,T]$. □

The major notion for the analysis of càdlàg functions is that of the *modified modulus of continuity* defined as

$$\tilde{w}(x,t,h) = \inf_{\Delta} \max_{k} \sup_{t_k \le r,s < t_{k+1}} \rho(x(r), x(s)), \tag{1.47}$$

where the infimum extends over all partitions $\Delta = (0 = t_0 < t_1 < \cdots < t_l < t)$ such that $t_{k+1} - t_k \ge h$ for $k = 1, \dots, l-1$.

It is easily seen that the definition of $\tilde{w}(x,t,h)$ is not changed if the infimum will be extended only over the partitions with $h \le t_{k+1} - t_k < 2h$. In particular, $\tilde{w}(x,t,h) \le \tilde{w}(x,t,2h)$ for all x.

Proposition 1.7.2. *If $x \in D([0,t], S)$, then* $\lim_{h\to 0} \tilde{w}(x,t,h) = 0$.

Proof. By Proposition 1.7.1, for an arbitrary δ there exists a partition $0 = t_0^0 < t_1^0 < \cdots < t_k^0 = t$ of $[0,t]$ such that inside the intervals $I_l = [t_l^0, t_{l+1}^0)$ of the partition there are no jumps with sizes exceeding δ. Let us make a further partition of each I_l, defining recursively

$$t_l^{j+1} = \min(t_{l+1}^0, \inf\{s > t_l^j : |x(s) - x(t_l^{j-1})| > 2\delta\})$$

with as many j as one needs to reach t_{l+1}^0. Clearly the new partition thus obtained is finite, and all the differences $t_l^j - t_l^{j+1}$ are strictly positive, so that

$$h = \min_{l,j}(t_l^j - t_l^{j+1}) > 0.$$

On the other hand, $\tilde{w}(x, t, h) \le 4\delta$. □

Looking for an appropriate topology on D is not an obvious task. The intuition arising from the study of random processes suggests that in a reasonable topology the convergence of the sizes and times of jumps should imply the convergence of paths. For example, the sequence of step-functions $\mathbf{1}_{[1+1/n,\infty)}$ should converge to $\mathbf{1}_{[1,\infty)}$ as $n \to \infty$ in $D([0,T], \mathbb{R}_+)$ with $T > 1$. On the other hand, the usual uniform distance

$$\|\mathbf{1}_{[1+1/n,\infty)} - \mathbf{1}_{[1,\infty)}\| = \sup_y |\mathbf{1}_{[1+1/n,\infty)}(y) - \mathbf{1}_{[1,\infty)}(y)|$$

equals one for all n, preventing convergence in the uniform topology. The main idea to make $\mathbf{1}_{[1+1/n,\infty)}$ and $\mathbf{1}_{[1,\infty)}$ close is by introducing a time change that may connect them. Namely, a *time change* on $[0,T]$ or $\mathbb{R}_+$ is defined as a monotone continuous bijection of $[0,T]$ or $\mathbb{R}_+$ on itself. One says that a sequence $x_n \in D([0,T], S)$ converges to $x \in D([0,T], S)$ in the *Skorohod topology* J_1 if there exists a sequence of time changes λ_n of $[0,T]$ such that

$$\sup_s |\lambda_n(s) - s| + \sup_{s \le t} \rho(x_n(\lambda_n(s)), x(s)) \to 0, \quad n \to \infty$$

for $t = T$ in case of a finite T or for all $t > 0$ in case $T = \infty$.

For example, for

$$\lambda_n = \begin{cases} (1 + 1/n)t, & t \le 1 \\ (1 - 1/n)(t - 1) + (1 + 1/n), & 1 \le t \le 2 \\ t, & t \ge 2 \end{cases} \tag{1.48}$$

one has $\mathbf{1}_{[1+1/n,\infty)}(\lambda_n(t)) = \mathbf{1}_{[1,\infty)}(t)$ for all t, so that

$$\sup_{s \le t}(|\lambda_n(s) - s| + |\mathbf{1}_{[1+1/n,\infty)}(\lambda_n(s)) - \mathbf{1}_{[1,\infty)}(s))|) = 1/n \to 0$$

as $n \to \infty$ for all $t \ge 2$, so that the step-functions $\mathbf{1}_{[1+1/n,\infty)}$ converge to $\mathbf{1}_{[1,\infty)}$ in the Skorohod topology of $D(\mathbb{R}_+, \mathbb{R}_+)$ as $n \to \infty$.

The following fundamental result due to Skorohod, Kolmogorov and Prohorov collects the main facts on the Skorohod topology.

Theorem 1.7.2. *Let S be a complete separable metric space and $T > 0$ or $T = \infty$.*

(i) *The Borel σ-algebra of $D([0,T],S)$ is generated by the evaluation maps $\pi_t : x \mapsto x(t)$ for all $t \le T$ (or $t < \infty$ in case $T = \infty$).*

(ii) *A set $A \subset D([0,T],S)$ is relatively compact in the J_1-topology if and only if $\pi_t(A)$ is relatively compact in S for each t and*

$$\lim_{h\to 0} \sup_{x\in A} \tilde{w}(x,t,h) = 0, \quad t > 0. \tag{1.49}$$

(iii) *There exists a metric d_P, called the Prohorov metric, that metrizes the J_1 Skorohod topology in such a way that $(D([0,T],S), d_P)$ becomes a complete separable metric space.*

A (standard by now, but rather involved) proof can be found e.g. in Jacod and Shiryaev [147], Billingsley [56] or Ethier and Kurtz [110].

Remark 6. If S is locally compact and T is finite, condition (ii) of the above theorem implies that there exists a compact set Γ_T such that $\pi_t(A) \subset \Gamma_T$ for all $t \in [0,T]$. In fact, if $\tilde{w}(x,T,h) < \epsilon$ let Γ be the union of the finite number of compact closures of $\pi_{hk/2}(A)$, $hk/2 \le T$, $k \in \mathbb{N}$. Then all intervals of any partition with $[t_k, t_{k+1}] \ge h$ contain a point $hk/2$, so that the whole trajectory belongs to the compact set $\bigcup \Gamma^{\epsilon}_{kh/2}$.

Measures often arise as dual of function spaces, by the following fundamental result. Recall that we denote by (f,μ) the usual pairing between functions and measures given by integration.

Theorem 1.7.3 (Riesz–Markov theorem). *If S is a locally compact topological space, the space of finite signed Borel measures μ on S with the total variation norm*

$$\|\mu\| = \sup_{f\in C_\infty(S):\|f\|\le 1} (f,\mu)$$

is the Banach dual to the Banach space $C_\infty(S)$. In particular, any positive bounded linear functional on $C_\infty(S)$ is specified by a (positive) finite Borel measure.

Apart from the norm topology, various weaker topologies on measures are of use. If S is a metric space, a sequence of finite Borel measures μ_n is said to converge *weakly* (resp. *vaguely*) to a measure μ as $n \to \infty$ if (f,μ_n) converges to (f,μ) for any $f \in C(S)$ (resp. for any $f \in C_c(S)$). If S is locally compact, the duality given by the Riesz–Markov theorem specifies the *weak-$\star$ topology* on measures, where the convergence μ_n to μ as $n \to \infty$ means that (f,μ_n) converges to (f,μ) for any $f \in C_\infty(S)$.

Example. Let $S = \mathbb{R}$. The sequence $\mu_n = n\delta_n$, $n \in \mathbb{N}$, in $\mathcal{M}(\mathbb{R})$ converges vaguely, but not weakly-$\star$. The sequence $\mu_n = \delta_n$ converges weakly-$\star$, but not weakly.

Proposition 1.7.3. *Suppose p_n, $n \in \mathbb{N}$, and p are finite Borel measures in $\mathbb{R}^d$.*

(i) *If p_n converge vaguely to p and $p_n(\mathbb{R}^d)$ converge to $p(\mathbb{R}^d)$ as $n \to \infty$, then p_n converge to p weakly (in particular, if p_n and p are probability laws, the vague and weak convergence coincide).*

(ii) *$p_n \to p$ weakly-$\star$ if and only if $p_n \to p$ vaguely and the sequence p_n is bounded.*

Proof. (i) Assuming p_n is not tight leads to a contradiction, since then $\exists\, \epsilon$: $\forall$ compact set K $\exists n : \mu_n(\mathbb{R}^d \setminus K) > \epsilon$, implying that

$$\liminf_{n\to\infty} p_n(\mathbb{R}^d) \geq p(\mathbb{R}^d) + \epsilon.$$

(ii) This is straightforward. □

Proposition 1.7.4. *If S is a separable metric space, then the space $\mathcal{M}(S)$ of finite Borel measures is separable in the weak (and hence also in the vague) topology.*

Proof. A dense countable set is given by the linear combinations with rational coefficients of the Dirac masses δ_{x_i}, where $\{x_i\}$ is a dense subset of S. □

The following general fact from the functional analysis is important for the analysis of measures.

Proposition 1.7.5. *Let B be a separable Banach space. Then the unit ball $B_1^\star$ in its dual Banach space $B^\star$ is weakly-$\star$ compact and there exists a complete metric in $B_1^\star$ compatible with this topology.*

Proof. Let $\{x_1, x_2, \dots\}$ be a dense subset in the unit ball of B. The formula

$$\rho(\mu, \eta) = \sum_{k=1}^{\infty} 2^{-k} |(\mu - \eta, x_k)|$$

specifies a complete metric in $B_1^\star$ compatible with the $\star$-weak convergence, i.e. $\rho(\mu_n, \mu) \to 0$ as $n \to 0$ if and only if $(\mu_n, x) \to (\mu, x)$ for any $x \in B$. Completeness and compactness follows easily. Say, to show compactness, we have to show that any sequence μ_n has a converging subsequence. To this end, we first chooses a subsequence μ_n^1 such that (μ_n^1, x_1) converges, then a further subsequence μ_n^2 such that (μ_n^2, x_2) converges, etc, and finally the diagonal subsequence μ_n^n converges on any of x_i and is therefore convergent. □

Remark 7. $B_1^\star$ is actually compact without the assumption of separability of B (the Banach–Alaoglu theorem).

Proposition 1.7.6. *If S is a separable locally compact metric space, then the set $\mathcal{M}_M(S)$ of Borel measures with norm bounded by M is a complete separable metric compact set in the vague topology.*

Proof. This follows from Propositions 1.7.5 and 1.7.4. □

To metrize the weak topology on measures one needs other approaches. Let S be a metric space with the distance d. For $P, Q \in \mathcal{P}(S)$ define the *Prohorov distance*

$$\rho_{\text{Proh}}(P, Q) = \inf\{\epsilon > 0 : P(F) \le Q(F^\epsilon) + \epsilon \ \forall \text{ closed } F\},$$

where $F^\epsilon = \{x \in S : \inf_{y \in F} d(x, y) < \epsilon\}$.

It is not difficult to show that

$$P(F) \le Q(F^\epsilon) + \beta \iff Q(F) \le P(F^\epsilon) + \beta,$$

leading to the conclusion that ρ_{Proh} is actually a metric.

Theorem 1.7.4. (i) *If S is separable, then $\rho(\mu_n, \mu) \to 0$ as $n \to \infty$ for $\mu, \mu_1, \mu_2, \dots \in \mathcal{P}(S)$ if and only if $\mu_n \to \mu$ weakly.*

(ii) *If S is separable (resp. complete and separable), then $(\mathcal{P}(S), \rho_{\text{Proh}})$ is separable (resp. complete and separable).*

Proof. See e.g. Chapter 3 of Ethier and Kurtz [110]. □

It is instructive to have a probabilistic interpretation of this distance.

One says that a measure $\nu \in \mathcal{P}(S \times S)$ is a *coupling* of the measures $\mu, \eta \in \mathcal{P}(S)$ if the margins of ν are μ and η, i.e. if $\nu(A \times S) = \mu(A)$ and $\nu(S \times A) = \eta(A)$ for any measurable A, or equivalently if

$$\int_{S \times S} (\phi(x) + \psi(y))\nu(dxdy) = (\phi, \mu) + (\psi, \eta) \tag{1.50}$$

for any $\phi, \psi \in C(S)$.

Theorem 1.7.5. *Let S be a separable metric space and $\mathbf{P}, \mathbf{Q} \in \mathcal{P}(S)$. Then*

$$\rho_{\text{Proh}}(\mathbf{P}, \mathbf{Q}) = \inf_\nu \inf\{\epsilon > 0 : \nu(x, y : d(x, y) \ge \epsilon) \le \epsilon\},$$

where $\inf_\nu$ *is taken over all couplings of of* $\mathbf{P}, \mathbf{Q}$.

Proof. See Chapter 3 of Ethier and Kurtz [110]. □

A family Π of measures on a complete separable metric space S is called *tight* if for any ϵ there exists a compact set $K \subset S$ such that $\mathbf{P}(S \setminus K) < \epsilon$ for all measures $\mathbf{P} \in \Pi$. The following fact is fundamental (a proof can be found e.g. in [110], Shiryayev [293] or Kallenberg [154]).

Theorem 1.7.6 (Prohorov's compactness criterion). *A family Π of measures on a complete separable metric space S is relatively compact in the weak topology if and only if it is tight.*

Another handy way to metricize the weak topology of measures is via Wasserstein–Kantorovich distances. Namely, let $\mathcal{P}^p(S)$ be the set of probability measures μ on S with the finite p-th moment, $p > 0$, i.e. such that

$$\int d^p(x_0, x)\mu(dx) < \infty$$

for some (and hence clearly for all) x_0.

The *Wasserstein–Kantorovich metrics* W_p, $p \geq 1$, on the set of probability measures $\mathcal{P}^p(S)$ are defined as

$$W_p(\mu_1, \mu_2) = \left(\inf_\nu \int d^p(y_1, y_2)\nu(dy_1 dy_2)\right)^{1/p}, \tag{1.51}$$

where the inf is taken over the class of probability measures ν on $S \times S$ that couple μ_1 and μ_2. Of course W_p depends on the metric d. It follows directly from the definition that

$$W_p^p(\mu_1, \mu_2) = \inf \mathbf{E}\, d^p(X_1, X_2), \tag{1.52}$$

where the inf is taken over all random vectors (X_1, X_2) such that X_i has the law μ_i, $i = 1, 2$. One can show (see e.g. Villani [314]) that W_p are actually metrics on $\mathcal{P}^p(S)$ (the only point not obvious being of course the triangle inequality), and that they are complete.

Proposition 1.7.7. *If S is complete and separable, the infimum in* (1.51) *is attained.*

Proof. In view of Theorem 1.7.6, in order to be able to pick a converging subsequence from a minimizing sequence of couplings, one needs to know that the set of couplings is tight. But this is straightforward: if K be a compact set in S such that $\mu_1(S \setminus K) \leq \delta$ and $\mu_2(S \setminus K) \leq \delta$, then

$$\nu(S \times S \setminus (K \times K)) \leq \nu(S \times (S \setminus K)) + \nu((S \setminus K) \times S) \leq 2\delta$$

for any coupling ν. □

The main result connecting weak convergence with the Wasserstein metrics is as follows.

Theorem 1.7.7. *If S is complete and separable, $p \geq 1$, $\mu, \mu_1, \mu_2, \dots$ are elements of $\mathcal{P}^p(S)$, then the following statements are equivalent:*

(i) $W_p(\mu_n, \mu) \to 0$ *as* $n \to \infty$,

(ii) $\mu_n \to \mu$ *weakly as* $n \to \infty$ *and for some (and hence any* x_0*)*

$$\int d^p(x, x_0)\mu_n(dx) \to \int d^p(x, x_0)\mu(dx).$$

For a proof see e.g. Chapter 7 of Villani [314].

Remark 8. If d is bounded, then of course $\mathcal{P}^p(S) = \mathcal{P}(S)$ for all p. Hence, changing the distance d to the equivalent $\tilde{d} = \min(d, 1)$ allows one to use Wasserstein metrics as an alternative way to metrize the weak topology of probability measures.

In case $p = 1$ the celebrated *Monge–Kantorovich theorem* states that

$$W_1(\mu_1, \mu_2) = \sup_{f \in \mathrm{Lip}} |(f, \mu_1) - (f, \mu_2)|,$$

where Lip is the set of continuous functions f such that $|f(x) - f(y)| \leq \|x - y\|$ for all x, y, see [314] or [268].

We shall need also the Wasserstein distances between the distributions in the spaces of paths (curves) $X : [0, T] \mapsto S$. Its definition depends of course on the way the distance between paths is measured. The most natural choices are the uniform and integral measures leading to the distances

$$\begin{aligned} W_{p,T,\mathrm{un}}(X^1, X^2) &= \inf(\mathbf{E} \sup_{t \leq T} d^p(X_t^1, X_t^2))^{1/p}, \\ W_{p,T,\mathrm{int}}(X_1, X_2) &= \inf\left(\mathbf{E} \int_0^T d^p(X_t^1, X_t^2)\, dt\right)^{1/p}, \end{aligned} \tag{1.53}$$

where the inf is taken over all couplings of the distributions of the random paths X_1, X_2. The estimates in $W_{p,T,\mathrm{int}}$ are usually easier to obtain, but the estimates in $W_{p,T,\mathrm{un}}$ are stronger. In particular, uniform convergence is stronger than the Skorohod convergence, implying that the limits in $W_{p,T,\mathrm{un}}$ preserve the Skorohod space of càdlàg paths, while the limits in $W_{p,T,\mathrm{int}}$ do not.

Exercise 1.7.1. Show that the Borel σ-field of the space of R^d-valued continuous functions on $[0, 1]$ is generated by the evaluations maps $f \mapsto f(t)$ for all $t \in [0, 1]$.

1.8 Fractional derivatives and pseudo-differential operators

Unlike other parts of this chapter (dealing mostly with probabilistic issues), this section summarizes some background in analysis related to pseudo-differential operators

(ΨDO) and equations (ΨDE). In stochastic analysis ΨDO appear naturally as the generators of Markov processes. The main aim here is to define the major class of these ΨDO, namely fractional derivatives, and deduce their basic properties. Several good books are available on the subject, see e.g. Samko, Kilbas and Marichev [286], Miller and Ross [247], Saichev and Woyczynski [284], or Jacob [142]. We shall explain here in an accessible way only the basic motivations and facts, needed for our analysis of Markov processes and random walks.

The most natural function space for our exposition is the Schwartz space $S = S(\mathbb{R}^d)$ of infinitely differentiable functions on $\mathbb{R}^d$ decreasing at infinity together with all their derivatives faster than any power, i.e.

$$S(\mathbb{R}^d) = \{f \in C^\infty(\mathbb{R}^d) : \forall k, l \in \mathbb{N},\ |x|^k \nabla^l f(x) \in C_\infty(\mathbb{R}^d)\}.$$

If not specified otherwise, all test functions in this section will be from S.

We shall start with the one-dimensional case. Let If be the integration operator

$$If(x) = \int_{-\infty}^x f(y)\,dy.$$

Straightforward integration by parts yields

$$I^2 f(x) = \int_{-\infty}^x (If)(y)\,dy = \int_{-\infty}^x (x-y) f(y)\,dy$$

and further by induction

$$I^k f(x) = \frac{1}{(k-1)!}\int_{-\infty}^x (x-y)^{k-1} f(y)\,dy, \quad k \in \mathbb{N}.$$

This formula motivates the following definition of the *fractional integral operator*:

$$I^\beta f(x) = \frac{1}{\Gamma(\beta)}\int_{-\infty}^x (x-y)^{\beta-1} f(y)\,dy, \quad \beta > 0. \tag{1.54}$$

As $\Gamma(k) = (k-1)!$ for $k \in \mathbb{N}$, this agrees with the above formula for integer β. Notice that (1.54) is not well defined for $\beta < 0$ (which should correspond to fractional derivatives). Hence, instead of using this formula for fractional derivatives we shall build the theory on the observation that the differentiation (usual) should decrease the order of integration by one. This leads to the following definition of *fractional derivative*:

$$\frac{d^\beta}{dx^\beta} f(x) = \frac{d^{[\beta+1]}}{dx^{[\beta+1]}} I^{[\beta+1]-\beta} f(x), \quad \beta > 0,\ \beta \notin \mathbb{N} \tag{1.55}$$

where $[\beta+1]$ denotes the integer part of the real number $\beta + 1$. In particular,

$$\frac{d^\beta}{dx^\beta} f(x) = \frac{d}{dx} I^{1-\beta} f(x) = (I^{1-\beta} f)'(x), \quad \beta \in (0,1). \tag{1.56}$$

Proposition 1.8.1. *If $\beta \in (0,1)$, the following explicit formulas hold true (at least for f from the Schwartz space):*

$$\frac{d^\beta}{dx^\beta} f(x) = \frac{1}{\Gamma(-\beta)} \int_{-\infty}^{x} (f(y) - f(x)) \frac{dy}{(x-y)^{1+\beta}}$$
$$= \frac{1}{\Gamma(-\beta)} \int_0^\infty (f(x-y) - f(x)) \frac{dy}{y^{1+\beta}}, \quad \beta \in (0,1). \tag{1.57}$$

Moreover,

$$(I^\beta f)' = I^\beta f'. \tag{1.58}$$

Proof. Using the integration by parts one rewrites (1.54) as

$$I^\beta f(x) = \frac{1}{\beta\Gamma(\beta)} \int_{-\infty}^{x} (x-y)^\beta f'(y)\, dy, \quad \beta > 0. \tag{1.59}$$

Differentiating leads to

$$(I^\beta f)'(x) = \frac{1}{\Gamma(\beta)} \int_{-\infty}^{x} (x-y)^{\beta-1} f'(y)\, dy, \quad \beta \in (0,1),$$

which firstly implies (1.58), and secondly rewrites (again integrating by parts) as

$$(I^\beta f)'(x) = \frac{1}{\Gamma(\beta)} \int_{-\infty}^{x} (x-y)^{\beta-1}\, d(f(y) - f(x))$$
$$= \frac{\beta - 1}{\Gamma(\beta)} \int_{-\infty}^{x} (x-y)^{\beta-2}(f(y) - f(x))\, dy,$$

yielding the first formula in (1.57) (taking also into account that $\Gamma(1-\beta) = -\beta\Gamma(-\beta)$). The second formula is obtained from the first one by change of variables. □

It is worth observing that for $\beta \in (0,1)$ one can now define the *fractional derivative* by formula (1.57) for all bounded and γ-Hölder continuous function with any $\gamma < \beta$.

Exercise 1.8.1. Show that for $\beta \in (0,1)$

$$\frac{d^\beta}{dx^\beta} e^{-ipx} = \exp\{-i\pi\beta \operatorname{sgn} p/2\} |p|^\beta e^{-ipx}. \tag{1.60}$$

Hint: by (1.57)

$$\frac{d^\beta}{dx^\beta} e^{-ipx} = \frac{1}{\Gamma(-\beta)} e^{-ipx} \int_0^\infty (e^{ipy} - 1) \frac{dy}{y^{1+\beta}},$$

then use (1.23).

Next, the observation that the operator

$$-\frac{d}{dx} = \frac{d}{d(-x)}$$

is adjoint to the derivative d/dx, motivates the following definition of the *adjoint fractional derivative*:

$$\begin{aligned}\frac{d^\beta}{d(-x)^\beta} f(x) &= \frac{1}{\Gamma(-\beta)} \int_x^\infty (f(y) - f(x)) \frac{dy}{(y-x)^{1+\beta}} \\ &= \frac{1}{\Gamma(-\beta)} \int_0^\infty (f(x+y) - f(x)) \frac{dy}{y^{1+\beta}} \quad \beta \in (0,1). \end{aligned} \tag{1.61}$$

Exercise 1.8.2. Check that the operator $d^\beta / d(-x)^\beta$ is adjoint to d^β / dx^β, i.e.

$$\int_{-\infty}^\infty \frac{d^\beta}{d(-x)^\beta} f(x) g(x)\, dx = \int_{-\infty}^\infty \frac{d^\beta}{dx^\beta} g(x) f(x)\, dx,$$

and that

$$\frac{d^\beta}{d(-x)^\beta} f(x) = \frac{d}{dx} (I^\star)^{1-\beta} f(x), \quad \beta \in (0,1), \tag{1.62}$$

where

$$(I^\star)^\beta f(x) = \frac{1}{\Gamma(\beta)} \int_x^\infty (y-x)^{\beta-1} f(y)\, dy, \quad \beta > 0. \tag{1.63}$$

Recall now that the Fourier transform

$$F[f](p) = Ff(p) = \frac{1}{(2\pi)^{d/2}} \int e^{-ipx} f(x)\, dx = \frac{1}{(2\pi)^{d/2}} (e^{-ip.}, f)$$

is a bijection on the Schwartz space $S(\mathbb{R}^d)$, with the inverse operator being the inverse Fourier transform

$$F^{-1} g(x) = \frac{1}{(2\pi)^{d/2}} \int e^{ipx} g(p)\, dp.$$

As one easily sees the Fourier transform takes the differentiation operator to the multiplication operator, i.e. $F(f')(p) = (ip)F(f)$. More generally,

$$F[\Phi(-i\nabla)\psi](y) = \Phi(y) F[\psi](y), \tag{1.64}$$

$$F[V(.)\psi(.)](y) = V(i\nabla) F[\psi](y) \tag{1.65}$$

for bounded continuous functions Φ and V.

Applied to fractional derivatives, this correspondence implies the following.

Proposition 1.8.2. *If $\beta \in (0,1)$, then*

$$F\left(\frac{d^\beta}{d(\pm x)^\beta} f\right)(p) = \exp\{\pm i\pi\beta \operatorname{sgn} p/2\}|p|^\beta F(f)(p), \tag{1.66}$$

$$F\left(\frac{d^\beta}{dx^\beta} + \frac{d^\beta}{d(-x)^\beta}\right)(p) = 2\cos(\pi\beta/2)|p|^\beta F(f)(p). \tag{1.67}$$

Proof. By Exercise 1.8.2 and the definition of F

$$F\left(\frac{d^\beta}{dx^\beta} f\right)(p) = \frac{1}{\sqrt{2\pi}} \int \left(\frac{d^\beta}{d(-x)^\beta} e^{-ipx}\right) f(x)\, dx.$$

Hence (1.60) proves (1.66). Equation (1.67) is a consequence of (1.66). □

Remark 9. Note that $\exp\{i\pi\beta \operatorname{sgn} p/2\}|p|^\beta$ is the value of the main branch of the analytic function $(ip)^\beta$ (if p is real). Thus Proposition 1.8.2 states that F takes fractional β-derivative to the multiplication by $(ip)^\beta$.

For stochastic processes one mostly needs fractional derivatives up to order 2. Let $\beta \in (1,2)$. From (1.55) and (1.58)

$$\frac{d^\beta}{dx^\beta} f(x) = \frac{d^2}{dx^2} I^{2-\beta} f(x) = I^{2-\beta} f''(x) = \frac{1}{\Gamma(2-\beta)} \int_{-\infty}^x (x-y)^{1-\beta} f''(y)\, dy,$$

which by integration by parts rewrites as

$$\frac{1}{\Gamma(1-\beta)} \int_{-\infty}^x (x-y)^{-\beta} (f'(y) - f'(x))\, dy,$$

and by yet another integration by parts as

$$\frac{1}{\Gamma(-\beta)} \int_{-\infty}^x (f(y) - f(x) - (y-x) f'(x)) \frac{dy}{(x-y)^{1+\beta}}.$$

So finally for $\beta \in (1,2)$

$$\frac{d^\beta}{dx^\beta} f(x) = \frac{1}{\Gamma(-\beta)} \int_0^\infty (f(x-y) - f(x) + y f'(x)) \frac{dy}{y^{1+\beta}}, \tag{1.68}$$

and similarly

$$\frac{d^\beta}{d(-x)^\beta} f(x) = \frac{1}{\Gamma(-\beta)} \int_0^\infty (f(x+y) - f(x) - y f'(x)) \frac{dy}{y^{1+\beta}}. \tag{1.69}$$

Consequently for $\beta \in (1,2)$

$$\frac{d^\beta}{dx^\beta} f(x) + \frac{d^\beta}{d(-x)^\beta} f(x) = \frac{1}{\Gamma(-\beta)} \int_{-\infty}^\infty (f(x+y) - f(x) - y f'(x)) \frac{dy}{|y|^{1+\beta}}. \tag{1.70}$$

Noting that the Fourier transform F takes $-d^2/dx^2$ to the operator of multiplication by p^2 one defines the *positive fractional derivative* $|d^2/dx^2|^{\beta/2} = |d/dx|^\beta$ as the operator that the Fourier transform takes to $|p|^\beta$.

Proposition 1.8.3.

$$\left|\frac{d}{dx}\right|^\beta f(x) = \begin{cases} -\frac{\beta}{2\Gamma(1-\beta)\cos(\pi\beta/2)} \int_{-\infty}^{\infty} (f(x+y) - f(x)) \frac{dy}{|y|^{1+\beta}}, \\ \hfill \beta \in (0,1) \\ \frac{\beta}{2\Gamma(\beta-1)\cos(\pi\beta/2)} \int_{-\infty}^{\infty} (f(x+y) - f(x) - yf'(x)) \frac{dy}{|y|^{1+\beta}}, \\ \hfill \beta \in (1,2) \end{cases} \tag{1.71}$$

Proof. Like Proposition 1.8.2, this follows from the definition and (1.26). □

So far we worked in one dimension. In finite dimensions we shall reduce the discussion only to symmetric fractional derivative. As one can weight differently the derivatives of different directions it is natural to consider a *symmetric fractional operator* in $\mathbb{R}^d$ of the form

$$\int_{S^{d-1}} |(\nabla, s)|^\beta \mu(ds),$$

where $\mu(ds)$ is an arbitrary centrally symmetric finite Borel measure on the sphere S^{d-1}. The most natural way to define this operator is via the Fourier transform, i.e. as the operator that multiplies the Fourier transform of a function by

$$\int_{S^{d-1}} |(p, s)|^\beta \mu(ds),$$

i.e. via the equation

$$F\left(\int_{S^{d-1}} |(\nabla, s)|^\beta \mu(ds) f\right)(p) = \int_{S^{d-1}} |(p, s)|^\beta \mu(ds) Ff(p).$$

Straightforward extension of Proposition 1.8.3 yields the following

Proposition 1.8.4.

$$\int_{S^{d-1}} |(\nabla, s)|^\beta \mu(ds) \tag{1.72}$$

$$= \begin{cases} -\frac{\beta}{\Gamma(1-\beta)\cos(\pi\beta/2)} \int_0^\infty \int_{S^{d-1}} (f(x+y) - f(x)) \frac{d|y|}{|y|^{1+\beta}} \mu(ds), \quad \beta \in (0,1) \\ \frac{\beta}{\Gamma(\beta-1)\cos(\pi\beta/2)} \int_0^\infty \int_{S^{d-1}} (f(x+y) - f(x) - (y, \nabla f(x))) \frac{d|y|}{|y|^{1+\beta}} \mu(ds), \\ \hfill \beta \in (1,2). \end{cases}$$

The last formula can be written equivalently as

$$\frac{\beta}{2\Gamma(\beta-1)\cos(\pi\beta/2)}\int_0^\infty\int_{S^{d-1}}(f(x+y)+f(x-y)-2f(x))\frac{d|y|}{|y|^{1+\beta}}\mu(ds)$$
$$=\frac{\beta}{\Gamma(\beta-1)\cos(\pi\beta/2)}\lim_{\epsilon\to0}\int_\epsilon^\infty\int_{S^{d-1}}(f(x+y)-f(x))\frac{d|y|}{|y|^{1+\beta}}\mu(ds). \tag{1.73}$$

Finally

$$\int_{S^{d-1}}|(\nabla,s)|\mu(ds)=-\frac{2}{\pi}\lim_{\epsilon\to0}\int_\epsilon^\infty\int_{S^{d-1}}(f(x+y)-f(x))\frac{d|y|}{|y|^2}\mu(ds). \tag{1.74}$$

In particular, if $\mu(ds)$ is the Lebesgue measure ds on S^{d-1}, then by (1.35)

$$\int_{S^{d-1}}|(p,s)|^\beta ds=2|p|^\beta\pi^{(d-1)/2}\frac{\Gamma((\beta+1)/2)}{\Gamma((\beta+d)/2)},$$

from which one can deduce the integral representation for the operator $\|\nabla\|^\beta$.

Fractional derivatives defined above via Fourier transform constitute a particular case of the so called pseudo-differential operators (ΨDO). For a function $\psi(p)$ in $\mathbb{R}^d$, called in this context a *symbol*, one defines the ΨDO $\psi(-i\nabla)$ as the operator that the Fourier transform takes to multiplication by ψ, i.e. via the equation

$$F(\psi(-i\nabla)f)(p)=\psi(p)(Ff)(p). \tag{1.75}$$

Comparing this definition with the above factional derivatives one sees that $\int_{S^{d-1}}|(\nabla,s)|^\alpha\mu(ds)$ is the ΨDO with symbol $\int_{S^{d-1}}|(p,s)|^\alpha\mu(ds)$.

The explicit formula for F^{-1} yields the explicit integral representation for the ΨDO with symbol ψ as

$$\psi(-i\nabla)f(x)=\frac{1}{(2\pi)^{d/2}}\int e^{ipx}\psi(p)(Ff)(p)\,dp.$$

This expression suggests the following extension. For a function $\psi(x,p)$ on $\mathbb{R}^d$ one defines the ΨDO $\psi(x,-i\nabla)$ *with symbol* ψ via the formula

$$\psi(x,-i\nabla)f(x)=\frac{1}{(2\pi)^{d/2}}\int e^{ipx}\psi(x,p)(Ff)(p)\,dp. \tag{1.76}$$

1.9 Propagators and semigroups

This section puts together in a systematic way those tools from functional analysis that are mostly relevant to random processes, namely the semigroups and propagators of linear operators. For completeness we recall the notion of unbounded operators (also fixing some notation) assuming however that the readers are familiar with such basic definitions for Banach and Hilbert spaces as convergence, bounded linear operators, dual spaces and operators.

Let us start by recalling the basic notions of semigroups and propagators. For a set S, a family of mappings $U^{t,r}$ from S to itself, parametrized by the pairs of numbers $r \le t$ (resp. $t \le r$) from a given finite or infinite interval is called a *propagator*[3] (resp. a *backward propagator*) in S, if $U^{t,t}$ is the identity operator in S for all t and the following *chain rule*, or *propagator equation*, holds for $r \le s \le t$ (resp. for $t \le s \le r$):

$$U^{t,s}U^{s,r} = U^{t,r}. \tag{1.77}$$

A family of mappings T^t from S to itself parametrized by non-negative numbers t is said to form a *semigroup* (of the transformations of S) if T^0 is the identity mapping in S and $T^t T^s = T^{t+s}$ for all t, s. If the mappings $U^{t,r}$ forming a backward propagator depend only on the differences $r - t$, then the family $T^t = U^{0,t}$ forms a semigroup.

A *linear operator* A on a Banach space B is a linear mapping $A : D \mapsto B$, where D is a subspace of B called the *domain of* A. We say that the operator A is *densely defined* if D is dense in B.

The operator A is called *bounded* if the norm $\|A\| = \sup_{x \in D} \|Ax\|/\|x\|$ is finite. If A is bounded and D is dense, then A has a unique bounded extension (with the same norm) to an operator with the whole of B as domain. It is also well known that a linear operator $A : B \to B$ is continuous if and only if it is bounded. For a continuous linear mapping $A : B_1 \to B_2$ between the two Banach spaces its *norm* is defined as

$$\|A\|_{B_1 \mapsto B_2} = \sup_{x \ne 0} \frac{\|Ax\|_{B_2}}{\|x\|_{B_1}}.$$

The space of bounded linear operators $B_1 \to B_2$ equipped with this norm is a Banach space itself, often denoted by $\mathcal{L}(B_1, B_2)$.

A sequence of bounded operators A_n, $n = 1, 2, \dots$, in a Banach space B is said to *converge strongly* to an operator A if $A_n f \to Af$ for any $f \in B$.

A linear operator on a Banach space is called a *contraction* if its norm does not exceed 1. A semigroup T_t of bounded linear operators on a Banach space B is called *strongly continuous* if $\|T_t f - f\| \to 0$ as $t \to 0$ for any $f \in B$.

Examples. (1) If A is a bounded linear operator on a Banach space, then

$$T_t = e^{tA} = \sum_{n=0}^{\infty} \frac{t^n}{n!} A^n$$

defines a strongly continuous semigroup.

[3] In alternative terminology, *semigroup with two parameters*, see e.g. Fleming and Soner [114].

(2) The shifts $T_t f(x) = f(x+t)$ form a strongly continuous group of contractions on $C_\infty(\mathbb{R})$, $L^1(\mathbb{R})$ or $L^2(\mathbb{R})$. However, it is not strongly continuous on $C(\mathbb{R})$. Observe also that if f is an analytic function, then

$$f(x+t) = \sum_{n=0}^{\infty} \frac{t^n}{n!}(D^n f)(x),$$

which can be written formally as $e^{tD} f(x)$.

(3) Let $\eta(y)$ be a complex-valued continuous function on $\mathbb{R}^d$ such that $\operatorname{Re}\eta \le 0$. Then

$$T_t f(y) = e^{t\eta(y)} f(y)$$

is a semigroup of contractions on the Banach spaces $L^p(\mathbb{R}^d)$, $L^\infty(\mathbb{R}^d)$, $B(\mathbb{R}^d)$, $C(\mathbb{R}^d)$ and $C_\infty(\mathbb{R}^d)$, which is strongly continuous on $L^p(\mathbb{R}^d)$ and $C_\infty(\mathbb{R}^d)$ but not on the other three spaces.

An operator A with domain D is called *closed* if its graph (which is defined as the space of the pairs (x, Ax) with $x \in D$) is a closed subset of $B \times B$, i.e. if $x_n \to x$ and $Ax_n \to y$ as $n \to \infty$ for a sequence $x_n \in D$, then $x \in D$ and $y = Ax$. A is called *closable* if a closed extension of A exists, in which case the *closure of* A is defined as the minimal closed extension of A, i.e. the operator with the graph being the closure of the graph of A. A subspace D of the domain D_A of a closed operator A is called a *core* for A if A is the closure of A restricted to D.

Let T_t be a strongly continuous semigroup of linear operators on a Banach space B. The *infinitesimal generator* or simply the *generator* of T_t is defined as the operator

$$Af = \lim_{t\to 0} \frac{T_t f - f}{t}$$

on the linear subspace $D_A \subset B$ (the *domain* of A), where this limit exists (in the topology of B). If the T_t are contractions, then the *resolvent* of T_t (or of A) is defined for any $\lambda > 0$ as the operator

$$R_\lambda f = \int_0^\infty e^{-\lambda t} T_t f \, dt.$$

For example, the generator A of the semigroup $T_t f = e^{t\eta} f$ from example (3) above is given by the multiplication operator $Af = \eta f$ on functions f such that $\eta^2 f \in C_\infty(\mathbb{R}^d)$ (or resp. $\eta^2 f \in L^p(\mathbb{R}^d)$).

Theorem 1.9.1 (Basic properties of generators and resolvents). *Let T_t be a strongly continuous semigroup of linear contractions on a Banach space B and let A be its generator. Then the following hold:*

(i) *$T_t D_A \subset D_A$ for each $t \ge 0$ and $T_t Af = AT_t f$ for each $t \ge 0$, $f \in D_A$.*

(ii) *$T_t f = \int_0^t AT_s f \, ds + f$ for $f \in D$.*

(iii) *R_λ is a bounded operator on B with $\|R_\lambda\| \le \lambda^{-1}$, for any $\lambda > 0$.*

(iv) *$\lambda R_\lambda f \to f$ as $\lambda \to \infty$.*

(v) *$R_\lambda f \in D_A$ for any f and $\lambda > 0$ and $(\lambda - A)R_\lambda f = f$, i.e. $R_\lambda = (\lambda - A)^{-1}$.*

(vi) *If $f \in D_A$, then $R_\lambda A f = A R_\lambda f$.*

(vii) *D_A is dense in B.*

(viii) *A is closed on D_A.*

Proof. (i) Observe that for $\psi \in D_A$

$$AT_t\psi = \left[\lim_{h\to 0}\frac{1}{h}(T_h - I)\right] T_t\psi = T_t\left[\lim_{h\to 0}\frac{1}{h}(T_h - I)\right]\psi = T_t A\psi.$$

(ii) Follows from (i).
(iii) $\|R_\lambda f\| \le \int_0^\infty e^{-\lambda t}\|f\|\,dt = \lambda^{-1}\|f\|$.
(iv) Follows from the equation

$$\lambda\int_0^\infty e^{-\lambda t}T_t f\,dt = \lambda\int_0^\infty e^{-\lambda t} f\,dt + \lambda\int_0^\epsilon e^{-\lambda t}(T_t f - f)\,dt + \lambda\int_\epsilon^\infty e^{-\lambda t}(T_t f - f)\,dt,$$

observing that the first term on the r.h.s. is f, the second (resp. third) term is small for small ϵ (resp. for any ϵ and large λ).

(v) By definition

$$\begin{aligned} AR_\lambda f &= \lim_{h\to 0}\frac{1}{h}(T_h - \mathbf{1})R_\lambda f = \frac{1}{h}\int_0^\infty e^{-\lambda t}(T_{t+h}f - T_t f)\,dt \\ &= \lim_{h\to 0}\left[\frac{e^{\lambda h}-1}{h}\int_0^\infty e^{-\lambda t}T_t f\,dt - \frac{e^{\lambda h}}{h}\int_0^h e^{-\lambda t}T_t f\,dt\right] = \lambda R_\lambda f - f. \end{aligned}$$

(vi) Follows from the definitions and (ii).
(vii) Follows from (iv) and (v).
(viii) If $f_n \to f$ as $n \to \infty$ for a sequence $f_n \in D$ and $Af_n \to g$, then

$$T_t f - f = \lim_{n\to\infty}\int_0^t T_s A f_n\,ds = \int_0^t T_s g\,ds.$$

Applying the fundamental theorem of calculus completes the proof. □

Remark 10. For all $\psi \in B$ the vector $\psi(t) = \int_0^t T_u\psi du$ belongs to D_A and $A\psi(t) = T_t\psi - \psi$. Moreover, $\psi(t) \to \psi$ as $t \to 0$ always, and $A\psi(t) \to A\psi$ for $\psi \in D_A$. This observation yields another insightful proof of statement (vii) of Theorem 1.9.1 (by-passing the resolvent).

Proposition 1.9.1. *Let an operator A with domain D_A generate a strongly continuous semigroup of linear contractions T_t. If D is a dense subspace of D_A of A that is invariant under all T_t, then D is a core for A.*

Proof. Let $\bar{D}$ be the domain of the closure of A restricted to D. We have to show that for $\psi \in D_A$ there exists a sequence $\psi_n \in \bar{D}$, $n \in \mathbb{N}$, such that $\psi_n \to \psi$ and $A\psi_n \to A\psi$. By the remark above it is enough to show this for $\psi(t) = \int_0^t T_u \psi \, du$. As D is dense there exists a sequence $\psi_n \in D$ converging to ψ and hence $A\psi_n(t) \to A\psi(t)$. To complete the proof it remains to observe that $\psi_n(t) \in \bar{D}$ by the invariance of D. □

An important tool for the construction of semigroups is *perturbation theory*, see e.g. V. P. Maslov [231] or Reed and Simon [274], which can be applied when a generator of interest can be represented as the sum of a well-understood operator and a term that is smaller (in some sense). Below we give the simplest result of this kind.

Theorem 1.9.2. *Let an operator A with domain D_A generate a strongly continuous semigroup T_t on a Banach space B, and let L be a bounded operator on B. Then*

(i) *$A+L$ with the same domain D_A also generates a strongly continuous semigroup Φ_t on B given by the series*

$$\Phi_t = T_t + \sum_{m=1}^{\infty} \int_{0 \le s_1 \le \cdots \le s_m \le t} T_{t-s_m} L T_{s_m - s_{m-1}} L \cdots L T_{s_1} \, ds_1 \cdots ds_m \tag{1.78}$$

converging in the operator norm;

(ii) *$\Phi_t f$ is the unique (bounded) solution of the integral equation*

$$\Phi_t f = T_t f + \int_0^t T_{t-s} L \Phi_s f \, ds, \tag{1.79}$$

with a given $f_0 = f$;

(iii) *if additionally D is an invariant core for A that is itself a Banach space under the norm $\|.\|_D$, the T_t are uniformly (for t from a compact interval) bounded operators $D \to D$ and L is a bounded operator $D \to D$, then D is an invariant core for $A + L$ and the Φ_t are uniformly bounded operators in D.*

Proof. (i) Clearly

$$\|\Phi_t\| \le \|T_t\| + \sum_{m=1}^{\infty} \frac{(\|L\| t)^m}{m!} \Big(\sup_{s \in [0,t]} \|T_s\| \Big)^{m+1},$$

implying the convergence of the series. Next,

$$\begin{aligned}
\Phi_t \Phi_\tau f &= \sum_{m=0}^{\infty} \int_{0\le s_1\le\cdots\le s_m\le t} T_{t-s_m} L T_{s_m-s_{m-1}} L\cdots L T_{s_1}\, ds_1\cdots ds_m \\
&\quad \times \sum_{n=0}^{\infty} \int_{0\le u_1\le\cdots\le u_n\le \tau} T_{\tau-u_n} L T_{u_n-u_{n-1}} L\cdots L T_{u_1}\, du_1\cdots du_n \\
&= \sum_{m,n=0}^{\infty} \int_{0\le u_1\le\cdots u_n\le\tau\le v_1\le\cdots\le v_m\le t+\tau} dv_1\cdots dv_m du_1\cdots du_n \\
&\qquad \times T_{t+\tau-v_m} L T_{v_m-v_{m-1}} L\cdots L T_{v_1-u_m} L\cdots L T_{u_1} \\
&= \sum_{k=0}^{\infty} \int_{0\le u_1\le\cdots\le u_k\le t+\tau} T_{t+\tau-u_k} L T_{u_k-u_{k-1}} L\cdots L T_{u_1}\, du_1\cdots du_k \\
&= \Phi_{t+\tau} f,
\end{aligned}$$

showing the main semigroup condition. Since the terms with $m > 1$ in (1.78) are of order $O(t^2)$ for small t,

$$\frac{d}{dt}\Big|_{t=0} \Phi_t f = \frac{d}{dt}\Big|_{t=0} \left(T_t f + \int_0^t T_{t-s} L T_s f\, ds\right) = \frac{d}{dt}\Big|_{t=0} T_t f + Lf,$$

so that $\frac{d}{dt}|_{t=0}\Phi_t f$ exists if and only if $\frac{d}{dt}|_{t=0} T_t f$ exists, and in this case

$$\frac{d}{dt}\Big|_{t=0} \Phi_t f = (A+L) f.$$

(ii) Equation (1.79) is a consequence of (1.78). On the other hand, if (1.79) holds, then substituting the l.h.s. of this equation into its r.h.s. recursively yields

$$\begin{aligned}
\Phi_t f &= T_t f + \int_0^t T_{t-s} L T_s f\, ds + \int_0^t ds_2 T_{t-s_2} L \int_0^{s_2} ds_1 T_{s_2-s_1} L \Phi_{s_1} f \\
&= T_t f + \sum_{m=1}^{N} \int_{0\le s_1\le\cdots\le s_m\le t} T_{t-s_m} L T_{s_m-s_{m-1}} L\ldots L T_{s_1} f\, ds_1\cdots ds_m \\
&+ \int_{0\le s_1\le\cdots\le s_{N+1}\le t} T_{t-s_{N+1}} L T_{s_{N+1}-s_N} L\cdots L T_{s_2-s_1} L \Phi_{s_1} f\, ds_1\cdots ds_m
\end{aligned}$$

for arbitrary N. As the last term tends to zero, the series representation (1.78) follows.

(iii) This is obvious, because the conditions on D ensure that the series (1.78) converges in the norm topology of D. □

Equation (1.79) is often called the *mild form* of the equation $\dot{\Phi} = (A + L)\Phi$. The possibility of obtaining this mild form is sometimes called the *Duhamel principle*.

For the analysis of time nonhomogeneous evolutions an extension of the notion of a generator to propagators is needed. A backward propagator $\{U^{t,r}\}$ of bounded linear operators on a Banach space B is called *strongly continuous* if the operators $U^{t,r}$ depend strongly continuously on t and r. The *principle of uniform boundedness* (well known in functional analysis) states that if a family T_α of bounded linear mappings from a Banach space X to another Banach space is such that the sets $\{\|T_\alpha x\|\}$ are bounded for each x, then the family T_α is uniformly bounded. This implies that if $U^{t,r}$ is a strongly continuous propagator of bounded linear operators, then the norms of $U^{t,r}$ are bounded uniformly for t, r from any compact interval.

Suppose $\{U^{t,r}\}$ is a strongly continuous backward propagator of bounded linear operators on a Banach space B with a common invariant domain D, which is itself a Banach space with the norm $\|.\|_D \geq \|.\|_B$. Let $\{A_t\}$, $t \geq 0$, be a family of bounded linear operators $D \to B$ depending strongly measurably on t (i.e. $A_t f$ is a measurable function $t \mapsto B$ for each $f \in D$). Let us say that the family $\{A_t\}$ *generates* $\{U^{t,r}\}$ on the invariant domain D if the equations

$$\frac{d}{ds}U^{t,s} f = U^{t,s} A_s f, \quad \frac{d}{ds}U^{s,r} f = -A_s U^{s,r} f, \quad t \leq s \leq r, \tag{1.80}$$

hold a.s. in s for any $f \in D$, that is there exists a set S of zero measure in $\mathbb{R}$ such that for all $t < r$ and all $f \in D$ equations (1.80) hold for all s outside S, where the derivatives exist in the Banach topology of B. In particular, if the operators A_t depend strongly continuously on t, this implies that equations (1.80) hold for all s and $f \in D$, where for $s = t$ (resp. $s = r$) it is assumed to be only a right (resp. left) derivative.

The next result extends Theorem 1.9.2 to propagators.

Theorem 1.9.3. *For $U^{t,r}$ a strongly continuous backward propagator of bounded linear operators in a Banach space B, suppose a dense subspace $D \subset B$ is itself a Banach space under the norm $\|.\|_D$ and $U^{t,r}$ are bounded operators $D \to D$. Suppose a family of linear operators $\{A_t\}$ generates this propagator on the common domain D (so that* (1.80) *holds). Let $\{L_t\}$ be a family of bounded operators both in B and in D that depend a.s. continuously on t in the strong topology of B, that is there exists a negligible set $S \subset \mathbb{R}$ such that $L_t f$ is a continuous function $t \to B$ for all $f \in B$ and all $s \notin S$. Then the family $\{A_t + L_t\}$ generates a strongly continuous propagator $\{\Phi^{t,r}\}$ in B, on the same invariant domain D, where*

$$\Phi^{t,r} = U^{t,r} + \sum_{m=1}^{\infty} \int_{t \leq s_1 \leq \cdots \leq s_m \leq r} U^{t,s_1} L_{s_1} U^{s_1,s_2} \cdots L_{s_m} U^{s_m,r} \, ds_1 \cdots ds_m \tag{1.81}$$

This series converges in the operator norms of both B and D. Moreover, $\Phi^{t,r} f$ is the unique bounded solution of the integral equation

$$\Phi^{t,r} f = U^{t,r} f + \int_t^r U^{t,s} L_s \Phi^{s,r} f \, ds, \tag{1.82}$$

with a given $f_r = f$.

Proof. This is a straightforward extension of Theorem 1.9.2. The only difference to note is that in order to conclude that

$$\frac{d}{dt}\bigg|_{t=r} \int_t^r U^{t,s} L_s \Phi^{s,r} f \, ds = \frac{d}{dt}\bigg|_{t=r} \int_t^r U^{t,s} L_r \Phi^{s,r} f \, ds = -L_r f$$

one uses the continuous dependence of L_s on s (since L_s are strongly continuous in s, the function $L_s \Phi^{s,r} f$ is continuous in s, because the family $\Phi^{s,r} f$ is compact as the image of a continuous mapping of the interval $[t, r]$). □

For a Banach space B or a linear operator A one usually denotes by $B^\star$ or $A^\star$ respectively its *Banach dual*. Sometimes the notations B' and A' are also in use.

Theorem 1.9.4 (Basic duality). *Let $U^{t,r}$ be a strongly continuous backward propagator of bounded linear operators in a Banach space B with a common invariant domain D, which is itself a Banach space with the norm $\|.\|_D \geq \|.\|_B$, and let the family $\{A_t\}$ of bounded linear operators $D \to B$ generate $U^{t,r}$ on D. Then the following hold.*

(i) *The family of dual operators $V^{s,t} = (U^{t,s})^\star$ forms a propagator of bounded linear operators in $B^\star$ (contractions if all $U^{t,r}$ are contractions), weakly continuous in s, t, such that*

$$\frac{d}{ds} V^{s,t} \xi = A_s^\star V^{s,t} \xi, \quad t \leq s \leq r, \tag{1.83}$$

weakly in $D^\star$, that is

$$\frac{d}{ds}(f, V^{s,t}\xi) = (A_s f, V^{s,t}\xi), \quad t \leq s \leq r, \quad f \in D, \tag{1.84}$$

for s outside a null set of $s \in \mathbb{R}$.

(ii) *$V^{s,t}\xi$ is the unique solution to the Cauchy problem of equation* (1.84), *i.e. if $\xi_t = \xi$ for a given $\xi \in B^\star$ and ξ_s, $s \in [t, r]$, is a weakly continuous family in $B^\star$ satisfying*

$$\frac{d}{ds}(f, \xi_s) = (A_s f, \xi_s), \quad t \leq s \leq r, \quad f \in D, \tag{1.85}$$

for s outside a null set of $\mathbb{R}$, then $\xi_s = V^{s,t}\xi$ for all $s \in [t, r]$.

(iii) $U^{s,r} f$ *is the unique solution to the inverse Cauchy problem of the second equation in* (1.80), *i.e. if* $f_r = f$, $f_s \in D$ *for* $s \in [t, r]$ *and satisfies the equation*

$$\frac{d}{ds} f_s = -A_s f_s, \quad t \leq s \leq r, \tag{1.86}$$

for s *outside a zero-measure subset of* $\mathbb{R}$ *(with the derivative existing in the norm topology of* B*), then* $f_s = U^{s,r} f$ *for all* $s \in [t, r]$.

Proof. Statement (i) is a direct consequence of duality and equation (1.80).

(ii) Let $g(s) = (U^{s,r} f, \xi_s)$ for a given $f \in D$. Writing

$$\begin{aligned}(U^{s+\delta,r} f, \xi_{s+\delta}) - (U^{s,r} f, \xi_s) &= (U^{s+\delta,r} f - U^{s,r} f, \xi_s) + (U^{s,r} f, \xi_{s+\delta} - \xi_s) \\ &\quad + (U^{s+\delta,r} f - U^{s,r} f, \xi_{s+\delta} - \xi_s)\end{aligned}$$

and using (1.80), (1.85) and the invariance of D, allows one to conclude that

$$\frac{d}{ds} g(s) = -(A_s U^{s,r} f, \xi_s) + (U^{s,r} f, A_s^\star \xi_s) = 0,$$

because a.s. in s

$$\left(\frac{U^{s+\delta,r} f - U^{s,r} f}{\delta}, \xi_{s+\delta} - \xi_s\right) \to 0,$$

as $\delta \to 0$ (since the family $\delta^{-1}(U^{s+\delta,r} f - U^{s,r} f)$ is relatively compact, being convergent, and ξ_s is weakly continuous). Hence $g(r) = (f, \xi_r) = g(t) = (U^{t,r} f, \xi_t)$, showing that ξ_r is uniquely defined.

(iii) As in (ii), this follows from the observation that

$$\frac{d}{ds}(f_s, V^{s,t} \xi) = 0.$$

□

We conclude with the following stability result.

Theorem 1.9.5 (Convergence of propagators). *Suppose we are given a sequence of propagators* $\{U_n^{t,r}\}$, $n = 1, 2, \ldots$, *generated by the families* $\{A_t^n\}$ *and a propagator* $\{U^{t,r}\}$ *generated by the family* $\{A_t\}$. *Suppose all these propagators satisfy the same conditions as* $U^{t,r}$ *and* A_t *from Theorem* 1.9.4 *with the same* D, B. *Suppose also that all* $U^{t,r}$ *are uniformly bounded as operators in* D.

(i) *Let*

$$\operatorname{ess\,sup}_{0 \leq t \leq r} \|A_t^n - A_t\|_{D \to B} \leq \epsilon_n. \tag{1.87}$$

Then

$$\sup_{0 \leq t \leq r} \|U_n^{t,r} g - U^{t,r} g\|_B = O(1)\epsilon_n \|g\|_D, \tag{1.88}$$

and $U_n^{t,r}$ *converge to* $U^{t,r}$ *strongly in* B.

(ii) *Let the families A_t^n and A_t depend càdlàg on t in the Banach topology of the space $\mathcal{L}(D, B)$ of bounded operators $D \to B$ and A_t^n converge to A_t as $n \to \infty$ in the corresponding Skorohod topology of the space $D([0, T], \mathcal{L}(D, B))$. Then $U_n^{t,r}$ converges to $U^{t,r}$ strongly in B.*

(iii) *Let the families A_t^n and A_t depend càdlàg on t in the strong topology of the space $\mathcal{L}(D, B)$, $A_t^n f$ converge to $A_t f$, as $n \to \infty$, in the Skorohod topology of the space $D([0, T], B)$ for any f and uniformly for f from any compact subset of D. Moreover, let the family $U^{t,r}$ is strongly continuous as a family of operators in D. Then again $U_n^{t,r}$ converges to $U^{t,r}$ strongly in B.*

Proof. By the density argument (taking into account that $U_n^{t,r} g$ are uniformly bounded), in order to prove the strong convergence of $U_n^{t,r}$ to $U^{t,r}$, it is sufficient to prove that $U_n^{t,r} g$ converges to $U^{t,r} g$ for any $g \in D$.

(i) If $g \in D$,

$$(U_n^{t,r} - U^{t,r})g = U_n^{t,s} U^{s,r} g|_{s=t}^r = \int_t^r U_n^{t,s}(A_s^n - A_s)U^{s,r} g \, ds, \tag{1.89}$$

and (1.87) implies (1.88).

(ii) In order to prove that expression (1.89) converges to zero as $n \to \infty$, it suffices to show that

$$\int_0^r \|A_s^n - A_s\|_{D \to B} \, ds \to 0$$

as $n \to \infty$. Since A_t^n converge to A_t in the Skorohod topology, there exists a sequence λ_n of the monotone bijections of $[0, r]$ such that

$$\sup_s |\lambda_n(s) - s| \to 0, \quad \sup_s \|A_s^n - A_{\lambda_n(s)}\|_{D \to B} \to 0,$$

as $n \to \infty$. Hence writing

$$\int_0^r \|A_s^n - A_s\|_{D \to B} \, ds = \int_0^r \|A_s^n - A_{\lambda_n(s)}\|_{D \to B} \, ds + \int_0^r \|A_s - A_{\lambda_n(s)}\|_{D \to B} \, ds,$$

we see that both terms tend to zero (the second one by the dominated convergence, as the limiting function vanishes a.s.).

(iii) The proof is similar to that of (ii). □

Chapter 2
Brownian motion (BM)

This chapter starts with an introductory section devoted to the basic definitions of the theory of random processes, including predictability, stopping times and martingales. Then we turn to the discussion of possibly the most beautiful object in stochastic analysis, the Brownian motion (BM). It unites in a remarkable synthesis intuitive and visual simplicity with deep structural complexity and almost universal practical applicability. No attempt is made here even to review the subject. The aim is merely to set necessary foundations for the future analysis of Markov processes. As fundamental treatises on BM one can mention Karatzas and Shreve [155], Revuz and Yor [278] or more financially oriented book of Jeanblanc, Yor and Chesney [149].

2.1 Random processes: basic notions

Let S be a complete metric space (for our main examples we need S to be either $\mathbb{R}^d$ or a closed or open subset of it). A *stochastic process* in S is a collection $X = (X_t)$, $t \geq 0)$ (or $t \in [0, T]$ for some $T > 0$) of S-valued random variables defined on the same probability space. The *finite-dimensional distributions* of such a process are defined as the collections of probability measures $p_{t_1,\dots,t_n}$ on S^n (parametrized by finite collections of pairwise different non-negative numbers $t_1, \dots, t_n$) defined as

$$p_{t_1,\dots,t_n}(H) = \mathbf{P}((X_{t_1}, \dots, X_{t_n}) \in H)$$

for each Borel subset H of S^n. These finite-dimensional distributions are (obviously) *consistent* (or satisfy *Kolmogorov's consistency criteria*): for any n, any permutation π of $\{1, \dots, n\}$, any sequence $0 \leq t_1 < \cdots < t_{n+1}$, and any collection of Borel subsets $H_1, \dots, H_n$ of S one has

$$\begin{aligned} p_{t_1,\dots,t_n}(H_1 \times \cdots \times H_n) &= p_{t_{\pi(1)},\dots,t_{\pi(n)}}(H_{\pi(1)} \times \cdots \times H_{\pi(n)}), \\ p_{t_1,\dots,t_n,t_{n+1}}(H_1 \times \cdots \times H_n \times S) &= p_{t_1,\dots,t_n}(H_1 \times \cdots \times H_n). \end{aligned}$$

A stochastic process is called *Gaussian* if all its finite-dimensional distributions are Gaussian.

Theorem 2.1.1 (Kolmogorov's existence theorem.). *Given a family of probability measures $p_{t_1,\dots,t_n}$ on S^n satisfying the Kolmogorov consistency criteria, there exists a probability space $(\Omega, \mathcal{F}, \mathbf{P})$ and a stochastic process X on it having $p_{t_1,\dots,t_n}$ as* *its finite-dimensional distributions. In particular, one can choose* Ω *to be the set* $S^{\bar{\mathbb{R}}_+}$ *of all mappings from* $\bar{\mathbb{R}}_+$ *to* S, $\mathcal{F}$ *to be the smallest* σ*-algebra containing all cylinder sets*

$$I^H_{t_1,\dots,t_n} = \{\omega \in \Omega : (\omega(t_1), \dots, \omega(t_n)) \in H\}, \quad H \in \mathcal{B}(S^n) \tag{2.1}$$

(in other words, $\mathcal{F}$ *is generated by all evaluation maps* $\omega \mapsto \omega(t)$*), and* X *to be the co-ordinate process* $X_t(\omega) = \omega(t)$.

A proof can be found e.g. in Chapter 6 of Kallenberg [154] or in Chapter 2 of Shiryayev [293].

The following two notions express the "sameness" between processes. Suppose two processes X and Y are defined on the same probability space $(\Omega, \mathcal{F}, \mathbf{P})$. Then

(i) X and Y are called *indistinguishable* if $\mathbf{P}(\forall t X_t = Y_t) = 1$;

(ii) X is called a *modification of* Y if for each t $\mathbf{P}(X_t = Y_t) = 1$.

As an illustrative example consider a positive r.v. ξ with a continuous distribution (i.e. such that $\mathbf{P}(\xi = x) = 0$ for any x). Put $X_t = 0$ for all t and let Y_t be 1 for $t = \xi$ and 0 otherwise. Then Y is a modification of X, but $\mathbf{P}(\forall t X_t = Y_t) = 0$.

Exercise 2.1.1. Suppose Y is a modification of X and both processes have right-continuous sample paths. Then X and Y are indistinguishable. Hint: show that if X is a modification of Y, then $\mathbf{P}(\forall t \in \mathbb{Q}\ X_t = Y_t) = 1$.

The process on the probability space $S^{\bar{\mathbb{R}}_+}$ described by Kolmogorov's existence theorem is called the *canonical process* for a given collection of finite-dimensional distributions, and the corresponding space $S^{\bar{\mathbb{R}}_+}$ is called the *canonical path space*. This space is tremendously big, as it contains all, even non-measurable, paths ω. Such paths are difficult to monitor or observe. Moreover, though the cylinder sets that can be visualized as gates through which the trajectories should pass are measurable in $S^{\bar{\mathbb{R}}_+}$, the *tunnels* defined as the sets of the form

$$I^H_{[t_1,t_2]} = \{\omega \in \Omega : \omega(t) \in H\ \forall\, t \in [t_1, t_2]\}, \quad H \in \mathcal{B}(S), \tag{2.2}$$

are not (as they represent uncountable intersections of gates). Similarly the set of all continuous paths is not measurable in $S^{\bar{\mathbb{R}}_+}$.

Therefore one often looks for modifications with possibly more regular paths. A process X, on a probability space $(\Omega, \mathcal{F}, \mathbf{P})$, is called *measurable, continuous, left or right-continuous* if its trajectories have the corresponding properties a.s. If a process is measurable then it becomes possible to measure the events of the form $\{t, \omega : X_t(\omega) \in H\}$ for Borel sets H and the indicators $\mathbf{1}_{\{X_t \in H\}}$ become measurable in t, allowing one, in particular, to measure the durations of time spent by a process in H: $\int_0^t \mathbf{1}_{\{X_s \in H\}}\, ds$. If a process is left or right-continuous, then the tunnels (2.2) become measurable.

According to the general definition, a random process is, after all, just a random variable taking values in a large space of paths. But what makes the theory of random processes really special within probability theory is the presence of the arrow of time, which leads to the possibility of observing the dynamics of processes and monitoring and storing information available up to present time t. Such historic information can be modeled by σ-algebras of sets measurable by time t. This idea gives rise to the following definition. Let $(\Omega, \mathcal{F})$ be a measurable space. A family $\mathcal{F}_t$, $t \geq 0$, of sub-σ-algebras of $\mathcal{F}$ is called a *filtration* if $\mathcal{F}_s \subset \mathcal{F}_t$ whenever $s \leq t$. By $\mathcal{F}_\infty$ one denotes the minimal σ-algebra containing all $\mathcal{F}_t$. A probability space $(\Omega, \mathcal{F}, \mathbf{P})$ with a filtration is said to be *filtered* and is denoted sometimes by the quadruple $(\Omega, \mathcal{F}, \mathcal{F}_t, \mathbf{P})$. The modern term for a filtered probability space is *stochastic basis*. A process $X = X_t$ defined on a filtered probability space $(\Omega, \mathcal{F}, \mathcal{F}_t, \mathbf{P})$ is called *adapted* or *non-anticipative* (more precisely $\mathcal{F}_t$-adapted or non-anticipative) if X_t is $\mathcal{F}_t$-measurable for each t. This means that at any moment of time we can assess the behavior of a random variable X_t on the basis of the information available up to time t. Any process X defines its own *natural filtration* $\mathcal{F}_t^X = \sigma\{X_s : 0 \leq s \leq t\}$ and X is clearly adapted to it.

The necessity of working with adapted and regular enough process gives rise to the following two fundamental concepts. The *predictable* or *previsible* σ-algebra in the product space $\Omega \times \mathbb{R}_+$ is defined as the σ-algebra generated by all left-continuous adapted processes. Processes that are measurable with respect to this σ-algebra are called *predictable*, or *previsible*. The *optional* (or *well-measurable* in old terminology) σ-algebra in the product space $\Omega \times \mathbb{R}_+$ is defined as the σ-algebra generated by all right-continuous adapted processes. Processes that are measurable with respect to the optional σ-algebra are called optional. Predictable and optional processes are widely studied in general stochastic analysis. Their properties are crucial for various applications, in particular, for filtering theory. In this book, for simplicity, we will not plunge deeply into the analysis of these processes, using just left-continuous or right-continuous process. However, we note the following simple but remarkable fact.

Proposition 2.1.1. *The predictable σ-algebra is generated also by continuous processes. In particular, it is a sub-algebra of the optional σ-algebra.*

Proof. Any left-continuous adapted process X_t can be approximated by its natural left-continuous discrete approximation

$$X_t^n = \sum_{k=1}^{\infty} X_{k/2^n} \mathbf{1}_{((k-1)/2^n, k/2^n]}.$$

Hence, to prove the statement it is enough to show that the process of the form $\xi \mathbf{1}_{(s,t]}$ with $\mathcal{F}_s$ measurable random variable ξ can be approximated (in the sense of a.s. convergence) by continuous adapted processes. And this clearly boils down to proving that the step processes of the form $\mathbf{1}_A \mathbf{1}_{(s,\infty)}(t)$ with $\mathcal{F}_s$-measurable set A can be approximated by continuous adapted processes. And for this process adapted continuous approximations can be defined as $Y^n(t) = \mathbf{1}_A \mathbf{1}_{(s,\infty)}(t)(1 \wedge n(t-s))$. □

Since continuous function are also right-continuous, Proposition 2.1.1 implies that the predictable σ-algebra is a sub-algebra of the optional sub-algebra. It is instructive to observe that the same continuous adapted approximations $Y^n(t) = \mathbf{1}_A \mathbf{1}_{(s,\infty)}(t)(1 \wedge n(t-s))$ converge a.s. to the right-continuous adapted step function $\mathbf{1}_A \mathbf{1}_{[s,\infty)}(t)$, where A is again assumed to be $\mathcal{F}_s$-measurable. Hence, for any number $s > 0$, the step function $\mathbf{1}_A \mathbf{1}_{[s,\infty)}(t)$ is right-continuous, but predictable, i.e. measurable with respect to the σ-algebra generated by adapted left-continuous processes. Does it prove that the optional σ-algebra coincides with the predictable σ-algebra? It does not, because a.s. convergence may be lost when approximating in the same way the step functions $\mathbf{1}_{[\tau,\infty)}$ with a random time τ. This discourse leads to the following fundamental definitions. A positive random variable is called a *stopping time* or a *Markov time*, if the step process $\mathbf{1}_{[\tau,\infty)}(t)$ is adapted. A stopping time is called *predictable* if the step process $\mathbf{1}_{[\tau,\infty)}(t)$ is also predictable. We shall use stopping times extensively when studying stopped processes and processes living on the domains with boundaries. As one may guess, predictability is crucial for forecasting, but its discussion lies beyond the scope of this book. Note that if τ is a stopping time, then the process $\mathbf{1}_{[\tau,\infty)}(t)$ is optional.[1]

Let us introduce a more general notion, which combines, in the most transparent way, the measurability of paths with some respect to the filtration. A process $X = X_t$ defined on a filtered probability space $(\Omega, \mathcal{F}, \mathcal{F}_t, \mathbf{P})$ is called *progressively measurable* (or *progressive*) if for each t the map $(s, \omega) \mapsto X_s(\omega)$ from $[0,t] \times \Omega$ into S is $\mathcal{B}([0,t]) \otimes \mathcal{F}_t$-measurable. In particular, for progressive processes the time $\int_0^t \mathbf{1}_{\{X_s \in H\}}\, ds$ spent in H up to time t can be measured on the basis of the information available at time t, which is of course quite natural. The link with adaptedness is given by the following.

Proposition 2.1.2. *Any progressive process is adapted. An adapted process with right or left-continuous paths is progressive.*

Proof. The first statement is clear, as projections of measurable mappings from a product space to its components are measurable. If X_t is adapted and right-continuous, define $X_0^{(n)}(\omega) = X_0(\omega)$ and

$$X_s^{(n)}(\omega) = X_{(k+1)t/2^n}(\omega) \quad \text{for } \frac{kt}{2^n} < s \leq \frac{k+1}{2^n} t,$$

[1] Motivated by this fact, some authors call stopping times *optional times*; we reserve the term optional for random variables such that $\mathbf{1}_{(\tau,\infty)}$ is adapted, see Section 3.10.

where $t > 0$, $n > 0$, $k = 0, 1, \ldots, 2^n - 1$. The map $(s, \omega) \mapsto X_s^{(n)}(\omega)$ is $\mathcal{B}([0,t]) \otimes \mathcal{F}_t$-measurable. Hence the same holds for X_s, since $X_s^{(n)} \to X_s$ by right-continuity. And if X_t is left-continuous, we can use its natural left-continuous discrete approximation from Proposition 2.1.1. □

Finally, working with filtration, allows us to monitor the dynamics of the Radon–Nikodym derivatives of measures, leading to one of the central notions of the whole theory of random processes, the notion of a martingale. Namely, if $(\Omega, \mathcal{F}, \mathcal{F}_t, \mathbf{P})$ is a filtered probability space and Q is a measure (possibly even signed) on it that is absolutely continuous with respect to $\mathbf{P}$ then the restriction Q_t of Q on the probability spaces $(\Omega, \mathcal{F}_t, \mathbf{P})$ are still absolutely continuous with respect to $\mathbf{P}$, so that one can define the Radon–Nikodym derivatives $M_t = dQ_t/d\mathbf{P}$, which are random variables on $(\Omega, \mathcal{F}_t, \mathbf{P})$. Hence, if $A_s \in \mathcal{F}_s$ and $s \leq t$, then

$$\int_{A_s} M_s(\omega)\mathbf{P}(d\omega) = \int_{A_s} M_t(\omega)\mathbf{P}(d\omega),$$

since both sides equal $\int_{A_s} Q(d\omega)$. By the definition of conditional expectation, this means that

$$\mathbf{E}(M_t|\mathcal{F}_s) = M_s. \tag{2.3}$$

The integrable processes satisfying (2.3) for any $s \leq t$ are called *martingales*. Thus we have shown the following.

Proposition 2.1.3. *An integrable process M_t, $t \in [0,T]$ with $0 < T < \infty$, is a martingale on the filtered probability space $(\Omega, \mathcal{F}, \mathcal{F}_t, \mathbf{P})$ if and only if there exists a measure Q on $(\Omega, \mathcal{F})$ (possibly signed, but finite) such that $M_t = dQ_t/d\mathbf{P}$ for all $t \leq T$.*

It is clear from this characterization of martingales that quite a special role belongs to non-negative martingales with $M_0 = 1$, as they describe the evolution of densities (or Radon–Nikodym derivatives) of probability measures. We shall study martingales in more detail in Section 3.9.

Exercise 2.1.2. Show that the process $\mathbf{E}(Y|\mathcal{F}_t)$, where Y is an arbitrary integrable r.v. on a filtered probability space, is a martingale. Hint: use Theorem 1.3.2 (iv). Such martingales are called *closed martingales*.

2.2 Definition and basic properties of BM

Brownian motion (briefly BM) is an acknowledged champion in stochastic modeling for a wide variety of processes in physics (statistical mechanics, quantum fields, etc), biology (e.g. population dynamics, migration, disease spreading), finance (e.g.

common stock prices). BM enjoys beautiful nontrivial properties deeply linked with various areas of mathematics. The general theory of modern stochastic process is strongly rooted in BM and was largely developed by extensions of its remarkable features. BM is simultaneously a continuous martingale, a Markov process, a Feller process, a Lévy process, a Gaussian process and a stable process. Learning the basic properties of BM, its potential and limitation as a modeling tool, understanding its place among general random processes and appreciating the methods of modern probability through its application to BM yields a handy entrance to the realm of modern stochastic analysis.

A *Brownian motion or a Wiener process (briefly BM) with variance* σ^2 (and without a drift) is defined as a Gaussian process B_t (on a probability space $(\Omega, \mathcal{F}, \mathbf{P})$) satisfying the following conditions:

(i) $B_0 = 0$ a.s.;

(ii) the increments $B_t - B_s$ have distribution $N(0, \sigma^2(t-s))$ for all $0 \le s < t$;

(iii) the r.v. $B_{t_2} - B_{t_1}$ and $B_{t_4} - B_{t_3}$ are independent whenever $t_1 \le t_2 \le t_3 \le t_4$;

(iv) the trajectories $t \mapsto B_t$ are a.s. continuous.

Notice that this is not obvious that the object called BM actually exists. There are several proofs of this existence, all of them quite ingenuous and each yielding specific insights in the remarkable features of BM. In the next sections we present four constructions and thus proofs of its existence. Yet another proof will appear from Theorem 3.11.6 of the next chapter. In this section we shall briefly discuss the basic sample path properties of BM.

Brownian motion with $\sigma = 1$ is called the *standard Wiener process or Brownian motion*. The corresponding induced measure on the space of continuous paths $t \mapsto B_t$ is called the *Wiener measure*. Closely connected with BM is the so called *Brownian bridge*, which is defined as a process on $t \in [0, 1]$ distributed like $X_t = B_t - tB_1$. Loosely speaking, Brownian bridge is BM forced to return to the origin at time one.

Exercise 2.2.1. Show that for any $n \in \mathbb{N}$ there exists a constant C_n s.t. $\mathbf{E}|X-\mu|^{2n} = C_n \sigma^{2n}$ for a random variable X with the normal distribution $N(\mu, \sigma^2)$. Hint: in the expression for the moments via the explicit normal density make the change of the variable x to $y = (x-\mu)/\sigma$.

Exercise 2.2.2 (Elementary transformations of BM). Let B_t be a BM. Then so are the processes

(i) $B_t^c = \frac{1}{\sqrt{c}} B_{ct}$ for any $c > 0$ (scaling),

(ii) $-B_t$ (symmetry),

(iii) $B_T - B_{T-t}, t \in [0, T]$ for any $T > 0$ (time reversal),

(iv) $tB_{1/t}$ (time inversion).

Hint: for (iv) in order to get continuity at the origin deduce from the law of large numbers that $B_t/t \to 0$ a.s. as $t \to \infty$.

As a simple example of the application of the transformations from the above exercise, let us show that a path B_t of a BM on $\mathbb{R}_+$ is a.s. unbounded. In fact, assuming that there exists a number $a > 0$ such that the subset A_a of Ω, where $\sup_t |B_t| < a$, has a positive measure p, it follows from transformation (i) that for any $n \in \mathbb{N}$ the set $A_{a/n} \subset A_a$ has the same measure p. By passing to the limit $n \to \infty$, this would imply that B_t vanishes identically on a set of positive measure, which is absurd. Alternatively, unboundedness of BM can be deduced from CLT applied to B_n, or from a more precise Theorem 2.2.2.

Theorem 2.2.1 (Quadratic variation of BM.). *Let B_t be a standard Brownian motion. Suppose a sequence of partitions*

$$\Delta_n = \{0 = t_{n,0} < t_{n,1} < \cdots < t_{n,k_n} = t\}$$

of a fixed interval $[0,t]$ is given such that

$$h_n = \max_k (t_{n,k} - t_{n,k-1}) \to 0$$

as $n \to \infty$. Then

$$\xi_n = \sum_k (B_{t_{n,k}} - B_{t_{n,k-1}})^2 \to t$$

in L^2. If $\sum h_n < \infty$, then this convergence holds also a.s.

Proof. Clearly $\mathbf{E}\xi_n = t$ for all n, so that $\mathbf{E}(\xi_n - t)^2 = \mathrm{Var}(\xi_n)$. By Exercise 2.2.1

$$\begin{aligned}
\mathrm{Var}(\xi_n) &= \sum_k \mathrm{Var}(B_{t_{n,k}} - B_{t_{n,k-1}})^2 \\
&= \sum_k [\mathbf{E}(B_{t_{n,k}} - B_{t_{n,k-1}})^4 - (t_{n,k} - t_{n,k-1})^2] \\
&= (C_4 - 1) \sum_k (t_{n,k} - t_{n,k-1})^2 \le (C_4 - 1) h_n t,
\end{aligned}$$

which tends to zero, implying the required convergence in L^2. Next, by Chebyshev's inequality

$$\mathbf{P}(|\xi_n - t| > \epsilon) \le \epsilon^{-2}(C_4 - 1) h_n t,$$

and consequently, if $\sum h_n < \infty$, the series

$$\sum_{n=1}^{\infty} \mathbf{P}(|\xi_n - t| > \epsilon)$$

converges for any ϵ. Hence by the Borel–Cantelli Lemma, see Theorem 1.1.1, there exists a.s. an $N = N(\epsilon)$ such that $|\xi_n - t| \leq \epsilon$ for all $n > N$. This implies the a.s. convergence. □

Corollary 3. *BM has a.s. unbounded variation on every interval* $[s, t]$.

This implies that a Brownian path is a.s. nowhere differentiable (a detailed deduction of this fact can be found e.g. in [278]).

Exercise 2.2.3 (p-th order variation of BM). Let B_t be a standard Brownian motion. Suppose a sequence of partitions

$$\Delta_n = \{0 = t_{n,0} < t_{n,1} < \cdots < t_{n,k_n} = t\}$$

of a fixed interval $[0, t]$ is given such that

$$h_n = \max_k (t_{n,k} - t_{n,k-1}) \to 0$$

as $n \to \infty$. Then

$$\xi_n(p) = \sum_k |B_{t_{n,k}} - B_{t_{n,k-1}}|^p$$

tends to 0 (resp. to infinity) in L^1 whenever $p > 2$ (resp. if $p \in [0, 2)$). If $\sum h_n < \infty$, this convergence holds also a.s. Hint: by Exercise 2.2.1,

$$\mathbf{E}\xi_n(p) = \sum_k |t_{n,k} - t_{n,k-1}|^{p/2}.$$

On the other hand, $\xi_n(p) \leq h_n^{p-2}\xi_n(2)$ for $p > 2$ and $\xi_n(p) \geq h_n^{-(2-p)}\xi_n(2)$ for $p \in (0, 2)$.

To complete this section let us say some words about the long time behavior of BM. First we formulate the famous law of iterated logarithm. However, we shall neither use nor prove it here. A standard proof (based on an ingenuous application of the Borel–Cantelli lemma) can be found basically in all textbooks on stochastic analysis, see e.g. Revuz and Yor [278] or Kallenberg [154].

Theorem 2.2.2. *For a standard Brownian motion* B *a.s.*

$$\limsup_{t \to 0} \frac{B_t}{\sqrt{2t \log\log(1/t)}} = \limsup_{t \to \infty} \frac{B_t}{\sqrt{2t \log\log t}} = 1.$$

Sometimes it is also useful to know how quickly a Brownian path goes to infinity (if it does), i.e. what is its *rate of escape*, see Section 5.10 for an application. The same question makes sense also for the integrals of a Brownian path that describe the movement of a particle with Brownian velocity.

Theorem 2.2.3 (Rate of escape for BM). *If $d > 2$, then for any $\beta < 1/2 - 1/d$*

$$\liminf_{t\to\infty} |B_t| t^{-\beta} = \infty$$

almost surely.

This is a corollary of a more general result from Dvoretsky–Erdős [105].

The same question makes sense also for the integral of a Brownian path that describes the movement of a particle with Brownian velocity.

Theorem 2.2.4 (Rate of escape for the integrated BM). *If $d > 1$, then for any $\beta < 3/2 - 1/d$*

$$\liminf_{t\to\infty} \left| \int_0^t B_s \, ds \right| t^{-\beta} = \infty$$

almost surely.

This theorem is proved in Kolokoltsov [174]. Some extensions can be found in Khoshnevisan [161]; see also Section 5.10.

2.3 Construction via broken-line approximation

We start here with arguably the most elementary and intuitively appealing construction of BM. The idea is to approximate a Brownian path by continuous piecewise-linear curves and then to recover the original path as the limit of this broken-line approximations.

In order to see how binary partitions are organized observe, that if $B(t)$ is the standard BM on $[0, 1]$, then one can write

$$B(t) = B(t/2) + \tilde{B}(t/2), \tag{2.4}$$

where $B(t/2)$ and $\tilde{B}(t/2) = B(t) - B(t/2)$ are independent $N(0, t/2)$ random variables. Hence

$$B(t/2) = \frac{1}{2} B(t) + \left(\frac{1}{2} B(t/2) - \frac{1}{2} \tilde{B}(t/2) \right).$$

But $(1/2)[B(t/2) - \tilde{B}(t/2)]$ is a $N(0, t/4)$ random variable independent of (2.4), which follows from the following simple fact.

Exercise 2.3.1. Check that if A, B are independent $N(0, \sigma^2)$ random variables, then $A + B$ and $A - B$ are independent $N(0, 2\sigma^2)$ random variables.

Exercise 2.3.2. This is a converse statement to the previous exercise. Let A, B be i.i.d. random variables with a characteristic function of the form $e^{\phi(p)}$. Show that if $A + B$ and $A - B$ are independent, then A and B are Gaussian. Hint: show that the assumed independence implies the functional equation $\phi(p + q) + \phi(p - q) = 2\phi(p) + \phi(q) + \phi(-q)$, which in turn implies that $\phi''(p) = 0$.

We conclude that $B(t)$ given, the value $B(t/2)$ can be obtained by adding an independent $N(0, t/4)$ random variable to $(1/2)B(t)$. This anticipates the following construction.

Let X_0 and Y_n^k, $n = 1, 2, \ldots$, $k = 1, 2, \ldots, 2^{n-1}$, be a family of independent random variables on some probability space such that X_0 is $N(0, 1)$ and each Y_n^k is $N(0, 2^{-(n+1)})$.

Exercise 2.3.3. Point out a probability space $(\Omega, \mathcal{F}, \mathbf{P})$, on which such a family can be defined. Hint: use countable products of probability spaces.

Let $X_0(t) = tX_0$. Next set

$$X_1(0) = 0, \quad X_1(1) = X_0, \quad X_1\left(\frac{1}{2}\right) = \frac{1}{2}X_0 + Y_1^1$$

and define the path $X_1(t)$ as a continuous function which is linear on the intervals $[0, 1/2]$ and $[1/2, 1]$, at the end of which it has the values prescribed above. Next, define $X_2(t)$ as a continuous function, which is linear on the intervals $[0, 1/4]$, $[1/4, 1/2]$, $[1/2, 3/4]$, $[3/4, 1]$ and such that

$$\begin{aligned} X_2(0) = X_1(0) = X_0(0) = 0, \quad & X_2\left(\frac{1}{2}\right) = X_1\left(\frac{1}{2}\right), \\ X_2(1) = X_1(1) = X_0(1) = X_0, \quad & X_2\left(\frac{1}{4}\right) = X_1\left(\frac{1}{4}\right) + Y_2^1, \\ & X_2\left(\frac{3}{4}\right) = X_1\left(\frac{3}{4}\right) + Y_2^2. \end{aligned}$$

Finally, define inductively $X_n(t)$ as a continuous broken-line random path which is linear between times $k2^{-n}$, $k = 0, \ldots, 2^n$, coincides with $X_{n-1}(t)$ at times $k2^{-(n-1)}$ and has values

$$X_n\left(\frac{2k-1}{2^n}\right) = X_n\left(\frac{k}{2^{n-1}} - \frac{1}{2^n}\right) = X_{n-1}\left(\frac{2k-1}{2^n}\right) + Y_n^k,$$

for $k = 1, \ldots, 2^{n-1}$. Thus the values of X_n at the middle of the interval $[(k-1)/2^{n-1}, k/2^{n-1}]$ is the sum of the values of the previous step approximation X_{n-1} at this point and an independent centered Gaussian random variable with an appropriate variance.

Theorem 2.3.1. *The processes $X_n(t)$ converge a.s. and in L_1 to a process $X(t)$ so that*

$$\mathbf{E} \sup_{t\in[0,1]} |X_n(t) - X(t)| \to 0, \quad n \to \infty \tag{2.5}$$

and also a.s.

$$\sup_{t\in[0,1]} |X_n(t) - X(t)| \to 0, \quad n \to \infty. \tag{2.6}$$

The limiting process $X(t)$ is a standard BM on $[0, 1]$.

Proof. As clearly

$$X_{n+m}\left(\frac{k}{2^n}\right) = X_n\left(\frac{k}{2^n}\right), \quad k = 0, \dots, 2^n,\ m \geq n,$$

the sequence $X_n(t)$ stabilizes after a finite number of steps and hence converges for all binary times $t = k2^{-n}$. Hence for these times the limiting trajectory $X(t)$ is defined. As the definition of X_n clearly implies that

$$X_n\left(\frac{k}{2^{n-1}} - \frac{1}{2^n}\right) - X_n\left(\frac{k-1}{2^{n-1}}\right) = \frac{1}{2}\left(X_{n-1}\left(\frac{k}{2^{n-1}}\right) - X_{n-1}\left(\frac{k-1}{2^{n-1}}\right)\right) + Y_n^k,$$

$$X_n\left(\frac{k}{2^{n-1}}\right) - X_n\left(\frac{k}{2^{n-1}} - \frac{1}{2^n}\right) = \frac{1}{2}\left(X_{n-1}\left(\frac{k}{2^{n-1}}\right) - X_{n-1}\left(\frac{k-1}{2^{n-1}}\right)\right) - Y_n^k,$$

it follows from Exercise 2.3.1 that the random variables

$$X_n\left(\frac{k}{2^{n-1}} - \frac{1}{2^n}\right) - X_n\left(\frac{k-1}{2^{n-1}}\right), \quad X_n\left(\frac{k}{2^{n-1}}\right) - X_n\left(\frac{k}{2^{n-1}} - \frac{1}{2^n}\right)$$

are independent, each having the law of $N(0, 1/2^n)$. This implies that generally for the binary rational times the differences $X(t)-X(s)$ are $N(0, t-s)$ with $X(t_4)-X(t_3)$ and $X(t_2) - X(t_1)$ independent for $t_1 < t_2 \leq t_3 < t_4$, as required in the definition of BM. Hence to complete the construction it is enough to show that the approximating paths converge a.s. in the norm topology, since then the limiting path $X(t)$ will be clearly continuous and will satisfy all the requirements.

But

$$\begin{aligned}\sup_{t\in[0,1]} |X_n(t) - X_{n-1}(t)| &= \max_{k=1,\dots,2^{n-1}} \left| X_n\left(\frac{2k-1}{2^n}\right) - X_{n-1}\left(\frac{2k-1}{2^n}\right)\right| \\ &= \max_{k=1,\dots,2^{n-1}} |Y_n^k| \leq \left(\sum_{k=1}^{2^{n-1}} |Y_n^k|^4\right)^{1/4},\end{aligned}$$

and hence by Jensen's inequality and Exercise 2.2.1

$$\begin{aligned}\mathbf{E} \sup_{t\in[0,1]} |X_n(t) - X_{n-1}(t)| &\leq \left(\mathbf{E}\sum_{k=1}^{2^{n-1}} |Y_n^k|^4\right)^{1/4} \\ &= (2^{n-1}2^{-(2n+2)}C_2)^{1/4} = C_2^{1/4}2^{-(n+3)/4},\end{aligned}$$

the sum

$$\sum_{n=1}^{\infty} \mathbf{E} \sup_{t\in[0,1]} |X_n(t) - X_{n-1}(t)|$$

is finite, implying that there exists a limiting process $X(t)$ such that (2.5) and (2.6) hold. □

Corollary 4. *Standard Brownian motion exists on* $\{t \geq 0\}$.

Proof. By Theorem 2.3.1 there exists a sequence $(\Omega_n, \mathcal{F}_n, \mathbf{P}_n)$, $n = 1, 2, \dots$, of probability spaces with Brownian motions W_n on each of them. Take the product probability space Ω and define B on it recursively by

$$B_t = B_n + W_{t-n}^{n+1}, \quad n \leq t \leq n+1. \qquad \square$$

2.4 Construction via Hilbert-space methods

Here we give an alternative construction of BM leading to a new proof of its existence.

Remark 11. The approximations used in this and previous sections actually coincide, only the methods of analysis are different. However, instead of the Haar functions used below, one can use an arbitrary orthogonal system on a finite interval. This would lead to different approximations (with a bit more involved calculations). For instance, in the original construction of Wiener, the trigonometric basis e^{ikx} was used, see e.g. Paley and Wiener [263].

The main bricks for the construction of BM on [0, 1] via Hilbert space methods are given by an orthogonal basis in $L_2([0,1])$. In the simplest version one works with the so called *Haar functions* H_k^n, $n = 1, 2, \dots$, $k = 0, 1, \dots, 2^{n-1} - 1$, on [0, 1]. By definition,

$$H_k^n(t) = \begin{cases} 2^{(n-1)/2}, & k/2^{n-1} \leq t < (k+1/2)/2^{n-1}, \\ -2^{(n-1)/2}, & (k+1/2)/2^{n-1} \leq t < (k+1)/2^{n-1}, \\ 0 & \text{otherwise.} \end{cases} \tag{2.7}$$

The integrals $S_k^n(t) = \int_0^t H_k^n(u)\,du$ are called the *Schauder functions*. The system of Haar functions is known to be an orthonormal basis in $L^2[0,1]$.

Exercise 2.4.1. Prove the latter statement in following steps. (a) Check the orthogonality condition:

$$(H_k^n, H_l^m) = \int_0^1 H_k^n(x) H_l^m(x)\,dx = \delta_m^n \delta_l^k.$$

Hint: the supports of H_k^n and H_l^n do not intersect for $k \neq l$. Notice that orthogonality implies linear independence.

(b) Comparing dimensions show that, for any $N \in \mathbb{N}$, the space generated by all Haar functions H_k^n with $n \leq N$ coincides with the space of piecewise-constant functions with discontinuities at the points $k/2^n$, $n \leq N$. Conclude that the space of finite linear combinations of Haar functions coincides with the space of piecewise-constant functions with discontinuities at binary-rational points. Hence any continuous function on $[0, 1]$ is a limit of a uniformly converging sequence of functions from this space.

(c) Conclude from (b) and the Stone–Weierstrass theorem that linear combinations of Haar functions are dense in $L^2([0, 1])$ and hence form an orthonormal basis.

Let ξ_k^n, $n = 1, 2, \dots, k = 0, 1, \dots, 2^{n-1} - 1$, be mutually independent $N(0, 1)$ r.v. on a probability space $(\Omega, \mathcal{F}, \mathbf{P})$. Such a family exists according to Exercise 2.3.3.

Consider the partial sums

$$B_t^m = \sum_{n=1}^{m} f_n(t, \omega), \quad f_n(t, \omega) = \sum_{k=0}^{2^{n-1}-1} \xi_k^n(\omega) S_k^n(t). \tag{2.8}$$

The main technical ingredient of the construction is the following

Lemma 2.4.1. *There exists a subset $\Omega_0 \subset \Omega$ such that B_t^m converges as $m \to \infty$ uniformly on $[0, 1]$ for all $\omega \in \Omega_0$ and $\mathbf{P}(\Omega_0) = 1$.*

Proof. Let

$$M_n(\omega) = \max\{|\xi_j^n| : 0 \le j \le 2^{n-1} - 1\}$$

Since

$$\mathbf{P}(M_n > a) \le \sum_{j=0}^{2^{n-1}-1} \mathbf{P}(|\xi_j^n| > a) = 2^n \frac{1}{\sqrt{2\pi}} \int_a^\infty e^{-x^2/2} dx$$

$$\le 2^n \frac{1}{\sqrt{2\pi}} \int_a^\infty \frac{x}{a} e^{-x^2/2} dx = 2^n \frac{1}{\sqrt{2\pi}} a^{-1} e^{-a^2/2},$$

one sees that

$$\sum_{n=1}^\infty \mathbf{P}(M_n > n) \le \frac{1}{\sqrt{2\pi}} \sum_{n=1}^\infty 2^n \frac{1}{n} e^{-n^2/2} < \infty.$$

Hence by Borel–Cantelli $\mathbf{P}(\Omega_0) = 1$, where

$$\Omega_0 = \{\omega : M_n(\omega) \le n \text{ for large enough } n\}.$$

Consequently for $\omega \in \Omega_0$

$$|f_n(t, \omega)| \le n \sum_{k=0}^{2^{n-1}-1} S_k^n(t) \le n 2^{-(n+1)/2}$$

for all large enough n, because $\max_t S_k^n(t) = 2^{-(n+1)/2}$ and the functions S_k^n have non-intersecting supports for different k. This implies that

$$\sum_{n=0}^\infty \max_{0 \le t \le 1} |f_n(t, \omega)| < \infty$$

on Ω_0, which clearly implies the claim of the lemma. □

Theorem 2.4.1. *Let B_t denote the limit of* (2.8) *for $\omega \in \Omega_0$ and let us put $B_t = 0$ for ω outside Ω_0. Then B_t is a standard Brownian motion on* $[0, 1]$.

Proof. Since B_t is continuous in t as a uniform limit of continuous functions, the conditions (i) and (iv) of the definition of BM hold. Moreover, the finite-dimensional distributions are clearly Gaussian and $\mathbf{E}B_t = 0$. Next, since

$$\sum (S_k^n)^2(t) = \sum (\mathbf{1}_{[0,t]}, H_k^n)^2 = (\mathbf{1}_{[0,t]}, \mathbf{1}_{[0,t]}) = t < \infty$$

(by Parseval) it follows that

$$\mathbf{E}[B_t - B_t^m]^2 = \sum_{n>m} \sum_{k=0}^{2^{n-1}-1} (S_k^n(t))^2 \to 0$$

as $m \to \infty$, and consequently B_t^m converge to B_t also in L_2. Hence one deduces that

$$\begin{aligned}\mathbf{E}(B_t B_s) = \lim_{m\to\infty} \mathbf{E}(B_t^m B_s^m) &= \sum_{n=0}^{\infty} \sum_{j=0}^{2^{n-1}-1} (\mathbf{1}_{[0,t]}, H_j^n)(\mathbf{1}_{[0,s]}, H_j^n)\\ &= (\mathbf{1}_{[0,t]}, \mathbf{1}_{[0,s]}) = \min(t, s).\end{aligned}$$

□

BM on the whole line can be now constructed as in the previous section.

2.5 Construction via Kolmogorov's continuity

Here we present yet another method of constructing BM, which together with its existence yields also a remarkable property of the a.s. Hölder continuity of its paths.

Theorem 2.5.1 (The Kolmogorov–Chentsov Continuity criterium). *Suppose a process X_t, $t \in [0, T]$, on a probability space $(\Omega, \mathcal{F}, \mathbf{P})$ satisfies the condition*

$$\mathbf{E}|X_t - X_s|^\alpha \le C|t - s|^{1+\beta}, \quad 0 \le s, t \le T,$$

for some positive constants α, β, C. Then there exists a continuous modification $\tilde{X}_t$ of X_t, which is a.s. locally Hölder continuous with exponent γ for every $\gamma \in (0, \beta/\alpha)$, i.e.

$$\mathbf{P}\left[\omega : \sup_{s,t\in[0,T]:|t-s|<h(\omega)} \frac{|\tilde{X}_t(\omega) - \tilde{X}_s(\omega)|}{|t - s|^\gamma} \le \delta\right] = 1,$$

where $h(\omega)$ is an a.s. positive r.v. and $\delta > 0$ is a constant.

Proof. Step 1. By the Chebyshev inequality,

$$\mathbf{P}(|X_t - X_s| \geq \epsilon) \leq \epsilon^{-\alpha}\mathbf{E}|X_t - X_s|^\alpha \leq C\epsilon^{-\alpha}|t-s|^{1+\beta},$$

and hence $X_s \to X_t$ in probability as $s \to t$.

Step 2. Setting $t = k/2^n$, $s = (k-1)/2^n$, $\epsilon = 2^{-\gamma n}$ in the above inequality yields

$$\mathbf{P}(|X_{k/2^n} - X_{(k-1)/2^n}| \geq 2^{-\gamma n}) \leq C 2^{-n(1+\beta-\alpha\gamma)}.$$

Hence

$$\mathbf{P}(\max_{1\leq k\leq 2^n} |X_{k/2^n} - X_{(k-1)/2^n}| \geq 2^{-\gamma n})$$
$$\leq \sum_{k=1}^{2^n} \mathbf{P}(|X_{k/2^n} - X_{(k-1)/2^n}| \geq 2^{-\gamma n}) \leq C 2^{-n(\beta-\alpha\gamma)}.$$

By Borel–Cantelli (by the assumption $\beta - \alpha\gamma > 0$) there exists Ω_0 with $\mathbf{P}(\Omega_0) = 1$ such that

$$\max_{1\leq k\leq 2^n} |X_{k/2^n} - X_{(k-1)/2^n}| < 2^{-\gamma n}, \quad \forall n \geq n^\star(\omega), \tag{2.9}$$

where $n^\star(\omega)$ is a positive, integer-valued random variable.

Step 3. For each $n \geq 1$ define $D_n = \{k/2^n : k = 0, 1, \ldots, 2^n\}$ and $D = \bigcup_{n=1}^\infty D_n$. For a given $\omega \in \Omega_0$ and $n \geq n^\star(\omega)$ we shall show that $\forall m > n$

$$|X_t(\omega) - X_s(\omega)| \leq 2\sum_{j=n+1}^{m} 2^{-\gamma j}, \quad \forall t, s \in D_m : 0 < t - s < 2^{-n}. \tag{2.10}$$

For $m = n+1$ necessarily $t - s = 2^{-(n+1)}$ and (2.10) follows from (2.9) with n replaced by $n+1$. Suppose inductively (2.10) is valid for $m = n+1, \ldots, M-1$. Take $s < t : s, t \in D_M$ and define the numbers

$$\tau_{\max} = \max\{u \in D_{M-1} : u \leq t\}, \quad \tau_{\min} = \min\{u \in D_{M-1} : u \geq s\},$$

so that

$$s \leq \tau_{\min} \leq \tau_{\max} \leq t; \quad \max(\tau_{\min} - s, t - \tau_{\max}) \leq 2^{-M}.$$

Hence from (2.9)

$$|X_{t_{\min}}(\omega) - X_s(\omega)| \leq 2^{-\gamma M}, \quad |X_{t_{\max}}(\omega) - X_t(\omega)| \leq 2^{-\gamma M},$$

and from (2.10) with $m = M-1$, completing the induction.

$$|X_{\tau_{\max}}(\omega) - X_{\tau_{\min}}(\omega)| \leq 2\sum_{j=n+1}^{M-1} 2^{-\gamma j},$$

which implies (2.10) with $m = M$.

Step 4. For $s, t \in D$ with

$$0 < t - s < h(\omega) = 2^{-n^\star(\omega)},$$

choose $n > n^\star(\omega)$ s.t.

$$2^{-(n+1)} \le t - s < 2^{-n}.$$

By (2.10),

$$|X_t(\omega) - X_s(\omega)| \le 2 \sum_{j=n+1}^{\infty} 2^{-\gamma j} \le 2(1-2^{-\gamma})^{-1} 2^{-(n+1)\gamma} \le 2(1-2^{-\gamma})^{-1} |t-s|^\gamma,$$

which implies the uniform continuity of X_t with respect to $t \in D$ for $\omega \in \Omega_0$.

Step 5. Define $\tilde{X}_t = \lim_{s \to t, s \in \mathbb{Q}} X_s$ for $\omega \in \Omega_0$ and zero otherwise. Then $\tilde{X}_t$ is continuous and satisfies (1) with $\delta = 2(1 - 2^{-\gamma})^{-1}$. Since $\tilde{X}_s = X_s$ for $s \in \mathbb{Q}$, it follows that $\tilde{X}_t = X_t$ a.s. for all t, because $X_s \to X_t$ in probability and $X_s \to \tilde{X}_t$ a.s. □

Corollary 5. *There exists a probability measure* **P** *on* $(\mathbb{R}^{[0,\infty)}, \mathcal{B}(\mathbb{R}^{[0,\infty)}))$ *and a stochastic process* W_t *on it which is a BM under* **P**. *The trajectories of this process are a.s. Hölder continuous with any exponent* $\gamma \in (0, 1/2)$.

Proof. By Kolmogorov's existence theorem $\exists$ **P** such that co-ordinate process X_t satisfies all properties of BM, but for continuity. By Kolmogorov's continuity theorem and the Exercise 2.2.1, for each T there exists an a.s. γ-Hölder continuous modification W_t^T of X_t on $[0, T]$ with any $\gamma < (n-1)/2n$ for a positive n, and hence with any $\gamma \in (0, 1/2)$. Set

$$\Omega_T = \{\omega : W_t^T(\omega) = X_t(\omega)\, \forall t \in [0, T] \cap \mathbb{Q}\}, \quad \Omega_0 = \bigcap_{T=1}^{\infty} \Omega_t.$$

As $W_t^T = W_t^S$ for $t \in [0, \min(T, S)]$ (continuous modifications of each other), their common values define a process on $t \ge 0$ of the required type. □

2.6 Construction via random walks and tightness

We shall start with the basic tightness criterion for distributions on the space of continuous functions $C([0, T], S)$, where (S, ρ) is a separable and complete metric space. Recall that by Prohorov's criterion tightness of a family of measures is equivalent to its relative compactness in the weak topology.

Theorem 2.6.1 (Tightness criterion in C). *Let* $X_1, X_2, \ldots$ *be a sequence of* $C([0, T], S)$*-valued random variables for a separable and complete metric space* (S, ρ). *Then this sequence is tight if*

(i) *the sequence* $\pi_t(X_n)$, $n = 1, 2, \ldots$, *is tight in* S *for* t *from a dense subset* $t = t_1, t_2, \ldots$ *of* $[0, T]$, *and*

(ii)

$$\lim_{h \to 0} \sup_n \mathbf{P}(w(X_n, T, h) > \epsilon) = 0 \tag{2.11}$$

for all ϵ.

Proof. Assuming (2.11) and given an $\epsilon > 0$, there exists a sequence $h_1, h_2, \ldots$ of positive numbers such that for all k

$$\sup_n \mathbf{P}(w(X_n, T, h_k) > 2^{-k}) \leq 2^{-k-1}\epsilon.$$

From the tightness of the families $\pi_{t_k}(X_n)$, one can choose a sequence of compact subsets $C_1, C_2, \ldots$ from S such that for all k

$$\sup_n \mathbf{P}(X_n(t_k) \in (S \setminus C_k)) \leq 2^{-k-1}\epsilon.$$

Then $\sup_n \mathbf{P}(X_n \in C([0, T], S) \setminus B) \leq \epsilon$ for

$$B = \bigcap_k \{x \in C([0, T], S) : x(t_k) \in C_k, w(x, T, h_k) \leq 2^{-k}\},$$

and by the Arzelà–Ascoli theorem B is relatively compact. □

Exercise 2.6.1. Condition (2.11) is equivalent to

$$\lim_{h \to 0} \sup_{n \to \infty} \mathbf{E} \min(1, w(X_n, T, h)) = 0. \tag{2.12}$$

Hint: use the same argument as in Proposition 1.1.1.

For our construction we shall need also the following auxiliary result.

Theorem 2.6.2 (Ottaviani's maximal inequality). *Let* $\xi_1, \xi_2, \ldots$ *be i.i.d. random variables with mean* 0 *and variance* 1. *Let* $S_n = \xi_1 + \cdots + \xi_n$ *and* $S_n^\star = \max_{k=1,\ldots,n} S_k$. *Then for any* $r > 2$

$$\mathbf{P}(S_n^\star \geq 2r\sqrt{n}) \leq \frac{\mathbf{P}(|S_n| \geq r\sqrt{n})}{1 - r^{-2}}, \tag{2.13}$$

and

$$\lim_{r \to \infty} r^2 \limsup_{n \to \infty} \mathbf{P}(S_n^\star \geq 2r\sqrt{n}) = 0. \tag{2.14}$$

Proof. Let $\tau = \min\{k : |S_k| \geq 2r\sqrt{n}\}$, so that the events $\{\tau \leq n\}$ and $\{S_n^\star \geq 2r\sqrt{n}\}$ coincide. Clearly

$$\begin{aligned}\mathbf{P}(|S_n| \geq r\sqrt{n}) &\geq \mathbf{P}(|S_n| \geq r\sqrt{n}, S_n^\star \geq 2r\sqrt{n}) \\ &\geq \mathbf{P}(\tau \leq n, |S_n - S_\tau| \leq r\sqrt{n}) = \mathbf{P}(\tau \leq n)P(|S_n - S_\tau| \leq r\sqrt{n}).\end{aligned}$$

The last equation comes from the formula

$$\begin{aligned}\mathbf{P}(\tau \leq n, |S_n - S_\tau| \leq r\sqrt{n}) &= \sum_{k=1}^{n} \mathbf{P}(\tau = k, |S_n - S_k| \leq r\sqrt{n}) \\ &= \sum_{k=1}^{n} \mathbf{P}(\tau = k)\mathbf{P}(|S_n - S_k| \leq r\sqrt{n}),\end{aligned}$$

that follows from the independence of ξ_j. Consequently

$$\begin{aligned}\mathbf{P}(|S_n| \geq r\sqrt{n}) &\geq \mathbf{P}(S_n^\star \geq 2r\sqrt{n}) \min_{k \leq n} \mathbf{P}(|S_n - S_k| \leq r\sqrt{n}) \\ &\geq \mathbf{P}(S_n^\star \geq 2r\sqrt{n}) \min_{k \leq n} \mathbf{P}(|S_k| \leq r\sqrt{n}).\end{aligned}$$

This implies (2.13), as by Chebyshev's inequality

$$\min_{k \leq n} \mathbf{P}(|S_k| \leq r\sqrt{n}) \geq \min_{k \leq n}(1 - k/r^2 n) = 1 - /r^2.$$

By the central limit theorem the laws of $S_n/\sqrt{n}$ converge to the standard normal. Hence (2.14) follows from (2.13). □

The main result of this section is the following.

Theorem 2.6.3 (Functional CLT). *Let $\xi_1, \xi_2, \dots$ be i.i.d. random variables with mean 0 and variance 1 and let*

$$X^n(t) = n^{-1/2}\Big(\sum_{k \leq nt} \xi_k + (nt - [nt])\xi_{[nt]+1}\Big), \quad t \in [0,1],\ n \in \mathbb{N},$$

be the sequence of linearly interpolated partial sums. Then the sequence X^n converges weakly in $C([0,1],\mathbb{R})$ and its limit is standard Brownian motion.

Proof. Clearly the finite-dimensional distributions of $X^n(t)$ converge to the distributions of the Brownian motion (by the central limit theorem). Hence to prove tightness one needs to establish (2.11). But (2.14) implies

$$\limsup_{n\to\infty} \sup_t \mathbf{P}(\sup_{0 \leq r \leq h} |X^n(t+r) - X^n(t)| \geq \epsilon) = o(h), \quad h \to 0,$$

because

$$(\sup_{0\le r\le h} |X^n(t+r) - X^n(t)| \ge \epsilon) = \left(S^*_{hn} \ge 2\frac{\epsilon}{\sqrt{h}}\sqrt{hn}\right),$$

where $S^\star_n$ is the same as in Theorem 2.6.2. Consequently, dividing the interval [0, 1] in subintervals of length h yields (2.11). □

Another way of proving Theorem 2.6.3 is via the following criterion of tightness in terms of moments.

Theorem 2.6.4. *Let $X_1, X_2, \dots$ be a sequence of $C([0,T],S)$-valued random variables for a separable and complete metric space (S,ρ). Then this sequence is tight if the sequence $\pi_t(X_n)$, $n = 1, 2, \dots$, is tight in S for t from a dense subset $t = t_1, t_2, \dots$ of $[0,T]$ and for some $\alpha, \beta > 0$*

$$\sup_n \mathbf{E}\rho^\alpha(X^n(s), X^n(t)) \le c|s-t|^{d+\beta}. \tag{2.15}$$

In this case the limiting process is a.s. γ-Hölder continuous with any $\gamma \in (0, \beta/\alpha)$.

A proof of this result could be obtained similarly to the proof of Theorem 2.5.1 taking into account Theorem 2.6.1; for details see e.g. Chapter 16 of Kallenberg [154].

2.7 Simplest applications of martingales

We shall discuss the important techniques arising from martingales and stopping times in the next chapter. However, it seems instructive to anticipate this development by illustrating the power of this technique on a simple example. To this end, we discuss here the exit times of BM from a fixed interval. From the theory of martingales (mentioned in Section 2.1) we shall use only the following particular case of the general optional sampling theorem proved in Section 3.10: if M_t is a martingale and τ is a bounded stopping time (i.e. $\tau \le a$ a.s. for some constant a), then $\mathbf{E}M_\tau = \mathbf{E}M_0$.

To begin with let us notice that standard Brownian motion B_t is a martingale with respect to its natural filtration $\mathcal{F}_t$. In fact, if $t > s$, then

$$\mathbf{E}(B_t|\mathcal{F}_s) = \mathbf{E}(B_t - B_s|\mathcal{F}_s) + \mathbf{E}(B_s|\mathcal{F}_s) = B_s$$

(the first term vanishes, because of the independence of increments of BM, and the second equals B_s because BM is adapted to its natural filtration). Similarly, $B_t^2 - t$ is a martingale, since

$$\begin{aligned}\mathbf{E}(B_t^2 - t|\mathcal{F}_s) &= \mathbf{E}(B_t^2 - B_s^2 - (t-s)|\mathcal{F}_s) + \mathbf{E}(B_s^2 - s|\mathcal{F}_s)\\ &= \mathbf{E}((B_t - B_s)^2 - (t-s) + 2B_s(B_t - B_s)|\mathcal{F}_s) + B_s^2 - s = B_s^2 - s.\end{aligned}$$

Let $a < 0 < b$ and τ denote the exit time of a standard BM from the interval (a, b):

$$\tau = \min\{s : B_s \notin (a, b)\} = \min\{s : B_s \in \{a, b\}\}.$$

It is straightforward to see that τ is a stopping time (according to the definition from Section 2.1). Our objective is to calculate the expectation $\mathbf{E}\tau$ together with the probabilities $p_a = \mathbf{P}(B_\tau = a)$ and $p_b = \mathbf{P}(B_\tau = b)$ that B_t exits the interval (a, b) via a and b respectively. To this end, let us show that

$$\mathbf{E}B_\tau = 0, \quad \mathbf{E}\tau = \mathbf{E}B_\tau^2. \tag{2.16}$$

Since BM is a.s. unbounded (see Section 2.2), τ is a.s. finite, but possibly not uniformly bounded, so that the above-mentioned property of martingales cannot be applied directly to τ. To circumvent this problem, one applies it first to the bounded stopping times $\tau \wedge t$ (for arbitrary $t > 0$), yielding the following equations:

$$\mathbf{E}B_{\min(t,\tau)} = 0, \quad \mathbf{E}(\min(t, \tau)) = \mathbf{E}B^2_{\min(t,\tau)}.$$

From these equations (2.16) is obtained by dominated and monotone convergence as $t \to \infty$.

Since τ is a.s. finite, we have $\mathbf{E}B_\tau = ap_a + bp_b$ and $p_a + p_b = 1$. Consequently, the first equation in (2.16) implies that

$$p_a = \frac{b}{b-a}, \quad p_b = \frac{|a|}{b-a}. \tag{2.17}$$

And from the second equation in (2.16) we deduce that $\mathbf{E}\tau = a^2 p_a + b^2 p_b$, and hence

$$\mathbf{E}\tau = b|a|. \tag{2.18}$$

In Section 4.2 we shall learn a systematic way of estimating exit times via Markov processes – the link with PDE. In the next section we apply equations (2.17), (2.18) to introduce the celebrated Skorohod embedding, that allows one to embed general random walks into BM by suitable stopping of the latter.

Exercise 2.7.1. Show that the following processes are martingales: (i) $B_t^3 - 3tB_t$, $B_t^4 - 6tB_t^2 + 3t^2$ and $M_t^u = \exp\{uB_t - u^2t/2\}$, $u > 0$ being an arbitrary parameter, for a standard BM B_t; (ii) $|B(t)|^2 - \mathrm{tr}(A)t$ for a d-dimensional Brownian motion $B(t)$ with covariance A.

Solution for M_t^u.

$$\mathbf{E}(M_t^u|\mathcal{F}_s) = \mathbf{E}(\exp\{u(B_t - B_s) - u^2t/2\}e^{uB_s}|\mathcal{F}_s)$$

Using Theorem 1.3.2 and the definition of BM this rewrites as

$$e^{uB_s}e^{-u^2t/2}\mathbf{E}\exp\{uB_{t-s}\},$$

which, applying the well known characteristic function or the moment generating function for the centered normal $N(0, t-s))$ random variable B_{t-s}, rewrites in turn as

$$e^{uB_s}e^{-u^2t/2}e^{u^2(t-s)/2} = M_s^u.$$

2.8 Skorohod embedding and the invariance principle

We discuss here the method of embedding the arbitrary sums of i.i.d. random variables into BM by choosing appropriate stopping times. Eventually it allows one to deduce properties of such sums from properties of BM. We shall demonstrate this idea by deducing a functional central limit for sums of i.i.d. random variables.

For $a \le 0 \le b$, let $\nu_{a,b}$ be the unique probability measure on the two-point set $\{a, b\}$ with mean zero so that $\nu_{a,b} = \delta_0$ for $ab = 0$ and

$$\nu_{a,b} = \frac{b\delta_a - a\delta_b}{b-a} \tag{2.19}$$

otherwise.

Proposition 2.8.1 (Randomization lemma). *For any distribution μ on $\mathbb{R}$ of zero mean, denote by $\mu_\pm$ its restriction on $\mathbb{R}_+ = \{x > 0\}$ and $\mathbb{R}_- = \{x < 0\}$ respectively and put $c = \int x\mu_+(dx) = -\int x\mu_-(dx)$. Then*

$$\mu = \int_{x\le 0\le y} \tilde{\mu}(dx\,dy)\nu_{x,y}, \tag{2.20}$$

where the distribution $\tilde{\mu}$ on $\bar{\mathbb{R}}_- \times \bar{\mathbb{R}}_+$ is given by

$$\tilde{\mu}(dx\,dy) = \mu(\{0\})\delta_{0,0}(dx\,dy) + c^{-1}(y-x)\mu_-(dx)\mu_+(dy).$$

Proof. For a continuous function f

$$\left(f, \int_{\{x\le 0\le y\}} \tilde{\mu}(dx\,dy)\nu_{x,y}\right) = f(0)\mu(\{0\}) + \int_{\{x<0<y\}} \tilde{\mu}(dx\,dy)\frac{yf(x) - xf(y)}{y-x}$$

$$= f(0)\mu(\{0\}) + \frac{1}{c}\int_{\{x<0<y\}} yf(x)\mu_-(dx)\mu_+(dy)$$

$$- \frac{1}{c}\int_{\{x<0<y\}} xf(y)\mu_-(dx)\mu_+(dy)$$

$$= f(0)\mu(\{0\}) + \int_{\{x<0\}} f(x)\mu_-(dx) + \int_{\{0<y\}} f(y)\mu_+(dy) = (f, \mu). \qquad \square$$

Proposition 2.8.2 (Embedding of random variables). *For a probability measure μ on $\mathbb{R}$ with mean zero choose a random pair (a, b) with distribution $\tilde{\mu}$ from Proposition* 2.8.1 *and an independent BM B_t. Then*

(i) *the random variable*

$$T = \min\{t : B_t \in \{a,b\}\} = \inf\{t : B_t \notin (a,b)\}$$

is a stopping time for filtration $\sigma\{a,b; B_s, s \le t\}$,

(ii) *the law of* B_T *is* μ,

(iii) *the expectation of* T *coincides with the second moment (variance) of* μ.

Proof. By (2.17) the random variable B_τ for fixed a,b would have the distribution (2.19). Hence

$$\mathbf{E}f(B_T) = \mathbf{E}\mathbf{E}(f(B_T)|a,b) = \iint f(z)\nu_{x,y}(dz)\tilde{\mu}(dx\,dy) = \int f(x)\mu(dx),$$

yielding (ii). Then (iii) follows from (2.16). □

Theorem 2.8.1 (Skorohod embedding). *Let* $\xi_1, \xi_2, \dots$ *be i.i.d. random variables with mean* 0 *and* $S_n = \xi_1 + \dots + \xi_n$. *Then there exist a filtered probability space with a BM* B_t *and stopping times* $0 = T_0 \le T_1 \le \dots$ *s.t. the differences* $\Delta T_n = T_n - T_{n-1}$ *are i.i.d. with* $\mathbf{E}\Delta T_n = \mathbf{E}\xi_1^2$ *and* B_{T_n} *are distributed like* S_n *for all* n.

Remark 12. $\tau_n = \inf\{t \ge \tau_{n-1} : B_t = S_n\}$ would give a trivial solution if the moment requirement were not imposed.

Proof. Let μ denote the common law of ξ_j. Take i.i.d. pairs (a_n, b_n), $n = 1, 2, \dots$, with the distribution $\tilde{\mu}$ from Proposition 2.8.1 and an independent BM. Everything follows from the recursive definition of random times $0 = T_0 \le T_1 \le T_2 \le \cdots$ by

$$T_n = \inf\{t \ge T_{n-1} : B_t - B_{T_{n-1}} \in \{a_n, b_n\}\}.$$ □

The following result is similar to Theorem 2.6.3, where we used linearly interpolated random walks to be able to work in the space of continuous functions only.

Theorem 2.8.2 (Functional CLT). *Let* $\xi_1, \xi_2, \dots$ *be i.i.d. random variables with mean* 0 *and variance* 1, *and let* $S_n = \xi_1 + \dots + \xi_n$. *Then there exists a BM* B_t *s.t.* $X_t = t^{-1/2}\sup_{s \le t}|S_{[s]} - B_s|$ *converges to zero in probability as* $t \to \infty$ *(*$[s]$ *denotes the integer part of* s*).*

Proof. Choose T_n and B as in Theorem 2.8.1. We then can take $S_n = B_{T_n}$ in the definition of X_t (as we are interested in the distribution of S_n, its realization is not relevant). Then $T_n/n \to 1$ a.s. as $n \to \infty$ by the law of large numbers (LLN). Hence $T_{[t]}/t \to 1$ a.s. as $t \to \infty$. Consequently $\delta_t/t \to 0$ a.s., where $\delta_t = \sup_{s \le t}|T_{[s]} - s|$. In fact, because a.s. there exists $M > 0$ such that $T_n/n \le M$ for all n, we have

$$\frac{\delta_t}{t} \le \sup_{s \le K}\frac{T_{[s]} - s}{t} + \sup_{K \le s \le t}\frac{T_{[s]} - s}{t} \le \frac{MK}{t} + \sup_{s \ge K}\frac{T_{[s]} - s}{s},$$

and this can be made arbitrary small by first choosing a large enough K and then a large t. Finally, for any t, h, ϵ

$$\mathbf{P}(X_t > \epsilon) = \mathbf{P}(X_t > \epsilon, \delta_t > th) + \mathbf{P}(X_t > \epsilon, \delta_t \leq th)$$
$$\leq \mathbf{P}(\delta_t > th) + \mathbf{P}(\sup_{u-v\leq th, u,v\leq t+th} |B_u - B_v| > \epsilon\sqrt{t}),$$

which by the scaling property of BM equals

$$= \mathbf{P}\left(\frac{\delta_t}{t} > h\right) + \mathbf{P}(\sup_{u-v\leq h, u,v\leq 1+h} |B_u - B_v| > \epsilon).$$

This can be made arbitrary small by choosing small h and large t. □

Corollary 6 (Functional CLT and invariance principle). (i) *For all $C, \epsilon > 0$ there exists N s.t. for all $n > N$ there exists a BM B_t (depending on n) s.t.*

$$\mathbf{P}\left(\sup_{t\leq 1}\left|\frac{S_{[tn]}}{\sqrt{n}} - B_t\right| > C\right) < \epsilon.$$

(ii) *Let F be a uniformly continuous function on the space $D[0,1]$ of càdlàg functions on $[0,1]$ equipped with the sup-norm topology. Then $F(\frac{S_{[tn]}}{\sqrt{n}})$ converges in distribution to $F(B)$ with $B = B_t$ standard BM.*

Proof. (i) Applying Theorem 2.8.2 with $t = n$ yields

$$\mathbf{P}(n^{-1/2}\sup_{s\leq n}|S_{[s]} - B_s| > C) \to 0$$

as $n \to \infty$ for any C. With $s = tn$ this rewrites as

$$\mathbf{P}\left(\sup_{t\leq 1}\left|\frac{S_{[tn]}}{\sqrt{n}} - \frac{B_{tn}}{\sqrt{n}}\right| > C\right) \to 0.$$

But by scaling $B_{tn}/\sqrt{n}$ is again a BM, and (i) follows.

(ii) One has to show that

$$\mathbf{E}\left(g\left(F\left(\frac{S_{[.n]}}{\sqrt{n}}\right)\right) - g(F(B_.))\right) \to 0 \tag{2.21}$$

as $n \to 0$ for any bounded uniformly continuous g. Choosing for each n a version of B from (i), one decomposes (2.21) into the sum of two terms with the function under the expectation multiplied by the indicators $\mathbf{1}_{Y_n>C}$ and $\mathbf{1}_{Y_n\leq C}$ respectively, where

$$Y_n = \sup_{t\leq 1}\left|\frac{S_{[tn]}}{\sqrt{n}} - B_t\right|.$$

Then the first term is small by (i) for any C and n large enough, and the second term is small for small C by the uniform continuity of F and g. □

Examples. 1. Applying statement (i) with $t = 1$ yields the usual CLT for random walks.

2. Applying (ii) with $F(h(.)) = \sup_{t\in[0,1]} h(t)$ and taking into account the distribution of the maximum of BM (obtained by the reflection principle in Theorem 3.12.1) yields

$$\mathbf{P}\left(\frac{\max\{S_k : k \leq n\}}{\sqrt{n}} \leq x\right) \to 2\mathbf{P}(N \leq x), \quad x \geq 0,$$

where N is a standard normal random variable.

2.9 More advanced Hilbert space methods: Wiener chaos and stochastic integral

Here we develop a more abstract version of the construction from Section 2.4 leading to new advanced insights. In particular, this development represents a starting-point for the celebrated Malliavin calculus. We shall use it in Section 9.6 to give a rigorous representation of the Feynman integral in terms of Wiener measure.

We start with the following fundamental definition. Let H be a separable real Hilbert space. An *isonormal Gaussian process on* H is a linear mapping from $h \in H$ to centered (i.e. with vanishing mean) Gaussian random variables $W(h)$ defined on a probability space $(\Omega, \mathcal{F}, \mathbf{P})$ such that

$$\mathrm{Cov}(W(h), W(g)) = \mathbf{E}(W(h)W(g)) = (h, g). \tag{2.22}$$

In other words, each $W(h)$ is normal $N(0, \|h\|^2)$ and $W(h)$ is an isometric linear inclusion of H into $L^2(\Omega, \mathcal{F}, \mathbf{P})$.

Existence of such a process is almost evident. In fact, choosing a basis $(e_1, e_2, \dots)$ in H and a family of i.i.d. $N(0, 1)$ random variables $\xi_1, \xi_2, \dots$ one sees straightforwardly that the mapping

$$W\Big(\sum_{i=1}^{\infty} h_i e_i\Big) = \sum_{i=1}^{\infty} h_i \xi_i, \tag{2.23}$$

satisfies all the requirements.

A closely related notion is that of a Gaussian random measure. Namely, let $(S, \mathcal{G}, \mu)$ be a measure space. A family of Gaussian random variables $M(A)$, $A \in \mathcal{G}$, defined on a certain probability space $(\Omega, \mathcal{F}, \mathbf{P})$, is called a *Gaussian random measure* or a *Gaussian white noise* on $(S, \mathcal{G})$ with the *control measure* μ if $\mathrm{Cov}(M(A), M(B)) = \mathbf{E}(M(A)M(B)) = \mu(A \cap B)$ (in particular, each $M(A)$ is a $N(0, \mu(A))$, $M(A)$, $M(B)$ are independent for $A \cap B = \emptyset$) and M is an additive function of A. Let us emphasize (to avoid possible confusion that may arise from the use of the term 'measure') that the function $M(A)$ is neither supposed to be positive, nor of finite variation, so that it is not a measure, or even not a signed measure of finite variation as defined in Section 1.1. Nevertheless, copying the standard method of the construction of the Lebesgue integral one can define the *stochastic integral* with respect to M as a (unique) continuous extension of the mapping

$$I\Big(\sum_{i=1}^{n} a_i \mathbf{1}_{A_i}\Big) = \sum_{i=1}^{n} a_i M(A_i)$$

from finite linear combinations of indicators to the linear mapping $I(h), h \in L^2(S, \mathcal{G}, \mu)$. Since the definition of the Gaussian measure implies that $\mathbf{E}(I(h)I(g)) = (h, g)$ for h, g being linear combinations of indicator functions, this extension is well defined and constitutes an isonormal Gaussian process on $L^2(S, \mathcal{G}, \mu)$. Thus any Gaussian random measure on $(S, \mathcal{G})$ with control measure μ defines a isonormal Gaussian process on $L^2(S, \mathcal{G}, \mu)$ via stochastic integration. Conversely, if $W(h)$ is an isonormal Gaussian process on $L^2(S, \mathcal{G}, \mu)$, then the mapping $A \mapsto W(\mathbf{1}_A)$ is obviously a Gaussian random measure with control measure μ. In particular such Gaussian measure always exists.

Remark 13. Not only Gaussian random measures are of interest. In Section 3.1 we shall introduce Poisson random measures.

Choosing an isonormal Gaussian process W on the Hilbert space $H = L^2([0, 1])$ (with Lebesgue measure) leads to the family of random variables $B_t = W(\mathbf{1}_{[0,t]})$, which satisfies all the properties of the Brownian motion, but for (possibly) continuity. To see how the construction of Section 2.4 fits to the present more abstract setting, notice that if $e_1, e_2, \dots$ is a basis in $L^2([0, 1])$, then

$$\mathbf{1}_{[0,t]} = \sum_{j=1}^{\infty} (\mathbf{1}_{[0,t]}, e_j) e_j = \sum_{j=1}^{\infty} \int_0^t e_j(s)\, ds\, e_j,$$

and according to (2.23)

$$B_t = W(\mathbf{1}_{[0,t]}) = \sum_{j=1}^{\infty} \int_0^t e_j(s)\, ds\, \xi_j,$$

which is precisely (2.8) in case of the Haar basis for $L^2([0, 1])$ used in Section 2.4.

Similarly one can defined BM on $L^2([0, \infty))$ by choosing an orthonormal basis there. Of course to complete the construction of BM by the present more abstract approach one still has to show the existence of a continuous modification, using either the method of Section 2.4 or Kolmogorov's continuity theorem. But what we have got here additionally is the stochastic integral $I(h)$, which in case of Brownian motion B_t

is usually denoted by $\int h(s)dB_s$. It is an isometric mapping from $L^2([0,\infty))$ to Gaussian random variables measurable with respect to $\sigma(B_t), t \in \mathbb{R}_+$, so that

$$\mathrm{Cov}\left(\int_0^t h(s)dB_s, \int_0^t g(s)dB_s\right) = \int_0^t h(s)g(s)\,ds = (h,g).$$

It is instructive to observe that the integral $\int_0^t h(s)dB_s$ cannot be defined in the usual Lebesgue–Stieltjes) sense, because B_t has unbounded variation (see Corollary 3).

Exercise 2.9.1. Show that equivalently the integral $\int h(s)dB_s$ can be constructed in the following way. If $h \in C_c^1(\mathbb{R}_+)$, then one can use the integration-by-parts formula to define

$$\int_0^\infty h(s)dB_s = -\int_0^\infty B(s)h'(s)\,ds.$$

Observing that the integral so defined is an isometry, one extends it to the whole $L^2([0,\infty))$.

Exercise 2.9.2. Find $\mathbf{E}\exp\{\int_0^t h(s)dB_s\}$, $h \in L^2([0,t])$. Hint: use the generating (or characteristic) function for the $N(0, \int_0^t h^2(s)\,ds)$ normal law.

The stochastic integral constructed above has a more or less straightforward multiple extension. Namely, let again $M(A)$, $A \in \mathcal{G}$, be a Gaussian random measure on $(S,\mathcal{G})$ with control measure μ. Let us say that a function f_n in $L^2(S^n, \mathcal{G}^{\otimes n}, \mu^n)$ is *simple* if it has the form

$$f_n(x_1,\ldots,x_n) = \sum_{i_1,\ldots,i_n=1}^{N} a_{i_1\ldots i_n}\mathbf{1}_{A_{i_1}}(x_1)\cdots\mathbf{1}_{A_{i_n}}(x_n),$$

where $A_1,\ldots,A_N$ is a collection of pairwise disjoint elements of the σ-algebra $\mathcal{G}$ and the only restriction on the coefficients $a_{i_1\ldots i_n}$ is that $a_{i_1\ldots i_n} = 0$ whenever at least two indices in the collection $\{i_1\ldots i_n\}$ coincide. For f_n of these form one defines the *multiple stochastic integral* by the formula

$$\begin{aligned} I_n(f_n) &= \int f_n(x_1,\ldots,x_n)M(dx_1)\cdots M(dx_n) \\ &= \sum_{i_1,\ldots,i_n=1}^{N} a_{i_1\ldots i_n}M(A_{i_1})\cdots M(A_{i_n}). \end{aligned}$$

The key assumption that $a_{i_1\ldots i_n} = 0$ whenever at least two indices coincide clearly implies that this integral is centered, that is $\mathbf{E}[I_n(f_n)] = 0$ for any simple function n.

It turns out that the integral does not change if the variables are permuted. To make this property precise, let us introduce the operator P_n that projects the space of functions of n variables on the space of symmetric functions (whose values do not change under any permutation of their arguments):

$$P_n f_n(x_1, \dots, x_n) = \frac{1}{n!} \sum_{\pi} f_n(x_{\pi(1)}, \cdots, x_{\pi(n)}),$$

where the sum is taken over all permutations of the set $\{1, \dots, n\}$. Clearly $P_n f_n$ is symmetric for any f_n and $P_n P_n = P_n$. Since

$$I_n(\mathbf{1}_{A_{i_1}} \cdots \mathbf{1}_{A_{i_n}}) = M(A_{i_1}) \cdots M(A_{i_n})$$

does not depend on the order of $\{i_1, \dots, i_n\}$, it follows that

$$I_n(P_n f_n) = I_n(f_n).$$

for any simple f_n.

The possibility of extending the multiple stochastic integral to more general functions depends crucially on its following *isometry property*.

Proposition 2.9.1. *For any symmetric simple f_n, g_m,*

$$\mathbf{E}(I_n(f_n) I_m(g_m)) = \delta_m^n n!(f_n, g_n), \tag{2.24}$$

where δ_m^n is the Kronecker symbol (1 for $m = n$ and zero otherwise) and (f_n, g_n) is the scalar product in $L^2(S^n, \mathcal{S}^{\otimes n}, \mu^n)$:

$$(f_n, g_n) = \int_{S^n} (f_n g_n)(x_1, \dots, x_n) \mu(dx_1) \cdots \mu(dx_n).$$

Proof. For simple functions f_n, g_m defined via families of subsets $A'_1, \dots, A'_{N'}$ and $A''_1, \dots, A''_{N''}$ one can equivalently represent them via the single family $A_1, \dots, A_N$ containing all possible nonempty intersections $A'_i \cap A''_j$. It is clear that

$$\mathbf{E}[I_n(\mathbf{1}_{A_{i_1}} \cdots \mathbf{1}_{A_{i_n}}) I_m(\mathbf{1}_{A_{j_1}} \cdots \mathbf{1}_{A_{j_m}})] = 0$$

if $n \neq m$ and each of the families $\{i_1, \dots, i_n\}$ and $\{j_1, \dots, j_m\}$ has no repeated indices. Hence it remains to prove (2.24) for $m = n$.

Next, if the sets $\{i_1, \dots, i_n\}$ and $\{j_1, \dots, j_n\}$ do not coincide as unordered sets, then clearly

$$\mathbf{1}_{A_{i_1} \times \cdots \times A_{i_n}} \mathbf{1}_{A_{j_1} \times \cdots \times A_{j_n}} = 0$$

and

$$\mathbf{E}[I_n(\mathbf{1}_{A_{i_1} \times \cdots \times A_{i_n}}) I_n(\mathbf{1}_{A_{j_1} \times \cdots \times A_{j_n}})] \\ = \mathbf{E}[M(A_{i_1}) \cdots M(A_{i_n}) M(A_{j_1}) \cdots M(A_{j_n})] = 0.$$

Finally, let the n-tuple $\{j_1, \dots, j_n\}$ be a permutation of the n-tuple $\{i_1, \dots, i_n\}$. Then

$$(P_n \mathbf{1}_{A_{i_1} \times \dots \times A_{i_n}}, P_n \mathbf{1}_{A_{j_1} \times \dots \times A_{j_n}}) = \frac{1}{n!} \mu(A_{i_1}) \cdots \mu(A_{i_n}),$$

and

$$\mathbf{E}[I_n((P_n \mathbf{1}_{A_{i_1} \times \dots \times A_{i_n}}) I_n(P_n \mathbf{1}_{A_{j_1} \times \dots \times A_{j_n}})] = \mu(A_{i_1}) \cdots \mu(A_{i_n}),$$

which completes the proof. □

It is not difficult to see that simple functions of n variables defined above are dense in $L^2(S^n, \mathcal{G}^{\otimes n}, \mu^n)$ for any n. Consequently Proposition 2.9.1 allows one to extend the stochastic integral to the isometric mappings $I_n : L^2(S^n, \mathcal{G}^{\otimes n}, \mu^n) \to L^2(\Omega, \mathcal{F}, \mathbf{P})$, called the *multiple stochastic integral* and denoted by

$$I_n(f_n) = \int_{S^n} f_n(x_1, \dots, x_n) dW_{s_1} \cdots dW_{s_n},$$

whose images for different n are orthogonal. Let us denote these images $\mathcal{H}_n$. The space $\mathcal{H}_n$ is called the *Wiener chaos of order n*.

Let us now define the *Fock space* built on the Hilbert space $L^2(S, \mathcal{G}, \mu)$ as the direct sum

$$F = \mathbb{C} \oplus L^2(S, \mathcal{G}, \mu) \oplus L^2(S^2, \mathcal{G}^{\otimes 2}, \mu^2) \oplus \cdots,$$

where the norm is defined by

$$\|(f_0, f_1, f_2, \dots)\|_F^2 = \sum_{n=0}^{\infty} n! \|f_n\|^2_{L^2(S^n, \mathcal{G}^{\otimes n}, \mu^n)}.$$

Proposition 2.9.1 implies that multiple stochastic integration defines an isometric linear inclusion $F \to L^2(\Omega, \mathcal{F}, \mathbf{P})$.

Recall that we previously denoted the mapping I_1 by W. The following celebrated result is called the *Wiener chaos decomposition.*

Theorem 2.9.1. (i) *Linear combinations of* $\{e^{W(h)}, h \in H\}$ *are dense in* $L^2(\Omega, \sigma(W), \mathbf{P})$.

(ii) *The space* $L^2(\Omega, \sigma(W), \mathbf{P})$ *is the orthogonal sum of* $\mathcal{H}_n$*:*

$$L^2(\Omega, \sigma(W), \mathbf{P}) = \bigoplus_{n=0}^{\infty} \mathcal{H}_n.$$

Proof. (i) Let $X \in L^2(\Omega, \sigma(W), \mathbf{P})$ and $\mathbf{E}(X e^{W(h)}) = 0$ for all $h \in H$. Hence

$$\mathbf{E}\Big(X \exp\Big\{\sum_{i=1}^m t_i W(h_i)\Big\}\Big) = \mathbf{E}\Big(X \exp\Big\{W\Big(\sum_{i=1}^m t_i h_i\Big)\Big\}\Big) = 0$$

for any $t_1, \dots, t_m \in \mathbb{R}$, $h_1, \dots, h_m \in H$. Hence by Exercise 1.1.2

$$0 = \mathbf{E}\Big(X \exp\Big\{\sum_{i=1}^m t_i W(h_i)\Big\}\Big) = \int_{\mathbb{R}^d} \exp\Big\{\sum_{i=1}^m t_i y_i\Big\} \nu(dy),$$

where $\nu(B) = \mathbf{E}[X \mathbf{1}_B(W(h_1), \dots, W(h_m))]$. Hence ν vanishes, as its Laplace transform vanishes. Consequently $\mathbf{E}(X \mathbf{1}_M) = 0$ for any $M \in \sigma(W)$, implying $X = 0$.

(ii) The spaces $\mathcal{H}_n$ and $\mathcal{H}_m$ are orthogonal for $n \neq m$, by Proposition 2.9.1. Hence one only needs to show that if $X \in L^2(\Omega, \sigma(W), \mathbf{P})$ and $\mathbf{E}(X H_n(W(h))) = 0$ for any $n \in \mathbb{N}$ and $h \in H$ (and hence $\mathbf{E}(X(W(h))^n) = 0$ for any $n \in \mathbb{N}$ and $h \in H$), then $X = 0$. But this follows from (i). □

Corollary 7. *The space $L^2(\Omega, \sigma(W), \mathbf{P})$ is isometric to the symmetric Fock space F.*

Finally, let us indicate some useful modifications, which are available when the space S reflects the structure of time. As an example, let us consider the basic BM and the corresponding Fock space built upon the Hilbert space $L^2(\mathbb{R}_+)$. In this case, the symmetric[2] functions on $\mathbb{R}_+^n$ are in one-to-one correspondence with the functions on the simplex

$$\mathrm{Sym}^n = \{(x^1, \dots, x^n) \in \mathbb{R}_+^n : x^1 \le x_2 \le \dots \le x^d\},$$

so that

$$\int_{\mathbb{R}_+^n} f_n(s_1, \dots, s_n) ds_1 \cdots ds_n = n! \int_{\mathrm{Sym}^n} f_n(s_1, \dots, s_n) ds_1 \cdots ds_n$$
$$= \int_0^\infty \int_0^{s_n} \cdots \int_0^{s_2} f_n(s_1, \dots, s_n) ds_1 \cdots ds_n,$$

and

$$I_n(f_n) = \int_{\mathbb{R}_+^n} f_n(s_1, \dots, s_n) dB_{s_1} \cdots dB_{s_n}$$
$$= n! \int_{\mathrm{Sym}^n} f_n(s_1, \dots, s_n) dB_{s_1} \cdots dB_{s_n} = n! J_n(f_n),$$

where the *iterated stochastic integral*

$$J_n(f_n) = \int_{\mathrm{Sym}^n} f_n(s_1, \dots, s_n) dB_{s_1} \cdots dB_{s_n}$$
$$= \int_0^\infty \int_0^{s_n} \cdots \int_0^{s_2} f_n(s_1, \dots, s_n) dB_{s_1} \cdots dB_{s_n}$$

[2]actually, as well as anti-symmetric

is defined by

$$J_n(f_n) = I_n(f_n \mathbf{1}_{\mathrm{Sym}^n}).$$

Instead of (2.24) we now get the simpler isometry relation

$$\mathbf{E}(J_n(f_n)J_m(g_m)) = \delta_m^n(f_n, g_n) = \delta_m^n \int_{\mathrm{Sym}^n} (f_n g_n)(s_1, \ldots, s_n) ds_1 \cdots ds_n. \tag{2.25}$$

Hence, defining the Fock space

$$F_{\mathrm{Sym}} = \mathbb{C} \oplus L^2(\mathbb{R}_+) \oplus L^2(\mathrm{Sym}^2) \oplus \cdots,$$

with the square norm

$$\|(f_0, f_1, f_2, \cdots)\|_F^2 = \sum_{n=0}^{\infty} \|f_n\|^2_{L^2(\mathrm{Sym}^n)},$$

the mapping $f_n \mapsto J_n(f_n)$ extends to an isometry $F_{\mathrm{Sym}} \to L^2(\Omega, \sigma(B), \mathbf{P})$.

Restricting the BM to a bounded interval $[0, t]$ leads similarly to the isometry of the corresponding space $L^2(\Omega, \sigma(B_{s \le t}), \mathbf{P})$ with the Fock space

$$F^t_{\mathrm{Sym}} = \mathbb{C} \oplus L^2([0, t]) \oplus L^2(\mathrm{Sym}_t^2) \oplus \cdots,$$

where $\mathrm{Sym}_t^n = \{x \in \mathrm{Sym}^n : x^d \le t\}$.

The Fock space turns out to be a meeting point for a remarkably wide variety of ideas and methods from classical and quantum probability, infinite-dimensional analysis and quantum and statistical physics.

2.10 Fock spaces, Hermite polynomials and Malliavin calculus

This section is an addendum to the previous one. It

(i) clarifies the structure of the Wiener chaos spaces, yielding its natural orthogonal basis via Hermite polynomials, and

(ii) introduces the Malliavin derivative as the image of the annihilation operator in Fock space.[3]

The *Hermite polynomials* $H_n(x)$, $x \in \mathbb{R}$, $n = 0, 1, \ldots$, are defined as

$$H_n(x) = \frac{(-1)^n}{n!} e^{x^2/2} \frac{d^n}{dx^n}(e^{-x^2/2}),$$

[3]The results of this section will not be used elsewhere in this book.

so that in particular $H_0(x) = 1$, $H_1(x) = x$, $H_2(x) = (x^2 - 1)/2$, and in general H_n is a polynomial of order n. These polynomials can be equivalently defined as the coefficients of the expansion in powers of t of the function $F(t,x) = \exp\{tx - t^2/2\}$, because by Taylor's formula

$$\begin{aligned} F(t,x) &= \exp\left\{\frac{x^2}{2} - \frac{1}{2}(x-t)^2\right\} \\ &= e^{x^2/2} \sum_{n=0}^{\infty} \frac{t^n}{n!} \left(\frac{d^n}{dt^n} e^{-(x-t)^2/2} \right) \Bigg|_{t=0} = \sum_{n=0}^{\infty} t^n H_n(x). \end{aligned}$$

The basic identities

$$\begin{aligned} H_n'(x) &= H_{n-1}(x), \\ (n+1)H_{n+1}(x) &= xH_n(x) - H_{n-1}(x), \\ H_n(-x) &= (-1)^n H_n(x), \end{aligned} \tag{2.26}$$

valid for all $n \geq 1$, follow respectively from the obvious equations $\frac{\partial F}{\partial x} = tF$, $\frac{\partial F}{\partial t} = (x-t)F$, $F(t,-x) = F(-t,x)$.

Finally, if X, Y are $N(0,1)$ random variables that are jointly Gaussian, then clearly

$$\mathbf{E}\left(\exp\left(sX - \frac{s^2}{2}\right) \exp\left(tY - \frac{t^2}{2}\right)\right) = \exp\{st\mathbf{E}(XY)\}$$

for all real s, t. Comparing the coefficients of the expansions of both sides of this equation in powers of t,s yields

$$\mathbf{E}(H_n(X)H_m(Y)) = \begin{cases} 0, & n \neq m, \\ \frac{1}{n!}(\mathbf{E}(XY))^n, & n = m. \end{cases} \tag{2.27}$$

In particular by choosing $X = Y$, this implies that $H_n(x)$ form an orthonormal system in $L^2(\mathbb{R}, e^{-x^2/2}/\sqrt{2\pi})$.

Let H be a separable (real) Hilbert space with an orthonormal basis $e_1, e_2, \ldots$. The *tensor power* $H^{\otimes n}$, $n = 1, 2, \ldots$ can be defined as the Hilbert space with the orthonormal basis $e_{i_1,\ldots,i_n}$, $i_1, i_2, \ldots, i_n \in \mathbb{N}$, denoted by

$$e_{i_1,\ldots,i_n} = e_{i_1} \otimes \cdots \otimes e_{i_n}.$$

It is more or less obvious that the mapping

$$e_{i_1} \times \cdots \times e_{i_n} \mapsto e_{i_1} \otimes \cdots \otimes e_{i_n}$$

extends to the n-linear (i.e. linear with respect to each of its n arguments) mapping

$$H^n = H \times \cdots \times H \to H^{\otimes n}$$

as

$$(f^1, \dots, f^n) \mapsto f^1 \otimes \cdots \otimes f^n = \sum_{i_1}^{\infty} \cdots \sum_{i_n}^{\infty} a^1_{i_1} \dots a^n_{i_n} e_{i_1} \otimes \cdots \otimes e_{i_n}$$

for

$$f^j = \sum_{i=1}^{\infty} a^j_i e_i, \quad j = 1, \dots, n,$$

called the *tensor product* of $f_1, \dots, f_n$. The orthogonal sum $H_0 \oplus H_1 \oplus H_2 \oplus \cdots$, where $H_0 = \mathbb{R}$, equipped with the product

$$((f_0, f_1, \dots), (g_0, g_1, \dots)) = \sum_{n=0}^{\infty} n!(f_n, g_n),$$

is called the *Fock space* based on H. Of course it extends the definition given above for H realized as a function space to the abstract setting. Namely, if $H = L^2(S, \mathcal{G}, \mu)$, then $H^{\otimes n}$ can be clearly identified with the space of functions of n variables: $H^{\otimes n} = L^2(S^n, \mathcal{G}^{\otimes n}, \mu^{\otimes n})$. In that case the tensor product $f^1 \otimes \cdots \otimes f^n$ is identified with the function $f^1(x_1) \cdots f^n(x_n)$.

For any $e_{i_1} \otimes \cdots \otimes e_{i_n}$ one defines the *symmetric tensor product* as

$$\text{symm}(e_{i_1} \otimes \cdots \otimes e_{i_n}) = \frac{1}{n!} \sum_{\pi} e_{i_{\pi(1)}} \otimes \cdots \otimes e_{i_{\pi(n)}}, \tag{2.28}$$

where the sum is taken over all permutations π of $\{1, \dots, n\}$. The image $H^{\hat{\otimes} n}$ of $H^{\otimes n}$ under this symmetrization operator is called the *symmetric tensor product* of H.

Exercise 2.10.1. Show that

$$\|\text{symm}(e_1^{n_1} \otimes \cdots \otimes e_k^{n_k})\|^2 = \frac{n_1! \cdots n_k!}{(n_1 + \cdots + n_k)!}. \tag{2.29}$$

Hint. In the sum on the r.h.s. of (2.28) there are only

$$\frac{(n_1 + \cdots + n_k)!}{n_1! \cdots n_k!}$$

different terms, so that

$$\|\text{symm}(e_1^{n_1} \otimes \cdots \otimes e_k^{n_k})\|^2 \\ = \frac{1}{[(n_1 + \cdots + n_k)!]^2} \frac{(n_1 + \cdots + n_k)!}{n_1! \cdots n_k!} \|n_1! \cdots n_k! e_1^{n_1} \otimes \cdots \otimes e_k^{n_k}\|^2,$$

yielding (2.29).

Let us return to the general isonormal Gaussian process $W : H \to L^2(\Omega, \mathcal{F}, \mathbf{P})$ on a Hilbert space H to construct a natural orthogonal basis for the space $L^2(\Omega, \sigma(W), \mathbf{P})$. Namely, let $\tilde{\mathcal{H}}_n$ denote the closed subspace of $L^2(\Omega, \sigma(W), \mathbf{P})$ generated by

$$\{H_n(W(h)), h \in H, \|h\| = 1\}.$$

In particular, $\tilde{\mathcal{H}}_1$ coincides with the image of H under W and $\mathcal{H}_0 = \mathbb{C}$ is the space of constants.

Proposition 2.10.1. *Each space $\tilde{\mathcal{H}}_n$ coincides with the Wiener chaos $\mathcal{H}_n$ of order n.*

Proof. The spaces $\tilde{\mathcal{H}}_n$ are orthogonal for different n, by (2.27). Next, it is easy to see that

$$\tilde{\mathcal{H}}_0 \oplus \tilde{\mathcal{H}}_1 \oplus \cdots \oplus \tilde{\mathcal{H}}_n = \mathcal{H}_0 \oplus \mathcal{H}_1 \oplus \cdots \oplus \mathcal{H}_n,$$

for any $n \in N$. The proof is then completed by trivial induction on n. □

Exercise 2.10.2. Observe that in the case of a one-dimensional H generated by a standard normal $N(0,1)$ the previous result reduces to the well-known fact from analysis that linear combinations of Hermite polynomials form a complete orthonormal system in $L^2(\mathbb{R}, \nu)$, ν being the law of $N(0,1)$.

We shall give now the infinite-dimensional version of this fact. Let $e_1, e_2, \dots$ be an orthonormal basis of the Hilbert space H.

Theorem 2.10.1. (i) *For any $n \in \mathbb{N}$, the family of random variables*

$$\Big\{\prod_{i=1}^{k} \sqrt{n_i!}H_{n_i}(W(e_i)),\ n_1 + \cdots + n_k = n,\ n_i \in \{0,1,\dots\},\ k \in \mathbb{N}\Big\} \tag{2.30}$$

form an orthonormal basis in $\mathcal{H}_n$.

(ii) *The mapping*

$$I_n\Big(\mathrm{symm}\Big(\prod_{i=1}^{k} e_i^{\otimes n_i}\Big)\Big) = \prod_{i=1}^{k} n_i! H_{n_i}(W(e_i)), \quad n_1 + \cdots + n_k = n$$

is an isometry between the symmetric tensor product $H^{\hat{\otimes} n}$ equipped with the norm $\sqrt{n!}\|.\|_{H^{\otimes n}}$ and the Wiener chaos $\mathcal{H}_n$.

Proof. (i) The family (2.30) is orthonormal by (2.27). Its linear combinations are dense, because any polynomial in $W(h)$ can be approximated by polynomials in $W(e_j)$, $j = 1, 2, \dots$.

(ii) This follows (2.29) and (i). □

Finally let us introduce the Malliavin derivative from the Fock-space setting. The Fock spaces form the basic scenery for the analysis of interacting particles, both classical and quantum. The main role in this analysis is played by the so called *creation* and *annihilation* operators, which for the Fock space F constructed from the Hilbert space $L^2(S,\mathcal{G},\mu)$ are defined as the operators $a_+ : L^2(S,F) \to F$ and $a_- : F \to L^2(S,F)$ (where $L^2(S,F)$ means the space of square-integrable functions from S to F) given by the formulas

$$(a_+ f^t)_n(x_1,\dots,x_n) = \frac{1}{n}\sum_{i=1}^{n} f_{n-1}^{x_i}(x_1,\dots,x_{i-1},x_{i+1},\dots,x_n),$$
$$(a_- f)_n^t(x_1,\dots,x_n) = (n+1) f_{n+1}(x_1,\dots,x_n,t).$$

It is straightforward to see that the operators a_- and a_+ are dual in the sense that

$$\int_S \mu(dt) n! \int_{S^n} (a_- f)_n^t(x_1,\dots,x_n) g^t(x_1,\dots,x_n)\mu(dx_1)\cdots\mu(dx_n)$$
$$= (n+1)! \int_{S^{n+1}} f_{n+1}(x_1,\dots,x_{n+1})(a_+ g^t)(x_1,\dots,x_{n+1}) dx_1\cdots dx_{n+1},$$

or, in more concise notations,

$$\int_S ((a_- f)^t, g^t)_F \mu(dt) = (f,(a_+ g^t))_F,$$

and that the product $a_+ a_-$ equals the so called *number of particles operator* N that acts as $(Nf)_n = nf_n$. The operator $-N$ clearly generates a semigroup in F given by the formula $(e^{-Nt} f)_n = e^{-nt} f_n$.

The isomorphism between the Fock space F and the space $L^2(\Omega,\sigma(W),\mathbf{P})$ allows one to transfer the operators a_+, a_- and N from F to $L^2(\Omega,\sigma(W),\mathbf{P})$, where they start playing a new role as the basic operators of the infinite-dimensional calculus of variations. The image of the annihilation operator a_- is called the *Malliavin derivative*, and the image of the creation operator a_- is called the *divergence operator*, which turns out to represent a natural extension of the Itô stochastic integral. The image of the semigroup e^{-Nt} is called the *Ornstein–Uhlenbeck semigroup*. For this development we refer to books on Malliavin calculus, e.g. Nualart [257] or [258], see also Applebaum [22].

2.11 Stationarity: OU processes and Holtzmark fields

Stationarity of a process, which means roughly speaking that its statistical characteristics do not change in time, is a crucial property for practical applications, as it allows one to estimate the parameters of the process by statistical methods applied to observed data evolving in time (time-series analysis). We give here only a basic definition and a couple of examples related to BM. One is the usual Ornstein–Uhlenbeck (OU) process, and the other a less standard example of moving Holtzmark fields.

A process ξ_t in $\mathbb{R}^d$, $t \in \mathbb{R}$ is called *stationary in the narrow sense* if $\mathbf{E}\xi_t = 0$ for all t and the covariance $\mathrm{Cov}(\xi_t, \xi_{t+s})$ does not depend on t for $s \geq 0$.

Of course BM is not stationary. But can we make it stationary by an appropriate change of time? More precisely, can we find monotone functions $f : \mathbb{R} \to \mathbb{R}_+$ and $g : \mathbb{R} \to \mathbb{R}_+$ such that the process $F_t = g(t)B_{f(t)}$ is stationary? Assume f is increasing. Since $\mathbf{E}F_t = 0$, we need to check only that

$$\mathrm{Cov}(F_t F_{t+s}) = g(t)g(t+s)f(t)$$

does not depend on t for $s \geq 0$. Equivalently, this means that $\ln g(t) + \ln g(t+s) + \ln f(t)$ is a function of s only. Differentiating with respect to t leads to the requirement that the derivative $(\ln g)'(t+s)$ does not depend on s, which means that the function $\ln g(t)$ is a constant. Hence there exists $a \in \mathbb{R}$ such that

$$g(t) = e^{-at}, \quad f(t) = e^{2at}.$$

If $a = 1$, the corresponding process $F_t = e^{-t}B_{e^{2t}}$ is called the standard *Ornstein–Uhlenbeck (OU) process* in $\mathbb{R}$. It is clearly a stationary Gaussian process.

As another example of stationarity we consider a slightly more nontrivial process, arising naturally from the Holtzmark distributions discussed in Section 1.5. At some moment of time, let the points in $\mathbb{R}^d$ be distributed like a Poisson point process with intensity $\lambda > 0$, i.e. in any bounded measurable $M \subset \mathbb{R}^d$ the number of points has the Poisson distribution with parameter $|M|$ (Lebesgue volume of M) and are independently and uniformly distributed in M (see Section 1.5 for a particular construction of this distribution and Chapter 3 for generalities). Each point acts on the origin with force (1.36), that is

$$F(x) = \gamma x|x|^{-m-1},$$

where $d/m < 2$. The total force equals the sum of forces arising from all points. Assuming now that each point moves according to an independent BM, we get the process Φ_t of the total force, which is zero mean, i.e. $\mathbf{E}\Phi_t = 0$ for all t, and at each time the distribution of Φ_t is the Holtzmark distribution described in Section 1.5.

Let us calculate the covariance matrix $\mathrm{Cov}(\Phi_t \Phi_{t+s})$. Following the same approach as in Section 1.5, and using the independence of all points that cancels the correlations between different points, we can write (upper index stands for the coordinate)

$$A_{ij}(s) = \mathrm{Cov}(\Phi_t^i \Phi_{t+s}^j) = \lim_{R\to\infty} \lambda |V_R| \mathbf{E} \frac{x^i(x^j + B_t^j)}{|x|^{m+1}|x+B_s|^{m+1}},$$

where the expectation arises from the uniform distribution of x in the ball B_R and an independent BM B_s in $\mathbb{R}^d$. Thus $A_{ij}(s)$ do not depend on t, as expected for a stationary process. It is easy to calculate this covariance. Namely, we have

$$A_{ij}(s) = \lambda \int_{\mathbb{R}^{2d}} (2\pi s)^{-d/2} \frac{x^i(x^j + y^j)}{|x|^{m+1}|x + y|^{m+1}} \exp\left\{-\frac{y^2}{2s}\right\} dx\, dy$$
$$= \lambda \int_{\mathbb{R}^{2d}} (2\pi s)^{-d/2} \frac{x^i z^j}{|x|^{m+1}|z|^{m+1}} \exp\left\{-\frac{|z - x|^2}{2s}\right\} dx\, dz.$$

Choosing $z/\sqrt{s}$ and $x/\sqrt{s}$ as the new variables of integration, this rewrites finally as

$$A_{ij}(s) = C_{ij} s^{-(2m-d)/2}, \tag{2.31}$$

where

$$C_{ij} = \lambda(2\pi)^{-d/2} \int_{\mathbb{R}^{2d}} \frac{x^i z^j}{|x|^{m+1}|z|^{m+1}} \exp\{-|z - x|^2/2\}\, dx\, dz. \tag{2.32}$$

Exercise 2.11.1. Check that the constants C_{ij} are finite for $m < d < 2m$. Hints:

(i) Condition $m < d$ is needed to integrate the singularity at the origin.

(ii) To estimate the integral for large x, z, return to the variables $x, y = z - x$ and decompose the domain of integration into the two parts, where $|y| \le |x|^\beta$ for sufficiently small $\beta > 0$ and otherwise.

Hence the covariance matrix $A(s)$ of the stationary process Φ_t is finite for all $s > 0$ whenever $m < d < 2m$ (which includes the classical case with $d = 3, m = 2$), and has a power singularity as $s \to 0$.

Remark 14. A model of this kind can be used as an intermediate regime between stationary (not moving) particles and very quick ones that are classical objects of studies in plasma diagnostics; see e.g. Lisitsa [223].

Chapter 3
Markov processes and martingales

This chapter discusses the two broad classes of random processes, where we can handle dependence well, that is martingales and Markov processes. Each lends itself to widespread use in stochastic modeling. We start slowly with Lévy processes, which are nothing else but spatially homogeneous Markov processes. More expanded introductions to the latter can be found e.g. in the recent books Applebaum [20], Kyprianou [215] or Sato [289]. Then we describe in some detail analytic and probabilistic facets or representations of Markov process. Interplay between these representations is central to this book. Afterwards the theory of martingales and stopping times is developed, the crucial link with Markov processes being supplied by Dynkin's martingales. We then turn to the strong Markov property, and conclude with several basic examples of the application of the methods developed to Brownian motion. Books on general Markov processes and martingales are numerous. To mention but a few: Dellacherie and Meyer [98], Chung and Walsh [87], Dynkin [106], Rogers and Williams [280], [281], Sharpe [292], Harlamov [126], Doob [102], Liptser and Shiryayev [222].

3.1 Definition of Lévy processes

Here we introduce the basic definitions related to the notion of a Lévy process.

A process $X = X_t$ in $\mathbb{R}^d$, $t \geq 0$, is said to have *independent increments* if for any collection of times $0 \leq t_1 < \cdots < t_{n+1}$ the random variables $X_{t_{j+1}} - X_{t_j}$, $j = 1, \dots, n$ are independent, and *stationary increments* if $X_t - X_s$ is distributed like $X_{t-s} - X_0$ for any $t > s$. A process X_t is called *stochastically continuous*, or *continuous in probability*, if X_t converges to X_s in probability as $t \to s$ for any s, i.e. if $\forall\, a > 0,\ s \geq 0$

$$\lim_{t \to s} \mathbf{P}(|X_t - X_s| > a) = 0.$$

A process $X = X_t, t \geq 0$, is called a *Lévy process* if $X_0 = 0$ a.s., X has stationary and independent increments and X is stochastically continuous.

If all other conditions are fulfilled, stochastic continuity is obviously equivalent to $\lim_{t\to 0} \mathbf{P}(|X_t| > a) = 0$ for all $a > 0$. By Proposition 1.1.2, this is equivalent to the weak continuity of X_t at $t = 0$. By Glivenko's theorem, this holds if the characteristic function $\phi_{X_t}(u)$ is continuous in t for each u.

To check the independence of increments, one usually shows that for all $s < t$ the random variable $X_t - X_s$ is independent of the σ-algebra generated by all X_τ with $\tau \leq s$.

An alternative version of the definition of the Lévy processes requires the a.s. right-continuity of paths instead of stochastic continuity. At the end of the day this leads to the same class of processes, because, on the one hand, a.s. right continuity implies stochastic continuity (since convergence a.s. implies convergence in probability), and on the other hand, any Lévy process as defined above has a right-continuous modification, as we shall see later. So we shall usually consider the right-continuous modifications of the Lévy processes.

Of course, Brownian motion is a Lévy process.

Remark 15. Note that our definition of BM required only the independence of pairs $B_t - B_s$ and $B_s - B_r$ for $r \leq s \leq t$. However since all distributions are Gaussian, a stronger version of independence in above definition follows easily.

In order to reveal the structure of Lévy processes we shall now exploit their connection with infinitely divisible distributions, and consequently with the Lévy–Khintchine formula.

Exercise 3.1.1. Let a right-continuous function $f : \mathbb{R}_+ \mapsto \mathbb{C}$ satisfy $f(t+s) = f(t)f(s)$ and $f(0) = 1$. Show that $f(t) = e^{t\alpha}$ with some α. Hint: consider first $t \in \mathbb{N}$, then $t \in \mathbb{Q}$, then use continuity.

Proposition 3.1.1. *If X is a Lévy process, then X_t is infinitely divisible for all t and $\phi_{X_t}(u) = e^{t\eta(u)}$, where $\eta(u)$ is the Lévy symbol of X_1:*

$$\eta(u) = i(b,u) - \frac{1}{2}(u, Gu) + \int_{\mathbb{R}^d} [e^{i(u,y)} - 1 - i(u,y)\mathbf{1}_{B_1}(y)]\nu(dy). \tag{3.1}$$

Proof. $\phi_{X_{t+s}}(u) = \phi_{X_t}(u)\phi_{X_s}(u)$ and $\phi_{X_0}(u) = 1$. Hence by the previous exercise $\phi_{X_t} = \exp\{t\alpha(u)\}$. But $\phi_{X_1} = \exp\{\eta(u)\}$, by the Lévy–Khintchine formula (1.18). □

It is worth noting that the mollifier $\mathbf{1}_{B_1}$ used in (3.1) can be replaced by any other measurable bounded function χ that equals one in a neighborhood of the origin and decreases at least linearly at infinity. For instance, another popular choice is $\chi(y) = 1/(1+y^2)$. The change of mollifier leads of course to a change in the drift b. If a Lévy measure has a finite outer first moment, i.e. if $\int_{|y|>1} |y|\nu(dy) < \infty$, one can clearly rewrite the characteristic exponent η without a mollifier (by adjusting the drift b if necessary) in the form

$$\eta(u) = i(b,u) - \frac{1}{2}(u, Gu) + \int_{\mathbb{R}^d} [e^{i(u,y)} - 1 - i(u,y)]\nu(dy). \tag{3.2}$$

Exercise 3.1.2. (i) Show that if $\int_{|y|>1} |y|^k \nu(dy) < \infty$, then $\mathbf{E}|Y_t|^k = O(t)$ for any integer $k > 1$ and small t.

(ii) Let Y_t be a Lévy process with the characteristic exponent of the form (3.2). Show that

$$\mathbf{E}(\xi + Y_t)^2 = (\xi + tb)^2 + t\left(\operatorname{tr} G + \int y^2 \nu(dy)\right) \tag{3.3}$$

for any $\xi \in \mathbb{R}^d$.

Hint: use characteristic functions.

A Lévy process X_t with characteristic exponent

$$\eta(u) = i(b,u) - \frac{1}{2}(u, Gu) \tag{3.4}$$

(where G is a positive definite $d \times d$-matrix, $b \in \mathbb{R}^d$) and with a.s. continuous paths is called the d-dimensional *Brownian motion with covariance G and drift b*. Recall that it is called *standard* if $G = I$, $b = 0$. Clearly if B_t is standard BM, then $B_t^{G,b} = bt + \sqrt{G} B_t$ is BM with covariance G and drift b.

3.2 Poisson processes and integrals

This section is devoted to Poisson random measures, compound Poisson processes and related integration.

Apart from BM, another basic example of a Lévy process is the *Poisson process of intensity* $c > 0$, defined as a right-continuous Lévy process N_t such that each random variable N_t is Poisson with parameter tc. We shall give two constructions (and thus two proofs of the existence) of a Poisson process.

The first method will be based on the notion of a Poisson random measure, which we introduce now and which plays a crucial rule also in the theory of the general Lévy processes.

Let μ be a σ-finite measure on a metric space S (we need only the case of S a Borel subset of $\mathbb{R}^d$). A random measure on S, i.e. the collection of random variables $\phi(B)$ parametrized by Borel subsets of S and such that $\phi(B)$ is a measure as a function of B, is called a *Poisson random measure with intensity measure* μ if each $\phi(B)$ is Poisson with parameter $\mu(B)$ whenever $\mu(B) < \infty$ and $\phi(B_1), \dots, \phi(B_n)$ are independent whenever $B_1, \dots, B_n$ are disjoint.

As it is often more convenient to work with centralized distributions, one defines the *compensated Poisson random measure* by $\tilde{\phi}(B) = \phi(B) - \mu(B)$, i.e. by subtracting from each $\phi(B)$ its mean. Clearly the (random) integrals of (deterministic) functions with respect to a compensated Poisson measure (defined in the sense of Lebesgue) always have zero mean.

Proposition 3.2.1. *For any σ-finite measure μ on a metric space S there exists a Poisson random measure with intensity μ.*

Proof. First assume that μ is finite. Let $\xi_1, \xi_2, \dots$ be a sequence of i.i.d. random variables with the common law $\mu/\|\mu\|$, and let N be a Poisson random variable with intensity $\|\mu\|$ independent of the sequence ξ_j. Then the random measure

$$\nu = \sum_{j=1}^{N} \delta_{\xi_j}$$

(where δ_x denotes as usual the Dirac point measure at the point x) is Poisson with intensity μ. In fact, $\nu(B)$ counts the number of points ξ_j lying in B, and its characteristic function is

$$\begin{aligned}
\mathbf{E}e^{ip\nu(B)} &= \sum_{n=0}^{\infty} \mathbf{E}(e^{ip\nu(B)}|N = n)\mathbf{P}(N = n) \\
&= \sum_{n=0}^{\infty} \mathbf{E}\exp\{ip[\mathbf{1}_B(\xi_1) + \cdots + \mathbf{1}_B(\xi_n)]\}\mathbf{P}(N = n) \\
&= \sum_{n=0}^{\infty} (\mathbf{E}e^{ip\mathbf{1}_B(\xi_1)})^n \frac{\|\mu\|^n}{n!} e^{-\|\mu\|} \\
&= \sum_{n=0}^{\infty} \left[e^{ip}\frac{\mu(B)}{\|\mu\|} + 1 - \frac{\mu(B)}{\|\mu\|}\right]^n \frac{\|\mu\|^n}{n!} e^{-\|\mu\|} = \exp\{(e^{ip} - 1)\mu(B)\},
\end{aligned}$$

which is the characteristic function of a Poisson random variable with parameter $\mu(B)$. Similarly, one shows that

$$\mathbf{E}e^{ip_1\nu(B_1)+\cdots+ip_k\nu(B_k)} = \prod_{j=1}^{k} \exp\{(e^{ip_j} - 1)\mu(B_j)\}$$

for disjoint collection of sets B_j, showing the required independence of $\nu(B_j)$.

Assume now that μ is σ-finite, so that there exist disjoint sets $\{A_j\}_{j=1}^{\infty}$ such that $S = \bigcup_j A_j$ and $\mu(A_j) < \infty$ for every j. If ϕ_n are independent Poisson random measures with the intensities $\mathbf{1}_{A_j}\mu$, then $\phi = \sum_{j=1}^{\infty} \phi_j$ is a Poisson random measure with intensity μ. □

Corollary 8. *A Poisson process of any given intensity $c > 0$ exists.*

Proof. Let ϕ be a Poisson random measure on $\mathbb{R}_+$ with the intensity being Lebesgue measure multiplied by c. Then $N_t = \phi([0, t])$ is a Poisson process with the intensity c. □

Alternatively, Poisson processes can be obtained by the following explicit construction. Let $\tau_1, \tau_2, \dots$ be a sequence of i.i.d. exponential random variables with parameter $c > 0$, i.e. $\mathbf{P}(\tau_i > s) = e^{-cs}$, $s > 0$. Introduce the partial sums $S_n = \tau_1 + \cdots + \tau_n$. These sums have the *Gamma* (c, n) distributions

$$\mathbf{P}(S_n \in ds) = \frac{c^n}{(n-1)!} s^{n-1} e^{-cs} ds$$

(which follows by induction, observing that the distribution of S_n is the convolution of the distributions S_{n-1} and τ_n). Define N_t as the right-continuous inverse to S_n, that is

$$N_t = \sup\{n \in \mathbb{N} : S_n \leq t\},$$

so that $\mathbf{P}(S_k \leq t) = \mathbf{P}(N_t \geq k)$ and

$$\begin{aligned}\mathbf{P}(N_t = n) &= \mathbf{P}(S_n \leq t, S_{n+1} > t) \\ &= \int_0^t \frac{c^n}{(n-1)!} s^{n-1} e^{-cs} e^{-c(t-s)}\, ds = e^{-ct} \frac{(ct)^n}{n!}.\end{aligned}$$

Exercise 3.2.1. Prove that the process N_t constructed above is in fact a Lévy process by showing that

$$\mathbf{P}(N_{t+r} - N_t \geq n, N_t = k) = \mathbf{P}(N_r \geq n)\mathbf{P}(N_t = k) = \mathbf{P}(S_n \leq r)\mathbf{P}(N_t = k). \tag{3.5}$$

Hint: take, say, $n > 1$ (the cases with $n = 0$ or 1 are even simpler) and observe that the l.h.s. of (3.5) is the probability of the event

$$\begin{aligned}&(S_k \leq t, S_{k+1} > t, S_{n+k} \leq t + r) \\ &\quad = (S_k = s \leq t, \tau_{k+1} = \tau > t - s, S_{n+k} - S_{k+1} = v \leq (t+r) - (s+\tau)),\end{aligned}$$

so that by independence the l.h.s. of (3.5) equals

$$\int_0^t \frac{c^k}{(k-1)!} s^{k-1} e^{-cs}\, ds \int_{t-s}^{\infty} ce^{-c\tau}\, d\tau \int_0^{(t+r)-(\tau+s)} \frac{c^{n-1}}{(n-2)!} v^{n-2} e^{-cv}\, dv,$$

which changing τ to $\tau + s$ and denoting it again by τ rewrites as

$$\int_0^t \frac{c^k}{(k-1)!} s^{k-1}\, ds \int_t^{\infty} ce^{-c\tau}\, d\tau \int_0^{t+r-\tau} \frac{c^{n-1}}{(n-2)!} v^{n-2} e^{-cv}\, dv.$$

By calculating the integral over ds and changing the order of v and τ this in turn rewrites as

$$\begin{aligned}&\frac{(ct)^k}{k!} \int_0^r \frac{c^{n-1}}{(n-2)!} v^{n-2} e^{-cv}\, dv \int_t^{t+r-v} ce^{-c\tau}\, d\tau \\ &\qquad = e^{-ct} \frac{(ct)^k}{k!} \int_0^r \frac{c^{n-1}}{(n-2)!} v^{n-2} (e^{-cv} - e^{-cr})\, dv.\end{aligned}$$

It remains to see that by integration by parts the integral in this expression equals

$$\int_0^r \frac{c^n}{(n-1)!} s^{n-1} e^{-cs}\, ds,$$

and (3.5) follows.

Exercise 3.2.2 (Law of large numbers for Poisson processes). Prove the law of large number for a Poisson process N_t of intensity c: $N_t/t \to c$ a.s. as $t \to \infty$. Hint: use the construction of N_t given above and the fact that $S_n/n \to 1/c$ as $n \to \infty$ according to the usual law of large numbers.

The next exercise shows that the standard Stieltjes integrals of the right and left modifications of a process should not coincide.

Exercise 3.2.3 (Poisson integrals). Recall first that right-continuous functions of bounded variation on $\mathbb{R}_+$ (or equivalently the differences of increasing functions) are in one-to-one correspondence with signed Radon measures on $\mathbb{R}_+$ according to the formulas $f_t = \mu([0,t])$, $\mu((s,t]) = f_t - f_s$, and the *Stieltjes integral* of a locally bounded Borel function g

$$\int_0^t g_s\, df_s = \int_{(0,t]} g_s\, df_s$$

is defined as the Lebesgue integral of g with respect to the corresponding measure μ. Let N_t be a Poisson process of intensity $c > 0$ with respect to a right-continuous filtration $\mathcal{F}_t$. Show that

$$\int_0^t N_s dN_s = \frac{1}{2} N_t(N_t + 1), \quad \int_0^t N_{s-} dN_s = \frac{1}{2} N_t(N_t - 1)$$

(integration in the sense of Stieltjes).

The rest of this section is devoted to compound Poisson processes and related integration. Let $Z(n)$, $n \in \mathbb{N}$, be a sequence of $\mathbb{R}^d$-valued i.i.d. random variables with law μ_Z. The *compound Poisson process (with distribution of jumps μ_Z and intensity λ)* is defined as

$$Y_t = Z(1) + \cdots + Z(N_t), \tag{3.6}$$

where N_t is a Poisson process of intensity λ. The corresponding *compensated compound Poisson process* is defined as

$$\tilde{Y}_t = Y_t - t\lambda \mathbf{E}(Z(1)).$$

From (1.17) it follows that Y_t is a Lévy process with Lévy exponent

$$\eta_Y(u) = \int (e^{i(u,y)} - 1)\lambda \mu_Z(dy) \tag{3.7}$$

and $\tilde{Y}_t$ is a Lévy process with Lévy exponent

$$\eta_{\tilde{Y}}(u) = \int (e^{i(u,y)} - 1 - i(u,y))\lambda\mu_Z(dy). \tag{3.8}$$

Remark 16. To check the stochastic continuity of Y_t, one can write

$$\mathbf{P}(|Y_t| > a) = \sum_{n=0}^{\infty} \mathbf{P}(|Z(1) + \cdots + Z(n)| > a)\mathbf{P}(N_t = n)$$

and use dominated convergence. Alternatively, this follows from the obvious right continuity of Y_t.

Consequently

$$\mathrm{Var}(Y_t) = \mathbf{E}\tilde{Y}_t^2 = -\frac{d}{du}\bigg|_{u=0} \mathbf{E}e^{t\eta_{\tilde{Y}}}\eta_{\tilde{Y}}(u) = t\int y^2\lambda\mu_Z(dy). \tag{3.9}$$

Let Y_t be the compound Poisson process (3.6). This process specifies a Poisson random measure on $\mathbb{R}_+ \times \mathbb{R}^d$ by the following prescription. $N((s,t] \times A)$ is defined as the number of jumps

$$\Delta Y_{t_k} = Y_{t_k} - Y_{t_k-} = Z(k)$$

of the processes Y_t of size $Z(k) \in A$ that occurred in the interval $(s,t]$. In other words, $N(dt\,dx)$ is the sum of Dirac delta masses at the points $t_k \times Z(k)$ (and there are a.s. only finite number of them in any compact set), so that for a bounded measurable f on $\mathbb{R}_+ \times \mathbb{R}^d$

$$\int_0^t \int f(s,z)N(ds\,dz) = \sum_{t_i \le t} f(t_i, Z(i)). \tag{3.10}$$

If f does not depend explicitly on t one can write also

$$\int_0^t \int f(s,z)N(ds\,dz) = \int f(z)N(t,\,dz),$$

where

$$N(t,A) = \int_0^t \int_A N(ds\,dx)$$

is a random measure on $\mathbb{R}^d$ for any fixed t. In particular,

$$Y_t = \int_0^t \int zN(ds\,dz) = \int zN(t,dz).$$

By Proposition 3.2.1, the measure $N(t,dz)$ is a Poisson measure on $\mathbb{R}^d$ with intensity $\lambda t\mu_Z$. Moreover, as one proves in the same way as Proposition 3.2.1, $N(dt\,dz)$ is a

Poisson measure on $\mathbb{R}_+ \times \mathbb{R}^d$ with intensity the product of the Lebesgue measure on $\mathbb{R}_+$ and $\lambda\mu_Z$ on $\mathbb{R}^d$.

Corresponding compensated Poisson random measure $\tilde{N}$ is

$$\tilde{N}((s,t] \times A) = N((s,t] \times A) - \lambda(t-s)\mu_Z(A),$$

so that for a bounded measurable $f(s,y)$ on $\mathbb{R}_+ \times \mathbb{R}^d$,

$$\int_0^t \int f(s,z)\tilde{N}(ds\,dz) = \int_0^t \int f(s,z)N(ds\,dz) - \lambda \int_0^t \int f(s,y)ds\mu_Z(dy). \tag{3.11}$$

In particular,

$$\tilde{Y}_t = \int_0^t \int z\tilde{N}(ds\,dz).$$

Let us stress again that the integrals with respect to a compensated Poisson measure have zero mean, implying that

$$\mathbf{E}\int_0^t \int f(s,z)N(ds\,dz) = \lambda \int_0^t \int f(s,y)ds\mu_Z(dy)$$

(whenever the r.h.s. is well defined) and consequently that

$$\frac{d}{dt}\mathbf{E}\int_0^t \int f(s,z)N(ds\,dz) = \lambda \int f(t,y)\mu_Z(dy)$$

when f depends continuously on s.

The following important statement is a direct consequence of the fact that $N(dt\,dz)$ is a Poisson measure.

Proposition 3.2.2. *Let Y_t be a compound Poisson process with Lévy exponent*

$$\eta_Y(u) = \int (e^{i(u,y)} - 1)\nu(dy)$$

for some bounded measure ν on $\mathbb{R}^d$, and let $\mathbb{R}^d$ be decomposed into the union of k non-intersecting Borel subsets $A_1, \dots, A_k$. Then Y_t can be represented as the sum $Y_t = Y_t^1 + \dots + Y_t^k$ of k independent compound Poisson processes with Lévy exponents

$$\eta_Y^j(u) = \int_{A_j} (e^{i(u,y)} - 1)\nu(dy), \quad j = 1, \dots, k.$$

Clearly for a measurable bounded f on $\mathbb{R}^d$ the process

$$\begin{aligned}\int_0^t \int f(z)N(ds\,dz) &= \int f(z)N(t,\,dz) \\ &= \sum_{t_i \le t} f(Z(i)) = f(Z(1)) + \dots + f(Z(N_t))\end{aligned}$$

is a compound Poisson process with distribution of jumps $f(Z(i))$ given by the probability law

$$\mu_Z^f(A) = \mu_Z(f^{-1}(A)) = \mu_Z\{y : f(y) \in A\},$$

which is the push forward of μ_Z by the mapping f. The corresponding compensated process is given by

$$\int_0^t \int f(z)\tilde{N}(ds\,dz) = \sum_{t_i \le t} f(Z(i)) - t\lambda \mathbf{E} f(Z(1)).$$

In particular, as follows from (3.9),

$$\begin{aligned} \mathrm{Var}\left(\int f(z)N(t\,dz)\right) &= \mathbf{E}\left|\int f(z)\tilde{N}(t\,dz)\right|^2 \\ &= t\lambda \int y^2 \mu_Z^f(dy) = t\lambda \int f^2(z)\mu_Z(dz). \end{aligned} \tag{3.12}$$

For a process X_t let a *nonlinear random integral* based on the noise dX_t be defined as a limit in probability

$$\int_0^t f(dX_s) = \lim_{\max_i (s_{i+1}-s_i)\to 0} \sum_{i=1}^n f(X(s_{i+1}) - X(s_i)) \tag{3.13}$$

(the limit is over finite partitions $0 = s_0 < s_1 < \dots . < s_n = t$ of the interval $[0,t]$). In particular, $\int_0^t (dY_s)^2$ is called the *quadratic variation* of Y. More generally, $\int_0^t (dY_s)^p$ is called the *p-th order variation* of Y.

Exercise 3.2.4. Let $Y, \tilde{Y}, N, \tilde{N}$ be as above. Show that

(i) if the integral (3.13) is defined (as a finite or infinite limit) for $X = Y$ and a measurable f (in particular this is the case for either bounded or positive f), then

$$\int_0^t f(dY_s) = \int_0^t \int f(z)N(ds\,dz) = f(Z(1)) + \dots + f(Z(N_t)); \tag{3.14}$$

in particular, the *quadratic variation* of a compound Poisson process equals the sum of the squares of its jumps;

(ii) if $f \in C^1(\mathbb{R}^d)$, then

$$\begin{aligned} \int_0^t f(d\tilde{Y}_s) &= \int_0^t \int f(z)N(dt\,dz) - t\lambda(\nabla f(0), \mathbf{E}Z(1)) \\ &= \int_0^t \int f(z)\tilde{N}(dt\,dz) + t\lambda[\mathbf{E}f(Z(1)) - (\nabla f(0), \mathbf{E}Z(1))]. \end{aligned} \tag{3.15}$$

Hint: since the number of jumps of $Y(t)$ is a.s. finite on each finite time interval, for partitions with small enough $\max_i(s_{i+1} - s_i)$, any interval $s_{i+1} - s_i$ will contain not more than one jump, implying that $\int_0^t f(d\tilde{Y})$ will equal $\int_0^t f(dY)$ plus the limit of the sums $\sum_{i=1}^n f(-\lambda \mathbf{E}Z(1)(s_{i+1}) - (s_i)))$.

3.3 Construction of Lévy processes

Here we explain the basic construction of the Lévy processes, revealing the celebrated Lévy–Itô decomposition that presents an arbitrary Lévy process as the sum of a scaled BM, a compound Poisson process and a centered Lévy process with bounded jumps and moments.

By $\Delta X_t = X_t - X_{t-}$ we shall denote the jumps of X_t.

Theorem 3.3.1 (Lévy–Itô decomposition and existence). *For any function $\eta(u)$ of form*

$$\eta(u) = i(b,u) - \frac{1}{2}(u, Gu) + \int_{\mathbb{R}^d} [e^{i(u,y)} - 1 - i(u,y)\mathbf{1}_{B_1}(y)]\nu(dy) \tag{3.16}$$

(where of course ν is a Lévy measure and G is a non-negative matrix), there exists a Lévy process X_t with characteristic exponent η. Moreover, X_t can be represented as the sum of three independent Lévy processes $X_t = X_t^1 + X_t^2 + X_t^3$, where X_t^1 is BM with drift specified by Lévy exponent (3.4),

$$X_t^2 = \sum_{s \le t} \Delta X_t \mathbf{1}_{|\Delta X_t| > 1}$$

is a compound Poisson process with exponent

$$\eta_2(u) = \int_{\mathbb{R}^d \setminus B_1} [e^{i(u,y)} - 1]\nu(dy) \tag{3.17}$$

obtained by summing the jumps of X_t of size exceeding 1, *and X_t^3 has exponent*

$$\eta_3(u) = \int_{B_1} [e^{i(u,y)} - 1 - i(u,y)]\nu(dy)$$

and is the limit, as $n \to \infty$, of compensated compound Poisson processes $X_t^3(n)$ with exponents

$$\eta_3^n(u) = \int_{B_1 \setminus B_{1/n}} [e^{i(u,y)} - 1]\nu(dy) - i(u, \int_{B_1 \setminus B_{1/n}} y\nu(dy)). \tag{3.18}$$

The process X_t^3 has jumps only of size not exceeding 1 *and has all moments $\mathbf{E}|X_t^3|^m$ finite, $m > 0$.*

Proof. If X^i, $i = 1, 2, 3$, are independent Lévy processes, as described above, then their sum is a Lévy process with exponent (3.16). Since we know already that the processes X_1 and X_2 exist (they are BM with a drift and a compound Poisson process), we only have to show the existence (and the required properties) of the process X_3 with exponent η_3.

To show the existence of X_t^3, one needs to show the existence of the limit of the compensated compound Poisson processes $X_t^3(n)$ with exponents (3.18). To this end, let $Y_t^1, Y_t^2, \dots$ be independent compound Poisson processes (defined on the same probability space) with intensities

$$c_n = \int_{2^{-n}<|y|\le 2^{-(n-1)}} \nu(dy)$$

and distributions of jumps given by the laws

$$\frac{1}{c_n}\mathbf{1}_{(2^{-n}, 2^{-(n-1)}]}(y)\nu(dy)$$

(for the probability space one can take, say, a product of probability spaces where Y^n are defined). Then

$$Z_t^n = Y_t^1 + \cdots + Y_t^n$$

are again compound Poisson processes, with intensities

$$z_n = \int_{2^{-n}<|y|\le 1} \nu(dy)$$

and distributions of jumps given by the law

$$\frac{1}{z_n}\mathbf{1}_{(2^{-n},1]}(y)\nu(dy).$$

The compensated Poisson processes

$$\tilde{Z}_t^n = Z_t^n - t\int_{2^{-n}<|y|\le 1} y\nu(dy)$$

clearly have Lévy exponents (3.18) (i.e. they are versions of processes $X_t^3(n)$ defined on the same probability space).

Now

$$\tilde{Z}_t^n - \tilde{Z}_t^m = \sum_{j=m+1}^{n} (Y_t^j - \tilde{Y}_t^j), \quad n > m,$$

are again compound Poisson processes, so that by (3.9)

$$\operatorname{Var}(Z_t^n - Z_t^m) = \mathbf{E}(\tilde{Z}_t^n - \tilde{Z}_t^m)^2 = t\int_{2^{-n}<|y|\le 2^{-(m+1)}} y^2\nu(dy),$$

implying that the sequence of random variables Z_t^n is Cauchy in the L^2-sense for any t (as $\int y^2 \nu(dy) < \infty$) and hence has a limit. However, one needs slightly more, namely convergence of the distributions on paths. From the above estimate for $\text{Var}(Z^n - Z^m)$ one obtains

$$\mathbf{E} \int_0^t |\tilde{Z}_s^n - \tilde{Z}_s^m|^2 \, ds \le t \int \sup_{s \le t} \mathbf{E} |\tilde{Z}_s^n - \tilde{Z}_s^m|^2 \le t \mathbf{E} (\tilde{Z}_t^n - \tilde{Z}_t^m)^2 \to 0, \quad m, n \to \infty,$$

implying the existence of a limit with respect to the norm $(\mathbf{E} \int_0^t \|Y_s\|^2 \, ds)^{1/2}$. The limiting process is stochastically continuous, because the characteristic function depends continuously on t.

Finally, the moments of X_t^3 are given by

$$\mathbf{E}|X_t^3|^{2k} = \left. \frac{d^{2k}}{du^{2k}} \right|_{u=0} \exp \left\{ t \int_{B_1} [e^{i(u,y)} - 1 - i(u, y)] \nu(dy) \right\},$$

so that

$$\mathbf{E}|X_t^3|^2 = \int_{B_1} |y|^2 \nu(dy)$$

and (as follows by induction)

$$\mathbf{E}|X_t^3|^{2k} = \int_{B_1} |y|^{2k} \nu(dy) + c_k, \quad k = 1, 2, \ldots,$$

with constants c_k depending on the moments $\int_{B_1} |y|^l \nu(dy)$, $l \le 2k - 1$. □

Remark 17. The above proof does not yield a càdlàg modification for the limiting process. The simplest way to obtain this is via martingale methods reviewed later in Section 3.9. These methods can be used in a variety of ways. For instance, from the regularity of martingales one can conclude that the limiting process X_t^3 (which is obviously a martingale as L_2 limit of martingales X_t^s) has a càdlàg modification. Alternatively, one can exploit the Doob maximum inequality. Namely, applying it to the martingales Z_t^n yields the estimate

$$\mathbf{E} \sup_{s \le t} |\tilde{Z}_s^n - \tilde{Z}_s^m|^2 \le 4 \mathbf{E} |\tilde{Z}_t^n - \tilde{Z}_t^m|^2 \to 0, \quad m, n \to \infty,$$

and hence the processes $\tilde{Z}_t^n$ converge with respect to the norm $(\mathbf{E} \sup_{s \le t} \|Y_s\|^2)^{1/2}$. This convergence clearly preserves càdlàg paths (each Z^n has càdlàg paths as a compensated compound Poisson process), implying that the limiting process is càdlàg.

Remark 18. Later on, we shall give an alternative proof of the existence of (right-continuous modification of) a Lévy process with given exponent by a more general procedure (applied to all Feller processes) in three steps:

(i) building finite-dimensional distributions via the Markov property,

(ii) using Kolmogorov's existence of a canonical process,

(iii) defining a right-continuous modification by martingale methods.

Corollary 9. *The only continuous Lévy processes are BM with drifts or deterministic processes (pure drifts).*

Corollary 10. *For any collection of disjoint Borel sets A_i, $i = 1, \dots, n$ not containing zero in their closures the processes*

$$X_t^{A_i} = \sum_{s \le t} \Delta X_t \mathbf{1}_{\Delta X_t \in A_i}$$

are independent compound Poisson process with characteristic exponents

$$\eta_{A_i}(u) = \int_{A_i} (e^{i(u,y)} - 1)\nu(dy), \tag{3.19}$$

and $X_t - \sum_{j=1}^n X_t^{A_j}$ is a Lévy process independent of all X^{A_j} with jumps only outside $\bigcup_j A_j$. Moreover, the processes $N(t, A_i)$ that count the number of jumps of X_t or $X_t^{A_i}$ in A_i up to time t are independent Poisson processes of intensity $\nu(A_i)$.

Corollary 11. *The collection of random variables $N((s,t], A) = N(t, A) - N(s, A)$ (notations from the previous corollary) counting the number of jumps of X_t of size A that occur in the time interval $(s,t]$ specifies a Poisson random measure on $(0,\infty) \times (\mathbb{R}^d \setminus \{0\})$ with intensity $dt \otimes \nu$.*

The following corollary shows that one can extend the integral (3.11) to the case of the Poisson random measure with unbounded intensity arising from a Lévy process.

Proposition 3.3.1. *Let $f(x)/\|x\|$ be a bounded Borel function and N be a Poisson random measure from Corollary 11 with the corresponding compensated measure defined by*

$$\tilde{N}((s,t], A) = N((s,t], A) - \mathbf{E}N((s,t], A) = N((s,t], A) - (t-s)\int_A \nu(dy)$$

for the Borel sets A bounded below. Then the integral

$$\int_0^t \int_{\{|x| \le 1\}} f(x)\tilde{N}(dsdx) = \int_{\{|x| \le 1\}} f(x)\tilde{N}(t, dx)$$

is well-defined as the L^2-limit of the approximations

$$\int_0^t \int_{\{\epsilon < |x| \le 1\}} f(x)\tilde{N}(dsdx)$$

in the sense of the Hilbert norm $(\mathbf{E}\sup_{s\le t}|Z_s|^2)^{1/2}$*, and represents a Lévy process with Lévy exponent*

$$\eta^f(u) = \int [e^{i(u,y)} - 1 - i(u,y)](\mathbf{1}_{B_1}\nu)^f(dy)$$
$$= \int_{|y|\le 1} [e^{i(u,f(y))} - 1 - i(u,f(y))]\nu(dy), \tag{3.20}$$

where ν^f *is the push forward of* ν *by the mapping* f*. Finally,*

$$\mathbf{E}\left(\int_0^t \int_{\{|x|\le 1\}} f(x)\tilde N(dsdx) \int_0^t \int_{\{|x|\le 1\}} g(x)\tilde N(dsdx)\right)$$
$$= t\int_{\{|x|\le 1\}} f(x)g(x)\nu(dx). \tag{3.21}$$

Proof. By (3.12), one has for the difference between two approximations that

$$\mathbf{E}\left(\int_0^t \int_{\{\epsilon_1<|x|\le\epsilon_2\}} f(x)\tilde N(dsdx)\right)^2 = t\int_{\{\epsilon_1<|x|\le\epsilon_2\}} f^2(x)\nu(dx),$$

implying the L^2-convergence uniformly for bounded t and (3.21) for $f = g$. The required stronger convergence is obtained as in the proof of the above theorem by Doob's maximum inequality. The form of the limiting exponent is straightforward from the exponents of the approximating compound Poisson processes. Finally, equation (3.21) is obtained by differentiation of the characteristic function of the Lévy process

$$\int_0^t \int_{|x|\le 1\}} (af(x) + bg(x))\tilde N(dsdx). \qquad \square$$

One can also define the *nonlinear random integral* $\int_0^t f(dX_s^3)$ as a limit of the integrals over the Poisson processes $\int_0^t f(dX_s^3(n))$ described in Exercise 3.2.4.

Exercise 3.3.1. Deduce from (3.15) that if $f \in C^2(\mathbb{R}^d)$, then

$$\int_0^t f(dX_s^3) = \lim_{n\to\infty}\int_0^t f(dX_s^3(n))$$
$$= \int_{|x|\le 1\}} f(x)\tilde N(t,dx) + t\int_{B_1} [f(x) - (x,\nabla f(0))]\nu(dx). \tag{3.22}$$

3.4 Subordinators

We introduce here the non-decreasing Lévy processes with values in $\mathbb{R}_+$, called *subordinators.* Their applications in stochastic analysis are numerous. Most importantly they are used as random time changes. In Chapter 8 we shall use their inverses as a time change.

Theorem 3.4.1. *A real-valued Lévy process X_t is a subordinator if its characteristic exponent has the form*

$$\eta(u) = ibu + \int_0^\infty (e^{iuy} - 1)\nu(dy), \tag{3.23}$$

where $b \geq 0$ and the Lévy measure ν has support in $\mathbb{R}_+$ and satisfies the additional condition

$$\int_0^1 x\nu(dx) < \infty. \tag{3.24}$$

Moreover

$$X_t = tb + \sum_{s \leq t} (\Delta X_s).$$

Proof. First, if X is positive, then it can only increase from $X_0 = 0$. Hence by the i.i.d. property it is a non-decreasing process and consequently the Lévy measure has support in $\mathbb{R}_+$ and X contains no Brownian part, e.g. $A = 0$ in (3.16). Next,

$$\sum_{s \leq t} (\Delta X_s) \mathbf{1}_{|X_s| \leq 1} = \sum_{s \leq t} |\Delta X_s| \mathbf{1}_{|X_s| \leq 1} \leq X_t^3,$$

where the notation X_t^3 is taken from Theorem 3.3.1, implying (by Theorem 3.3.1, X_t^3 has finite moments) that

$$\mathbf{E} \sum_{s \leq t} (\Delta X_s) \mathbf{1}_{|X_s| \leq 1} \leq \mathbf{E} X_t^3 < \infty.$$

But

$$\mathbf{E} \sum_{s \leq t} (\Delta X_s) \mathbf{1}_{|X_s| \leq 1} = \lim_{\epsilon \to 0} \mathbf{E} \sum_{s \leq t} (\Delta X_s) \mathbf{1}_{\epsilon \leq |X_s| \leq 1} = \int_0^1 x\nu(dx),$$

implying (3.24). □

Clearly for a subordinator X_t the Laplace transform is well-defined and

$$\mathbf{E} e^{-\lambda X_t} = \exp\{-t\Phi(\lambda)\}, \tag{3.25}$$

where

$$\Phi(\lambda) = \eta(i\lambda) = -b\lambda + \int_0^\infty (1 - e^{-\lambda y})\nu(dy) \tag{3.26}$$

is called the *Laplace exponent* or *cumulant*.

A subordinator X_t is called a *one-sided stable process* or *one-sided stable Lévy motion* if to each $a \geq 0$ there corresponds a constant $b(a) \geq 0$ such that aX_t and $X_{tb(a)}$ have the same law.

Exercise 3.4.1 (Exponents of stable subordinators). 1. Show that $b(a)$ in this definition is continuous and satisfies the equation $b(ac) = b(a)b(c)$, hence deduce that $b(a) = a^\alpha$ with some $\alpha > 0$, called the *index of stability* or *stability exponent*.

2. Deduce further that $\Phi(a) = b(a)\Phi(1)$, and hence

$$\mathbf{E}e^{-uX_t} = \exp\{-tru^\alpha\} \tag{3.27}$$

with a constant $r > 0$, called the *rate*. Taking into account that Φ from (3.26) is concave, deduce that necessarily $\alpha \in (0, 1)$.

3. From the equation

$$\int_0^\infty (1 - e^{-uy}) \frac{dy}{y^{1+\alpha}} = \frac{\Gamma(1-\alpha)}{\alpha} u^\alpha$$

that holds for $\alpha \in (0, 1)$, deduce that stable subordinators with index α and rate r described by (3.27) have the Laplace exponent (3.26) with Lévy measure

$$\nu(dy) = \frac{\alpha r}{\Gamma(1-\alpha)} y^{-(1+\alpha)}. \tag{3.28}$$

Stable subordinators represent a particular case of a class of Lévy processes called *stable Lévy motions*. A comprehensive presentation of this class is given in Samorodnitski and Taqqu [288]. The importance of these processes is due to the fact that these processes describe the limits of homogeneous random walks.

For simplicity, in dimension $d > 1$ we shall work mostly with *symmetric stable Lévy motions*, defined via their exponents of the form

$$\eta(u) = -\int |(u, s)|^\alpha \mu(ds), \quad \alpha \in (0, 2),$$

where μ is an arbitrary centrally symmetric Borel measure on S^{d-1}, or simply

$$\eta(u) = -\sigma |u|^\alpha$$

with a constant σ in case of the uniform μ. Scaling properties of their distributions and the connection with the Lévy–Khintchine representation were discussed in Section 1.4. The properties of the corresponding stable densities

$$S(x, \alpha, \sigma t) = \frac{1}{(2\pi)^d} \int_{\mathbb{R}^d} \exp\{-\sigma t |p|^\alpha\} e^{-ipx}\, dp \tag{3.29}$$

will be studied later.

3.5 Markov processes, semigroups and propagators

In this section we introduce Markov processes, the main object of analysis in this book. We also discuss various formulations of Markovianity and the fundamental connection with the theory of semigroups.

Let us start by defining transition kernels. A *transition kernel* from a measurable space $(X, \mathcal{F})$ to a measurable space $(Y, \mathcal{G})$ is a function of two variables $\mu(x, A)$, $x \in X, A \in \mathcal{G}$, which is $\mathcal{F}$-measurable as a function of x for any A and is a measure in $(Y, \mathcal{G})$ for any x. A transition kernel is said to be *bounded* if $\sup_x \|\mu(x, .)\| < \infty$. It is then called a *transition probability kernel* or simply a *probability kernel* or a *stochastic kernel* if all measures $\mu(x, .)$ are probability measures. In particular, a *random measure* on a measurable space $(X, \mathcal{F})$ is a transition kernel from a probability space to $(X, \mathcal{F})$. We shall distinguish also the *Lévy kernels* from a measurable space $(X, \mathcal{F})$ to $\mathbb{R}^d$, which are defined as above, but with each $\mu(x, .)$ being a Lévy measure on $\mathbb{R}^d$, i.e. a (possibly unbounded) Borel measure such that $\mu(x, \{0\}) = 0$ and $\int \min(1, y^2)\mu(x, dy) < \infty$. In Section 9.1 we shall extend the notion of a kernel to complex measures.

The following definition is fundamental. An adapted process $X = X_t$ on a filtered probability space $(\Omega, \mathcal{F}, \mathcal{F}_t, \mathbf{P})$ with values in a metric space S (our main example is $S = \mathbb{R}^d$) is said to satisfy the *Markov property*, if

$$\mathbf{E}(f(X_t)|\mathcal{F}_s) = \mathbf{E}(f(X_t)|X_s) \quad \text{a.s.} \tag{3.30}$$

for all $f \in B(S)$, $0 \le s \le t$. If the filtration is not specified one means the filtration generated by the process itself.

For instance, a deterministic curve X_t in S enjoys this property.

Let us define a *Markov process* in S as a family of processes $X_t^{s,x}$, $t \ge s \ge 0$, depending on $s \in \mathbb{R}^+$ and $x \in S$ as parameters (starting point x at time s) if there exists a family of probability kernels $p_{s,t}(x, A)$ from S to S, $t \ge s \ge 0$, called *transition probabilities*, such that

$$\mathbf{E}(f(X_t^{s,x})|\mathcal{F}_u) = \mathbf{E}(f(X_t^{s,x})|X_u^{s,x}) = \int_S f(y)p_{u,t}(X_u^{s,x}, dy) \quad \text{a.s.} \tag{3.31}$$

for all $f \in B(S)$, $0 \le s \le u \le t$. The operator

$$\Phi^{s,t} f(x) = \int_S f(y)p_{s,t}(x, dy) \tag{3.32}$$

from the r.h.s. of (3.31) is called the *transition operator* of the Markov process $X_t^{s,x}$. Of course, transition probabilities are expressed in terms of the transition operators as

$$p_{s,t}(x, A) = (\Phi^{s,t}\mathbf{1}_A)(x).$$

One can also start this process from any initial law μ by defining

$$X_t^{s,\mu} = \int_S X_t^{s,x} \mu(dx).$$

To shorten the formulas, the upper indices for $X_t^{s,x}$ are often omitted if their particular values are not relevant.

A Markov process is called (time) *homogeneous* if $\Phi^{s,t}$ and $p_{s,t}(x, A)$ depend on the difference $t-s$ only. If this is the case we shall write Φ_{t-s} for $\Phi^{s,t}$ and $p_{t-s}(x, A)$ for $p_{s,t}(x, A)$.

Theorem 3.5.1. *Any Lévy process X (e.g. Brownian motion) is a time-homogeneous Markov with respect to its natural filtration. Moreover*

$$\mathbf{E}(f(X_t)|\mathcal{F}_s^X) = \int_{\mathbb{R}^d} f(X_s + z) p_{t-s}(dz) \tag{3.33}$$

for $f \in B(\mathbb{R}^d)$, $0 \leq s < t$, *where* p_t *is the law of* X_t.

Proof. By the properties of conditioning (see formula (1.12))

$$\mathbf{E}(f(X_t)|\mathcal{F}_s^X) = \mathbf{E}(f(X_t - X_s + X_s)|\mathcal{F}_s^X) = G_f(X_s),$$

where

$$G_f(y) = \mathbf{E}(f(X_t - X_s + y)) = \mathbf{E}(f(X_{t-s} + y)) = \int f(z + y) p_{t-s}(dz),$$

and (3.33) follows. Similarly the r.h.s. of (3.31) equals the r.h.s. of (3.33) implying (3.31) with respect to the filtration $\mathcal{F}_t^X$. □

A Lévy process X_t on a probability space $(\Omega, \mathcal{F}, \mathbf{P})$ equipped with a filtration $\mathcal{F}_t$ is called an $\mathcal{F}_t$*-Lévy process* if it is $\mathcal{F}_t$-adapted and the increments $X_t - X_s$ are independent of $\mathcal{F}_s$ for all $0 \leq s < t$.

Let us point out some properties of the transition operators $\Phi^{s,t}$. For this purpose it is convenient to introduce the general concept of a Markov propagator.

A linear operator L on a functional space (e.g. $B(S)$ or $C(S)$) is called *positive* if $f \geq 0 \Rightarrow Lf \geq 0$. Recall that a linear contraction is a linear operator in a Banach space with a norm not exceeding 1. A backward propagator (respectively a semigroup) of positive linear contractions in either $B(S)$, or $C(S)$, or $L_p(\mathbb{R}^d)$ is said to be a *sub-Markov backward propagator* (resp. a *sub-Markov semigroup*). It is called a *Markov backward propagator* (resp. a *Markov semigroup*), if additionally all these contractions are *conservative*, i.e. they take any constant function to itself.

Theorem 3.5.2. *The family of the transition operators $\Phi^{s,t}$ of a Markov process in S defined by* (3.32) *forms a Markov propagator in $B(S)$. In particular, if this Markov process is time-homogeneous, the family*

$$\Phi_t f(x) = \mathbf{E}_x f(X_t) \tag{3.34}$$

forms a Markov semigroup.

Proof. Only the propagator equation (1.77) is not obvious. But by (3.30)

$$\begin{aligned}\Phi^{r,t} f(x) &= \mathbf{E}(f(X_t)|X_r = x)\\ &= \mathbf{E}(\mathbf{E}(f(X_t)|\mathcal{F}_s)|X_r = x) = \mathbf{E}(\mathbf{E}(f(X_t)|X_s)|X_r = x)\\ &= \mathbf{E}(\Phi^{s,t} f(X_s)|X_r = x) = (\Phi^{r,s}(\Phi^{s,t} f))(x). \qquad \square\end{aligned}$$

A Markov process in $\mathbb{R}^d$ is said to have *transition densities* whenever the measures $p_{s,t}(x,.)$ have densities, say $\rho_{s,t}(x,y)$, so that $p_{s,t}(x,A) = \int_A \rho_{s,t}(x,y)\,dy$.

Proposition 3.5.1 (The Chapman–Kolmogorov equation). *If X is a Markov process, then for any Borel A*

$$p_{r,t}(x,A) = \int_S p_{s,t}(y,A)p_{r,s}(x,dy).$$

Proof. Apply the propagator equation $\Phi^{r,s}\Phi^{s,t} = \Phi^{r,t}$ to the indicator function $\mathbf{1}_A$. □

If a Markov process in $\mathbb{R}^d$ has transition densities, the Chapman–Kolmogorov equation clearly rewrites as

$$\rho_{r,t}(x,z) = \int_{\mathbb{R}^d} \rho_{r,s}(x,y)\rho_{s,t}(y,z)\,dy.$$

A family of probability kernels $\{p_{s,t} : 0 \leq s \leq t < \infty\}$ from S to S is called a *Markov transition family* if the Chapman–Kolmogorov equation hold.

Theorem 3.5.3 (Markov property in terms of transition families). *A process X is Markov with a Markov transition family $p_{s,t}$ if and only if for any $0 = t_0 < t_1 < \dots . < t_k$ and positive Borel f_i, $i = 0,\dots,k$,*

$$\mathbf{E}\prod_{i=0}^k f_i(X_{t_i}) = f_0(x_0)\int p_{0,t_1}(x_0,dx_1)f_1(x_1)\cdots\int p_{t_{k-1},t_k}(x_{k-1},dx_k)f_k(x_k). \tag{3.35}$$

Proof. Let X be Markov with Markov transition family $p_{s,t}$. Then

$$\mathbf{E}\prod_{i=0}^{k} f_i(X_{t_i}) = \mathbf{E}\Big(\prod_{i=0}^{k-1} f_i(X_{t_i})\mathbf{E}(f_k(X_{t_k}|\mathcal{F}_{t_{k-1}}))\Big)$$
$$= \mathbf{E}\Big(\prod_{i=0}^{k-1} f_i(X_{t_i})\Phi^{t_{k-1},t_k} f_k(X_{t_{k-1}})\Big)$$
$$= \mathbf{E}\Big(\prod_{i=0}^{k-1} f_i(X_{t_i})\int p_{t_{k-1},t_k}(X_{t_{k-1}}, dx_k) f_k(x_k)\Big),$$

and repeating this inductively one arrives at the r.h.s. of (3.35). Conversely, since $\mathcal{F}_s$ is generated by the sets $\prod_{i=1}^{k} \mathbf{1}_{X_{t_i}\in A_i}$, $t_1 < \cdots < t_k \le s$, and taking into account (3.32), to prove that X is Markov one has to show that for any $t_1 < \cdots < t_k \le s < t$ and Borel functions $f_0, \ldots, f_k, g$

$$\mathbf{E}\Big(\prod_{i=0}^{k} f_i(X_{t_i})g(X_t)\Big) = \mathbf{E}\Big(\prod_{i=0}^{k} f_i(X_{t_i})\Phi^{s,t} g(X_t)\Big).$$

But this follows from (3.35). □

When a kernel is given, we can naturally define the process $X_t = X_t^{0,\mu}$ started from any probability law μ via its distributions given by the equation

$$\mathbf{E}_\mu \prod_{i=0}^{k} f_i(X_{t_i}) = \int \mu(dx_0) f_0(x_0) \int p_{0,t_1}(x_0, dx_1) f_1(x_1) \cdots \times \int p_{t_{k-1},t_k}(x_{k-1}, dx_k) f_k(x_k). \tag{3.36}$$

Theorem 3.5.4 (Existence of Markov processes). *Let $\{p_{s,t} : 0 \le s \le t < \infty\}$ be a transition family and μ a probability measure on a complete metric space S. Then there exists a probability measure P on the measure space $S^{\bar{\mathbb{R}}_+}$ equipped with its canonical filtration $\mathcal{F}_t^0 = \sigma(X_u : u \le t)$ generated by the co-ordinate process X_t such that the co-ordinate process X_t is Markov with initial distribution μ and Markov transition family $p_{s,t}$.*

Proof. On cylinder sets, define

$$p_{t_0,t_1,\ldots,t_n}(A_0 \times A_1 \times \cdots \times A_n) = \int_{A_0} \mu(dx_0) \int_{A_1} p_{0,t_1}(x_0, dx_1) \int_{A_2} p_{t_1,t_2}(x_1, dx_2) \cdots \times \int_{A_n} p_{t_{n-1},t_n}(x_{n-1}, dx_n). \tag{3.37}$$

The Chapman–Kolmogorov equation implies consistency, which implies (Kolmogorov's theorem) the existence of a process X_t with such finite-dimensional distributions. Clearly X_0 has law μ and X_t is adapted to its natural filtration. Theorem 3.5.3 ensures that this process is Markov. □

A Markov process constructed in the above theorem is called the *canonical process* corresponding to transition family $p_{s,t}$.

Remark 19. It is important to stress that canonical Markov processes built from the same transition family but with different initial points live on different probability spaces (even if you combine the underlying measurable spaces for processes starting at various points in a common joint space of all paths, the probability measure would be different). Hence it makes no sense to compare the trajectories X_t^x starting at various points x, say, the expression $\mathbf{E}|X_t^x - X_t^y|$ is not defined. This constitutes a crucial difference with the case of Lévy processes, where trajectories starting at various points are deterministically related (they are linked by a shift). A big advantage of constructing Markov processes via SDE driven by a certain given process of noise lies mostly in the fact that such a construction yields a natural coupling for trajectories with different initial data.

Remark 20. It is natural to ask when a Markov propagator gives rise to a Markov process. Let us recall that a sequence of measurable functions f_n is said to converge to a function f, as $n \to \infty$, in bp-topology (bp stands for bounded pointwise) if the family f_n is uniformly bounded and $f_n(x) \to f(x)$ for any $x \in S$. One easily sees that a positive linear functional F on $B(S)$ is given by an integral with respect to a measure if and only if it is bp-continuous (bp-continuity allows one to obtain the σ-additivity property of the corresponding function $F(\mathbf{1}_A)$ on the Borel subsets $A \in \mathcal{B}(S)$). Consequently, a Markov backward propagator $\Phi^{s,t}$ in $B(S)$ is given by equation (3.32) with a certain Markov transition family $p_{s,t}$ if and only if the linear functionals $f \mapsto (\Phi^{s,t} f)(x)$ are bp-continuous.

Exercise 3.5.1. Assume X is a canonical Markov process and Z is a $\mathcal{F}_\infty^0$-measurable bounded (or positive) function on $(\mathbb{R}^d)^{\bar{\mathbb{R}}_+}$. Then the map $x \mapsto E_x(Z)$ is (Borel) measurable and

$$\mathbf{E}_\nu(Z) = \int \nu(dx)\mathbf{E}_x(Z)$$

for any probability measure ν (initial distribution of X). Hint: extend by the monotone class theorem from the mappings Z being indicators of cylinders, for which this is equivalent to (3.37).

Theorem 3.5.5 (Markov property via shifts). *The coordinate process on* $((\mathbb{R}^d)^{\bar{\mathbb{R}}_+}$, $\mathcal{F}_\infty^0 \mathbf{P})$ *is Markov* $\Longleftrightarrow$ *for any bounded (or positive) random variable* Z *on* $(\mathbb{R}^d)^{\bar{\mathbb{R}}_+}$, *every $t > 0$ and a initial measure ν*

$$\mathbf{E}_\nu(Z \circ \theta_t | \mathcal{F}_t^0) = \mathbf{E}_{X_t}(Z) \quad P_\nu\text{-a.s.},$$

where θ is the canonical shift operator $X_s(\theta_t(\omega)) = X_{t+s}(\omega)$.

Proof. One needs to show that

$$\mathbf{E}_\nu((Z \circ \theta_t)Y) = \mathbf{E}_\nu(\mathbf{E}_{X_t}(Z)Y)$$

for arbitrary $\mathcal{F}_t^0$-measurable r.v. Y (see (1.11)). By the usual extension arguments it is enough to do it for $Y = \prod_{i=1}^k f_i(X_{t_i})$ and $Z = \prod_{j=1}^n g_j(X_{s_j})$, where $t_i \le t$ and f_i, g_j are positive Borel. Thus one has to show that

$$\mathbf{E}_\nu\Big(\prod_{j=1}^n g_j(X_{s_j+t}) \prod_{i=1}^k f_i(X_{t_i})\Big) = \mathbf{E}_\nu\Big(\mathbf{E}_{X_t}\Big(\prod_{j=1}^n g_j(X_{s_j})\Big) \prod_{i=1}^k f_i(X_{t_i})\Big).$$

But the l.h.s. equals

$$\mathbf{E}_\nu\Big(\mathbf{E}\Big(\prod_{j=1}^n g_j(X_{s_j+t}) | \mathcal{F}_t^0\Big) \prod_{i=1}^k f_i(X_{t_i})\Big),$$

which coincides with the r.h.s. by the homogeneous Markov property. □

3.6 Feller processes and conditionally positive operators

In this section we introduce Feller processes and semigroups, stressing the interplay between analytic (semigroups as solutions to certain evolution equations) and probabilistic (semigroups as expectations of averages over random paths) interpretations.

A strongly continuous semigroup of positive linear contractions on $C_\infty(S)$, where S is a locally compact metric space, is called a *Feller semigroup*.

A (homogeneous) Markov process in a locally compact metric space S is called a *Feller process* if its Markov semigroup reduced to $C_\infty(S)$ is a Feller semigroup, i.e. it preserves $C_\infty(S)$ and is strongly continuous there.

Theorem 3.6.1. *For an arbitrary Feller semigroup Φ_t in $C_\infty(S)$ there exists a (uniquely defined) family of positive Borel measures $p_t(x, dy)$ on S with norm not exceeding one, depending vaguely continuous on x, i.e.*

$$\lim_{x_n \to x} \int f(y) p_t(x_n, dy) = \int f(y) p_t(x, dy), \quad f \in C_\infty(S),$$

and such that

$$\Phi_t f(x) = \int p_t(x, dy) f(y). \tag{3.38}$$

Proof. Representation (3.38) follows from the Riesz–Markov theorem. Other mentioned properties of $p_t(x, dy)$ follow directly from the definition of a Feller semigroup. □

Formula (3.38) allows us to extend the operators Φ_t to contraction operators in $B(S)$. This extension clearly forms a sub-Markov semigroup in $B(S)$.

Let $K_1 \subset K_2 \subset \cdots$ be an increasing sequence of compact subsets of S exhausting S, i.e. $S = \bigcup_n K_n$. Let χ_n be any sequence of functions from $C_c(S)$ with values in $[0, 1]$ and such that $\chi(x) = 1$ for $|x| \in K_n$. Then for any $f \in B(S)$ one has (by monotone or dominated convergence)

$$\begin{aligned}\Phi_t f(x) &= \int p_t(x, dy) f(y) \\ &= \lim_{n\to\infty} \int p_y(x, dy)\chi_n(y) f(y) = \lim_{n\to\infty} (\Phi_t(\chi_n f))(x) \qquad (3.39)\end{aligned}$$

(for positive f the limit is actually the supremum over n). This simple equation is important, as it allows one to define the *minimal extension of* Φ_t *to* $B(S)$ directly via Φ_t by-passing the explicit reference to $p_t(x, dy)$.

Theorem 3.6.2. *If Φ_t is a Feller semigroup, then uniformly for x from a compact set*

$$\lim_{t\to 0} \Phi_t f(x) = f(x), \quad f \in C(\mathbb{R}^d),$$

where Φ_t denote the extension (3.39).

Proof. By linearity and positivity it is enough to show this for $0 \le f \le 1$. In this case, for any compact set K and a nonnegative function $\phi \in C_\infty(\mathbb{R}^d)$ that equals 1 in K

$$(f - \Phi_t f)\mathbf{1}_K \le (f\phi - \Phi_t(f\phi))\mathbf{1}_K,$$

and similarly

$$(1 - f - \Phi_t(1 - f))\mathbf{1}_K \le ((1 - f)\phi - \Phi_t((1 - f)\phi))\mathbf{1}_K.$$

The second inequality implies

$$\begin{aligned}(\Phi_t f - f)\mathbf{1}_K &\le (\Phi_t \mathbf{1} - \mathbf{1} + \mathbf{1} - f - \Phi_t((1 - f)\phi))\mathbf{1}_K \\ &\le ((1 - f)\phi - \Phi_t((1 - f)\phi))\mathbf{1}_K.\end{aligned}$$

Consequently

$$|f - \Phi_t f|\mathbf{1}_K \le |f\phi - \Phi_t(f\phi)|\mathbf{1}_K + |(1 - f)\phi - \Phi_t((1 - f)\phi)|\mathbf{1}_K,$$

which implies the required convergence on the compact set K by the strong continuity of Φ_t. □

Corollary 12. *If Φ is a Feller semigroup, then the dual semigroup $\Phi_t^\star$ on $\mathcal{M}(X)$ is a positivity-preserving semigroup of contractions depending continuously on t in both vague and weak topologies.*

Proof. Everything is straightforward from definitions except weak continuity, which follows from the previous theorem, since

$$\begin{aligned}(f, \Phi_t^\star \mu - \mu) &= (\Phi_t f - f, \mu) \\ &= \int_{|x|<K} (\Phi_t f - f)(x)\mu(dx) + \int_{|x|\geq K} (\Phi_t f - f)(x)\mu(dx),\end{aligned}$$

and for $f \in C(\mathbb{R}^d)$ the second integral can be made arbitrarily small by choosing large enough K, and then the first integral is small for small t by Theorem 3.6.2. □

A Feller semigroup Φ_t is called *conservative* if all measures $p_t(x,.)$ in the representation (3.38) are probability measures, or equivalently if the natural extension of Φ_t to $B(S)$ given by (3.39) preserves constants and hence forms a Markov semigroup in $B(S)$.

Apart from the Feller property, another useful link between Markovianity and continuity is stressed in the following modification of this property. A *C-Feller semigroup* in $C(S)$ is a sub-Markov semigroup in $C(S)$, i.e. it is a semigroup of contractions Φ_t in $C(S)$ such that $0 \leq u \leq 1$ implies $0 \leq \Phi_t u \leq 1$. Note that, on the one hand, this definition does not include the strong continuity, and on the other hand, it applies to any topological space S, not necessarily locally compact or even metric. Of course, a Feller semigroup Φ_t is *C-Feller*, if the space $C(S)$ is invariant under the natural extension (3.39), and a C-Feller semigroup Φ_t is Feller if $C_\infty(S)$ is invariant under all Φ_t and the corresponding restriction is strongly continuous. It is worth stressing that a Feller semigroup may not be C-Feller and vice versa; see examples at the end of Section 4.1.

Feller semigroups arising from Markov processes are obviously conservative. Conversely, any conservative Feller semigroup is the semigroup of a certain Markov process, which follows from representation (3.38) for the kernels p_t and a basic construction of Markov processes based on Kolmogorov's existence theorem.

Proposition 3.6.1. *A Feller semigroup is C-Feller if and only if Φ_t applied to a constant is a continuous function. In particular, any conservative Feller semigroup is C-Feller.*

Proof. By Proposition 1.7.3 the vague and weak continuity of $p_t(x, dy)$ with respect to x coincide under the condition of continuous dependence of the total mass $p_t(x, S)$ on x. □

Theorem 3.6.3. *If X_t^x is a Feller process in $\mathbb{R}^d$ with starting point x, then*

(i) *$X_t^x \to X_t^y$ weakly as $x \to y$ for any t, and*

(ii) *$X_t^x \to x$ in probability as $t \to 0$.*

Proof. Statement (i) follows from Proposition 1.7.3, and (ii) follows from the last bit of Proposition 1.1.2. □

Theorem 3.6.4. *Let X_t be a Lévy process with characteristic exponent*

$$\eta(u) = i(b,u) - \frac{1}{2}(u, Gu) + \int_{\mathbb{R}^d} [e^{i(u,y)} - 1 - i(u,y)\mathbf{1}_{B_1}(y)]\nu(dy). \tag{3.40}$$

Then X_t is a Feller process with semigroup Φ_t s.t.

$$\Phi_t f(x) = \int f(x+y)p_t(dy), \quad f \in C(\mathbb{R}^d), \tag{3.41}$$

where p_t is the law of X_t. This semigroup is translation-invariant, i.e.

$$(\Phi_t f)(x+z) = (\Phi_t f(.+z))(x).$$

Proof. Formula (3.41) follows from the definition of Lévy processes as time-homogeneous and translation-invariant Markov process. Notice that any $f \in C_\infty(\mathbb{R}^d)$ is uniformly continuous. For any such f

$$\begin{aligned}\Phi_t f(x) - f(x) &= \int (f(x+y) - f(x))p_t(dy) \\ &= \int_{|y|>K} (f(x+y) - f(x))p_t(dy) \\ &\quad + \int_{|y|\le K} (f(x+y) - f(x))p_t(dy),\end{aligned}$$

and the first (resp. the second) term is small for small t and any K by stochastic continuity of X (resp. for small K and arbitrary t by uniform continuity of f). Hence $\|\Phi_t f - f\| \to 0$ as $t \to 0$. To see that $\Phi_t f \in C_\infty(\mathbb{R}^d)$ for $f \in C_\infty(\mathbb{R}^d)$ one writes similarly

$$\Phi_t f(x) = \int_{|y|>K} f(x+y)p_t(dy) + \int_{|y|\le K} f(x+y)p_t(dy)$$

and observes that the second term clearly belongs to $C_\infty(\mathbb{R}^d)$ for any K and the first one can be made arbitrarily small by choosing K large enough. □

Remark 21. The Fourier transform takes the semigroup Φ_t to a multiplication semigroup:

$$\Phi_t f(x) = F^{-1}(e^{t\eta} Ff), \quad f \in S(\mathbb{R}^d),$$

because

$$\begin{aligned}(F\Phi_t f)(p) &= \frac{1}{(2\pi)^{d/2}} \int e^{-ipx} \int f(x+y) p_t(dy) \\ &= \frac{1}{(2\pi)^{d/2}} \int \int e^{-ipz+ipy} \int f(z) p_t(dy) = (Ff)(p) e^{t\eta(p)}.\end{aligned}$$

This yields another proof of the Feller property of the semigroup Φ_t.

Theorem 3.6.5. *If X_t is a Lévy process with characteristic exponent* (3.40), *its generator (that is, the generator of the corresponding Feller semigroup) is given by*

$$\begin{aligned}Lf(x) = &\sum_{j=1}^d b_j \frac{\partial f}{\partial x_j} + \frac{1}{2} \sum_{j,k=1}^d G_{jk} \frac{\partial^2 f}{\partial x_j \partial x_k} \\ &+ \int_{\mathbb{R}^d} \Big[f(x+y) - f(x) - \sum_{j=1}^d y_j \frac{\partial f}{\partial x_j} \mathbf{1}_{B_1}(y) \Big] \nu(dy). \end{aligned} \tag{3.42}$$

on the Schwartz space S of rapidly decreasing smooth functions. Moreover, the Lévy exponent is expressed via the generator by the formula

$$\eta(u) = e^{-iux} L e^{iux}. \tag{3.43}$$

Each space $C^k_\infty(\mathbb{R}^d)$ with $k \geq 2$ is an invariant core for L.

Proof. Let us first check (3.42) on the exponential functions. Namely, for $f(x) = e^{i(u,x)}$

$$\Phi_t f(x) = \int f(x+y) p_t(dy) = e^{i(u,x)} \int e^{i(u,y)} p_t(dy) = e^{i(u,x)} e^{t\eta(u)}.$$

Hence

$$Lf(x) = \frac{d}{dt}\Big|_{t=0} \Phi_t f(x) = \eta(u) e^{i(u,x)}$$

is given by (3.42) due to the elementary properties of the exponent. By linearity this extends to functions of the form $f(x) = \int e^{i(u,x)} g(u) du$ with $g \in S$. But this class coincides with S, by Fourier's theorem. To see that $C^k_\infty(\mathbb{R}^d)$ is invariant under Φ_t for any $k \in \mathbb{N}$ it is enough to observe that the derivative $\nabla_l \Phi_t f$ for a function $f \in C^l_\infty(\mathbb{R}^d)$ satisfies the same equation as $\Phi_t f$ itself. Finally $Lf \in C_\infty(\mathbb{R}^d)$ for any $f \in C^2_\infty(\mathbb{R}^d)$. □

By a straightforward change of variable one obtains that the operator $L^\star$ given by

$$L^\star f(x) = -\sum_{j=1}^{d} b_j \frac{\partial f}{\partial x_j} + \frac{1}{2}\sum_{j,k=1}^{d} G_{jk}\frac{\partial^2 f}{\partial x_j \partial x_k}$$
$$+ \int_{\mathbb{R}^d} \Big[f(x-y) - f(x) + \sum_{j=1}^{d} y_j \frac{\partial f}{\partial x_j} \mathbf{1}_{B_1}(y) \Big] \nu(dy) \qquad (3.44)$$

is adjoint to (3.42), in the sense that

$$\int Lf(x)g(x)\,dx = \int f(x)L^\star g(x)\,dx$$

for f, g from the Schwartz space S.

Remark 22. Operator (3.42) is a ΨDO (see Section 1.8) with symbol $\eta(p)$, where η is the characteristic exponent (3.40). In fact, by (1.75) one has to check that $(FLf)(p) = \eta(p)(Ff)(p)$. Since

$$(FLf)(p) = \frac{1}{(2\pi)^{d/2}}(e^{-ip\cdot}, Lf) = \frac{1}{(2\pi)^{d/2}}(L^\star e^{-ip\cdot}, f),$$

this follows from the equation

$$L^\star e^{-ipx} = \eta(p)e^{-ipx},$$

which in turn is a direct consequence of the properties of exponents.

The following are the basic definitions related to the generators of Markov processes. One says that an operator A in $C(\mathbb{R}^d)$ defined on a domain D_A

(i) is *conditionally positive*, if $Af(x) \geq 0$ for any $f \in D_A$ s.t. $f(x) = 0 = \min_y f(y)$;

(ii) satisfies the *positive maximum principle (PMP)*, if $Af(x) \leq 0$ for any $f \in D_A$ s.t. $f(x) = \max_y f(y) \geq 0$;

(iii) is *dissipative* if $\|(\lambda - A)f\| \geq \lambda\|f\|$ for $\lambda > 0$, $f \in D_A$;

(iv) is *local* if $Af(x) = 0$ whenever $f \in D_A \cap C_c(\mathbb{R}^d)$ vanishes in a neighborhood of x;

(v) is *locally conditionally positive* if $Af(x) \geq 0$ whenever $f(x) = 0$ and has a local minimum there;

(vi) satisfies a *local PMP*, if $Af(x) \leq 0$ for any $f \in D_A$ having a local non-negative maximum at x.

For example, the operator of multiplication $u(x) \mapsto c(x)u(x)$ on a function $c \in C(\mathbb{R}^d)$ is always conditionally positive, but it satisfies PMP only in the case of non-negative c.

The importance of these notions lie in the following fact.

Theorem 3.6.6. *Let A be a generator of a Feller semigroup Φ_t. Then*

(i) *A is conditionally positive,*

(ii) *satisfies the PMP on D_A,*

(iii) *is dissipative.*

If moreover A is local and D_A contains C_c^∞, then it is locally conditionally positive and satisfies the local PMP on C_c^∞.

Proof. This is very simple. For (i), note that

$$Af(x) = \lim_{t\to 0} \frac{\Phi_t f(x) - f(x)}{t} = \lim_{t\to 0} \frac{\Phi_t f(x)}{t} \geq 0$$

by positivity preservation. For (ii) note that if $f(x) = \max_y f(y)$, then $\Phi_t f(x) \leq f(x)$ for all t implying $Af(x) \leq 0$. For (iii) choose x to be the maximum point of $|f|$. By passing to $-f$ if necessary we can consider $f(x)$ to be positive. Then

$$\|(\lambda - A)f\| \geq \lambda \|f\| \geq \lambda f(x) - Af(x) \geq \lambda f(x)$$

by PMP. □

Let us observe that if S is compact and a Feller semigroup in $C(S)$ is conservative, then obviously the constant unit function $\mathbf{1}$ belongs to the domain of its generator A and $A\mathbf{1} = 0$. Hence it is natural to call such generators *conservative*. In case of noncompact $S = \mathbb{R}^d$, we shall say that a generator of a Feller semigroup A is *conservative* if $A\phi_n(x) \to 0$ for any x as $n \to \infty$, where $\phi_n(x) = \phi(x/n)$ and ϕ is an arbitrary function from $C_c^2(\mathbb{R}^d)$ that equals one in a neighborhood of the origin and has values in $[0, 1]$. We shall see at the end of the next section, that conservativity of a semigroup implies the conservativity of the generator with partial inverse being given in Theorem 4.1.3.

We shall recall now the basic structural result about the generators of Feller processes by formulating the following fundamental fact, due to Courrège.

Theorem 3.6.7. *If the domain of a conditionally positive operator L (in particular, the generator of a Feller semigroup) in $C_\infty(\mathbb{R}^d)$ contains the space $C_c^2(\mathbb{R}^d)$, then it has the following Lévy–Khintchine form with variable coefficients:*

$$Lf(x) = \frac{1}{2}(G(x)\nabla, \nabla)f(x) + (b(x), \nabla f(x)) + c(x)f(x) \tag{3.45}$$

$$+ \int [f(x+y) - f(x) - (\nabla f(x), y)\mathbf{1}_{B_1}(y)]\nu(x, dy), \quad f \in C_c^2(\mathbb{R}^d),$$

with $G(x)$ being a symmetric non-negative matrix and $\nu(x,.)$ being a Lévy measure on $\mathbb{R}^d$, i.e.

$$\int_{\mathbb{R}^n} \min(1,|y|^2)\nu(x,dy) < \infty, \quad \nu(\{0\}) = 0, \tag{3.46}$$

depending measurably on x. If additionally L satisfies PMP, then $c(x) \le 0$ everywhere.

The proof of this theorem is based only on standard calculus, though requires some ingenuity (the last statement being of course obvious). It can be found in [90], [63] [143] and will not be reproduced here. Let us only indicate the main strategy, showing how the Lévy kernel comes into play. Namely, as follows from conditional positivity, $Lf(x)$, for any x, is a positive linear functional on the space of continuous functions with support in $\mathbb{R}^d \setminus \{0\}$, hence by the Riesz–Markov theorem for these functions

$$Lf(x) = \tilde{L}f(x) = \int f(y)\tilde{\nu}(x,dy) = \int f(x+y)\nu(x,dy)$$

with some kernel ν such that $\nu(x,\{x\}) = 0$. Next one deduces from conditional positivity that L should be continuous as a mapping from $C_c^2(\mathbb{R}^d)$ to bounded Borel functions. This in turn allows us to deduce the basic moment condition (3.46) on ν. One then observes that the difference between L and $\tilde{L}$ should be a second-order differential operator. Finally one shows that this differential operator should be also conditionally positive.

Remark 23. Actually when proving Theorem 3.6.7 (see Bony, Courrège and Priouret [63]) one obtains the characterization not only for conditionally positive operators, but also for conditionally positive linear functionals obtained by fixing the arguments. Namely, it is shown that if a linear functional $(Ag)(x) : C_c^2 \mapsto \mathbb{R}^d$ is *conditionally positive* at x, i.e. if $Ag(x) \ge 0$ whenever a non-negative g vanishes at x, then $Ag(x)$ is continuous and has form (3.45) (irrespectively of the properties of $Ag(y)$ at other points y).

Corollary 13. *If the domain of the generator L of a conservative Feller semigroup Φ_t in $C_\infty(\mathbb{R}^d)$ contains C_c^2, then it has form (3.45) with vanishing $c(x)$. In particular, L is conservative.*

Proof. By Theorems 3.6.6 and 3.6.7 L has form (3.45) on $C_c^2(\mathbb{R}^d)$ with non-positive $c(x)$. Conservativity of L means that $L\phi_n(x) \to 0$ for any x as $n \to \infty$, where $\phi_n(x) = \phi(x/n)$ and ϕ is an arbitrary function from $C_c^2(\mathbb{R}^d)$ that equals one in a neighborhood of the origin and has values in [0, 1]. Clearly $\lim_{n\to\infty} L\phi_n(x) = c(x)$. So conservativity is equivalent to $c(x) = 0$ identically. Since Φ_t is a conservative Feller semigroup it corresponds to a certain Markov (actually Feller) process X_t. □

The inverse question of whether a given operator of form (3.45) (or its closure) actually generates a Feller semigroup, which roughly speaking means the possibility to have regular solutions to the equation $\dot{f} = Lf$ (see the next section), is nontrivial and has attracted much attention. We shall deal with it in the next few chapters.

Remark 24. C-Feller semigroups T_t are usually not strongly continuous in $C(S)$. There are several reasonable ways to define the generator in such a case. One approach is to introduce the subspace $B_0 \subset C(S)$ of those functions f that $\|T_t f - f\| \to 0$ as $t \to 0$. It is straightforward to see that B_0 is T_t invariant and closed in $C(S)$, so that one can define the generator of T_t in the usual way. The second approach is by introducing the *generalized generator* of T_t as $A_t f = \lim[(T_t f - f)/t]$, where the limit is understood in the sense of uniform convergence on compact sets. Probabilistically, the most natural way to link the infinitesimal operator with a process is via the concept of the martingale problem solution, see Section 3.9.

Exercise 3.6.1. This exercise is meant to demonstrate why strong continuity of the semigroups of contractions, and not continuity in the Banach norm, is natural. Show that if T_t is a Feller semigroup with a transition density, i.e. such that $T_t f(x) = \int f(y)p(x,y)\,dy$ with some measurable $p(x,y)$, then $\|T_t - \mathbf{1}\| = 2$ for all t, where $\mathbf{1}$ is the identity operator. (Hint: the total variance norm between any Dirac measure and any probability measure with a density always equals 2.) Note however, that if T_t is regularizing enough, for instance, if this is the semigroup of Brownian motion, then $\|T_{s+t} - T_s\| \to 0$ as $t \to 0$ for any $s > 0$.

Exercise 3.6.2. (i) Show that the resolvent of the standard BM is given by the formula

$$R_\lambda f(x) = \int_{-\infty}^{\infty} R_\lambda^1(|x-y|)f(y)\,dy = \frac{1}{\sqrt{2\lambda}} \int_{-\infty}^{\infty} e^{-\sqrt{2\lambda}|y-x|} f(y)\,dy. \tag{3.47}$$

Hint: Check this identity for the exponential functions $f(x) = e^{i\theta x}$ using the known ch.f. of the normal r.v. $N(0,t)$.

(ii) Show that for standard BM in $\mathbb{R}^3$

$$R_\lambda f(x) = \int_{\mathbb{R}^3} R_\lambda^3(|x-y|)f(y)\,dy = \int_{\mathbb{R}^3} \frac{1}{2\pi|x-y|} e^{-\sqrt{2\lambda}|y-x|} f(y)\,dy. \tag{3.48}$$

Hint: observe that

$$R_\lambda^3(|z|) = \int_0^{\infty} e^{-\lambda t}(2\pi t)^{-3/2} e^{-|z|^2/(2t)}\,dt = -\frac{1}{2\pi|z|}(R_\lambda^1)'(|z|).$$

Finally let us describe how Markov processes and their generators are transformed under homeomorphisms (continuous bijections with a continuous inverse) of a state space. Let S be a locally compact space, X_t a C-Feller process in S with the semigroup Φ_t and Ω_t a family (depending continuously on time) of homeomorphisms of S. Then the transformed process $Y_t = \Omega_t(X_t)$ is of course also Markov, though not time-homogeneous any more. The corresponding averaging operators

$$U^{s,t} f(y) = \mathbf{E}(f(Y_t)|Y_s = y), \quad s \le t,$$

form a backward Markov propagator, each $U^{s,t}$ being a conservative contraction in $C(S)$ that can be expressed in terms of Φ_t as

$$U^{s,t} f(y) = \mathbf{E}(f(\Omega_t(Z_t^x))|\Omega_s(Z_s^x) = y) = \Phi_{t-s}[f \circ \Omega_t](\Omega_s^{-1}(y)).$$

Lifting the transformations Ω_t to functions as the operators

$$\tilde{\Omega}_t f(y) = (f \circ \Omega_t)(y) = f(\Omega_t(y)),$$

we can write equivalently that

$$U^{s,t} f = \tilde{\Omega}_s^{-1} \Phi_{t-s} \tilde{\Omega}_t f, \quad f \in C(\Omega_t(S)), \quad s \le t. \tag{3.49}$$

Exercise 3.6.3. Suppose $S = \mathbb{R}^d$, the semigroup Φ_t is Feller with a generator A and Ω_t is linear:

$$\Omega_t(z) = (z - \xi_t)/a,$$

where $a \neq 0$ is a constant and ξ_t a given differentiable curve in $\mathbb{R}^d$. Show that the propagator (3.49) is generated (in the sense of the definition given before Theorem 1.9.3) by the family of operators

$$A_t f = \tilde{\Omega}_t^{-1} A \tilde{\Omega}_t f - \frac{1}{a}(\dot{\xi}_t, \nabla f) \tag{3.50}$$

(so that equations (1.80) hold).

Hint:

$$\frac{d}{dt}\tilde{\Omega}_t f(y) = \frac{d}{dt} f(\Omega_t(y)) = -\frac{1}{a}(\dot{\xi}_t, \nabla f(\Omega_t(y))) = -\frac{1}{a}[\tilde{\Omega}_t(\dot{\xi}_t, \nabla f)](y).$$

Consequently

$$\frac{d}{dt} U^{s,t} f = \tilde{\Omega}_s^{-1} \Phi_{t-s} A \tilde{\Omega}_t f - \frac{1}{a} \tilde{\Omega}_s^{-1} \Phi_{t-s} \tilde{\Omega}_t (\dot{\xi}_t, \nabla f),$$

and

$$\frac{d}{ds} U^{s,t} f = -\tilde{\Omega}_s^{-1} A \Phi_{t-s} \tilde{\Omega}_t f + \tilde{\Omega}_s^{-1}(\dot{\xi}_s, \nabla(\Phi_{t-s}\tilde{\Omega}_t f)),$$

implying equations (1.80).

If, in particular,

$$Af(x) = \int [f(x+y) - f(x)]\nu(x, dy)$$

in the above exercise, then

$$\begin{aligned} A_t f &= \int [f(\Omega_t(\Omega_t^{-1}(x) + y)) - f(x)]\nu(\Omega_t^{-1}(x), dy) - \frac{1}{a}(\dot{\xi}_t, \nabla f) \\ &= \int \left[f\left(x + \frac{y}{a}\right) - f(x) \right] \nu(ax + \xi_t, dy) - \frac{1}{a}(\dot{\xi}_t, \nabla f). \end{aligned} \tag{3.51}$$

3.7 Diffusions and jump-type Markov processes

In this section we consider in more detail two classes of conditionally positive operators, namely local and bounded, leading to the two basic classes of Markov processes, respectively diffusions and pure jump processes.

Theorem 3.7.1. (i) *If L is a locally conditionally positive operator $C_c^2(\mathbb{R}^d) \mapsto C(\mathbb{R}^d)$, then for $f \in C_c^\infty$*

$$Lf(x) = c(x)f(x) + \sum_{j=1}^d b_j(x)\frac{\partial f}{\partial x_j} + \frac{1}{2}\sum_{j,k=1}^d a_{jk}(x)\frac{\partial^2 f}{\partial x_j \partial x_k}, \quad f \in C_c^2(\mathbb{R}^d), \tag{3.52}$$

with certain $c, b_i, a_{ij} \in C(\mathbb{R}^d)$ such that $A = (a_{ij})$ is a positive definite matrix.

(ii) *If additionally L satisfies local PMP, then $c(x) \le 0$ in* (3.52).

(iii) *If L is local and generates a conservative Feller semigroup Φ_t with transition family $p_t(x, dy)$, then it has representation* (3.52) *with vanishing $c(x)$.*

Proof. (i) To shorten the formulas, assume $d = 1$. Let χ be a smooth function $\mathbb{R} \mapsto [0, 1]$ that equals 1 (resp. 0) for $|x| \le 1$ (resp. $|x| > 2$). For an $f \in C_c^\infty$ one can write

$$f(y) = \left(f(x) + f'(x)(y - x) + \frac{1}{2}f''(x)(y - x)^2 \right) \chi(y - x) + g_x(y),$$

where $g_x(y) = o(1)(y - x)^2$ as $y \to x$. Hence

$$Lf(x) = c(x)f(x) + b(x)f'(x) + \frac{1}{2}a(x)f''(x) + (Lg_x)(x)$$

with

$$b(x) = L[(. - x)\chi(. - x)](x) = \lim_{t \to 0} \frac{1}{t} \int (y - x)\chi(y - x)p_t(x, dy),$$

$$a(x) = L[(. - x)^2\chi(. - x)](x) = \lim_{t \to 0} \frac{1}{t} \int (y - x)^2\chi(y - x)p_t(x, dy).$$

However, $\pm g_x(y) + \epsilon(y-x)^2$ vanishes at $y = x$ and has a local minimum there for any ϵ so that

$$L[\pm g_x(.) + \epsilon(. - x)^2](x) \geq 0$$

for any ϵ and hence $Lg_x(x) = 0$.

(ii) Applying PMP to a non-negative $f \in C_c^2(\mathbb{R}^d)$ that equals one in a neighborhood of x yields $c(x) \leq 0$.

(iii) By (i), (ii) and Theorem 3.6.6 L has form (3.52) with non-positive c. Choosing f as in (i) and using conservativity yields

$$c(x) = Lf(x) = \lim_{t\to 0} \frac{1}{t} \int (f(y) - f(x)) p_t(x, dy).$$

This integral taken over a neighborhood of x, where f equals one, obviously vanishes. And the integral over the compliment to any neighborhood of x tends to zero as $t \to 0$ by locality. □

A Feller process with a generator of type (3.52) is called a *(Feller) diffusion*. Later on we shall address the question of the existence of a Feller process specified by a generator of form (3.52).

Exercise 3.7.1. Show that the coefficients b_j and a_{ij} can be defined as

$$b_j(x) = \lim_{t\to 0} \frac{1}{t} \int (y-x)_j \mathbf{1}_{\{|y-x|\leq\epsilon\}}(y) p_t(x, dy), \tag{3.53}$$

$$a_{ij}(x) = \lim_{t\to 0} \frac{1}{t} \int (y-x)_i (y-x)_j \mathbf{1}_{\{|y-x|\leq\epsilon\}}(y) p_t(x, dy) \tag{3.54}$$

for any $\epsilon > 0$. Conversely, if these limits exist and are independent of ϵ, then the generator is local, so that the process is a diffusion.

Exercise 3.7.2. If the generator L of a (conservative) Feller semigroup Φ_t with transition family $p_t(x, dy)$ is such that $C_c^\infty \subset D_L$ and

$$p_t(x; \{y : |y-x| \geq \epsilon\}) = o(t), \quad t \to 0,$$

for any $\epsilon > 0$, then L is local (and hence of diffusion type).

Corollary 14. *A BM (possibly with drift) can be characterized as*

(i) *a diffusion with i.i.d. increments or as*

(ii) *a Lévy process with a local generator.*

The next important class of Markov processes is that of *pure jump* processes, which can be defined as Markov processes with bounded generators.

Proposition 3.7.1. *Let S be a locally compact metric space and L be a bounded conditionally positive operator from $C_\infty(S)$ to $B(S)$. Then there exists a bounded transition kernel $\nu(x,dy)$ in S with $\nu(x,\{x\}) = 0$ for all x, and a function $a(x) \in B(S)$ such that*

$$Lf(x) = \int_S f(z)\nu(x,dz) - a(x)f(x). \tag{3.55}$$

Conversely, if L is of this form, then it is a bounded conditionally positive operator $C(S) \mapsto B(S)$.

Proof. If L is conditionally positive in $C_\infty(S)$, then $Lf(x)$ is a positive functional on $C_\infty(S \setminus \{x\})$, and hence by the Riesz–Markov theorem there exists a measure $\nu(x,dy)$ on $S \setminus \{x\}$ such that $Lf(x) = \int_S f(z)\nu(x,dz)$ for $f \in C_\infty(S \setminus \{x\})$. As L is bounded, these measures are uniformly bounded. As any $f \in C_\infty(S)$ can be written as $f = f(x)\chi + (f - f(x)\chi)$ with χ an arbitrary function with a compact support and with $\chi(x) = 1$, it follows that

$$Lf(x) = f(x)L\chi(x) + \int (f - f(x)\chi)(z)\nu(x,dz)$$

which clearly has the form (3.55). The inverse statement is obvious. □

Remark 25. Condition $\nu(x,\{x\}) = 0$ is natural for the probabilistic interpretation (see below). From the analytic point of view it makes representation (3.55) unique.

We shall now describe analytic and probabilistic constructions of pure jump processes, reducing attention to the most important case of continuous kernels.

Theorem 3.7.2. *Let $\nu(x,dy)$ be a weakly continuous uniformly bounded transition kernel in a complete metric space S such that $\nu(x,\{x\}) = 0$ and $a \in C(S)$. Then operator* (3.55) *has $C(S)$ as its domain and generates a strongly continuous semigroup T_t in $C(S)$ that preserves positivity and is given by transition kernels $p_t(x,dy)$:*

$$T_t f(x) = \int p_t(x,dy)f(y).$$

In particular, if $a(x) = \|\nu(x,.)\|$, then $T_t\mathbf{1} = \mathbf{1}$ and T_t is the semigroup of a Markov process that we shall call a pure jump or jump-type Markov process.

Proof. Since L is bounded, it generates a strongly continuous semigroup. As it can be written in the integral form

$$Lf(x) = \int_S f(z)\tilde{\nu}(x,dz)$$

with the signed measure $\tilde{\nu}(x,.)$ coinciding with ν outside $\{x\}$ and with $\tilde{\nu}(x,\{x\}) = -a(x)$, it follow from the convergence in norm of the exponential series for $T_t = e^{tL}$

that all T_t are integral operators. To see that these operators are positive we observe that T_t are bounded below by the resolving operators of the equation $\dot{f}(x) = -a(x)f(x)$ which are positive. Application of the standard construction of Markov process (via Kolmogorov's existence theorem) yields the existence of the corresponding Markov process. □

Remark 26. An alternative analytic proof can be given by perturbation theory (Theorem 1.9.2) when considering the integral part of (3.55) as a perturbation. This approach leads directly to the representation (3.58) obtained below probabilistically. From this approach the positivity is straightforward.

A characteristic feature of pure jump processes is the property that their paths are a.s. piecewise constant, as the following result on a probabilistic interpretation of these processes shows.

Theorem 3.7.3. *Let $\nu(x, dy)$ be a weakly continuous uniformly bounded transition kernel on S (S being a metric space) such that $\nu(x, \{x\}) = 0$. Let $a(x) = \nu(x, S)$. Define the following process X_t^x. Starting at a point x it stays there a random $a(x)$-exponential time τ (i.e. distributed according to $\mathbf{P}(\tau > t) = \exp(-ta(x))$) and then jumps to a point $y \in S$ distributed according to the probability law $\nu(x, .)/a(x)$. Then the same repeats starting from y, etc. Let N_t^x denote the number of jumps of this process during the time t when starting from a point x. Then*

$$\mathbf{P}(N_t^x = k) = \int_{0<s_1<\cdots<s_k<t} \int_{S^k} e^{-a(y_k)(t-s_k)} \nu(y_{k-1}, dy_k) \times e^{-a(y_{k-1})(s_k-s_{k-1})} \cdots e^{-a(y_1)(s_2-s_1)} \nu(x, dy_1) e^{-s_1 a(x)} ds_1 \cdots ds_k, \tag{3.56}$$

$$\mathbf{P}(N_t^x > k) = \int_{0<s_1<\cdots<s_k<t} \int_{S^k} (1 - e^{-a(y_k)(t-s_k)}) \nu(y_{k-1}, dy_k) \times e^{-a(y_{k-1})(s_k-s_{k-1})} \cdots e^{-a(y_1)(s_2-s_1)} \nu(x, dy_1) e^{-s_1 a(x)} ds_1 \cdots ds_k, \tag{3.57}$$

and N_t^x is a.s. finite. Moreover, for a bounded measurable f

$$\begin{aligned} \mathbf{E} f(X_t^x) &= \sum_{k=0}^{\infty} \mathbf{E} f(X_t^x) \mathbf{1}_{N_t^x = k} \\ &= \sum_{k=0}^{\infty} \int_{0<s_1<\cdots<s_k<t} \int_{S^k} e^{-a(y_k)(t-s_k)} \nu(y_{k-1}, dy_k) \cdots \\ &\quad \times e^{-a(y_1)(s_2-s_1)} \nu(x, dy_1) e^{-s_1 a(x)} f(y_k) ds_1 \cdots ds_k, \end{aligned} \tag{3.58}$$

and there exists (in the sense of the sup norm) the derivative

$$\left.\frac{d}{dt}\right|_{t=0} \mathbf{E} f(X_t^x) = \int_S f(z) \nu(x, dz) - a(x) f(x).$$

Proof. Let $\tau_1, \tau_2, \dots$ denote the (random) sequence of the jump times. By the definition of the exponential waiting time,

$$\mathbf{P}(N_t^x = 0) = \mathbf{P}(\tau_1 > t) = e^{-a(x)t}.$$

Next, by conditioning,

$$\begin{aligned}
\mathbf{P}(N_t^x = 1) &= \mathbf{P}(\tau_2 > t - \tau_1, \tau_1 \leq t) \\
&= \int_0^t \mathbf{P}(\tau_2 > t - \tau_1 | \tau_1 = s) a(x) e^{-sa(x)}\, ds \\
&= \int_0^t \int_S \mathbf{P}(\tau_2 > t - s | \tau_1 = s, X(s) = y) \nu(x, dy) e^{-sa(x)}\, ds \\
&= \int_0^t \int_S e^{-a(y)(t-s)} \nu(x, dy) e^{-sa(x)}\, ds,
\end{aligned}$$

and

$$\mathbf{P}(N_t^x > 1) = \mathbf{P}(\tau_2 \leq t - \tau_1, \tau_1 \leq t) = \int_0^t \int_S (1 - e^{-a(y)(t-s)}) \nu(x, dy) e^{-sa(x)}\, ds,$$

and similarly one obtains (3.56), (3.57) with arbitrary k. Denoting $M = \sup_x a(x)$ and taking into account the elementary inequality $1 - e^{-a} \leq a$, $a > 0$, one obtains from (3.57)

$$\mathbf{P}(N_t^x > k) \leq M^{k+1} t \int \int_{0 < s_1 < \cdots < s_k < t} ds_1 \cdots ds_k \leq (Mt)^{k+1}/k!,$$

implying the convergence of the series $\sum_{k=0}^{\infty} \mathbf{P}(N_t^x > k)$. Hence by the Borel–Cantelli lemma N_t^x is a.s. finite. In particular, the first equation in (3.58) holds. Next,

$$\begin{aligned}
\mathbf{E} f(X_t^x) \mathbf{1}_{N_t^x = 1} &= \int_0^t \int_S f(y) \nu(x, dy) e^{-sa(x)} \mathbf{P} \tau_2 > t - s | X_s = y)\, ds \\
&= \int_0^t \int_S e^{-a(y)(t-s)} f(y) \nu(x, dy) e^{-sa(x)} ds.
\end{aligned}$$

Similarly one computes the other terms of the series (3.58). The equation for the derivative then follows straightforwardly as only the first two terms of series (3.58) contribute to the derivative (other terms being of order at least t^2). □

Exercise 3.7.3. If S in Theorem 3.7.2 is locally compact and a bounded ν (depending weakly continuous on x) is such that $\lim_{x \to \infty} \int_K \nu(x, dy) = 0$ for any compact set K, then L of form (3.55) preserves the space $C_\infty(S)$ and hence generates a Feller semigroup.

Finally, let us put into the general context the usual *Markov chains*, defined as Markov processes with finite (or more generally denumerable) state spaces. Any Markov process with a finite state space is of course pure jump and Feller. Let us describe its generator. A $d \times d$-matrix $Q = (Q_{mn})$ is called an *infinitesimally stochastic matrix* or *Q-matrix* if $Q_{nm} \geq 0$ for $m \neq n$ and

$$Q_{nn} = -\sum_{m \neq n} Q_{nm} \tag{3.59}$$

for all n. To any such matrix there corresponds a linear operator in $\mathbb{R}^d$ (which we denote by the same letter) acting as

$$(Qf)_n = \sum_m Q_{nm} f_m.$$

Taking into account the properties of Q, one can rewrite this in two other useful forms:

$$(Qf)_n = \sum_{m \neq n} Q_{nm}(f_m - f_n) = \sum_m Q_{nm}(f_m - f_n). \tag{3.60}$$

It is easy to see that any such Q generates a (unique) Feller semigroup in $\mathbb{R}^d$ that defines a Markov chain on a finite state space of d points. Conversely, any Markov process with a finite state space has a generator of this form. Markov chains represent the simplest examples of pure jump Markov processes described above. The intensity of jumps from a site n equals $|Q_{nn}|$.

3.8 Markov processes on quotient spaces and reflections

One is often interested in whether the Markov property of a process is preserved under a transformation reducing the state space. Let us start with the following typical problem. Given a Markov process X_t^x in $\mathbb{R}^d$, is the magnitude process $|X_t^{|x|}|$ again Markov? Clearly this is not always the case, as one sees already looking at the deterministic process $|x - t|$. This process is not Markov, as its position $x > 0$ does not tell us whether we are moving right (that is, before reflection) or left (after reflection).

To proceed, let us say that a function f on $\mathbb{R}^d$ is *invariant under rotations*, if $f(Ox) = f(x)$ for any $x \in \mathbb{R}^d$ and any orthogonal $d \times d$-matrix O, or equivalently, if $f(x) = h(|x|)$ for some function h on $\mathbb{R}_+$.

Proposition 3.8.1. *Let X_t^x be a Markov process in $\mathbb{R}^d$ with semigroup of transition operators Φ_t. Then the process $Y_t^y = |X_t^x|$, $y = |x|$, in $\bar{\mathbb{R}}_+$ is Markov if and only if the semigroup $\{\Phi_t\}$ preserves the space of functions invariant under rotations. The latter property holds, in particular, if all Φ_t commute with rotations, that is $\Phi_t \tilde{O} = \tilde{O}\Phi_t$ for all $t > 0$ and all orthogonal $d \times d$-matrix O, where the operator $\tilde{O}$ is defined via the formula $\tilde{O} f(x) = f(Ox)$.*

Proof. For a bounded measurable f

$$\mathbf{E}(h(|X^x_{s+t}|)|\mathcal{F}_s) = \mathbf{E}(f(X^x_{s+t})|X^x_s) = \Phi_t f(X^x_s),$$

where $f(z) = h(|z|)$. The operators Φ_t preserve the space of functions that are invariant under rotations if and only if $\Phi_t f(X^x_s)$ depend only on the magnitude of X^x_s, that is

$$\mathbf{E}(h(Y^y_{s+t})|\mathcal{F}_s) = \mathbf{E}(h(Y^y_{s+t})|Y^x_s),$$

which is precisely the Markov property for Y. The last statement is obvious. □

Another typical example of a transformation is a reflection. Namely, let $R_i : \mathbb{R}^d \to \mathbb{R}^d$ denote the *reflection* of the i-th coordinate, that is

$$R_i(x^1, \ldots, x^d) = (x^1, \ldots, x^{i-1}, -x^i, x^{i+1}, \ldots, x^d),$$

and let $\tilde{R}_i$ be the corresponding linear transformation on functions: $\tilde{R}_i f(x) = f(R_i(x))$. A function f on $\mathbb{R}^d$ will be called invariant under the reflection R_i if $\tilde{R}_i f = f$.

Let $\mathbb{R}^d_i$ denote the half-space

$$\mathbb{R}^d_i = \{(x^1, \ldots, x^d) \in R^d : x^i > 0\}$$

and $\bar{\mathbb{R}}^d_i$ its closure. And let the transformation $\bar{R}_i : \mathbb{R}^d \to \bar{\mathbb{R}}^d_i$ be defined by the formula

$$\bar{R}_i(x^1, \ldots, x^d) = (x^1, \ldots, x^{i-1}, |x^i|, x^{i+1}, \ldots, x^d).$$

For a process X_t, the process $\bar{R}_i(X_t)$ is said to be obtained from X_t by the *reflection* at the ith coordinate hyperplane $\{x^i = 0\}$ (or shortly $\bar{R}_i(X_t)$ is the reflected process of X_t). Similarly to Proposition 3.8.1 one obtains the following.

Proposition 3.8.2. *Let X^x_t be a Markov process in $\mathbb{R}^d$ with the semigroup of transition operators Φ_t. Then the reflected process $Y^y_t = \bar{R}_i(X^x_t)$, $y = \bar{R}_i(x)$, in $\bar{\mathbb{R}}^d_i$ is Markov whenever all Φ_t commute with the operator $\tilde{R}_i$.*

One can easily modify this result to include several reflections. But let us instead formulate a more general fact, whose proof is a straightforward extension of the proof of Proposition 3.8.1. Let us recall that if G is a compact topological group of continuous transformations of a metric space S, then one defines the corresponding *quotient space* S_G of S as the collection of the *orbits* of the action of G, where the orbit of $x \in S$ is defined as the subset $G_x = \{g(x) : g \in G\}$ of S. It is easy to see that S_G has a natural structure as a metric space and the projection $P_G : S \to S_G$ that assigns to each $x \in S$ its orbit is a continuous mapping.

Proposition 3.8.3. *Let X_t^x be a Markov process in a metric space S, and let G be a compact topological group of continuous transformations of S. Then the process $Y_t^y = P_G(X_t^x)$, $y = P_G(x)$, in S_G is Markov whenever $\Phi_t \tilde{g} = \tilde{g}\Phi_t$ for all $g \in G$, $t > 0$, where the linear transformation $\tilde{g}$ on $C(S)$ is defined by $(\tilde{g} f)(x) = f(g(x))$.*

In Propositions 3.8.1 and 3.8.2 we dealt with the case of $S = \mathbb{R}^d$ and the groups of transformations generated respectively by all rotations and by the reflection $\mathbb{R}_i$. Other natural examples are given by the group generated by all reflections $\mathbb{R}_i$ in $\mathbb{R}^d$, $i = 1, \dots, d$, yielding reflected processes in the octant $\bar{\mathbb{R}}_+^d$, or by the group generated by the reflections with respect to all hyper-planes $\{x^i = 0\}$ and $\{x^i = 1\}$, yielding reflected processes in the box of vectors in $\mathbb{R}^d$ with all coordinates from the interval $[0, 1]$.

Finally, if a reflected process is not Markov it can be lifted to a Markov one by adding orientation as an additional state parameter. For instance, as we mentioned above, the deterministic process $|x - t|$ in $\bar{R}_+$ is not Markov. But we can define the new state space as the collection of pairs $(x, +)$ or $(x, -)$, $x \geq 0$, where $\pm$ indicates the direction of the movement, and the corresponding Markov semigroup acting on the pairs (f_+, f_-) of functions from $C(\bar{\mathbb{R}}_+)$ with values coinciding at the boundary by the formulas

$$[T_t(f_\pm)]_+(x) = f_+(x + t)$$

and

$$[T_t(f_\pm)]_-(x) = \begin{cases} f_-(x - t), & x \geq t, \\ f_+(t - x), & x \leq t. \end{cases}$$

3.9 Martingales

An adapted integrable process on a filtered probability space is called a *submartingale* if, for all $0 \leq s \leq t < \infty$,

$$\mathbf{E}(X_t | \mathcal{F}_s) \geq X_s,$$

a *supermartingale*, if the reverse inequality holds, and a *martingale* if

$$\mathbf{E}(X_t | \mathcal{F}_s) = X_s.$$

A filtration $\mathcal{F}_t$ is said to satisfy the *usual hypotheses* (or *usual conditions*) if (i) (completeness) $\mathcal{F}_0$ contains all sets of P-measure zero (all P-negligible sets), (ii) (right continuity) $\mathcal{F}_t = \mathcal{F}_{t+} = \bigcap_{\epsilon>0} \mathcal{F}_{t+\epsilon}$. Adding to all $\mathcal{F}_t$ (of an arbitrary filtration) all P-negligible sets leads to a new filtration called the *augmented filtration*.

The following fact about martingales is fundamental.

Theorem 3.9.1 (Regularity of submartingales). *Let M be a submartingale.*

(i) *The following left and right limits exist and are a.s. finite for each $t > 0$:*

$$M_{t-} = \lim_{s\in\mathbb{Q}, s\to t, s<t} M_s; \quad M_{t+} = \lim_{s\in\mathbb{Q}, s\to t, s>t} M_s.$$

(ii) *If the filtration satisfies the usual hypotheses and if the map $t \mapsto \mathbf{E}M_t$ is right-continuous, then M has a càdlàg (right-continuous with finite left limits everywhere) modification.*

Let us quote also the following convergence result.

Theorem 3.9.2. *Let X_t be a right-continuous non-negative supermartingale. Then a.s. there exists a limit $\lim_{t\to\infty} X_t$.*

The proof of the above two theorems can be found in any text on martingale theory and will be omitted here as it is not much relevant to the content of this book.

Theorem 3.9.3. *If X is a Lévy process with Lévy symbol η, then $\forall u \in \mathbb{R}^d$, the process*

$$M_u(t) = \exp\{i(u, X_t) - t\eta(u)\}$$

is a complex $\mathcal{F}_t^X$-martingale.

Proof. $\mathbf{E}|M_u(t)| = \exp\{-t\eta(u)\} < \infty$ for each t. Next, for $s \le t$

$$M_u(t) = M_u(s)\exp\{i(u, X_t - X_s) - (t-s)\eta(u)\}.$$

Then

$$\mathbf{E}(M_u(t)|\mathcal{F}_s^X) = M_u(s)\mathbf{E}(\exp\{i(u, X(t-s))\})\exp\{-(t-s)\eta(u)\} = M_u(s). \quad \square$$

Exercise 3.9.1. Show that the following processes are martingales:

(i) $\exp\{(u, B(t)) - t(u, Au)/2\}$ for d-dimensional Brownian motion $B(t)$ with covariance A and any u;

(ii) the compensated Poisson process $\tilde{N}_t = N_t - \lambda t$ with an intensity λ and the process $\tilde{N}_t^2 - \lambda t$.

As an instructive example let us describe the transformation of martingales by the multiplication on deterministic functions.

Proposition 3.9.1. *Let M_t be a càdlàg $\mathcal{F}_t$-martingale and f_t a continuously differentiable function. Then*

$$\mathbf{E}(M_t f_t - M_s f_s|\mathcal{F}_s) = (f_t - f_s)M_s = \mathbf{E}\left(\int_s^t M_\tau \frac{d}{d\tau} f_\tau \, d\tau \,\Big|\, \mathcal{F}_s\right) \tag{3.61}$$

for any $s < t$. In particular, the process

$$M_t f_t - \int_0^t M_\tau \frac{d}{d\tau} f_\tau \, d\tau$$

is also an $\mathcal{F}_t$-martingale.

Proof. Writing

$$M_t f_t - M_s f_s = (M_t - M_s) f_t + M_s (f_t - f_s)$$

yields the first equation in (3.61). Writing

$$\begin{aligned} \mathbf{E}(M_t f_t - M_s f_s | \mathcal{F}_s) &= \mathbf{E}\Big(\sum_{j=0}^{n-1} (M_{s+(j+1)\tau} f_{s+(j+1)\tau} - M_{s+j\tau} f_{s+j\tau}) \Big| \mathcal{F}_s \Big) \\ &= \mathbf{E}\Big(\sum_{j=0}^{n-1} M_{s+j\tau} (f_{s+(j+1)\tau} - f_{s+j\tau}) \Big| \mathcal{F}_s \Big) \end{aligned}$$

with $\tau = (t-s)/n$ and passing to the limit $n \to \infty$ yields the second equation in (3.61). The last statement is a direct consequence of this equation. □

We turn now to the basic connection between martingales and Markov processes.

Theorem 3.9.4 (Dynkin's formula). *Let $f \in D$, the domain of a Feller process X_t. Then the process*

$$M_t^f = f(X_t) - f(X_0) - \int_0^t Af(X_s) \, ds, \quad t \geq 0, \tag{3.62}$$

is a martingale (with respect to the same filtration, for which X_t is a Markov process) under any initial distribution ν. It is often called Dynkin's martingale.

Proof.

$$\begin{aligned} &\mathbf{E}(M_{t+h}^f | \mathcal{F}_t) - M_t^f \\ &= \mathbf{E}\Big(f(X_{t+h}) - \int_0^{t+h} Af(X_s) \, ds \Big| \mathcal{F}_t \Big) - \Big(f(X_t) - \int_0^t Af(X_s) \, ds \Big) \\ &= \Phi_h f(X_t) - \mathbf{E}\Big(\int_t^{t+h} Af(X_s) \, ds | \mathcal{F}_t \Big) - f(X_t) \\ &= \Phi_h f(X_t) - f(X_t) - \int_0^h A\Phi_s f(X_t) \, ds = 0. \end{aligned}$$

□

Theorem 3.9.5. *A Feller process X_t admits a càdlàg modification.*

Remark 27. (i) Unlike martingales, we do not need the right continuity of the filtration here.

(ii) Proving the result for Lévy processes only one often utilizes the special martingales $M_u(t) = \exp\{i(u, X_t) - t\eta(u)\}$ instead of Dynkin's one used in the proof below.

Proof. Let f_n be a sequence in C_∞^+ that separates points. By Dynkin's formula and the regularity of martingales, there exists a set Ω of full measure s.t. $f_n(X_t)$ has right and left limits on it for all n along all rational numbers $\mathbb{Q}$. Hence X_t has right and left limits on Ω. Define

$$\tilde{X}_t = \lim_{s \to t, s > t, s \in \mathbb{Q}} X_s.$$

Then $X_s \to \tilde{X}_t$ a.s. and $X_s \to X_t$ weakly (by Feller property) and in probability (by Theorem 3.6.3). Hence $\tilde{X}_t = X_t$ a.s. □

The previous two theorems motivate the following important definition. Let L be a linear operator $L : D \mapsto B(S)$, $D \in C(S)$ with a metric space S. One says that a S-valued process X_t with càdlàg paths (or the corresponding probability distribution on the Skorohod space) solves the (L, D)*-martingale problem* with the initial distribution μ if X_0 is distributed according to μ and processes (3.62) are martingales for any $f \in D$ with respect to its own filtration. If $\mathcal{F}_t$ is a given filtration and processes (3.62) are $\mathcal{F}_t$-martingales, we say that X_t solves the (L, D)*-martingale problem* with respect to $\mathcal{F}_t$. The (L, D)-martingale problem is called *well posed* if for any initial μ there exists a unique X_t solving it. We shall say that the (L, D)-martingale problem is *measurably well posed* if the distribution of the unique solution with the Dirac initial law δ_x depends measurably on x. If the (L, D)-martingale problem is measurably well posed, then clearly the solution depends linearly on the initial measure in the sense that the measure P_μ solving the problem with initial law μ can be expressed as $P_\mu = \int P_x \mu(dx)$ in terms of the solutions with the Dirac initial conditions. In the most reasonable situations well posedness implies measurable well posedness, see Theorem 4.10.2. The following result is a direct consequence of Theorems 3.9.4 and 3.9.5. It will be used later for the constructions of Markov semigroups.

Remark 28. Some authors define solutions to the martingale problem without requiring the paths to be càdlàg, but in the most examples of interest these solutions have càdlàg modifications anyway, which justifies the definition adopted here.

Proposition 3.9.2. (i) *A Feller process on a locally compact metric space X_t solves the (L, D)-martingale problem, where L is the generator of X_t and D is any subspace of its domain.*

(ii) *If the (L, D)-martingale problem is well posed, there can exist no more than one Feller process with a generator being an extension of L.*

The next result suggests a more general form of Dynkin's martingale, which will be used in Section 4.10.

Proposition 3.9.3. *Let X_t solve the (L, D)-martingale problem and ϕ be a bounded continuously differentiable function. Then*

$$S_t = f(X_t)\phi_t - \int_0^t \left[\frac{d}{ds}\phi_s f(X_s) + \phi_s Lf(X_s)\right] ds$$

is a martingale for any $f \in D$. In particular, choosing $\phi_s = e^{-\lambda s}$ with $= \lambda > 0$ yields the martingale

$$f(X_t)e^{-\lambda t} - \int_0^t e^{-\lambda s}(\lambda - L)f(X_s)\, ds.$$

Proof. Integration by parts implies that

$$S_t = \left[f(X_t) - \int_0^t Lf(X_s)\, ds\right]\phi_t - \int_0^t \left[f(X_s) - \int_0^s Lf(X_u)du\right]\frac{d}{ds}\phi_s\, ds,$$

which is a martingale by Proposition 3.9.1. □

The following technical result shows that the usual hypotheses for a filtration are in fact natural for the analysis of Feller processes.

Theorem 3.9.6. *Let X_t be a canonical Feller process (see Theorem 3.5.4) on the measure space $(\mathbb{R}^d)^{\mathbb{R}^+}$ equipped with its canonical filtration $\mathcal{F}_t^0 = \sigma(X_u : u \le t)$. Then the augmented filtration $\mathcal{F}_t^\nu$ of the canonical filtration $\mathcal{F}_t^0$ (augmented with respect to the probability measure $\mathbf{P}_\nu$ of this process started with an arbitrary initial distribution ν) is right-continuous.*

Proof. Because $\mathcal{F}_t^\nu$ and $\mathcal{F}_{t+}^\nu$ are $\mathbf{P}_\nu$-complete, it is enough to show

$$\mathbf{E}_\nu(Z|\mathcal{F}_t^\nu) = \mathbf{E}_\nu(Z|\mathcal{F}_{t+}^\nu) \quad \mathbf{P}_\nu\text{-a.s.}$$

for $\mathcal{F}_\infty^0$-measurable and positive Z. By the monotone class theorem, it suffices to show this for $Z = \prod_{i=1}^n f_i(X_{t_i})$ with $f \in C_\infty$ and $t_1 < \cdots < t_n$. We shall use the observation that

$$\mathbf{E}_\nu(Z|\mathcal{F}_t^\nu) = \mathbf{E}_\nu(Z|\mathcal{F}_t^0) \quad \mathbf{P}_\nu\text{-a.s.}$$

For a $t > 0$ choose an integer k: $t_{k-1} \le t < t_k$ so that for $h < t_k - t$

$$\mathbf{E}_\nu(Z|\mathcal{F}_{t+h}^\nu) = \prod_{i=1}^{k-1} f_i(X_{t_i})g_h(X_{t+h}) \quad \mathbf{P}_\nu\text{-a.s.,}$$

where

$$g_h(x) = \int p_{t_k-t-h}(x, dx_k) f_k(x_k) \times \int p_{t_{k+1}-t_k}(x_k, dx_{k+1}) f_{k+1}(x_{k+1}) \cdots \int p_{t_n-t_{n-1}}(x_{n-1}, dx_n) f_n(x_n).$$

As $h \to 0$, g_h converges uniformly (Feller!) to

$$g(x) = \int p_{t_k-t}(x, dx_k) f_k(x_k) \times \int p_{t_{k+1}-t_k}(x_k, dx_{k+1}) f_{k+1}(x_{k+1}) \cdots \int p_{t_n-t_{n-1}}(x_{n-1}, dx_n) f_n(x_n).$$

Moreover, $X_{t+h} \to X_t$ a.s. (right continuity!), and by Theorem 1.3.3

$$\mathbf{E}_\nu(Z|\mathcal{F}^\nu_{t+}) = \lim_{h\to 0} \mathbf{E}_\nu(Z|\mathcal{F}^\nu_{t+h}) = \prod_{i=1}^{k-1} f_i(X_{t_i}) g(X_t) = \mathbf{E}_\nu(Z|\mathcal{F}^\nu_t). \qquad \square$$

Remark 29. The Markov property is preserved by augmentation (we shall not address this important but technical issue in detail).

Exercise 3.9.2. Show that if a random process X_t is left-continuous (e.g. is a Brownian motion), then its natural filtration $\mathcal{F}^X_t$ is left-continuous. Hint: $\mathcal{F}^X_t$ is generated by the sets $\Gamma = \{(X_{t_1}, \dots, X_{t_n}) \in B\}, 0 \le t_1 < \dots < t_n = t$.

Exercise 3.9.3. Let X_t be a Markov chain on $\{1, \dots, n\}$ with transition probabilities $q_{ij} > 0$, $i \ne j$, which can be defined via the semigroup of stochastic matrices Φ_t with generator

$$(Af)_i = \sum_{j \ne i} (f_j - f_i) q_{ij}.$$

Let $N_t = N_t(i)$ denote the number of transitions during time t of a process starting at some point i. Show that $N_t - \int_0^t q(X_s)\, ds$ is a martingale, where $q(l) = \sum_{j\ne l} q_{lj}$ denote the *intensity of the jumps*. Hint: to check that $\mathbf{E}N_t = \mathbf{E}\int_0^t q(X_s)\, ds$ show that the function $\mathbf{E}N_t$ is differentiable and

$$\frac{d}{dt}\mathbf{E}(N_t) = \sum_{j=1}^n \mathbf{P}(X_t = j) q_j.$$

Exercise 3.9.4 (Further Poisson integrals). Let N_t be a Poisson process of intensity $c > 0$ with respect to a right-continuous filtration $\mathcal{F}_t$ and let H be a left-continuous bounded adapted process. Show that the processes

$$M_t = \int_0^t H_s\, dN_s - c\int_0^t H_s\, ds, \quad M_t^2 - c\int_0^t H_s^2\, ds \tag{3.63}$$

are martingales. Hint: check this first for simple left-continuous processes $H_s = \xi(\omega)\mathbf{1}_{(a,b]}(s)$, where from adaptedness ξ is $\mathcal{F}_t$-measurable for any $t \in (a,b]$ and hence $\mathcal{F}_a$-measurable by right continuity. Then, say

$$M_t = \xi[(N_{\min(t,b)} - N_a) - c(\min(t,b) - a)], \quad t \geq a,$$

and one concludes that $\mathbf{E}M_t = 0$ by the independence of ξ and $N_{a+u} - N_a$ and the properties of the latter.

3.10 Stopping times and optional sampling

We discuss here in more detail the notion of stopping times introduced briefly in Section 2.1. Let $(\Omega, \mathcal{F}, \mathcal{F}_t, \mathbf{P})$ be a filtered probability space. A *stopping time* (respectively *optional time*) is defined as a random variable $T : \Omega \mapsto [0,\infty]$ such that $\forall t \geq 0$, $(T \leq t) \in \mathcal{F}_t$ (respectively $(T < t) \in \mathcal{F}_t$). Equivalently, $T : \Omega \mapsto [0,\infty]$ is a stopping time (respectively *optional time*) if the step random process $\mathbf{1}_{[T,\infty)}$ is $\mathcal{F}_t$-adapted. Loosely speaking, stopping times mark the events whose occurrence or nonoccurrence until any given time can be judged from this time. Clearly, if T is a stopping time, then so is the random variable $T + a$ for any $a > 0$, because $(T + a \leq t) = (T \leq t - a) \in \mathcal{F}_{t-a} \in \mathcal{F}_t$ (at any given time we can say whether the event occurred or not, say, a minutes ago), but not for $a < 0$ (at a given time we can predict whether the event will occur a minutes).

Proposition 3.10.1. (i) *T is a stopping time $\Rightarrow$ T is an optional time.*

(ii) *if $\mathcal{F}_t$ is right-continuous, the two notions coincide.*

Proof. (i) $\{T < t\} = \bigcup_{n=1}^{\infty}\{T \leq t - 1/n\} \in \mathcal{F}_{t-1/n} \subset \mathcal{F}_t$.
(ii) $\{T \leq t\} = \bigcap_{n=m}^{\infty}\{T < t + 1/n\} \in \mathcal{F}_{t+1/m}$. Hence $\{T \leq t\} \in \mathcal{F}_{t+}$. □

Hitting time is defined as the random variable $T_A = \inf\{t \geq 0 : X_t \in A\}$, where X_t is a process and A is a Borel set.

Exercise 3.10.1. Show that if X is a $\mathcal{F}_t$-adapted and right-continuous and A is (i) open, or (ii) closed, then T_A is (i) an optional or (ii) a stopping time respectively. Hint:

(i) $$\{T < t\} = \bigcup_{s<t, s\in\mathbb{Q}} \{X_s \in A\} \subset \mathcal{F}_t,$$

(ii) $$\{T > t\} = \bigcap_{s\leq t, s\in\mathbb{Q}} \{X_s \notin A\} \subset \mathcal{F}_t.$$

Proposition 3.10.2. *If T, S are stopping times, then so are*

(i) $\min(T, S)$,

(ii) $\max(T, S)$ *and*

(iii) $T + S$.

Proof. (i) $\{\min(T, S) \le t\} = \{T \le t\} \cup \{S \le t\}$.

(ii) $\{\max(T, S) \le t\} = \{T \le t\} \cap \{S \le t\}$.

(iii) Observe that

$$\begin{aligned}\{T + S > t\} &= \{T = 0, S > t\} \cup \{T > t, S = 0\} \\ &\quad \cup \{T \ge t, S > 0\} \cup \{0 < T < t, T + S > t\}.\end{aligned}$$

The first three events are in $\mathcal{F}_t$ trivially or by Proposition 3.10.1 see that the same holds for the last one, which can be written as

$$\bigcup_{r \in (0,t) \cap \mathbb{Q}} \{t > T > r, S > t - r\}.$$

□

If T is a stopping time and X is a adapted process, the *stopped σ-algebra* $\mathcal{F}_T$ (of events determined prior to T) is defined by

$$\mathcal{F}_t = \{A \in \mathcal{F} : A \cap \{T \le t\} \in \mathcal{F}_t, \forall t \ge 0\}$$

and the *stopped r.v.* X_T is $X_T(\omega) = X_{T(\omega)}$.

Exercise 3.10.2. (i) Convince yourself that $\mathcal{F}_T$ is a σ-algebra.

(ii) Show that if S, T are stopping times s.t. $S \le T$ a.s., then $\mathcal{F}_S \subset \mathcal{F}_T$.

Exercise 3.10.3. If X is an adapted process and T a stopping time taking finitely many values, then X_T is $\mathcal{F}_T$-measurable. Hint: if the range of T is $t_1 < \cdots < t_n$, then

$$\begin{aligned}\{X_T \in B\} \cap \{T \le t_j\} &= \bigcup_{k=1}^{j} \{X_T \in B\} \cap \{T = t_k\} \\ &= \bigcup_{k=1}^{j} \{X_{t_k} \in B\} \cap \{T = t_k\} \in \mathcal{F}_{t_j}.\end{aligned}$$

Exercise 3.10.4. (i) If T_n is a sequence of $\mathcal{F}_t$ stopping times, then $\sup_n T_n$ is a stopping time. Hint: $\{\sup T_n \le t\} = \bigcap \{T_n \le t\}$.

(ii) If $\mathcal{F}_t$ is right-continuous, then $\inf_n T_n$ is a stopping time. Hint: $\{\inf T_n < t\} = \bigcup \{T_n < t\} \in \mathcal{F}_t$ and use Proposition 3.10.1 (ii).

(iii) If additionally T_n is decreasing and converging to T, then $\mathcal{F}_T = \bigcap_n \mathcal{F}_{T_n}$.

The following fact characterizes predictability in terms of stopping times:

Exercise 3.10.5. Let $\mathcal{F}_t$ satisfy the usual hypothesis. Show that the predictable σ-algebra is generated by each of the following collections of sets:

(i) $\mathcal{F}_0 \times \mathbb{R}_+$ and the sets $A \times (t, \infty)$ with $A \in \mathcal{F}_t$;

(ii) $\mathcal{F}_0 \times \mathbb{R}_+$ and the intervals (T, ∞) with stopping times T;

Hint: It is shown when proving Proposition 2.1.1 that the first collection generates the predictable σ-algebra. It remains to observe that any set of the first collection belongs to the second one, because for given $t > 0$ and $A \in \mathcal{F}_t$

$$A \times (t, \infty) = \{(\omega, s) : s > t, \omega \in A\} = (\tau_A, \infty),$$

where τ_A is the stopping time that equals t on A and ∞ outside it.

Proposition 3.10.3. *If X is progressive and T is a stopping time, then the stopped r.v. X_T is $\mathcal{F}_T$-measurable on $\{T < \infty\}$.*

Proof. The r.v. $(s, \omega) \mapsto X_s(\omega)$ is $\mathcal{B}([0,t]) \otimes \mathcal{F}_t$ measurable, and the mapping $\omega \mapsto (T(\omega), \omega)$ is $\mathcal{F}_t, \mathcal{B}([0,t]) \otimes \mathcal{F}_t$ measurable, and so is its restriction on the set $\{T \leq t\}$. Hence the composition $X_T(\omega)$ of these maps is $\mathcal{F}_t$-measurable on the set $\{T \leq t\}$, which means $\{\omega : X_T(\omega) \in B, T \leq t\} \in \mathcal{F}_t$ for a Borel set B, as required. □

It is often easier to work with stopping times that may take only discrete values. Hence a problem of approximation arises. This should be done with a certain care, as say the most natural discrete approximation to a stopping time T would be $[T]$ (the integer part of T), but this need not to be a stopping time. However, if for an optional time T we define the sequence (T_n), $n \in \mathbb{N}$, of decreasing random times as

$$T_n(\omega) = \begin{cases} \infty, & \text{if } T(\omega) = \infty, \\ k/2^n, & \text{if } (k-1)/2^n \leq T(\omega) < k/2^n, \end{cases} \tag{3.64}$$

then all T_n are stopping times converging monotonically to T.

Anticipating further construction of stochastic integrals, let us introduce now its simplified discrete version. Even this elementary notion turns out to be useful. We shall use it to prove Doob's optional sampling theorem. We start with the discrete analog of predictability. A process $H_n, n = 1, 2, \ldots$ is called *predictable* with respect to a discrete filtration $\mathcal{F}_n$, $n = 0, 1, \ldots$, if H_n is $\mathcal{F}_{n-1}$-measurable for all n. Let (X_n), $n = 0, 1, \ldots$ be a stochastic process adapted to $\mathcal{F}_n$, and H_n a positive bounded predictable process. The process $H \circ X$ defined inductively by

$$(H \circ X)_0 = X_0, \quad (H \circ X)_n = (H \circ X)_{n-1} + H_n(X_n - X_{n-1})$$

is called the *transform of X by H* and a *martingale transform* if X is a martingale.

Proposition 3.10.4. *$(H \circ X)$ is a (sub)martingale whenever X so is.*

Proof. Follows from

$$\mathbf{E}((H \circ X)_n | \mathcal{F}_{n-1}) = (H \circ X)_{n-1} + H_n \mathbf{E}(X_n - X_{n-1} | \mathcal{F}_{n-1}). \qquad \square$$

Exercise 3.10.6. Let T, S be bounded stopping times s.t. $S \le T \le M$. Let

$$H_n = \mathbf{1}_{n \le T} - \mathbf{1}_{n \le S} = \mathbf{1}_{S < n \le T}.$$

Show that H is predictable and $(H \circ X)_n - X_0 = X_T - X_S$ for $n > M$. Hint:

$$(H \circ X)_n - X_0 = \mathbf{1}_{S < 1 \le T}(X_1 - X_0) + \cdots + \mathbf{1}_{S < n \le T}(X_n - X_0).$$

Proposition 3.10.5 (Discrete optional sampling and martingale characterization). *Let (X_n), $n = 0, 1, \ldots$ be a $\mathcal{F}_n$-adapted integrable process. The following three statements are equivalent:*

1. *X_t is a submartingale (respectively a martingale),*
2. *for any bounded stopping times $S \le T$*

$$\mathbf{E}(X_S) \le \mathbf{E}(X_T) \tag{3.65}$$

   (respectively with the equality sign),

3. *for any bounded stopping times $S \le T$*

$$X_S \le \mathbf{E}(X_T | \mathcal{F}_s) \quad \text{a.s.} \tag{3.66}$$

   (respectively with equality).

Proof. (3) $\Longrightarrow$ (1) is obvious. (1) $\Longrightarrow$ (2) follows from Exercise 3.10.6 and Proposition 3.10.4. Finally, to get (2) $\Longrightarrow$ (3) one applies (3.65) to the stopping times

$$S^B = S\mathbf{1}_B + M(1 - \mathbf{1}_B), \quad T^B = T\mathbf{1}_B + M(1 - \mathbf{1}_B)$$

with $B \in \mathcal{F}_S$ (check they are stopping times!) yielding

$$\mathbf{E}(X_S \mathbf{1}_B + X_M(1 - \mathbf{1}_B)) \le \mathbf{E}(X_T \mathbf{1}_B + X_M(1 - \mathbf{1}_B)),$$

which implies $\mathbf{E}(X_S \mathbf{1}_B) \le \mathbf{E}(X_T \mathbf{1}_B)$ and hence (3.66). $\square$

As an easy application one gets the following fundamental estimates, which are called *Doob's maximum inequalities*:

Proposition 3.10.6. (i) *If X_n is a submartingale, $n = 1, \ldots, N$, then*

$$\lambda P(\sup |X_n| \ge \lambda) \le \mathbf{E}(|X_N| \mathbf{1}_{\sup_n |X_n| \ge \lambda}) \le \mathbf{E}(|X_N|).$$

(ii) *If X_t is a right-continuous submartingale on $t \in [0, T]$ or $t \geq 0$, then*

$$\lambda P(\sup_t |X_t| \geq \lambda) \leq \sup_t \mathbf{E}(|X_t|).$$

Proof. (i) As $|X_n|$ is again a submartingale, it is enough to consider the case of positive X. Define a stopping time S being equal to N if $\sup_n X_n < \lambda$ and $S = \inf\{n : X_n \geq \lambda\}$ otherwise. Then

$$\mathbf{E}(X_N) \geq \mathbf{E}(X_S) = \mathbf{E}(X_S \mathbf{1}_{\sup_n |X_n| \geq \lambda}) + \mathbf{E}(X_S \mathbf{1}_{\sup_n |X_n| < \lambda})$$

$$\geq \lambda \mathbf{P}(\sup X_n \geq \lambda) + \mathbf{E}(X_N \mathbf{1}_{\sup_n |X_n| < \lambda}),$$

and the required estimate follows by subtraction.

(ii) From a finite index set one directly extends it to a countable index set, and then uses right continuity to obtain the general estimate. □

Theorem 3.10.1 (Doob's optional stopping (or sampling) theorem). *If X is a right-continuous (sub)martingale, $S \leq T$ are two stopping times and either*

(i) *T is bounded or*

(ii) *the family X_τ with τ running through all stopping times is uniformly integrable (the latter occurs e.g. if $X_t = \mathbf{E}(X_\infty | \mathcal{F}_t)$ for some integrable X_∞),*

then X_S and X_T are integrable with

$$X_S \leq \mathbf{E}(X_T | \mathcal{F}_S),$$

with equality in case X is a martingale.

Proof. Let $S_n \leq T_n$ be a sequences of decreasing stopping times with countably many values converging to S and T. Then

$$\int_A X_{S_n} d\mathbf{P} \leq \int_A X_{T_n} d\mathbf{P} \tag{3.67}$$

for all $A \in \mathcal{F}_{S_n}$, and in particular for $A \in \mathcal{F}_S$. By right continuity X_{T_n} (respectively X_{S_n}) converge to X_T (respectively X_S) point-wise and by uniform integrability also in L^1 (use Theorem 1.3.3 in case (i)). Hence (3.67) implies

$$\int_A X_S d\mathbf{P} \leq \int_A X_T d\mathbf{P}$$

for $A \in \mathcal{F}_S$, i.e. the required result. □

Example (Violation of optional sampling). Suppose (η_n), $n \in \mathbb{N}$, are i.i.d. Bernoulli random variables such that η_n equals 1 (success) or -1 (loss) with probability p and $q = 1 - p$. The player's stake at n-th turn is V_n. Naturally V_n is $\mathcal{F}_{n-1} = \sigma(\eta_1, \ldots, \eta_{n-1})$-measurable. Then the total gain is

$$X_n = \sum_{i=1}^{n} V_i \eta_i = \sum_{i=1}^{n} V_i \Delta Y_i = (V \circ Y)_n,$$

where $Y_n = \eta_1 + \cdots + \eta_n$. The game is *fair* (or *favorable*, or *unfavorable*) if $p = q$ (or $p > q$, or $p < q$) $\Leftrightarrow$ $(X_n, \mathcal{F}_n)$ is a martingale (or submartingale, or supermartingale).

Consider a strategy V (called the *martingale strategy*) s.t. $V_1 = 1$ and further

$$V_n = \begin{cases} 2^{n-1}, & \text{if } \eta_1 = \cdots = \eta_{n-1} = -1, \\ 0, & \text{otherwise.} \end{cases} \tag{3.68}$$

Thus if $\eta_1 = \cdots = \eta_n = -1$, the total loss after n turns will be $\sum_{i=1}^{n} 2^{i-1} = 2^n - 1$ and if then $\eta_{n+1} = 1$, $X_{n+1} = 2^n - (2^n - 1) = 1$. Denoting $T = \inf\{n : X_n = 1\}$ and assuming $p = q = 1/2$ yields $\mathbf{P}(T = n) = (1/2)^n$, and hence $\mathbf{P}(T < \infty) = 1$, by the Borel–Cantelli lemma. Consequently

$$\mathbf{E} X_T = \mathbf{P}(X_T = 1) = 1 > X_0 = 0,$$

though X_n is a martingale and $\mathbf{E} X_n = 0$ for all n.

3.11 Strong Markov property; diffusions as Feller processes with continuous paths

The following definition is fundamental. A time-homogeneous Markov process with transition family p_t is called *strong Markov* if

$$\mathbf{E}_\nu(f(X_{S+t}) | \mathcal{F}_S) = (\Phi_t f)(X_S) \quad \mathbf{P}_\nu\text{-a.s.} \quad \text{on } \{S < \infty\} \tag{3.69}$$

for any $\{\mathcal{F}_t\}$-stopping time S, initial distribution ν and positive Borel f.

Exercise 3.11.1. (i) If (3.69) holds for bounded stopping times, then it holds for all stopping times. Hint: For any n and a stopping time S

$$\mathbf{E}_\nu(f(X_{\min(S,n)+t}) | \mathcal{F}_{\min(S,n)}) = (\Phi_t f)(X_{\min(S,n)}) \quad \mathbf{P}_\nu\text{-a.s.}$$

Hence by locality (Theorem 1.3.5)

$$\mathbf{E}_\nu(f(X_{S+t}) | \mathcal{F}_S) = (\Phi_t f)(X_S) \quad \mathbf{P}_\nu\text{-a.s.} \quad \text{on } \{S \le n\}.$$

To complete the proof take $n \to \infty$ thus exhausting the set $\{S < \infty\}$.

(ii) A Markov process X_t is strong Markov $\Longleftrightarrow$ for all a.s. finite stopping time T the process $Y_t = X_{T+t}$ is a Markov process with respect to $\mathcal{F}_{T+t}$ with the same transition family. Hint: strong Markov $\Longleftrightarrow$

$$\mathbf{E}_\nu(f(X_{T+t+s})|\mathcal{F}_{T+t}) = (\Phi_s f)(X_{T+t}) \quad \mathbf{P}_\nu\text{-a.s.}$$

(iii) A Markov process X_t is strong Markov $\Longleftrightarrow$ for arbitrary times $t_1 < \cdots < t_n$, bounded measurable functions $f_1, \ldots, f_n$ and stopping times T

$$\mathbf{E}_\nu\Big(\prod_{i=1}^n f_i(X_{T+t_i})\Big|\mathcal{F}_T\Big) = \mathbf{E}_{X_T}\prod_{i=1}^n f_i(X_{t_i}). \tag{3.70}$$

Hint: for $n = 1$ this coincides with the definition of the strong Markovianity. Use conditioning and induction to complete the proof.

(iv) A canonical Markov process is strong Markov $\Leftrightarrow$

$$\mathbf{E}_\nu(Z \circ \theta_t|\mathcal{F}_S) = \mathbf{E}_{X_S}(Z) \quad \mathbf{P}_\nu\text{-a.s.}$$

for any $\{\mathcal{F}_t\}$-optional time S, initial distribution ν and $\mathcal{F}_\infty$-measurable r.v. Z, where θ is the canonical shift. Hint: see Theorem 3.5.5.

Theorem 3.11.1. *Any Feller process X_t is strong Markov.*

Proof. Let T take values on a countable set D. Then

$$\begin{aligned}\mathbf{E}_\nu(f(X_{T+t})|\mathcal{F}_T) &= \sum_{d\in D} \mathbf{1}_{T=d}\mathbf{E}_\nu f(X_{d+t})|\mathcal{F}_d) \\ &= \sum_{d\in D} \mathbf{1}_{T=d}\Phi_t f(X_d) = \Phi_t f(X_T).\end{aligned}$$

For a general T take a decreasing sequence of T_n with only finitely many values converging to T. Then

$$\mathbf{E}_\nu(f(X_{T_n+t})|\mathcal{F}_{T_n}) = \Phi_t f(X_{T_n}),$$

for all n, i.e.

$$\int_A f(X_{T_n+t})\mathbf{P}(d\omega) = \int_A \Phi_t f(X_{T_n})\mathbf{P}(d\omega)$$

for all $A \in \mathcal{F}_{T_n}$, in particular for $A \in \mathcal{F}_T$, as $\mathcal{F}_T \subset \mathcal{F}_{T_n}$. Hence, by right continuity of X_t and dominated convergence, passing to the limit $n \to \infty$ yields

$$\int_A f(X_{T+t})\mathbf{P}(d\omega) = \int_A \Phi_t f(X_T)\mathbf{P}(d\omega)$$

for all $A \in \mathcal{F}_T$, as required. □

Theorem 3.11.2. *If X is a Lévy process and T is a stopping time, then the process $X_T(t) = X_{T+t} - X_T$ is again a Lévy process, which is independent of $\mathcal{F}_T$, and its law under $\mathbf{P}_\nu$ is the same as that of X under $\mathbf{P}_0$.*

Proof. First method (as a corollary of the strong Markov property of Feller processes). For a positive Borel functions f_i, by (3.70)

$$\mathbf{E}_\nu\Big(\prod_i f_i(X_{T+t_i} - X_T)\Big|\mathcal{F}_T\Big) = \mathbf{E}_{X_T}\Big(\prod_i f_i(X_{t_i} - X_0)\Big),$$

but this is a constant not depending on X_T.

Second method (direct using special martingales $M_u(t) = \exp\{i(u, X_t) - t\eta(u)\}$). Assume T is bounded. Let $A \in \mathcal{F}_T$, $u_i \in \mathbb{R}^d$, $0 = t_0 < t_1 < \cdots < t_n$. Then

$$\mathbf{E}\Big(\mathbf{1}_A \exp\Big\{i\sum_{j=1}^n (u_j, X_T(t_j) - X_T(t_{j-1}))\Big\}\Big) = \mathbf{E}\Big(\mathbf{1}_A \prod_{j=1}^n \frac{M_{u_j}(T+t_j)}{M_{u_j}(T+t_{j-1})}\Big)\prod_{j=1}^n \phi_{t_j - t_{j-1}}(u_j),$$

where $\phi_t(u) = \mathbf{E}e^{i(u,X_t)} = e^{t\eta(u)}$. By conditioning for $s < t$,

$$\mathbf{E}\Big(\mathbf{1}_A\frac{M_u(T+t)}{M_u(T+s)}\Big) = \mathbf{E}\Big(\frac{\mathbf{1}_A}{M_u(T+s)}\mathbf{E}(M_u(T+t)|\mathcal{F}_{T+s})\Big) = \mathbf{P}(A).$$

Repeating this argument yields

$$\mathbf{E}\Big(\mathbf{1}_A \exp\Big\{i\sum_{j=1}^n (u_j, X_T(t_j) - X_T(t_{j-1}))\Big\}\Big) = \mathbf{P}(A)\prod_{j=1}^n \phi_{t_j-t_{j-1}}(u_j),$$

which implies the statement of the Theorem by means of the following fact. □

Exercise 3.11.2. Suppose X is a r.v. on $(\Omega, \mathcal{F}, \mathbf{P})$, $\mathcal{G}$ is a sub-σ-algebra of $\mathcal{F}$ and

$$\mathbf{E}(e^{i(u,X)}\mathbf{1}_A) = \phi_p(u)\mathbf{P}(A)$$

for any $A \in \mathcal{G}$, where ϕ_p is the ch.f. of a probability law p. Then X is independent of $\mathcal{G}$ and the distribution of X is p.

Theorem 3.11.3 (Blumenthal's zero-one law). *Let X_t be a Feller process in $\mathbb{R}^d$ defined on a filtered probability space with a filtration $\mathcal{F}_t$ satisfying the usual hypothesis. Then $\mathbf{P}_x(A) = 0$ or 1 for any $x \in \mathbb{R}^d$ and $A \in \mathcal{F}_0$. In particular, $\mathbf{P}_x(\tau = 0) = 0$ or 1 for a $\mathcal{F}$-stopping time τ.*

Proof. A.s. $\mathbf{1}_A = \mathbf{E}(\mathbf{1}|\mathcal{F}_0) = \mathbf{P}_x(A)$. □

Denote

$$\tau_h(x) = \inf\{t \geq 0 : |X_t^x - x| > h\}, \quad h > 0.$$

A point x is called *absorbing* if $\tau_h = \infty$ a.s. for every h.

Theorem 3.11.4. *Let X_t^x be a Feller process with a transition family $p_t(x, dy)$.*

(i) *Then $\mathbf{E}_x(\tau_h) > 0$ for all x and h.*

(ii) *If x is not absorbing, then $\mathbf{E}_x(\tau_h) < \infty$ for all sufficiently small h.*

Proof. Statement (i) follows from stochastic continuity of X_t (Theorem 3.6.3). Assume x is not absorbing. Then $p_t(x, B_\epsilon(x)) < p < 1$ for some t, ϵ, where $B_\epsilon(x)$ is the ball centered at x of radius ϵ. By weak continuity of the family p_t, there exists a small $h < \epsilon$ such that

$$p_t(y, B_h(x)) \leq p_t(y, B_\epsilon(x)) \leq p$$

for $y \in B_h(x)$. Then

$$\mathbf{P}(\tau_h(x) \geq nt) \leq \mathbf{P}\Big[\bigcap_{k \leq n} (X_{kt}^x \in B_h(x))\Big] \leq p^n$$

for all natural n. Finally,

$$\mathbf{E}\tau_h(x) = \int_0^\infty \mathbf{P}(\tau_h(x) \geq t)\, dt \leq t \sum_{n=0}^\infty \mathbf{P}(\tau_h(x) \geq nt) \leq \frac{t}{1-p} < \infty. \qquad \square$$

Theorem 3.11.5 (Dynkin's formula for generators). *Let X be a Feller process with a generator A and $f \in D_A$. Then*

$$Af(x) = \lim_{h \to 0} \frac{\mathbf{E}_x f(X_{\tau_h}) - f(x)}{\mathbf{E}_x \tau_h} \tag{3.71}$$

if x is not absorbing, and $Af(x) = 0$ otherwise.

Proof. By Dynkin's martingale and optional stopping,

$$\mathbf{E}_x f(X_{\min(t,\tau_h)}) - f(x) = \mathbf{E}_x \int_0^{\min(t,\tau_h)} Af(X_s)\, ds, \quad t, h > 0.$$

As $\mathbf{E}_x \tau_h < \infty$ for small h, this extends to $t = \infty$ by dominated convergence. This implies (3.71) by continuity of Af, taking also into account that $\mathbf{E}_x(\tau_h) > 0$. □

The next beautiful result designates the distinguished place of diffusions among other Feller processes.

Theorem 3.11.6. *Let A be a generator of a Feller process X_t s. t. $C_c^\infty \subset D_A$. Then X_t is a.s. continuous $\mathbf{P}_\nu$ for every ν if and only if A is local on C_c^∞ and hence X_t is a diffusion.*

Proof. In one direction, namely that continuity implies locality, the statement is a consequence of (3.71). Assume that A is local. Choose arbitrarily small h. Then $Af(y) = 0$ for $y \in B_h(x)$ whenever the support of f does not contain $B_h(x)$. For such f it follows that $f(X_{\min(t,\tau_h)})$ is a martingale for each t, and hence by dominated convergence $\mathbf{E}_x f(X_{\tau_h}) = 0$. Consequently $\mathbf{P}_x(|X_{\tau_h} - x| \le h) = 1$ (otherwise one would be able to find a non-negative $f \in C_c^\infty$, vanishing in $B_h(x)$, but with positive $\mathbf{E}_x f(X_{\tau_h})$). By the Markov property,

$$\mathbf{P}_\nu(\sup_{t\in\mathbb{Q}} |X_{t+\tau_h(X_t)} - X_t| \le h) = 1,$$

and hence by right continuity,

$$\mathbf{P}_\nu(\sup_t |X_{t+\tau_h(X_t)} - X_t| \le h) = 1,$$

implying

$$\mathbf{P}_\nu(\sup_t |X_{t+} - X_{t-}| \le h) = 1.$$

As this holds for any $h > 0$, the trajectories X_t are a.s. continuous. □

Remark 30. Theorem 3.11.6 yields another proof that BM (possibly with drift) is the only Lévy process with continuous paths. It also implies another construction and proof of the existence of BM (additional to the four methods exposed in Chapter 2): define it as a Markov process with required Gaussian transition probabilities, deduce that it is Feller with a local generator, hence the trajectories are continuous.

3.12 Reflection principle and passage times

This section presents some basic applications of the strong Markov property to the analysis of BM. They are given as an illustration of the general theory and will not be used further.

Let B be a Brownian motion on $(\Omega, \mathcal{F}, \mathbf{P})$. The *passage time* or the *hitting time* T_b to a level b is defined as

$$T_b(\omega) = \inf\{t \ge 0 : B_t(\omega) = b\}. \tag{3.72}$$

The (intuitively clear) equation

$$\mathbf{P}(T_b < t, B_t \ge b) = \mathbf{P}(T_b < t, B_t < b)$$

for $b > 0$ is called the *reflection principle*. Since

$$\mathbf{P}(T_b < t) = \mathbf{P}(T_b < t, B_t \geq b) + \mathbf{P}(T_b < t, B_t < b),$$

and $\mathbf{P}(T_b < t, B_t \geq b) = \mathbf{P}(B_t \geq b)$, this implies

$$\mathbf{P}(T_b < t) = 2\mathbf{P}(B_t \geq b) = \sqrt{2/(t\pi)} \int_b^\infty e^{-x^2/2t}\, dx = 2/\pi \int_{bt^{-1/2}}^\infty e^{-x^2/2}\, dx.$$

Differentiating yields the density

$$\mathbf{P}(T_b \in dt) = \frac{|b|}{\sqrt{2\pi t^3}} e^{-b^2/2t}\, dt. \tag{3.73}$$

The necessity to justify the reflection principle and hence these calculations was one reason to introduce the strong Markov property.

Theorem 3.12.1 (Reflection principle). *For a BM B_t*

$$\mathbf{P}(T_b \leq t) = \mathbf{P}(M_t \geq b) = 2\mathbf{P}(B_t \geq b) = \mathbf{P}(|B_t| \geq b), \tag{3.74}$$

where $M_t = \inf\{b : T_b \geq t\} = \sup\{B_s : s \leq t\}$. In particular, formula (3.73) *holds true.*

Proof.

$$\begin{aligned}\mathbf{P}(M_t \geq b, B_t < b) &= \mathbf{P}(T_b \leq t, B_{T_b+(t-T_b)} - B_{T_b} < 0)\\ &= \mathbf{P}(T_b \leq t)\mathbf{P}(B_s < 0) = \frac{1}{2}\mathbf{P}(T_b \leq t) = \frac{1}{2}\mathbf{P}(M_t \geq b),\end{aligned}$$

and the result follows as

$$\mathbf{P}(M_t \geq b) = \mathbf{P}(B_t \geq b) + \mathbf{P}(M_t \geq b, B_t < b).$$ □

Theorem 3.12.2. *The process T_a is a left-continuous non-decreasing Levy process (i.e. it is a subordinator), and $T_{a+} = \inf\{t : B_t > a\}$ is its right-continuous modification.*

Proof. Since $T_b - T_a = \inf\{t \geq 0 : B_{T_a+t} - B_{T_a} \geq b-a\}$, this difference is independent of $\mathcal{F}_{T_a}$ by the strong Markov property of BM. Stochastic continuity follows from the density (3.73). Clearly the process T_a (respectively T_{a+}) is non-decreasing and left-continuous (respectively right continuous) and $T_{a+} = \lim_{s\to 0, s>0} T_{a+s}$. Lastly, it follows from the continuity of BM that $T_a = T_{a+}$ a.s. □

Theorem 3.12.3. *For the process T_a,*

$$\mathbf{E}e^{-uT_a} = e^{-a\sqrt{2u}}, \tag{3.75}$$

which implies by (3.27), (3.28) *that T_a is a stable subordinator with index $\alpha = 1/2$ and Lévy measure $\nu(dx) = (2\pi x^3)^{-1/2}dx$.*

Proof. We shall give three proofs, as they all are instructive.

(a) Compute the l.h.s. of (3.75) directly from density (3.73) using the integral calculated in (3.48).

(b) As $M_s(t) = \exp\{sB_t - s^2t/2\}$ is a martingale, one concludes from optional sampling that

$$1 = \mathbf{E}\exp\{sB_{T_a} - s^2T_a/2\} = e^{sa}\mathbf{E}\exp\{-s^2T_a/2\},$$

and (3.75) follows by substituting $u = s^2/2$. (Remark. As Doob's theorem is stated for bounded stopping times, in order to be precise here one has to consider first the stopping times $\min(n, T_a)$ and then let $n \to \infty$.)

(c) For any $a > 0$ the process $\frac{1}{b}T_{a\sqrt{b}}$ is the first hitting time of the level a for the process $b^{-1/2}B_{bt}$. As by the scaling property of BM the latter is again a BM, $\frac{1}{b}T_{a\sqrt{b}}$ and T_a are identically distributed, and thus the subordinator T_a is stable. Comparing expectations one identifies the rate leading again to (3.75). □

Theorem 3.12.4. *The joint distribution of B_t and M_t is given by the density*

$$\phi(t,a,b) = \frac{2(2b-a)}{\sqrt{2\pi t^3}}\exp\left\{-\frac{(2b-a)^2}{2t}\right\}. \tag{3.76}$$

Proof. Let $a \le b$. Then

$$\begin{aligned}\mathbf{P}(B_t < a, M_t \ge b) &= \mathbf{P}\left(M_t \ge b, B_{T_b+(t-T_b)} - B_{T_b} < -(b-a)\right)\\ &= \mathbf{P}(M_t \ge b, B_{T_b+(t-T_b)} - B_{T_b} \ge b-a)\\ &= \mathbf{P}(M_t \ge b, B_t \ge 2b-a) = \mathbf{P}(B_t \ge 2b-a).\end{aligned}$$

Hence (3.76) follows by differentiation, as

$$\phi(t,a,b) = -\frac{\partial^2}{\partial a\partial b}\mathbf{P}(B_t < a, M_t \ge b) = -\frac{\partial}{\partial b}p_t(2b-a), \tag{3.77}$$

where p_t is a probability density function of BM at time t. □

Theorem 3.12.5. *The reflected Brownian motion $|B_t|$ and the process $Y_t = M_t - B_t$ are both Markov with the same probability density*

$$p_t^+(x,y) = p_t(x-y) + p_t(x+y), \tag{3.78}$$

where $p_t(x-y)$ is the transition density of the standard BM.

Proof. For $|B_t|$ the claim follows from a more general result from Section 3.8. Turning to Y_t let $m = M_t > 0$, $b = B_t < m$ and $r = m - b$. Then, by the strong Markov property,

$$\begin{aligned}\mathbf{P}(M_{t+h} - B_{t+h} < \xi|\mathcal{F}_t) &= \mathbf{P}(M_{t+h} - B_{t+h} < \xi|B_t = b, M_t = m) \\ &= \mathbf{E}(\mathbf{1}_{M_{t+h}-B_{t+h}<\xi}\mathbf{1}_{M_{t+h}=m}|B_t = b, M_t = m) \\ &\quad + \mathbf{E}(\mathbf{1}_{M_{t+h}-B_{t+h}<\xi}\mathbf{1}_{M_{t+h}>m}|B_t = b, M_t = m) \\ &= \mathbf{E}(\mathbf{1}_{r-B_h<\xi}\mathbf{1}_{M_h<r}) + \mathbf{E}(\mathbf{1}_{M_h-B_h<\xi}\mathbf{1}_{M_h\geq r}),\end{aligned}$$

which is the integral of $\phi(t, x, y)$ from (3.77) over the domain $r-\xi < x < y < x+\xi$, i.e. it equals

$$\int_{r-\xi}^{\infty} dx \int_{x}^{x+\xi} dy\phi(t, x, y) = -\int_{r-\xi}^{\infty} p_t(2y - x)|_{y=x}^{y=x+\xi}.$$

Hence

$$\begin{aligned}\mathbf{P}(M_{t+h} - B_{t+h} < \xi|\mathcal{F}_t) &= \int_{r-\xi}^{\infty} p_t(x)\,dx - \int_{r-\xi}^{\infty} p_t(2\xi + x)\,dx \\ &= \int_{r-\xi}^{\infty} p_t(x)\,dx - \int_{r+\xi}^{\infty} p_t(y)\,dy.\end{aligned}$$

Differentiating with respect to ξ yields (3.78). □

The *arcsin law* is defined as the distribution of $\xi = \sin^2 X$ when X is $U(0, 2\pi)$ (uniformly distributed on $[0, 2\pi]$). Clearly

$$\mathbf{P}(\xi \leq t\} = \mathbf{P}\{|\sin X| \leq \sqrt{t}\} = \frac{2}{\pi}\arcsin\sqrt{t}, \quad t \in [0, 1]. \tag{3.79}$$

Theorem 3.12.6. *Let B_t be a Brownian motion on $[0, 1]$, with maximum process M_t. Then the random times $\tau = \inf\{t : B_t = M_1\}$ (when B_t first attains its maximum), $\tilde{\tau} = \sup\{t : B_t = M_1\}$ (when B_t for the last time attains its maximum) and the time $\theta = \sup\{t : B_t = 0\}$ of the last exit from the origin all obey the arcsin law. In particular, as $\tau \leq \tilde{\tau}$ this implies that $\tau = \tilde{\tau}$ a.s.*

Proof.

$$\begin{aligned}\mathbf{P}(\tilde{\tau} \leq t) = \mathbf{P}(\tau \leq t) &= \mathbf{P}(\sup_{s\leq t}(B_s - B_t) \geq \sup_{s\geq t}(B_s - B_t)) = \mathbf{P}(|B_t| \geq |B_1 - B_t|) \\ &= \mathbf{P}(t\xi^2 \geq (1-t)\eta^2) = \mathbf{P}\left(\frac{\eta^2}{\xi^2 + \eta^2} \leq t\right) = \mathbf{P}(\sin^2 X \leq t),\end{aligned}$$

where ξ, η are independent $N(0,1)$ r.v. and X is uniformly distributed on $[0, 2\pi]$. (Use Theorems 3.12.1 and 3.12.5 to explain the reasoning behind all these equivalences!). Also

$$\begin{aligned}\mathbf{P}(\theta < t) &= \mathbf{P}(\sup_{s\geq t} B_s < 0) + \mathbf{P}(\inf_{s\geq t} B_s > 0) = 2\mathbf{P}(\sup_{s\geq t}(B_s - B_t) < -B_t) \\ &= 2\mathbf{P}(|B_1 - B_t| < B_t) = \mathbf{P}(|B_1 - B_t| < |B_t|) = \mathbf{P}(\tau \leq t). \qquad \square\end{aligned}$$

Exercise 3.12.1. Show (either directly or by applying the scaling transformation to (3.79)) that for $\tau_t = \inf\{s \in (0,t) : B_s = M_t\}$,

$$\mathbf{P}(\tau_t \leq r) = \int_0^r \frac{dy}{\pi\sqrt{y(t-y)}} = \frac{2}{\pi} \arcsin \sqrt{\frac{r}{t}}.$$

Chapter 4
SDE, ΨDE and martingale problems

The evolution of the averages of functions over a random position of a Markov process usually satisfies an evolution pseudo-differential equation (ΨDE) $\dot{f} = Lf$ with a generator L of Lévy–Khintchine type. The global dynamics are then described by a Markov or sub-Markov semigroup, while infinitesimal changes are specified by the generator L, or equivalently by the corresponding bilinear form, called a Dirichlet form. This dynamics can be analyzed in various functional spaces revealing the functional analytic description of a Markov dynamics. One can also look at the random trajectories themselves. Their dynamics can often be described by an appropriate stochastic differential equation (SDE), yielding a probabilistic interpretation (and an underlying probabilistic structure) for the dynamics given by a ΨDE. The solutions to the corresponding stationary equations $Lf = 0$ can be obtained probabilistically via stopped or killed processes. This chapter is devoted to a closer look at these two sides of analysis. Moreover, a very handy intermediate notion is that of a martingale problem, which can be looked at as a kind of generalized solution for both SDE and ΨDE, providing a convenient technical link between these two descriptions.

We start with a relatively elementary introduction to this circle of ideas on the example of diffusion processes (Sections 4.1–4.4), followed by the analysis of Markov processes that can be constructed from the usual SDE driven by Lévy processes and Poisson random measures (Sections 4.5–4.6). Then, in Section 4.7, the theory of stochastic integration for martingales with continuous predictable quadratic variation (continuity in this sense being the main simplification as compared with the general theory) is developed. Finally we plunge into the deep (and technically demanding) general theory of the convergence of martingale problems, revealing the basic link with Markov processes. Presenting this theory we used Ethier and Kurtz [110], Joffe and Métivier [150], Kallenberg [154] and Stroock [300]. However, the proofs were simplified and streamlined essentially.

4.1 Markov semigroups and evolution equations

This section can be considered as a continuation of Section 3.6. It details the link between Feller semigroups and the solutions to the corresponding evolution ΨDE. In particular, it reveals the crucial role played by positivity preservation in the analysis of these ΨDE.

From the definition of the generator and the invariance of its domain it follows that if Φ_t is the Feller semigroup of a process X_t with a generator L and domain D_L, then

$\Phi_t f(x)$ solves the Cauchy problem

$$\frac{d}{dt} f_t(x) = L f_t(x), \quad f_0 = f, \tag{4.1}$$

whenever $f \in D_L$, the derivative being taken in the sense of the sup-norm of $C(\mathbb{R}^d)$. Formula (3.34) yields the probabilistic interpretation of this solution and an explicit formula.

In the theory of linear differential equations the solution $G(t, x, x_0)$ of (4.1) with $f_0 = \delta_{x_0} = \delta(. - x_0)$ (i.e. satisfying (4.1) for $t > 0$ and the limiting condition in the weak form

$$\lim_{t \to 0} \int G(t, x, x_0) g(x)\, dx = g(x_0)$$

for any $g \in C_c^\infty$), is called the *Green function* or (usually for parabolic type equations) the *heat kernel* of the problem (4.1) (whenever it exists of course, which may not be the case in general). In its probabilistic interpretation, the Green function $G(t, x, x_0)$ represents the density at x_0 of the distribution of X_t started at x.[1]

In particular, if the distribution of a Lévy process X_t has a density $\omega(t, y)$, then $\Phi_t \delta_{x_0}(x) = \omega(t, x_0 - x)$, as follows from (3.41), so that $\omega(t, x_0 - x)$ is the Green function $G(t, x, x_0)$ in this case.

For example, the Green function for the pseudo-differential (fractional parabolic) equation

$$\frac{\partial u}{\partial t} = (A, \nabla u(x)) - a|\Delta u|^{\alpha/2}$$

(see the end of Section 1.8 for fractional derivatives or Laplace operator) is given by the stable density (see (3.29))

$$S(x_0 - At - x; \alpha, at) = (2\pi)^{-d} \int_{\mathbb{R}^d} \exp\{-at|p|^\alpha + ip(x + At - x_0)\}\, dp.$$

Together with the existence of a solution one is usually interested in its uniqueness and continuous dependence on initial data. The next statement shows how naturally this issue is settled via the PMP.

Theorem 4.1.1. *Let a subspace $D \subset C(\mathbb{R}^d)$ contain constant functions, and let an operator $L : D \mapsto C(\mathbb{R}^d)$ satisfying PMP be given. Let $T > 0$ and $u(t, x) \in C([0, T] \times R^d$. Assume $u(0, x)$ is everywhere non-negative, $u(t, .) \in C_\infty(\mathbb{R}^d) \cap D$ for all $t \in [0, T]$, is differentiable in t for $t > 0$ and satisfies the evolutionary equation*

$$\frac{\partial u}{\partial t} = Lu, \quad t \in (0, T].$$

[1] Note the difference in the direction of time: x_0 is the initial point for the evolution equation, but a final point for the corresponding process. This difference is often not revealed in the analysis of time-homogeneous processes, but becomes explicit for time-nonhomogeneous ones.

Then $u(t, x) \geq 0$ *everywhere, and*

$$\max\{u(t, x) : t \in [0, T], x \in \mathbb{R}^d\} = \max\{u(0, x) : x \in \mathbb{R}^d\}. \tag{4.2}$$

Proof. First suppose $\inf u = -\alpha < 0$. For a $\delta < \alpha/T$ consider the function

$$v_\delta = u(t, x) + \delta t.$$

Clearly this function also has a negative infimum. Since v tends to a positive constant δt as $x \to \infty$, v has a global negative minimum at some point (t_0, x_0), which lies in $(0, T] \times \mathbb{R}^d$. Hence $(\partial v/\partial t)(t_0, x_0) \leq 0$ (with equality if $t_0 < T$), and by the PMP $Lv(t_0, x_0) \geq 0$. Consequently

$$\left(\frac{\partial v}{\partial t} - Lv\right)(t_0, x_0) \leq 0.$$

On the other hand, from the evolution equation one deduces that

$$\left(\frac{\partial v}{\partial t} - Lv\right)(t_0, x_0) = \left(\frac{\partial u}{\partial t} - Lu\right)(t_0, x_0) + \delta = \delta.$$

This contradiction completes the proof of the first statement. Similarly, assume that

$$\max\{u(t, x) : t \in [0, T], x \in \mathbb{R}^d\} > \max\{u(0, x) : x \in \mathbb{R}^d\}.$$

Then there exists a $\delta > 0$ such that the function $v_\delta = u(t, x) - \delta t$ also attains its maximum at a point (t_0, x_0) with $t_0 > 0$. Hence $(\partial v/\partial t)(t_0, x_0) \geq 0$ (with equality if $t_0 < T$), and by the PMP $Lv(t_0, x_0) \leq 0$. Consequently,

$$\left(\frac{\partial v}{\partial t} - Lv\right)(t_0, x_0) \geq 0.$$

But from the evolution equation,

$$\left(\frac{\partial v}{\partial t} - Lv\right)(t_0, x_0) = \left(\frac{\partial u}{\partial t} - Lu\right)(t_0, x_0) - \delta = -\delta.$$

This is again a contradiction. □

Corollary 15. *Under the condition on* D *and* L *as in the above theorem, assume* $f \in C([0, T] \times \mathbb{R}^d)$, $g \in C_\infty(\mathbb{R}^d)$. *Then the Cauchy problem*

$$\frac{\partial u}{\partial t} = Lu + f, \quad u(0, x) = g(x), \tag{4.3}$$

can have at most one solution $u \in C([0, T] \times \mathbb{R}^d)$ *such that* $u(t, .) \in C_\infty(\mathbb{R}^d)$ *for all* $t \in [0, T]$.

We shall touch now upon the problem (developed extensively later on) of reconstructing a Feller semigroup (and hence the corresponding process) from a rich enough class of solutions to the Cauchy problem (4.1).

Theorem 4.1.2. *Let L be a conditionally positive operator in $C_\infty(\mathbb{R}^d)$ satisfying PMP, and let D be a dense subspace of $C_\infty(\mathbb{R}^d)$ containing $C_c^2(\mathbb{R}^d)$ and belonging to the domain of L. Let $\{U_t\}$, $t \ge 0$, be a family of bounded (uniformly for $t \in [0,T]$ for any $T > 0$) linear operators in $C_\infty(\mathbb{R}^d)$ such that U_t preserves D and $U_t f$ for any $f \in D$ is a classical solution of* (4.1) *(i.e. it holds for all $t \ge 0$, the derivative being taken in the sense of the sup-norm of $C(\mathbb{R}^d)$). Then $\{U_t\}$ is a strongly continuous semigroup of positive operators in C_∞ defining a unique classical solution $U_t \in C_\infty(\mathbb{R}^d)$ of* (4.1) *for any $f \in D$.*

Proof. Uniqueness and positivity follows from the previous theorem, if one takes into account that by Courrège's Theorem 3.6.7 the operator L naturally extends to constant functions preserving the PMP. On the other hand, uniqueness implies the semigroup property, because U_{t+s} and $U_t U_s$ solves the same Cauchy problem. Finally, to prove strong continuity, observe that if $\phi \in D$, then (as L and U_s commute by Theorem 1.9.1)

$$U_t\phi - \phi = \int_0^t LU_s\phi\, ds = \int_0^t U_s L\phi\, ds,$$

and

$$\|U_t\phi - \phi\| \le t \sup_{s\le t} \|U_s\|\|L\phi\|.$$

Since D is dense, arbitrary ϕ are dealt with by the standard approximation procedure. □

The next result gives a simple analytic criterion for conservativity. It also introduces a very important formula (4.4) for the solution of non-homogeneous equations that is called sometimes the *Duhamel principle*.

Theorem 4.1.3. (i) *Under the assumption of the previous theorem assume additionally that D is itself a Banach space under a certain norm $\|\phi\|_D \ge \|\phi\|$ such that L is a bounded operator $D \mapsto C_\infty(\mathbb{R}^d)$ and the operators U_t are bounded (uniformly for t on compact sets) as operators in D. Then the function*

$$u = U_t g + \int_0^t U_{t-s} f_s\, ds \tag{4.4}$$

is the unique solution to equation (4.3) *in $C_\infty(\mathbb{R}^d)$.*

(ii) *Let L be uniformly conservative in the sense that $\|L\phi_n\| \to 0$ as $n \to \infty$ for $\phi_n(x) = \phi(x/n)$, $n \in \mathbb{N}$, and any $\phi \in C_c^2(\mathbb{R}^d)$ that it equals one in a neighborhood of the origin and has values in $[0,1]$. Then U_t is a conservative Feller semigroup.*

Proof. (i) Uniqueness follows from Theorem 4.1.1. Since U_t are uniformly bounded in D it follows that the function u of form (4.4) is well defined and belongs to D for all t. Next, straightforward formal differentiation shows that u satisfies (4.3). To prove the existence of the derivative one writes

$$\frac{\partial g}{\partial t} = LU_f + \lim_{\delta \to 0} \frac{1}{\delta} \int_0^t (U_{t+\delta-s} - U_{t-s})\phi_s \, ds + \lim_{\delta \to 0} \frac{1}{\delta} \int_t^{t+\delta} U_{t+\delta-s}\phi_s \, ds.$$

The first limit here exists and equals $L \int_0^t U_{t-s}\phi_s \, ds$. On the other hand,

$$\lim_{\delta \to 0} \frac{1}{\delta} \int_t^{t+\delta} U_{t+\delta-s}\phi_s \, ds = \phi_t + \lim_{\delta \to 0} \frac{1}{\delta} \int_t^{t+\delta} (U_{t+\delta-s}\phi_s - \phi_t) \, ds,$$

and the second limit vanishes.

(ii) Clearly the function ϕ_n solves the problem

$$\frac{\partial u}{\partial t} = Lu - L\phi_n, \quad u(0,x) = \phi(x),$$

and hence by (i)

$$\phi_n(x) = U_t\phi_n + \int_0^t U_{t-s} L\phi_n \, ds.$$

As $n \to \infty$ the integral on the r.h.s. of this equation tends to zero in $C_\infty(\mathbb{R}^d)$ and $\phi_n(x)$ tends to one for each x. Hence

$$\lim_{n \to \infty} U_t\phi_n(x) = 1, \quad x \in \mathbb{R}^d,$$

implying that in the representation of type (3.38) for U_t (that exists due to the positivity of U_t), all measures $p_t(x, dy)$ are probability measures. □

We conclude this section with a couple of simple examples illustrating various versions of the Feller property.

Exercise 4.1.1. Let X_t be a deterministic process in $\mathbb{R}$ solving the ODE $\dot{x} = x^3$. Show that

(i) the solution to this equation with the initial condition $X(0) = x$ is

$$X_x(t) = \operatorname{sgn}(x) \left(\frac{1}{-2t + \frac{1}{x^2}} \right)^{1/2}, \quad |x| < \frac{1}{\sqrt{2t}},$$

(ii) the corresponding semigroup has the form

$$\Phi_t f(x) = \begin{cases} f(X_x(t)), & |x| < \frac{1}{\sqrt{2t}}, \\ 0, & |x| \geq \frac{1}{\sqrt{2t}} \end{cases} \tag{4.5}$$

in $C_\infty(\mathbb{R})$ and is Feller,

(iii) the corresponding measures from representation (3.38) are

$$p_t(x,dy) = \begin{cases} \delta(X_x(t) - y)), & |x| < \frac{1}{\sqrt{2t}}, \\ 0, & |x| \geq \frac{1}{\sqrt{2t}}, \end{cases} \tag{4.6}$$

implying that this Feller semigroup is not conservative, as its minimal extension takes the constant one to the indicator function of the interval $(-1/\sqrt{2t}, 1/\sqrt{2t})$. (It is instructive to see where the criterion of conservativity of Theorem 4.1.3 breaks down in this example.)

Exercise 4.1.2. Let X_t be a deterministic process in $\mathbb{R}$ solving the ODE $\dot{x} = -x^3$. Show that

(i) the solution to this equation with initial condition $X(0) = x$ is

$$X_x(t) = \operatorname{sgn}(x)\left(\frac{1}{2t + \frac{1}{x^2}}\right)^{1/2},$$

(ii) the corresponding semigroup is conservative and C-Feller, but not Feller, as it does not preserve the space $C_\infty(\mathbb{R}^d)$.

Exercise 4.1.3. Let X_t be a deterministic process in $\mathbb{R}_+$ solving the ODE $\dot{x} = -1$ and *killed at the boundary* $\{x = 0\}$, i.e. it vanishes at the boundary at the moment it reaches it. Show that the corresponding semigroup on $C_\infty(\mathbb{R}_+)$ (which is the space of continuous functions on $\mathbb{R}_+$ tending to zero both for $x \to \infty$ and $x \to 0$) is given by (3.38) with

$$p_t(x,dy) = \begin{cases} \delta(x - t - y), & x > t, \\ 0, & x \leq t \end{cases} \tag{4.7}$$

and is Feller, but not conservative, as its minimal extension to $C(\mathbb{R}_+)$ (that stands for killing at the boundary) takes the constant one to the indicator $\mathbf{1}_{[t,\infty)}$. On the other hand, if instead of a killed process, one defines the corresponding *stopped process* that sticks to the boundary $\{x = 0\}$ once it reaches it, the corresponding semigroup is given on $C_\infty(\bar{\mathbb{R}}_+)$ by (3.38) with

$$p_t(x,dy) = \begin{cases} \delta(x - t - y), & x > t, \\ \delta(y), & x \leq t. \end{cases} \tag{4.8}$$

This is a conservative Feller semigroup on $C_\infty(\bar{\mathbb{R}}_+)$ that is an extension (but not a minimal one) of the previously constructed semigroup of the killed process.

Exercise 4.1.4. This exercise aims to show that the stopped process from the previous one does not give a unique extension of a Feller semigroup on $C_\infty(\mathbb{R}_+)$ to

$C_\infty(\bar{\mathbb{R}}_+)$. Namely, consider a mixed killed and stopped process, where a particle moves according to the equation $\dot{x} = -1$ until it reaches the boundary, where it stays a θ-exponential random time and then vanishes. Show that such a process specifies a non-conservative Feller semigroup on $C_\infty(\bar{\mathbb{R}}_+)$ given by

$$\Phi_t f(x) = \begin{cases} f(x-t), & x > t, \\ f(0)e^{-\theta(t-x)}, & x \le t. \end{cases} \tag{4.9}$$

Exercise 4.1.5. Consider the deterministic process $X_t^x = e^{-t}x$ in $\mathbb{R}_+$. Clearly the corresponding process stopped at the origin coincides with $X_t^x = e^{-t}x$ restricted to $\mathbb{R}_+$ and also coincides with the reflected process $|X_t^x|$. Show that this process is Feller in $\mathbb{R}_+$, that is its semigroup is strongly continuous in $C_\infty(\mathbb{R}_+)$ (the space of continuous functions on $\mathbb{R}_+$ vanishing at the origin and at infinity), and the domain of the generator consists of functions f from $C_\infty(\mathbb{R}_+)$ such that $f'(x)$ exists and is continuous for $x > 0$ and the function $xf'(x)$ vanishes at the origin and at infinity.

4.2 The Dirichlet problem for diffusion operators

In this section we describe the crucial link between stationary equations and stopped Markov processes for the example of diffusions. The "simplicity" used here is crucially linked with the continuity of the diffusion trajectories.

Assume a_{ij}, b_j are continuous bounded function s.t. the matrix (a_{ij}) is positive definite and the operator

$$Lf(x) = \sum_{j=1}^{d} b_j(x)\frac{\partial f}{\partial x_j} + \frac{1}{2}\sum_{j,k=1}^{d} a_{jk}(x)\frac{\partial^2 f}{\partial x_j \partial x_k}$$

generates a Feller diffusion X_t. Assume Ω is an open subset of $\mathbb{R}^d$ with the boundary $\partial\Omega$ and closure $\bar{\Omega}$. The *Dirichlet problem for L in Ω* consists of finding a $u \in C_b(\bar{\Omega}) \cap C_b^2(\Omega)$ s.t.

$$Lu(x) = f(x), x \in \Omega, \quad u|_{\partial\Omega} = \psi \tag{4.10}$$

for given $f \in C_b(\Omega)$, $\psi \in C_b(\partial\Omega)$. A fundamental link between probability and PDE is given by the following

Theorem 4.2.1. *Let Ω be bounded and $\mathbf{E}_x\tau_\Omega < \infty$ for all $x \in \Omega$, where $\tau_\Omega = \inf\{t \ge 0 : X_t \in \partial\Omega\}$ (e.g. if X is a BM, see Section* 2.7*), and let $u \in C_b(\bar{\Omega}) \cap C^2(\Omega)$ be a solution to* (4.10)*. Then*

$$u(x) = \mathbf{E}_x\left[\psi(X_{\tau_\Omega}) - \int_0^{\tau_\Omega} f(X_t)\,dt\right]. \tag{4.11}$$

In particular, such a solution u is unique.

Proof. (i) Assume first that u can be extended to the whole $\mathbb{R}^d$ as a function $u \in C_c(\mathbb{R}^d) \cap C_b^2(\mathbb{R}^d)$. Then $u \in D_L$, and applying the stopping time τ_Ω to Dynkin's martingale yields

$$\mathbf{E}\left[u(X_{\tau_\Omega}) - u(x) - \int_0^{\tau_\Omega} Lu(X_t)\,dt\right] = 0, \tag{4.12}$$

implying (4.11).

(ii) In the general case choose an expanding sequence of domains $\Omega_n \subset \Omega$ with smooth boundaries tending to Ω as $n \to \infty$. Clearly the function u solves the problem

$$Lu_n(x) = f(x), x \in \Omega_n, \quad u_n|_{\partial\Omega_n} = u.$$

As it can be extended to $\mathbb{R}^d$ as is required in (i), it is a unique solution given by

$$u(x) = u_n(x) = \mathbf{E}_x\left[u(X_{\tau_{\Omega n}}) - \int_0^{\tau_{\Omega n}} f(X_t)\,dt\right], \quad x \in \Omega_n.$$

Taking the limit as $n \to \infty$ yields (4.11), because $\tau_{\Omega_n} \to \tau_\Omega$ a.s., as $n \to \infty$, and hence $X_{\tau_{\Omega n}} \to X_{\tau_\Omega}$ by the continuity of diffusion paths (proved in Theorem 3.11.6). □

For example, we can now extend the results of Section 2.7 on the exit times of BM to arbitrary one-dimensional diffusions. Namely, let us take $\Omega = (\alpha, \beta) \subset \mathbb{R}$ and

$$L = \frac{1}{2}a(x)\frac{d^2}{dx^2} + b(x)\frac{d}{dx}$$

with $a, b \in C(\bar{\Omega})$, $a > 0$. Then $u(x) = \mathbf{P}_x(X_{\tau_\Omega} = \beta)$ is the probability that X_t starting at a point $x \in (\alpha, \beta)$ reaches β before α. By (4.11), this u represents a solution to the problem

$$\frac{1}{2}a(x)\frac{d^2u(x)}{dx^2} + b(x)\frac{du(x)}{dx} = 0, \quad x \in (\alpha, \beta), \quad u(\alpha) = 0,\ u(\beta) = 1. \tag{4.13}$$

On the other hand, $u(x) = \mathbf{E}_x \tau_\Omega$ is the mean exit time from Ω that solves the problem

$$\frac{1}{2}a(x)\frac{d^2u(x)}{dx^2} + b(x)\frac{du(x)}{dx} = -1, \quad x \in (\alpha, \beta), \quad u(\alpha) = u(\beta) = 0. \tag{4.14}$$

Exercise 4.2.1. (i) Solve problem (4.13) analytically showing that

$$\mathbf{P}_x(X_{\tau_\Omega} = \beta) = \int_\alpha^x \exp\{g(y)\}\,dy\left(\int_\alpha^\beta \exp\{g(y)\}\,dy\right)^{-1}, \tag{4.15}$$

where $g(x) = -\int_\alpha^x (2b/a)(y)\,dy$. In particular, for a standard BM B_t starting at x this gives $\mathbf{P}_x(B_{\tau_\Omega} = \beta) = (x - \alpha)/(\beta - \alpha)$.

(ii) Solve (4.14) with $b = 0$ showing that in this case

$$\mathbf{E}_x \tau_\Omega = 2\frac{x-\alpha}{\beta-\alpha}\int_\alpha^\beta \frac{\beta-y}{a(y)}\,dy - 2\int_\alpha^x \frac{x-y}{a(y)}\,dy. \tag{4.16}$$

In particular, in case of BM this turns to $(x-\alpha)(\beta-x)$. Hint: show first that the solution to the Cauchy problem

$$\frac{1}{2}a(x)u''(x) = -1, \quad u(\alpha) = 0$$

is given by the formula

$$u(x) = \omega(x-\alpha) - 2\int_\alpha^x (x-y)a^{-1}(y)\,dy$$

with a constant ω.

Exercise 4.2.2. Check that $\Delta\phi = h''(|x|) + \frac{d-1}{|x|}h'(|x|)$ for $\phi(x) = h(|x|)$. Deduce that if such ϕ is harmonic (i.e. satisfies the Laplace equation $\Delta\phi = 0$) in $\mathbb{R}^d$, then

$$h(r) = \begin{cases} A + Br^{-(d-2)}, & d > 2 \\ A + B\ln r, & d = 2 \end{cases} \tag{4.17}$$

for some constants A, B.

Exercise 4.2.3. Solve the equation $\Delta\phi = 0$ in the shell $S_{r,R} = \{x \in \mathbb{R}^d : r < |x| < R\}$ subject to the following boundary conditions: $\phi(x)$ equals 1 (respectively zero) on $|x| = R$ (resp. $|x| = r$). Hence compute the probability that standard Brownian motion started from a point $x \in S_{r,R}$ leaves the shell via the outer part of the boundary. Hint: choosing appropriate A, B from (4.17) one finds

$$\phi(x) = \begin{cases} \frac{|x|^{2-d} - r^{2-d}}{R^{2-d} - r^{2-d}}, & d > 2, \\ \frac{\ln|x| - \ln r}{\ln R - \ln r}, & d = 2. \end{cases} \tag{4.18}$$

This describes the required probability due to Theorem 4.2.1.

Exercise 4.2.4. Calculate the probability of the Brownian motion W_t ever hitting the ball B_r if started at a distance $a > r$ from the origin. Hint: Let T_R (resp. T_r) be the first time $\|W_t\| = R$ (resp. r). By letting $R \to \infty$ in (4.18)

$$\mathbf{P}_x(T_r < \infty) = \lim_{R\to\infty} \mathbf{P}_x(T_r < T_R) = \begin{cases} (r/a)^{d-2}, & d > 2, \\ 1, & d = 2. \end{cases} \tag{4.19}$$

Exercise 4.2.5. Use the Borel–Cantelli lemma and the previous exercise to deduce that for $d > 2$ and any starting point $x \neq 0$ there exists a.s. a positive $r > 0$ s.t. W_t^x starting at x never hits the ball B_r. Hint: For any $r < a$ let A_n be the event that W_t^x ever hits the ball $B_{r/2^n}$. Then $\sum P(A_n) < \infty$.

Exercise 4.2.6. Show that BM in dimension $d > 2$ is *transient*, i.e. that a.s. $\lim_{t\to\infty} \|W_t\| = \infty$. Hint: As W_t is a.s. unbounded (see Section 2.2), the event that W_t does not tend to infinity means that there exists a ball B_r s.t. infinitely many events A_n occur, where A_n means that the trajectory returns to B_r after being outside $B_{2^n r}$. This leads to a contradiction by the Borel–Cantelli lemma and (4.19).

The next result is a more abstract version of Exercise 4.2.6.

Theorem 4.2.2. *Let L be a generator of a Feller diffusion X_t. Given a domain $\Omega \subset \mathbb{R}^d$, assume that there exists a twice continuously differentiable function $f \geq 0$ in $\mathbb{R}^d \setminus \Omega$ s.t. $Lf(x) \leq 0$ and for some $a > 0$ and a point $x_0 \in \mathbb{R}^d \setminus \Omega$ one has*

$$f(x_0) < a < \inf\{f(x) : x \in \partial\Omega\}.$$

Then for the process X_t started at x_0 there is a positive probability of never hitting Ω. (this actually means that the diffusion X_t is transient).

Remark 31. General arguments show that the conclusion of the Theorem implies that the diffusion X_t is *transient*, that is $X_t \to \infty$ a.s., as $t \to \infty$, see e.g. Freidlin [117].

Proof. Let $N > \|x_0\|$, and let τ_Ω and τ_N denote the hitting times of Ω and the sphere $\|y\| = N$ respectively. Put $T_N = \min(\tau_N, \tau_\Omega)$. From Dynkin's martingale it follows that

$$\mathbf{E}_{x_0} f(X_{T_n}) \leq f(x_0) < a.$$

Hence

$$a > \inf\{f(x) : x \in \partial\Omega\}\mathbf{P}_{x_0}(\tau_\Omega < \tau_N) > a\mathbf{P}_{x_0}(\tau_\Omega < \tau_N).$$

Passing to the limit as $N \to \infty$ yields

$$a > a\mathbf{P}_{x_0}(\tau_\Omega < \infty),$$

implying $\mathbf{P}_{x_0}(\tau_\Omega < \infty) < 1$. □

This result is an example of the application of the method of Lyapunov functions that will be extensively developed in Chapter 6.

4.3 The stationary Feynman–Kac formula

Recall that the equation

$$\lambda g = Ag + f, \tag{4.20}$$

where A is the generator of a Feller semigroup Φ_t, $f \in C_\infty(\mathbb{R}^d)$, $\lambda > 0$, is solved uniquely by the formula

$$g(x) = R_\lambda f(x) = \mathbf{E}_x \int_0^\infty e^{-\lambda t} f(X_t)\, dt.$$

This suggests a guess that a solution to the more general equation

$$(\lambda + k)g = Ag + f, \tag{4.21}$$

where the additional letter k denotes a bounded continuous function could look like

$$g(x) = \mathbf{E}_x \int_0^\infty \exp\left\{-\lambda t - \int_0^t k(X_s)\, ds\right\} f(X_t)\, dt. \tag{4.22}$$

This is the *stationary Feynman–Kac formula* that we are going to discuss now. The fastest way of proving it (at least for diffusions) is by means of Itô's formula. We did not introduce this tool and hence choose a different method (following essentially [155]) by first rewriting it in terms of the resolvents (thus rewriting the differential equation (4.21) in an integral form).

Theorem 4.3.1. *Let X_t be a Feller process with semigroup Φ_t and generator A. Suppose $f \in C_\infty(\mathbb{R}^d)$, k is a continuous bounded non-negative function and $\lambda > 0$. Then $g \in D_A$ and satisfies* (4.21) *iff $g \in C_\infty(\mathbb{R}^d)$ and*

$$R_\lambda(kg) = R_\lambda f - g. \tag{4.23}$$

Proof. Applying R_λ to both sides of (4.21) and using $R_\lambda(\lambda - A)g = g$ yields (4.23). Conversely, subtracting the resolvent equations for f and kg

$$AR_\lambda f = \lambda R_\lambda f - f, \quad AR_\lambda(kg) = \lambda R_\lambda(kg) - kg, \tag{4.24}$$

and using (4.23) yields (4.21). □

Theorem 4.3.2. *Under the assumptions of Theorem* 4.3.1 *the function* (4.22) *yields a solution to* (4.23) *and hence to* (4.21).

Proof. Using the Markov property one writes

$$\begin{aligned} R_\lambda(kg) &= \mathbf{E}_x \int_0^\infty e^{-\lambda s} k(X_s) g(X_s)\, ds \\ &= \mathbf{E}_x \int_0^\infty e^{-\lambda s} k(X_s) \int_0^\infty \exp\left\{-\lambda t - \int_0^t k(X_{u+s})\, du\right\} f(X_{t+s})\, dt\, ds. \end{aligned}$$

Changing the variables of integration t, u to $\tilde{t} = s + t$ and $\tilde{u} = s + u$ and denoting them again by t and u respectively leads to

$$R_\lambda(kg) = \mathbf{E}_x \int_0^\infty e^{-\lambda t} f(X_t) \int_0^t k(X_s) \exp\left\{-\int_s^t k(X_u)\,du\right\} ds\,dt.$$

Since

$$\int_0^t k(X_s) \exp\left\{-\int_s^t k(X_u)\,du\right\} ds = 1 - \exp\left\{-\int_0^t k(X_s)\,ds\right\}$$

(Newton–Leibniz formula), $R_\lambda(kg)$ rewrites as

$$\mathbf{E}_x \int_0^\infty e^{-\lambda t} f(X_t)\left[1 - \exp\left\{-\int_0^t k(X_s)\,ds\right\}\right] ds\,dt = (R_\lambda f - g)(x),$$

as required. □

In many interesting situations the validity of formula (4.22) can be extended beyond the general conditions of Theorem 4.3.2. Let us consider one of these extensions for a one-dimensional BM.

Theorem 4.3.3. *Assume $k \geq 0$ and f are piecewise-continuous bounded functions on $\mathbb{R}$ with the finite sets of discontinuity being* Disc_k *and* Disc_f. *Then the (clearly bounded) function g given by* (4.22) *with X_t being a BM B_t is continuously differentiable, has a piece-wise continuous second derivative and satisfies*

$$(\lambda + k)g = \frac{1}{2}g'' + f \quad \text{outside } \mathrm{Disc}_k \cup \mathrm{Disc}_f. \tag{4.25}$$

Proof. The calculations in the proof of Theorem 4.3.2 remain valid for all bounded measurable f and k, showing that g satisfies (4.23). Moreover, for piecewise continuous f and k, one sees from dominated convergence that this g is continuous. Next, from formula (3.47) one finds that

$$R_\lambda f(x) = \frac{1}{\sqrt{2\lambda}}\left[\int_{-\infty}^x e^{\sqrt{2\lambda}(y-x)} f(y)\,dy + \int_x^\infty e^{\sqrt{2\lambda}(x-y)} f(y)\,dy\right].$$

Hence $R_\lambda f$ is continuously differentiable for any bounded measurable f with

$$(R_\lambda f)'(x) = \int_x^\infty e^{\sqrt{2\lambda}(x-y)} f(y)\,dy - \int_{-\infty}^x e^{\sqrt{2\lambda}(y-x)} f(y)\,dy.$$

This implies in turn that $(R_\lambda f)''$ is piecewise continuous for a piecewise continuous f and the resolvent equations (4.24) hold outside $\mathrm{Disc}_f \cup \mathrm{Disc}_k$. Hence one shows as in Theorem 4.3.2 that g satisfies (4.25), which by integration implies the continuity of g'. □

Exercise 4.3.1. Show that for $\alpha, \beta > 0$, $x \geq 0$ and a BM B_t

$$\mathbf{E}_x \int_0^\infty \exp\left\{ -\alpha t - \beta \int_0^t \mathbf{1}_{(0,\infty)}(B_s)\, ds \right\} dt$$
$$= \frac{1}{\alpha+\beta}\left[1 + \frac{\sqrt{\alpha+\beta}-\sqrt{\alpha}}{\sqrt{\alpha}} e^{-\sqrt{2(\alpha+\beta)}x} \right]. \quad (4.26)$$

Hint: by Theorem 4.3.3 the function $z(x)$ on the l.h.s. of (4.26) is a bounded solution of the equation

$$\begin{cases} \alpha z(x) = \frac{1}{2} z''(x) - \beta z(x) + 1, & x > 0 \\ \alpha z(x) = \frac{1}{2} z''(x) + 1, & x < 0 \end{cases} \quad (4.27)$$

with the boundary conditions

$$z(0_+) = z(0_-), \quad z'(0_+) = z'(0_-).$$

Bounded solutions to (4.27) have the form

$$z(x) = \begin{cases} A \exp\{-\sqrt{2(\alpha+\beta)}x\} + \frac{1}{\alpha+\beta}, & x > 0, \\ B \exp\{\sqrt{2\alpha}x\} + \frac{1}{\alpha}, & x < 0. \end{cases} \quad (4.28)$$

Theorem 4.3.4 (Arcsin law for the occupation time). *The law for the occupation time $O_t = \int_0^t \mathbf{1}_{(0,\infty)}(B_s)\, ds$ of $(0,\infty)$ by a standard BM B_t has the density*

$$\mathbf{P}(O_t \in dy) = \frac{dy}{\pi\sqrt{y(t-y)}}. \quad (4.29)$$

Proof. By the uniqueness of the Laplace transform it is enough to show that

$$\mathbf{E} e^{-\beta O_t} = \int_0^t e^{-\beta y} \frac{dy}{\pi\sqrt{y(t-y)}}. \quad (4.30)$$

But from (4.26)

$$\int_0^\infty e^{-\alpha t} \mathbf{E} e^{-\beta O_t}\, dt = z(0) = \frac{1}{\sqrt{\alpha(\alpha+\beta)}},$$

and on the other hand

$$\int_0^\infty e^{-\alpha t} \int_0^t e^{-\beta y} \frac{dy}{\pi\sqrt{y(t-y)}} dy dt = \frac{1}{\pi} \int_0^\infty \frac{e^{-(\alpha+\beta)y}}{\sqrt{y}} dy \int_0^\infty \frac{e^{-\alpha s}}{\sqrt{s}} ds$$
$$= \frac{1}{\sqrt{\alpha(\alpha+\beta)}},$$

which implies (4.30) again by uniqueness of the Laplace transform. □

Exercise 4.3.2. From formula (3.48) yielding the solution to equation $(\lambda - \Delta)g = f$, $\lambda > 0$, in $\mathbb{R}^3$, deduce that the solution to the Poisson equation $\Delta g = -f$ in $\mathbb{R}^3$ is given by formula

$$g = \frac{1}{2\pi} \int \frac{f(y)}{|x-y|} dy,$$

whenever f decreases quickly enough at infinity.

4.4 Diffusions with variable drift, Ornstein–Uhlenbeck processes

In order to solve probabilistically equations involving second-order differential operators, one has to know that these operators generate Markov (Feller) semigroups. Here, anticipating further development of SDE, we show how BM can be used to construct processes with generators of the form

$$Lf(x) = \frac{1}{2}\Delta f(x) + \left(b(x), \frac{\partial f}{\partial x}\right), \quad x \in \mathbb{R}^d. \tag{4.31}$$

Let b be a bounded Lipschitz continuous function, i.e. $|b(x) - b(y)| \le C|x-y|$ with a constant C. Let B_t be a $\mathcal{F}_t$-BM on a filtered probability space with a filtration $\mathcal{F}_t$ satisfying usual conditions. Then, for any x, the equation

$$X_t = x + \int_0^t b(X_s)\, ds + B_t \tag{4.32}$$

has a unique global continuous solution $X_t(x)$ depending continuously on x and t, which is proved by fixed-point arguments in literally the same way as for usual ODE. Clearly $X_t(x)$ is a $\mathcal{F}_t$-Markov process starting at x, because the behavior of the process after reaching any position x is uniquely specified by the corresponding solution to (4.32).

Theorem 4.4.1. *X_t is a Feller process with generator* (4.31).

Proof. Clearly $\Phi_t f(x) = \mathbf{E} f(X_t(x))$ is a semigroup of positive contractions on $C_b(\mathbb{R}^d)$. Let $f \in C_c^\infty(\mathbb{R}^d)$. Then

$$\Phi_t f(x) - f(x) = \mathbf{E}\frac{\partial f}{\partial x}(x)\left(B_t + \int_0^t b(X_s)\, ds\right)$$
$$+ \frac{1}{2}\mathbf{E}\left(\frac{\partial^2 f}{\partial x^2}(x)\left(B_t + \int_0^t b(X_s)\, ds\right), B_t + \int_0^t b(X_s)\, ds\right) + \cdots,$$

where dots denote the correcting term of the Maclaurin (or Taylor) series. As $\mathbf{E}|B_t^k| = O(t^{k/2})$, it follows that the r.h.s. of this expression is

$$\left(\frac{\partial f}{\partial x}(x), \mathbf{E}(B_t + tb(x))\right) + \frac{1}{2}\mathbf{E}\left(\frac{\partial^2 f}{\partial x^2}(x)B_t, B_t\right) + o(t), \quad t \to 0,$$

so that

$$\frac{1}{t}[\Phi_t f(x) - f(x)] \to Lf(x), \quad t \to 0.$$

Hence any $f \in C_c^\infty(\mathbb{R}^d)$ belongs to the domain of the generator L, and Lf is given by formula (4.31). As clearly $\Phi_t f \to f$ for any such f, $t \to 0$, it follows that the same holds for all $f \in C_\infty(\mathbb{R}^d)$, by density arguments. □

Exercise 4.4.1. Convince yourself that the assumption that b is bounded can be dispensed with (only Lipschitz continuity is essential).

Example 1. The solution to the *Langevin* equation

$$v_t = v - b\int_0^t v_s\,ds + B_t$$

with a given constant $b > 0$ defines a Feller process called the *Ornstein–Uhlenbeck (velocity) process* with generator

$$Lf(v) = \frac{1}{2}\Delta f(v) - b\left(v, \frac{\partial f}{\partial v}\right), \quad v \in \mathbb{R}^d. \tag{4.33}$$

This process was already introduced in Section 2.11. The pair $(v_t, x_t = x_0 + \int v_s\,ds)$ describes the evolution of a *Newtonian particle subject to white noise driving force and friction*, and is also called the *Ornstein–Uhlenbeck process*.

Example 2. The solution to the system

$$\begin{cases} \dot{x}_t &= y_t, \\ y_t &= -\int_0^t \frac{\partial V}{\partial x}(x_s)\,ds - b\int_0^t y_s\,ds + B_t \end{cases} \tag{4.34}$$

describes the evolution of a *Newtonian particle in a potential field V subject to friction and white noise driving force*.

Exercise 4.4.2. Assume that $b = 0$ and the potential V is bounded below, say $V \geq 1$ everywhere, and is increasing to ∞ as $|x| \to \infty$.

1. Write down the generator L of the pair process (x_t, y_t). Answer:

$$Lf(x, y) = \left(y, \frac{\partial f}{\partial x}\right) - \left(\frac{\partial V}{\partial x}, \frac{\partial f}{\partial y}\right) + \frac{1}{2}\Delta_y f,$$

where Δ_y denotes the Laplacian with respect to the variable y.

2. Check that $L(H^{-\alpha}) \leq 0$ for $0 < \alpha < (d/2)-1$, where $H(x, y) = V(x)+y^2/2$ is the energy function (Hamiltonian).
3. Applying Dynkin's formula with $f = H^{-\alpha}$ for the process starting at (x, y) with the stopping time

$$\tau_h = \inf\{t \geq 0 : H(x_t, y_t) = h\}$$

with $h < H(x, y)$, show that $\mathbf{E}_{x,y} f((x, y)(\tau_h)) < f(x, y)$, and consequently

$$\mathbf{P}_{x,y}(\tau_h < \infty) \leq (h/H(x, y))^\alpha.$$

4. Follow the same reasoning as in Exercise 4.2.6 to establish that the process (x_t, y_t) is transient in dimension $d \geq 3$ (this result is remarkable, as it holds for all smooth enough V).

4.5 Stochastic integrals and SDE based on Lévy processes

Stochastic integrals belong to the major tools of stochastic analysis. The modern theory of integration with respect to general semimartingales is technically rather involved. We shall sketch here a more elementary theory for Lévy integrators that would form an appropriate platform for nonlinear extensions developed further.

As we shall see, solving SDEs yields a remarkable probabilistic method for the construction of Markov processes, both linear and nonlinear.

We shall start with the simplest stochastic integral of the form $\int_0^t \alpha_s dY_s$, where Y_t is a Lévy process. Notice that if Y_t is a compound Poisson process, its paths are functions of finite variation (more precisely they are piecewise constant), and the above integral is defined in the usual Lebesgue or Riemann sense depending on the regularity of α. Therefore, since in the Lévy–Itô decomposition the term X_t^2 (see Theorem 3.3.1) with large jumps is a compound Poisson process, one would not lose much generality in discussing stochastic integrals by discarding this term. Thus let us concentrate on the centered (with zero mean) Lévy process Y_t in $\mathbb{R}^d$ adapted to a filtration $\mathcal{F}_t$ satisfying the usual conditions on a given probability space $(\Omega, \mathcal{F}, \mathbf{P})$ and having generator

$$Lf(x) = \frac{1}{2}(G\nabla, \nabla)f(x) + \int [f(x + y) - f(x) - (y, \nabla)f(x)]\nu(dy), \quad (4.35)$$

where the Lévy measure ν has a finite second moment $\int |y|^2\nu(dy) < \infty$ (for instance, this is always the case for ν with bounded support) and G is a nonnegative $d \times d$-matrix. In particular, if G is the unit matrix and ν vanishes, Y is a BM and our stochastic integral would become the standard Itô integral with respect to the Wiener process.

Exercise 4.5.1. Show that under the above assumptions

$$\mathbf{E}(Y_t)^2 = t\left[\operatorname{tr} G + \int |y|^2 \nu(dy)\right], \tag{4.36}$$

$$\mathbf{E}(Y_t^i Y_t^j) = t\left[G_{ij} + \int y^i y^j \nu(dy)\right]. \tag{4.37}$$

Hint: use the characteristic function.

In defining the integral it is convenient to start with piecewise constant integrands α. To simplify the notation we shall assume that they are constant on intervals with binary rational bounds, i.e. have the form

$$\alpha_t = \sum_{j=0}^{[t/\tau_k]} \alpha^j \mathbf{1}_{(j\tau_k,(j+1)\tau_k]}, \tag{4.38}$$

where $\tau_k = 2^{-k}$ and α^j are $\mathcal{F}_{j\tau_k}$-measurable and square-integrable $\mathbb{R}^d$-valued random variables. The stochastic integral for such α is defined by

$$\int_0^t \alpha_s dY_s = \sum_{j=0}^{[t/\tau_k]} \alpha^j (Y_{\min(t,(j+1)\tau_k)} - Y_{j\tau_k}). \tag{4.39}$$

As Y_t has zero mean and is square-integrable, this process is clearly a right-continuous $\mathbb{R}^d$-valued $\mathcal{F}_t$-martingale. Moreover,

$$\mathbf{E}\left(\int_0^t \alpha_s dY_s\right)^2 = \mathbf{E}\left(\sum_{j=0}^{[t/\tau_k]} (\alpha^j)^2 (Y_{\min(t,(j+1)\tau_k)} - Y_{j\tau_k})^2\right),$$

since

$$\begin{aligned}
&\mathbf{E}[\alpha^i \alpha^j (Y_{(i+1)\tau_k} - Y_{i\tau_k})(Y_{(j+1)\tau_k} - Y_{j\tau_k})] \\
&\quad = \mathbf{E}\mathbf{E}(\alpha^i \alpha^j (Y_{(i+1)\tau_k} - Y_{i\tau_k})(Y_{(j+1)\tau_k} - Y_{j\tau_k}) \mid \mathcal{F}_{j\tau_k}) \\
&\quad = \mathbf{E}[\mathbf{E}(\alpha^i \alpha^j (Y_{(i+1)\tau_k} - Y_{i\tau_k}) \mid \mathcal{F}_{j\tau_k})\mathbf{E}(Y_{(j+1)\tau_k} - Y_{j\tau_k})] = 0
\end{aligned}$$

for $i < j$. Consequently, again by conditioning and (4.37),

$$\mathbf{E}\left(\int_0^t \alpha_s dY_s\right)^2 = \mathbf{E}\int_0^t (\alpha_s, \tilde{G}\alpha_s)\, ds, \tag{4.40}$$

where $\tilde{G}_{ij} = G_{ij} + \int y^i y^j \nu(dy)$.

Hence the mapping from α_t to the integrals $\int_0^t \alpha_s \, dY_s$ is a bounded (in the L_2 sense) linear mapping from simple left-continuous square-integrable $\mathcal{F}_t$-adapted processes to the right-continuous square-integrable martingales.

We can now define the *stochastic integral driven by Lévy noise* $\int_0^t \alpha_s dY_s$ for any left-continuous adapted processes α, which is square-integrable in the sense that $\mathbf{E}\int_0^t \alpha_s^2\, ds < \infty$, as the L_2-limit of the corresponding integrals over the simple approximations of α, since all these processes α can be approximated in L^2 (on each bounded interval $[0, t]$) by simple left-continuous processes.

Remark 32. To prove the last claim, observe that any such process can be approximated in L^2 by bounded left-continuous adapted processes. And any bounded left-continuous adapted processes α_t is approximated in L^2 and a.s. by its natural left-continuous binary approximations

$$\alpha_t^k = \sum_{j=0}^{[t/\tau_k]} \alpha_{j\tau_k} \mathbf{1}_{(j\tau_k,(j+1)\tau_k]}.$$

Thus for any left-continuous adapted square-integrable process α the stochastic integral $\int_0^t \alpha_s dY_s$ is well defined and represents a square-integrable right continuous martingale satisfying (4.40). Of course one can naturally extend the integral to vector-valued integrands.

The next object of our analysis is the simplest *stochastic differential equation (SDE) driven by Lévy noise*:

$$X_t = x + \int_0^t a(X_{s-})dY_s + \int_0^t b(X_{s-})\, ds. \tag{4.41}$$

Proposition 4.5.1. *Let Y_t satisfy the above conditions and let b, a be bounded Lipschitz continuous functions from $\mathbb{R}^n$ to $\mathbb{R}^n$ and to $n \times d$-matrices respectively, with Lipschitz constant κ. Finally, let x be a $\mathcal{F}_0$-measurable $\mathbb{R}^n$-valued random variable independent of Y_s. Then there exists a unique right-continuous $\mathcal{F}_t$-adapted square-integrable process X_t on $(\Omega, \mathcal{F}, \mathbf{P})$ satisfying* (4.41).

Proof. This is a consequence of the contraction principle and is omitted being a particular case of the next proposition. □

Next we consider an extension of the above integral, where the integration is carried over the random Poisson measure underlying the Lévy process.

Let Y_t be the same Lévy process as above and $N(dsdx)$ be the corresponding Poisson measure with the (unbounded) intensity $dt \times \nu$ introduced in Corollary 11. Let $\tilde{N}(dsdx)$ be the corresponding compensated measure defined as

$$\tilde{N}((s,t], A) = N((s,t], A) - \mathbf{E}N((s,t], A) = N((s,t], A) - (t-s)\int_A \nu(dy)$$

for the Borel sets A bounded below. We now aim to define the stochastic integral

$$\int_0^t \int f(s,z)\tilde{N}(dsdz)$$

for a random function f extending the integral from Proposition 3.3.1 to the case of random integrands. Notice directly that due to the assumption of the finiteness of the second moment of ν the restriction to the domain $\{|x| \le 1\}$ is not needed.

The approach to the definition of the integral will be as above. Namely, we shall start with simple functions f which will be defined as random process of the form

$$f(s,x) = \sum_{j=0}^{[t/\tau_k]} \alpha^j \mathbf{1}_{(j\tau_k,(j+1)\tau_k]}(s)\phi_j(x), \tag{4.42}$$

where α_j are bounded $\mathcal{F}_{j\tau_k}$-measurable random variables and ϕ_j are Borel functions on $\mathbb{R}^d$ such that $\phi_j(x)/|x|$ are bounded. For these processes the integral is defined by the formula

$$\int_0^t \int f(s,x)\tilde{N}(dsdx) = \sum_{j=0}^{[t/\tau_k]} \alpha^j \int_{j\tau_k}^{\min(t,(j+1)\tau_k)} \int \phi_j(x)\tilde{N}(dsdx), \tag{4.43}$$

where the last integral is given by Corollary 3.3.1. Acting as above one concludes that

$$\mathbf{E}\left(\int_0^t \int f(s,x)\tilde{N}(dsdx)\right)^2 = \int_0^t ds \int \mathbf{E} f^2(s,x)\nu(dx). \tag{4.44}$$

This again allows for extensions. Namely, first one extends the integral to processes of the form

$$f(s,x) = \sum_{j=0}^{[t/\tau_k]} \phi^j(x)\mathbf{1}_{(j\tau_k,(j+1)\tau_k]}(s), \tag{4.45}$$

where $\phi^j(x)$ are $\mathcal{F}_{j\tau_k} \otimes \mathcal{B}(\mathbb{R}^d)$-measurable random variables on $\Omega \times \mathbb{R}^d$ such that $\phi^j(x)/|x|$ is bounded, using the fact that bounded L^2-functions of two variables can be approximated in L^2 by linear combinations of bounded products of functions of one variable. One then extends the integral to all left-continuous $\mathcal{F}_t \otimes \mathcal{B}(\mathbb{R}^d)$-adapted processes on $\Omega \times \mathbb{R}^d$ such that the r.h.s. of (4.44) is finite.

Remark 33. It is possible to extend the definition of stochastic integral to left-continuous processes $f(s,x)$ such that

$$\int_0^t ds \int f^2(s,x)\nu(dx) < \infty$$

a.s. To this end, one approximates such processes by processes f_n with a finite r.h.s. of (4.44) in the sense that

$$\int_0^t ds \int |f_n - f|^2(s,x)\nu(dx) \to 0, \quad n \to \infty,$$

a.s. Then one defines the corresponding stochastic integral of f as the limit in probability of the approximating integrals of f_n, see details e.g. in [20]. We shall not use this extension.

To connect with the previously defined integral $\int \alpha_s dY_s$ it is instructive to observe that

$$\int_0^t \alpha_s dY_s = \int_0^t \alpha_s dB_s^G + \int_0^t \int \alpha_s x \tilde{N}(dsdx),$$

where B^G is the BM with covariance matrix G.

Generalizing equation (4.41), one can analyze now SDE of the form

$$X_t = x + \int_0^t \sigma(X_{s-})dB_s + \int_0^t b(X_{s-})ds + \int_0^t \int F(X_{s-}, z)\tilde{N}(dsdz), \quad (4.46)$$

where B_t is standard Brownian motion, F is a measurable mapping $\mathbb{R}^n \times \mathbb{R}^d \mapsto \mathbb{R}^n$ and σ maps $\mathbb{R}^n$ to $n \times d$-matrices.

Theorem 4.5.1. *Let Y_t satisfies the above conditions, that is its generator has form* (4.35) *and ν has finite second moment. Moreover, let*

$$|b(y_1) - b(y_2)|^2 + \|\sigma(y_1) - \sigma(y_2)\|^2 \\ + \int |F(y_1, w) - F(y_2, w)|^2 \nu(dw) \le \kappa^2 |y_1 - y_2|^2 \quad (4.47)$$

and

$$|b(y)|^2 + \|\sigma(y)\|^2 + \int |F(y, w)|^2 \nu(dw) \le \kappa^2 |y|^2$$

with some constant κ. Finally, let x be an $\mathcal{F}_0$-measurable random variable independent of Y_s. Then there exists a unique càdlàg $\mathcal{F}_t$-adapted square-integrable process X_t on $(\Omega, \mathcal{F}, \mathbf{P})$ satisfying (4.46).

Proof. By (4.47), (4.44) and (4.40) applied to B one has

$$\left\| \int_0^t \left((\sigma(X_{s-}^1) - \sigma(X_{s-}^2))dB_s + \int (F(X_{s-}^1, z) - F(X_{s-}^2, z))\tilde{N}(dsdz) \right) \right. \\ \left. + \int_0^t (b(X_{s-}^1) - b(X_{s-}^2))\, ds \right\|^2 \le c\kappa^2 \mathbf{E} \int_0^t \|X_{s-}^1 - X_{s-}^2\|^2\, ds,$$

implying for the mapping $X \mapsto \Phi(X)$ ($\Phi(X)$ being the r.h.s. of equation (4.46)) the estimate

$$\mathbf{E} \int_0^t \|\Phi(X_{s-}^1) - \Phi(X_{s-}^2)\|^2\, ds \le ct\kappa^2 \mathbf{E} \int_0^t \|X_{s-}^1 - X_{s-}^2\|^2\, ds.$$

This means that for $t < 1/c\kappa^2$ this mapping is a contraction on the space of càdlàg adapted processes distributed like x at $t=0$, taken with the L^2-norm $(\mathbf{E}\int_0^t |X_s|^2 ds)^{1/2}$. This implies well-posedness for this range of time. As usual in the theory of ODE, for large times the result is obtained by iterating this procedure. □

Adding terms with compound Poisson integrators to equations (4.46) does not involve much trouble. Notice that for a bounded Poisson measure $N(dsdx)$ the integral of a left-continuous process $f(s,x)$ can be defined by the same formula (3.10) as for a deterministic f (alternatively, an extended definition from Remark 33 can be used).

Proposition 4.5.2. *Let Y_t be a general Lévy process, initial condition x and functions $b(x), \sigma(x)$ and $F(y,z)\mathbf{1}_{|z|\leq 1}$ satisfy the assumptions of Proposition* 4.5.1, *and let $G(y,z)$ be Lipschitz continuous with respect to the first argument. Then the equation*

$$X_t = x + \int_0^t \sigma(X_{s-})dB_s + \int_0^t b(X_{s-})ds + \int_0^t \int_{\{|z|\leq 1\}} F(X_{s-},z)\tilde{N}(dsdz) + \int_0^t \int_{\{|z|>1\}} G(X_{s-},z)N(dsdz) \tag{4.48}$$

has a unique càdlàg adapted solution.

Proof. The simplest way is to observe (as in e.g. [20]) that this solution can be constructed via the *interlacing procedure*. Namely, let τ_n be the (random) jump times of the compound Poisson process $P_t = \int_{\{|z|>1\}} zN(t,dz)$ (by ΔP we shall denote its jumps), and let Z_t denote the solution to equation (4.46). By definition of compound Poisson integrals, the solution to (4.48) is given by the following 'explicit' recursive formulas

$$\begin{aligned} Y_t &= Z_t, \quad 0 \leq t < \tau_1, \\ Y_{\tau_1} &= Z_{\tau_1 -} + G(Z_{\tau_1 -}, \Delta P(\tau_1)), \\ Y_t &= Y_{\tau_1} + Z_t^1 - Z_{\tau_1}^1, \quad \tau_1 < t < \tau_2, \\ Y_{\tau_2} &= Z_{\tau_2 -} + G(Z_{\tau_2 -}, \Delta P(\tau_2)), \end{aligned} \tag{4.49}$$

and so on, where Z^1 is the unique solution to (4.46) with $Z_0^1 = Y_{\tau_1}$ instead of x. □

4.6 Markov property and regularity of solutions

Theorem 4.6.1. *Let the assumptions of Theorem* 4.5.1 *hold. Then*

(i) *for any x, y, t*

$$\mathbf{E}|X_t^x - X_t^y|^2 \leq e^{\kappa t}|x-y|^2; \tag{4.50}$$

(ii) *if f is Lipschitz continuous with a constant ω, then the function $\mathbf{E}f(X_t^x)$ is also Lipschitz continuous with the Lipschitz constant $\omega e^{\kappa t}$;*

(iii) *X_t is a C-Feller Markov process and for any $f \in C_c^2(\mathbb{R}^d)$*

$$\mathbf{E}\left[f(X_t^x) - \int_0^t Lf(X_s^x)\,ds\right] - f(x) = 0, \tag{4.51}$$

where

$$\begin{aligned} Lf(x) = {} & \frac{1}{2}((\sigma(x)\sigma(x)^T)\nabla, \nabla) f(x) + (b(x), \nabla f(x)) \\ & + \int [f(x + F(x,y)) - f(x) - (F(x,y), \nabla f(x))]\nu(dy); \end{aligned} \tag{4.52}$$

(iv) *if the coefficients are bounded, i.e.*

$$|b(y)|^2 + \|\sigma(y)\|^2 + \int |F(y,w)|^2 \nu(dw) \le \kappa^2$$

with some constant κ, then

$$\mathbf{E}|X_t^x - x|^2 \le \kappa^2 t \tag{4.53}$$

and the process X_t^x is Feller with the generator L having a domain containing $C_c^2(\mathbb{R}^d)$.

Proof. (i) From (4.46) and (4.47) it follows that

$$\mathbf{E}|X_t^x - X_t^y|^2 \le |x - y|^2 + \kappa \int_0^t \mathbf{E}|X_s^x - X_s^y|^2\,ds,$$

implying statement (i) by Gronwall's lemma.

(ii) This is a direct consequence of (i).

(iii) As by uniqueness, X_{t+s} is the solution of the corresponding SDE started at X_t at time t, it follows that $\mathbf{E}(f(X_{t+s})|\mathcal{F}_t) = \mathbf{E}(f(X_{t+s})|X_t)$, giving the Markov property. Then statement (ii) implies the C-Feller property. Next, by definition of stochastic integrals, the process X_t solving (4.46) is the limit $X_t = \lim_{n\to\infty} X_t(n)$ (it is sufficient to have a limit in L_2-sense for any t locally uniformly in t), where

$$\begin{aligned} X_t^x(k) = {} & x + \int_0^t \sigma(X_{[s2^k]2^{-k}})dB_s + \int_0^t b(X_{[s2^k]2^{-k}})ds \\ & + \int_0^t \int F(X_{[s2^k]2^{-k}}, z)\tilde{N}(dsdz) \end{aligned} \tag{4.54}$$

(we do not need to write $X_{[s2^k]2^{-k}-}$, because X_t is stochastically continuous meaning that for a fixed countable set of binary rational it is a.s. continuous on this set). But the r.h.s. of (4.52) is the sum of simple integrals that by Proposition 3.3.1 are Lévy processes. Hence (applying Dynkin's martingale), we can write

$$\mathbf{E}\Big[f(X_t^x(k)) - \int_{t_k}^t L_{X_{t_k}^x(k)} f(X_s^x(k))\,ds - f(X_{t_k}^x(k))\Big|\mathcal{F}_{[t2^k]2^{-k}}\Big] = 0,$$

$$\mathbf{E}\Big[f(X_{t_k}^x(k)) - \int_{t_k-2^{-k}}^{t_k} L_{X_{t_k-2^{-k}}^x}(k) f(X_s^x(k))\,ds - f(X_{t_k-2^{-k}}^x(k))\Big|\mathcal{F}_{t_k-2^{-k}}\Big] = 0,$$

$$\ldots,\quad \mathbf{E}\Big[f(X_{2^{-k}}^x(k)) - \int_0^{2^{-k}} L_x f(X_s^x(k))\,ds - f(x)\Big] = 0,$$

where $t_k = [t2^k]2^{-k}$ and

$$L_x f(y) = \frac{1}{2}((\sigma(x)\sigma(x)^T)\nabla,\nabla)f(y) + (b(x),\nabla f(y))$$
$$+ \int [f(y+F(x,z)) - f(y) - (F(x,z),\nabla f(y))]\nu(dz).$$

Summing up these equations yields

$$\mathbf{E}\Big[f(X_t^x(k)) - \int_0^t L_{X_{[s2^k]2^{-k}}^x}(k) f(X_s^x(k))\,ds - f(x)\Big] = 0,$$

which implies (4.51) by letting $k\to\infty$.

(iv) Estimate (4.53) follows from (4.46) and the assumed boundedness of the coefficients. This estimate clearly implies that the space $C_\infty(\mathbb{R}^d)$ is preserved by the Markov semigroup of the process X_t^x, i.e. it is a Feller process. It remains to identify the generator on the space $C_c^2(\mathbb{R}^d)$, which follows from (4.51). □

As we mentioned earlier, the possibility of representing a Markov process as a solution of an SDE of type (4.46) gives much more than is available for general Markov processes, as it allows one to compare trajectories started at different points (the corresponding processes are defined on the same filtered probability space), in particular, to obtain estimates of type (4.50). Even more, it allows one to analyze the regularity of the trajectories with respect to initial points, as differentiating equation (4.50) with respect to x yields the equation for the derivative $\partial X_t^x/\partial x$ of the same type. This leads naturally to the regularity of the corresponding Feller semigroup. For instance, one can obtain the following.

Recall that we denote by C_{Lip}^k (resp. C_∞^k) the subspace of functions from $C^k(\mathbb{R}^d)$ with a Lipschitz continuous derivative of order k (resp. with all derivatives up to order k vanishing at infinity).

Theorem 4.6.2. *Assume that the conditions of Theorem* 4.6.1 (iv) *hold.*

(i) *Assume that* $b, \sigma \in C^2_{\mathrm{Lip}}(\mathbb{R}^d)$,

$$\sup_z \int \left\| \frac{\partial^2}{\partial z^2} F(z, w) \right\|^2 \nu(dw) < \infty, \tag{4.55}$$

and

$$\sup_z \int \left\| \frac{\partial}{\partial z} F(z, w) \right\|^\beta \nu(dw) < \infty \tag{4.56}$$

holds with $\beta = 4$. *Then the solutions* X_t^x *of* (4.46) *are a.s. differentiable with respect to* x *and the spaces* C^1_{Lip} *and* $C^1_{\mathrm{Lip}} \cap C^1_\infty$ *are invariant under the semigroup* T_t.

(ii) *Assume further that* $b, \sigma \in C^3_{\mathrm{Lip}}(\mathbb{R}^d)$,

$$\sup_z \int \left\| \frac{\partial^3}{\partial z^3} F(z, w) \right\|^2 \nu(dw) < \infty, \tag{4.57}$$

and (4.56) *holds with* $\beta = 6$. *Then the solutions* X_t^x *of* (4.46) *are a.s. twice differentiable with respect to* x,

$$\sup_{s \le t} \mathbf{E} \left\| \frac{\partial^2 X_s^x}{\partial x^2} \right\|^2 \le c(t), \tag{4.58}$$

the spaces C^2_{Lip} *and* $C^2_{\mathrm{Lip}} \cap C^2_\infty$ *are invariant under the semigroup* T_t, *and the latter represents an invariant core for* T_t. *Moreover, in this case the Markov semigroup* T_t *and the corresponding process are uniquely defined by the generator* L.

Proof. We omit the proof referring to [196] for full detail; see also [302] for a slightly different presentation. □

Of course not any Lévy–Khintchine-type operator can be written in form (4.52) with regular enough F, which puts natural limitations on the method of analysis of Markov processes based on SDEs of type (4.46). In Section 5.8 we shall develop a theory of SDE driven by a nonlinear Lévy noise that can be applied to a more general class of Markov processes.

Notice that the integral part of (4.52) can be rewritten equivalently as

$$\int [f(x+z) - f(x) - (z, \nabla f(x))] \nu^{F_x}(dy),$$

where $\nu^{F_x}(dy)$ is the push forward of ν via the mapping $F_x : y \mapsto F(x, y)$. Hence, in order to represent a Lévy–Khintchine-type operator in form (4.52) one has to express the Lévy kernel as a push forward of a given Lévy measure with respect to a family of transformation. This can often be done in case of a common star shape of the measures $\nu(x, .)$, i.e. if they can be represented as

$$\nu(x, dy) = \nu(x, s, dr)\,\omega(ds), \quad y \in \mathbb{R}^d,\ r = |y| \in \mathbb{R}_+,\ s = y/r \in S^{d-1}, \tag{4.59}$$

with a certain measure ω on S^{d-1} and a family of measures $\nu(x, s, dr)$ on $\mathbb{R}_+$. In this case the above representation via pushing is reduced to a one-dimensional problem. Namely, the measure ν^F is the pushing forward of a measure ν on $\mathbb{R}_+$ by a mapping $F : \mathbb{R}_+ \mapsto \mathbb{R}_+$ whenever

$$\int f(F(r))\nu(dr) = \int f(u)\nu^F(du)$$

for a sufficiently rich class of test functions f, say for the indicators of intervals. Suppose we are looking for a family of monotone continuous bijections $F_{x,s} : \mathbb{R}_+ \mapsto \mathbb{R}_+$ such that $\nu^{F_{x,s}} = \nu(x, s, .)$. Choosing $f = \mathbf{1}_{[F(z),\infty)}$ as a test function in the above definition of pushing yields

$$G(x, s, F_{x,s}(z)) = \nu([z, \infty)) \tag{4.60}$$

for $G(x, s, z) = \nu(x, s, [z, \infty)) = \int_z^\infty \nu(x, s, dy)$. Clearly if all $\nu(x, s, .)$ and ν are unbounded (though bounded on any interval separated from the origin), have no atoms and do not vanish on any open interval, then this equation defines a unique continuous monotone bijection $F_{x,s} : \mathbb{R}_+ \mapsto \mathbb{R}_+$ with also continuous inverse. Hence we arrive at the following criterion.

Proposition 4.6.1. *Suppose the Lévy kernel $\nu(x, .)$ can be represented in the form* (4.59) *and ν is a Lévy measure on $\mathbb{R}_+$ such that all $\nu(x, s, .)$ and ν are unbounded, have no atoms and do not vanish on any open interval. Then the family $\nu(x, .)$ can be represented as the push forward of the measure $\omega(ds)\nu(dr)$ via the family of mappings $F_{x,s}(r)$, that is*

$$\int f(F_{x,s}(r))\nu(dr)\omega(ds) = \int f(u)\nu(x, s, du),$$

where $F_{x,s}(z)$ is the unique continuous solution $F_{x,s}(z)$ to (4.60)*. If $F_{x,s}(z)$ is Lipschitz continuous in x with a constant $\kappa_F(z, s)$ satisfying the condition*

$$\int_{\mathbb{R}_+} \int_{S^{d-1}} \kappa_F^2(r, s)\omega(ds)\nu(dr) < \infty, \tag{4.61}$$

then

$$\int_{\mathbb{R}_+} \int_{S^{d-1}} (F_{x,s}(r) - F_{y,s}(r))^2 \omega(ds)\nu(dr) \le c(x - y)^2. \tag{4.62}$$

Proof. It is clear that (4.61) implies (4.62). □

The following is the basic example.

Corollary 16. *Let*

$$\nu(x;dy) = a(x,s)r^{-(1+\alpha(x,s))}\,dr\,\omega(ds),$$
$$y \in \mathbb{R}^d,\ r = |y| \in \mathbb{R}_+,\ s = y/r \in S^{d-1}, \quad (4.63)$$

where a, α are $C^1(\mathbb{R}^d)$ functions of the variable x, depend continuously on s and take values in $[a_1, a_2]$ and $[\alpha_1, \alpha_2]$ respectively, where $0 < a_1 \le a_2$, $0 < \alpha_1 \le \alpha_2 < 2$. Then condition (4.62) holds for the family of measures $\mathbf{1}_{B_K}(y)\nu(x, dy)$.

Proof. Choose $\nu(z, K]) = 1/z - 1/K$. Since now

$$G(x,s,z) = \int_z^K a(x,s)r^{-(1+\alpha(x,s))}\,dr = \frac{a(x,s)}{\alpha(x,s)}(z^{-\alpha(x,s)} - K^{-\alpha(x,s)}),$$

it follows that the solution to (4.60) is given by

$$F_{x,s}(z) = \left[K^{-\alpha} + \frac{\alpha}{a}\left(\frac{1}{z} - \frac{1}{K}\right)\right]^{-1/\alpha}(x,s),$$

implying that $F(1) = 1$, $F_{x,s}(z)$ is of order $(az/\alpha)^{1/\alpha}$ for small z and $|\nabla_x F|$ is bounded by $O(1)z^{1/\alpha}\log z$. Hence condition (4.61) rewrites as the integrability around the origin of the function

$$z^{2(\alpha_2^{-1}-1)}\log^2 z,$$

which clearly holds. □

This leads to the construction of possibly *degenerate diffusions* combined with mixtures of possibly degenerate *stable-like process*. Namely, let

$$Lf(x) = \frac{1}{2}\mathrm{tr}(\sigma(x)\sigma^T(x)\nabla^2 f(x)) + (b(x), \nabla f(x)) \quad (4.64)$$
$$+ \int_P (dp) \int_0^K d|y| \int_{S^{d-1}} a_p(x,s)\frac{f(x+y) - f(x) - (y, \nabla f(x))}{|y|^{\alpha_p(x,s)+1}}d|y|\omega_p(ds),$$

where $s = y/|y|$, $K > 0$ and (P, dp) is a Borel space with a finite measure dp and ω_p are certain finite Borel measures on S^{d-1}.

Proposition 4.6.2. (i) *Let σ, b be Lipschitz continuous, a_p, α_p be $C^1(\mathbb{R}^d)$ functions of the variable x (uniformly in s, p) that depend continuously on s, p and take values in compact subintervals of $(0,\infty)$ and $(0,2)$ respectively. Then an extension of L defined on $C_c^2(\mathbb{R}^d)$ by (4.64) generates a Feller process.*

(ii) *Suppose additionally that for a $k > 2$ one has $\sigma, b \in C^k_{\text{Lip}}(\mathbb{R}^d)$, a, α are of class $C^k(\mathbb{R}^d)$ as functions of x uniformly in s. Then for each $= l = 2, \dots, k-1$, the space $C^l_{\text{Lip}} \cap C^l_\infty$ is an invariant domain for the Feller semigroup and this semigroup is uniquely defined.*

Proof. The result follows from Theorem 4.6.2 taking into account Corollary 16. □

4.7 Stochastic integrals and quadratic variation for square-integrable martingales

This section introduces more advanced stochastic calculus, which will be crucial for the theory of convergence of solutions of martingale problems, developed further. Here we will define the integrals $\int Z_s \, dX_s$ when X_t is a general square-integrable martingale and Z a left-continuous process. For this integral to be defined in the usual Riemann–Stieltjes sense, the process X_t should be of finite variation. But already Brownian motion does not enjoy this property. Hence some more sophisticated approach is required.

We start by introducing the basic discrete approximations to stochastic integrals and covariances motivating the crucial stochastic integration by parts formula (4.70), and then discuss the limiting procedures leading to the general construction of the integral.

We assume that a filtered probability space $(\Omega, \mathcal{F}, \mathbf{P})$ with filtration $\mathcal{F}_t$, satisfying the usual conditions, is chosen, where all our processes and random variables are defined. We shall denote by D and L the spaces of adapted càdlàg processes and, respectively, adapted left-continuous processes with right limits. The *left-continuous version* of a process X_t from D, denoted X_{t-}, obviously belongs to L. If a filtration is right-continuous (as we always assume), then, conversely, the *right-continuous version* of a process X_t from L, denoted X_{t+} belongs to D.

A *partition* of an interval $[0, T]$ is a finite sequence $\sigma = \{0 = t_0 < \cdots < t_k < t_{k+1} = T\}$. For $T = \infty$, by a partition of $[0, \infty)$ we mean a finite or countable sequence $0 = t_0 < t_1 < \dots$ such that any compact interval contains not more than finite number of points. For $t < T$ let us denote by σ_t the corresponding partition of $[0, t]$. By the *size* $|\sigma|$ of a partition we mean the maximum increment $\max_k (t_{k+1} - t_k)$. It is often handy to work with random partitions, where all t_k are assumed to be stopping times. Let us say that the size of a family of partition tends to zero locally if $|\sigma_t| \to 0$ a.s. for any finite $t \le T$.

A *simple left-continuous adapted process* (shortly SLA) on $[0, T]$ is a process of the form

$$Z_t = \sum_k Z^k \mathbf{1}_{(t_k, t_{k+1}]}(t), \tag{4.65}$$

where $\sigma = \{0 = t_0 < t_1 < \dots\}$ is a partition of $[0, T]$ (possibly random) and Z^k are $\mathcal{F}_{t_k}$-measurable random variables. For another random process X_t it is natural to define the integral of Z_t with respect to X_t as

$$\int_0^t Z_s dX_s = \sum_k Z^k (X_{t_{k+1}\wedge t} - X_{t_k}) \mathbf{1}_{t_k < t}. \tag{4.66}$$

If X_t is a càdlàg martingale and all Z^k in (4.65) are integrable, then the integral (4.66) is also a càdlàg martingale. In fact, the càdlàg property is obvious and the martingale property follows from the equations

$$\mathbf{E}(Z^k (X_{t_{k+1}\wedge t} - X_{t_k}) | \mathcal{F}_{t_k}) = Z^k \mathbf{E}(X_{t_{k+1}\wedge t} - X_{t_k} | \mathcal{F}_{t_k}) = 0,$$

where the first equation is due to the $\mathcal{F}_{t_k}$-measurability of Z^k and the second to the martingale property of X_t.

For a process Z_t on $(\Omega, \mathcal{F}, \mathbf{P})$ and a partition σ (deterministic or random) we can define a natural *left-continuous approximation* of Z_t by

$$Z_t^\sigma = \sum_k Z_{t_k} \mathbf{1}_{(t_k, t_{k+1}]}(t). \tag{4.67}$$

If there exists a limit in probability of the simple integrals $\int_0^t Z_s^\sigma dX_s$, as the size of partitions σ tends to zero locally, it is natural to call this limit the *stochastic integral* of the left-continuous modification of Z_t with respect to X_t and to denote it $\int_0^t Z_{s-} dX_s$.

For two process Y and X and a partition σ, let us introduce the process

$$[Y, X]_t^\sigma = \sum_k (Y_{t_{k+1}\wedge t} - Y_{t_k})(X_{t_{k+1}\wedge t} - X_{t_k}) \mathbf{1}_{t_k < t},$$

which is càdlàg whenever both Y and X are càdlàg. If there exists a limit in probability of these processes, as the size of partitions tends to zero locally, it is called the *covariance* of Y and X and is denoted $[Y, X]_t$. The covariance $[X, X]_t$ is called the *quadratic variation* of X and is often denoted briefly by $[X]_t$.

The problem of the existence of the limits of $\int_0^t Z_s^\sigma dX_s$ and $[X, Y]_t^\sigma$ are closely connected, due to the following crucial identity obtained by re-arrangement of sums:

$$\begin{aligned} (XY)_t - (XY)_0 &= \sum_{j=1}^n (X_{t_j \wedge t} - X_{t_{j-1}}) Y_{t_j \wedge t} + \sum_{j=1}^n (Y_{t_j \wedge t} - Y_{t_{j-1}}) X_{t_{j-1}} \qquad (4.68) \\ &= [X, Y]_t^\sigma + \sum_{j=1}^n (Y_{t_j \wedge t} - Y_{t_{j-1}}) X_{t_{j-1}} + \sum_{j=1}^n (X_{t_j \wedge t} - X_{t_{j-1}}) Y_{t_{j-1}}, \end{aligned}$$

where $t_{n-1} < t \le t_n$, or briefly

$$(XY)_t - (XY)_0 = [X, Y]_t^\sigma + \int_0^t Y_s^\sigma dX_s + \int_0^t X_s^\sigma dY_s. \tag{4.69}$$

If the limits defining integrals and covariance exist, this equation implies the following fundamental formula of stochastic *integration by parts*:

$$(XY)_t - (XY)_0 = [X,Y]_t + \int_0^t Y_{s-}\,dX_s + \int_0^t X_{s-}\,dY_s. \tag{4.70}$$

Moreover, (4.69) implies the following crucial fact. If X and Y are martingales, the last two terms in (4.69) are martingales, and consequently the expectation of $[X,Y]_t^\sigma$ does not depend on the partition:

$$\mathbf{E}[X,Y]_t^\sigma = \mathbf{E}(XY)_t - \mathbf{E}(XY)_0. \tag{4.71}$$

This preliminary discussion suggests that one can start rigorous analysis either by constructing stochastic integrals (i.e., proving the convergence of the approximations $\int_0^t Z_s^\sigma dX_s$) and then defining the covariance from (4.70), or, vice versa, by proving the existence of covariance and then using it to construct the integrals. In fact, both approaches are widely presented in the literature. The second approach is particularly appealing when the covariance $[X,X]_t$ can be explicitly calculated. In particular, this is the case for X a Lévy process, which we studied in the previous section (where $[X,X]_t$ was a deterministic process). Here, for integration over general square-integrable martingales, we show how one can exploit the first route, following mainly the approach from Kurtz and Protter [212].

Depending on Z and X the convergence of the approximations $\int_0^t Z_s^\sigma dX_s$ can be obtained in various topologies. The usual strategy is to gradually extend the class of integrands using continuity with different topologies. The strongest topology is the uniform one, for which the following holds.

Proposition 4.7.1. *If X_t is a càdlàg square-integrable martingale on $[0,T]$ with $X_0 = 0$ and Z is a bounded SLA process, then the integral* (4.66) *is also a càdlàg square-integrable martingale with*

$$\mathbf{E}\left|\int_0^t Z_s dX_s\right|^2 = \mathbf{E}\sum_{t_k<t}(Z^k)^2(X_{t_{k+1}\wedge t} - X_{t_k})^2 \le \sup_{s\in[0,t],\omega\in\Omega} |Z_s(\omega)|^2 \mathbf{E}|X_t|^2. \tag{4.72}$$

Proof. If Z is given by (4.65),

$$\left(\int_0^t Z_s dX_s\right)^2 = \sum_{t_k<t}(Z^k)^2(X_{t_{k+1}\wedge t} - X_{t_k})^2 + 2\sum_{t_i<t_j<t} Z^i Z^j (X_{t_{i+1}\wedge t} - X_{t_i})(X_{t_{j+1}\wedge t} - X_{t_j}).$$

The expectation of the second term vanishes by conditioning. And for the expectation of the first term we get the required estimate by (4.71). □

From (4.72) the following continuity property of the integral (4.66) follows. Let X_t be a càdlàg square-integrable martingale on $[0, T]$. If a sequence of SLA processes Z^n is fundamental (or Cauchy) in the uniform topology, i.e.

$$\sup_{t\in[0,T],\omega\in\Omega} |Z_t^n(\omega) - Z_t^m(\omega)| \to 0, \quad n,m \to \infty,$$

the sequence of the integrals $\int_0^t Z_s^n dX_s$ is fundamental in L_2:

$$\mathbf{E}\left|\int_0^t Z_s^n dX_s - \int_0^t Z_s^m dX_s\right|^2 \to 0, \quad m,n \to \infty.$$

This property allows one to extend the integral by continuity, with respect to the uniform topology, to the completion of the set of bounded SLA processes. The next statement shows that this completion contains all left-continuous bounded processes. Here the use of random partitions is crucial.

Proposition 4.7.2. *For any $t, \epsilon > 0$ and any bounded process Y from L there exists an SLA process Z^ϵ such that*

$$\sup_{s\le t,\omega\in\Omega} |Y_s - Z_s^\epsilon| \le \epsilon.$$

Proof. Let $Z = Y_+$ be the right modification of Y and let T_n^ϵ be defined recursively by

$$T_{n+1}^\epsilon = \inf\{t : t > T_n^\epsilon, \ |Z_t - Z_{T_n^\epsilon}| > \epsilon/2\}$$

starting with $T_0^\epsilon = 0$. They are stopping times, because Z is adapted. Moreover, since Z is càdlàg, $T_{n+1}^\epsilon - T_n^\epsilon > 0$, and the number of T_n^ϵ not exceeding t is finite for any t. The process

$$Z^\epsilon = Y_0 \mathbf{1}_{\{0\}} + \sum_n Z_{T_n^\epsilon} \mathbf{1}_{(T_n^\epsilon, T_{n+1}^\epsilon]}$$

enjoys all the properties required. □

Remark 34. Notice the necessity of combining cleverly the right and left modifications. If we defined

$$T_{n+1}^\epsilon = \inf\{t : t > T_n^\epsilon, \ |Y_t - Y_{T_n^\epsilon}| > \epsilon/2\},$$

we would have $T_{n+1}^\epsilon = T_n^\epsilon$ for $\epsilon < |Y_{T_n^\epsilon} - Z_{T_n^\epsilon}|$. It is also instructive to observe the following property of approximations (4.67): $((Z^\sigma)_+)^\sigma = Z^\sigma$, but usually $(Z^\sigma)^\sigma \neq Z^\sigma$.

Our aim is to extend the integral with respect to the martingale X_t to all Z from L. For this purpose the uniform topology is of course too strong. A convenient weaker version is the topology of *uniform convergence on compacts in probability* (shortly *ucp*): a sequence of processes Y^n converges to Y in ucp if

$$(Y^n - Y)_t^* = \sup_{s \le t} |Y_s^n - Y_s| \to 0, \quad n \to \infty,$$

in probability for any t.

Proposition 4.7.3. *For any Y from L there exists a sequence of bounded process Z^n from L converging to Y in ucp.*

Proof. This is of course done by stopping. Let $T_n = \inf\{t : |Y_t| > n\}$. Then each $Y_{t \wedge T_n}$ is bounded and

$$\mathbf{P}(\sup_{s \le t} |Y_{s \wedge T_n} - Y_s| > 0) = \mathbf{P}(\sup_{s \le t} |Y_s| > n),$$

which tends to zero as $n \to \infty$, because $\sup_{s \le t} |Y_s|$ is finite a.s. for any Y from L or D. □

Hence to extend the integral from bounded to arbitrary Z from L, we only need the continuity in ucp, which is settled in the following statement.

Proposition 4.7.4. *If X_t is a càdlàg square-integrable martingale on $[0, T]$, the mapping $Z \mapsto \int_0^t Z_s dX_s$ given by* (4.66) *is a continuous mapping from SLA to D if both spaces are equipped with the ucp topology.*

Proof. By (4.72), if Z_n converge to Z uniformly, then the corresponding integrals converge in L^2, and hence, by Doob's maximum inequality, also in ucp. Now, let Z_n converge to Z in ucp and let $\epsilon > 0$ is given. Denote $Y_t^n = \int_0^t Z_s^n dX_s$ and $Y_t = \int_0^t Z_s dX_s$. We have for any η:

$$\mathbf{P}((Y^n - Y)_t^* > \epsilon) \le \mathbf{P}((Y^n - Y)_t^* > \epsilon, (Z^n - Z)_t^* \le \eta) + \mathbf{P}((Z^n - Z)_t^* > \eta).$$

Choosing η small enough we can make the first term arbitrary small uniformly in n. Then by choosing n large we can make the second term small. □

Consequently, using continuity in ucp topology, we can define the *stochastic integral* $\int_0^t Z_s dX_s$ for any Z from L and any càdlàg square-integrable martingale X_t. This makes the stochastic integral a well-defined mapping from L to D, continuous in ucp.

However, the natural approximations (4.67) to a process Z from L need not to converge to Z in ucp. Nevertheless, as we shall show, the integrals still converge. The main step in the argument is the following.

Proposition 4.7.5. *Let S_t be a left-continuous step function with unit jump at $s_0 > 0$, i.e. $S = \mathbf{1}_{(s_0,\infty)}$, and let X_t be a right-continuous process. Then*

$$\max_{t\geq 0}\left|\int_0^t (S - S^\sigma)_s\, dX_s\right| \to 0 \tag{4.73}$$

a.s., as the size $|\sigma|$ of the partitions $\sigma = \{0 = t_0 < t_1 < \cdots\}$ tends to zero locally (the approximations S^σ are defined by (4.67)*).*

Proof. First observe that

$$\int_0^t S_s\, dX_s = \begin{cases} 0, & t \leq s_0, \\ X_t - X_{s_0}, & t \geq s_0 \end{cases}$$

(notice that for $t = s_0$ both expression coincide by right-continuity of X_t). As $S_s^\sigma = \mathbf{1}_{(t_j,\infty)}$, where $t_j = \min\{t_k : t_k > s_0\}$, we have

$$\int_0^t S_s^\sigma\, dX_s = \begin{cases} 0, & t \leq t_j, \\ X_t - X_{t_j}, & t \geq t_j. \end{cases}$$

Consequently,

$$\int_0^t (S_s - S_s^\sigma)\, dX_s = \begin{cases} 0, & t \leq s_0, \\ X_{t\wedge t_j} - X_{s_0}, & t \geq s_0, \end{cases}$$

and hence

$$\max_{t\geq 0}\left|\int_0^t (S - S^\sigma)_s\, dX_s\right| \leq \sup_{s\in[s_0,t_j]} |X_s - X_{s_0}|.$$

This clearly implies the required convergence. □

We can now obtain the main result of this section.

Proposition 4.7.6. *Let Z_s be a process from L and X_t be a càdlàg square-integrable martingale. Then the integrals of the approximations Z_s^σ (given by* (4.67)*) with respect to X_t converge in ucp to the integral of Z itself.*

Proof. Approximating Z by bounded SLA process and using Proposition 4.7.4 reduces the statement to the SLA processes. And for these processes the required convergence follow from linearity and Proposition 4.7.5. □

The next statement is a direct consequence of Proposition 4.7.6 and equation (4.69):

Proposition 4.7.7. *If X_t and Y_t are two square-integrable càdlàg martingales, the limit of $[X,Y]_t^\sigma$ exists in ucp, as the size of partitions tends to zero locally, and the limit $[X,Y]_t$ satisfies* (4.70). *In particular, $(XY)_t - [X,Y]_t$ is a martingale. For $Y = X$ this implies $\mathbf{E}X_t^2 = \mathbf{E}[X]_t$.*

An important particular case is that of continuous square-integrable martingales X_t. In this case, integrals (4.66) of SLA processes are continuous. And hence the integrals of all processes from the class L are continuous as well. Hence $[X]_t$ is continuous.

An easy, but important generalization of the above theory deals with processes X_t of the form

$$X_t = M_t + V_t, \tag{4.74}$$

where M_t is a square-integrable martingale and V_t is a process of finite variation. One easily sees that the integral (4.66) remains continuous in ucp as a mapping from L to D. This implies the existence of the covariance $[X,Y]_t$ for processes X, Y of this type and the validity of the integration by parts formula (4.70) for them. Because the covariance $[X,Y]$ is bilinear and positive definite in the sense that $[X] \geq 0$, we deduce, as for Hilbert space, the Cauchy inequality

$$[X,Y]_t \leq \sqrt{[X]_t}\sqrt{[Y]_t} \tag{4.75}$$

for processes X and Y of this class. Let us observe also for the future use that from the integration by parts formula (4.70) it follows that if

$$X_t^1 = M_t^1 + \int_0^t b_s^1 ds, \quad X_t^2 = M_t^2 + \int_0^t b_s^2 ds,$$

where M_t^1, M_t^2 are square-integrable martingales and b_s^1, b_s^2 are bounded càdlàg processes, then

$$(X^1X^2)_t - [X^1, X^2]_t - \int_0^t (X_s^1 b_s^2 + X_s^2 d_s^1) ds \tag{4.76}$$

is a martingale.

To move forward, we need the following technical result.

Proposition 4.7.8. *Let A_s be an integrable continuous process of finite variation on a finite interval $[0,t]$. For a partition $\sigma = \{0 = t_0 < \cdots < t_j < t_{j+1} = t\}$ of $[0,t]$ let*

$$A_t(\sigma) = \sum_j \mathbf{E}(A_{t_{j+1}} - A_{t_j} | \mathcal{F}_{t_j}). \tag{4.77}$$

Then $\mathbf{E}|A_t(\sigma) - A_t| \to 0$, as $|\sigma| \to 0$.

Proof. By localization, i.e. by introducing stopping times $\sigma_n = \{s : \mathrm{Var}(A_s) \geq n\}$ and working with $A_{t\wedge\sigma_n}$ the problem is reduced to the case of A_t with uniformly bounded variation. Assume this is the case. Denote $\delta_j = A_{t_{j+1}} - A_{t_j}$. We have by sequential conditioning

$$\mathbf{E}|A_t(\sigma) - A_t|^2 = \mathbf{E}\Big[\sum_j (\delta_j - \mathbf{E}(\delta_j|\mathcal{F}_{t_j}))\Big]^2 = \mathbf{E}\sum_j [\delta_j - \mathbf{E}(\delta_j|\mathcal{F}_{t_j})]^2$$
$$= \mathbf{E}\sum_j (\delta_j^2 - [\mathbf{E}(\delta_j|\mathcal{F}_{t_j})]^2) \leq \mathbf{E}\sum_j \delta_j^2 \leq \mathbf{E}|\Delta|\,\mathrm{Var}(A_t),$$

which tends to zero by dominated convergence, as $|\sigma| \to 0$ a.s. □

Proposition 4.7.9. *If A_t is a continuous submartingale of finite variation, then A_t is an increasing process. In particular, if X_t is a martingale of finite variation, then it is constant.*

Proof. For given $s < t$ consider the partitions σ of $[0, t]$ that contain s as a partition point. By the submartingale property, all terms in (4.77) are nonnegative. Hence $A_t(\sigma) \geq A_s(\sigma)$, implying the required statement by passing to the limit as $|\sigma| \to 0$. □

Proposition 4.7.10. *If a process X_t is of form* (4.74) *and Y_t is a process of finite variation, then*

$$[X, Y]_t = \sum_{s\leq t} \Delta X_t \Delta Y_t, \tag{4.78}$$

where ΔZ for a càdlàg process Z denotes the corresponding process of jumps: $\Delta Z_s = Z_s - Z_{s-}$.

Proof. Recall that for any function of finite variation Y_t it can be decomposed into the sum $Y_t = Y_t^c + Y_t^d$ of its continuous and jump (discrete) parts. This is called the *Lebesgue decomposition*.

Observe now that if Z is of finite variation and W_t is continuous, then $[Z, W]_t = 0$, because

$$[Z, W]_t^\sigma \leq w(Z, t, |\sigma|)\,\mathrm{Var}(W_t),$$

where $w(Z, t, h)$ denotes, as usual, the modulus of continuity of Z. Hence, if X is also of finite variation, we deduce (4.78) by applying the Lebesgue decomposition to both X and Y. Hence it remains to analyze the case when X is a square-integrable martingale. As $[Y^c]_t = 0$, we obtain $[X, Y^c]_t = 0$ by the Cauchy inequality (4.75). Thus it remains to show that

$$[X, Y^d]_t = \sum_{s\leq t} \Delta X_t \Delta Y_t.$$

Let $Y^d = Y^{d,\epsilon} + Y^{d,\bar{\epsilon}}$, where $Y^{d,\bar{\epsilon}}$ is the sum of jumps of size larger than ϵ. Since $Y^{d,\bar{\epsilon}}$ contains only finite number of jumps, it easy to see that

$$[X, Y^{d,\bar{\epsilon}}]_t = \sum_{s \le t} \Delta X_t \Delta Y^{d,\bar{\epsilon}}.$$

To complete the proof it remains to observe that by the Cauchy inequality

$$[X, Y^{d,\epsilon}]_t \le \epsilon \sqrt{[X]_t} \operatorname{Var}(Y_t^{d,\epsilon}),$$

which approaches zero as $\epsilon \to 0$. □

As we mentioned, if X_t and Y_t are continuous, the covariance $[X, Y]$ is also continuous. In other cases, a further decomposition turns out to be useful. If there exists a continuous adapted process of finite variation $\langle X, Y\rangle_t$ such that $\langle X, Y\rangle_0 = 0$ and $[X, Y]_t - \langle X, Y\rangle_t$ is a martingale, then $\langle X, Y\rangle_t$ is called the *predictable covariance* of X_t and Y_t. It follows from Proposition 4.7.9 that (i) if $\langle X, Y\rangle_t$ exists, then it is uniquely defined and (ii) $\langle X\rangle_t$ is always a nondecreasing process. The process $\langle X, X\rangle_t$ is usually denoted $\langle X\rangle_t$ and is called the *predictable quadratic variation* or the *Meyer increasing process*.

Remark 35. If instead of continuity one requires only predictability of $\langle X\rangle_t$, then a deep result of stochastic analysis, the Doob–Meyer decomposition, states that for any submartingale X_t this predictable quadratic variation exists. However, we will not use this result.

Let us show how one calculates the predictable covariance for functions of a Markov process.

Proposition 4.7.11. *Let X_t be a solution to the martingale problem for the operator L with domain D that forms an algebra under pointwise multiplication. Then for any function $f, g \in D$*

$$\langle f(X), g(X)\rangle_t = \langle M^f, M^g\rangle_t = \int_0^t (L(fg) - fLg - gLf)(X_s)\, ds, \qquad (4.79)$$

where

$$M_t^f = f(X_t) - f(X_0) - \int_0^t Lf(X_s)\, ds, \quad M_t^g = g(X_t) - g(X_0) - \int_0^t Lg(X_s)\, ds.$$

Proof. The processes M_t^f, M_t^g and

$$M_t^{fg} = (fg)(X_t) - (fg)(X_0) - \int_0^t L(fg)(X_s)\, ds$$

are well-defined martingales. Notice now that

$$[f(X), g(X)]_t = [M^f, M^g]_t,$$

by Proposition 4.7.10, implying the first equation in (4.79). From equation (4.76) it follows that

$$[f(X), g(X)]_t = f(X_t)g(X_y) - \int_0^t [f(X_s)Lg(X_s) + g(X_s)Lf(X_s)]ds + M_t$$

with M_t a martingale, and consequently

$$[f(X), g(X)]_t = \int_0^t [L(fg)(X_s) - f(X_s)Lg(X_s) - g(X_s)Lf(X_s)]ds + \tilde{M}_t$$

with $\tilde{M}_t$ another martingale, implying the second part of (4.79). □

4.8 Convergence of processes and semigroups

Here we discuss the basic criteria for tightness (or, equivalently, relative compactness) for distributions on Skorohod spaces, or, in other words, for càdlàg stochastic processes. These are the cornerstones of the modern theory of the convergence of martingales and Markov processes touched upon at the end of the chapter.

Everywhere in this section S denotes a complete separable metric space with distance d.

Let X^α be a family of S-valued random processes, each defined on its own probability space with a fixed filtration $\mathcal{F}_t^\alpha$ with respect to which it is adapted. One says that the family X^α enjoys *the compact containment condition* if for any $\eta, T > 0$ there exists a compact set $\Gamma_{\eta,T} \subset S$ such that

$$\inf_\alpha \mathbf{P}(X_\alpha(t) \in \Gamma_{\eta,T} \; \forall\, t \in [0, T]) \geq 1 - \eta. \tag{4.80}$$

The following is the basic criterion of compactness for distributions on Skorohod spaces (see (1.47) for the definition of the modulus of continuity $\tilde{w}$), which is a close analogue of the criterion for compactness of distributions on spaces of continuous functions (see Theorem 2.6.1). As usual, π_t denotes the evaluation map: $\pi_t f = f(t)$.

Theorem 4.8.1 (Basic criterion for tightness in D). *Let X^α be a family of random processes with sample paths in $D([0, T], S)$, $T > 0$ or $T = \infty$. Then $\{X^\alpha\}$ is relatively compact if*

(i) *the family $\pi_t(X^\alpha)$ is tight in S for rational t, that is, by Prohorov's criterion, for every $\eta > 0$ and a rational $t \geq 0$ there exists a compact set $\Gamma_{\eta,t} \subset S$ such that*

$$\inf_\alpha \mathbf{P}(X_\alpha(t) \in \Gamma_{\eta,t}) \geq 1 - \eta, \tag{4.81}$$

and

(ii) *for all* $\epsilon > 0$, $t > 0$,

$$\lim_{h\to 0} \sup_\alpha \mathbf{P}(\tilde{w}(X^\alpha, t, h) > \epsilon) = 0. \tag{4.82}$$

Moreover, if (i), (ii) *hold and* S *is locally compact, then the compact containment condition holds.*

Proof. It suffices to prove the theorem for any finite $T > 0$. It is then almost literally the same as for Theorem 2.6.1. Namely, by (4.82), given $\epsilon > 0$, there exists a sequence $h_1, h_2, \dots$ of positive numbers such that for all k

$$\sup_\alpha \mathbf{P}(\tilde{w}(X^\alpha, T, h_k) > 2^{-k}) \le 2^{-k-1}\epsilon.$$

By tightness of the families $\pi_{t_k}(X^\alpha)$, for all rational numbers $t_1, t_2, \dots$ one can choose a sequence of compact subsets $C_1, C_2, \dots$ from S such that for all k

$$\sup_\alpha \mathbf{P}(X^\alpha(t_k) \in (S \setminus C_k)) \le 2^{-k-1}\epsilon.$$

Then $\sup_\alpha \mathbf{P}(X^\alpha \in D([0,T], S) \setminus B) \le \epsilon$ for

$$B = \bigcap_k \{x \in D([0,T], S) : x(t_k) \in C_k, \tilde{w}(x, T, h_k) \le 2^{-k}\}.$$

By Theorem 1.7.2, B is relatively compact. Hence the required relative compactness follows from Prohorov's criterion for distributions in $D([0,T], S)$ (to apply Prohorov's criterion, one has to know that $D([0,T], S)$ is a complete separable metric space, which is the main content of Theorem 1.7.2).

The last statement follows from Remark 6. □

As the conditions of Theorem 4.8.1 are not easy to check, more concrete criteria were developed.

A sequence X^n of S-valued random processes (each defined on its own probability space with a fixed filtration $\mathcal{F}^n$ with respect to which it is adapted) is said to enjoy the *Aldous condition* [A] if $d(X^n_{\tau_n}, X^n_{\tau_n+h_n}) \to 0$ in probability, as $n \to \infty$, for any sequence of bounded $\mathcal{F}^n$-stopping times $\{\tau_n\}$ and any sequence of positive numbers $h_n \to 0$.

Because of its importance, let us give several handy reformulations of this property.

Proposition 4.8.1. *Condition* [A] *for a sequence* X^n *of* S*-valued* $\mathcal{F}^n$*-adapted random processes is equivalent to any of the conditions* [A$_1$]–[A$_4$] *given below.*

[A$_1$] *For each* $t, \epsilon, \eta > 0$ *there exists* $\delta > 0$ *and* n_0 *such that for any sequence of* $\mathcal{F}^n$*-stopping times* $\{\tau_n\}$ *with* $\tau_n \le t$

$$\sup_{n \ge n_0} \sup_{\theta \le \delta} \mathbf{P}(d(X^n_{\tau_n}, X^n_{\tau_n+\theta}) \ge \eta) \le \epsilon.$$

[A$_2$] *For each $t, \epsilon, \eta > 0$ there exists $\delta > 0$ and n_0 such that for any sequence of pairs of $\mathcal{F}^n$-stopping times $\{\sigma_n, \tau_n\}$ with $\sigma_n \le \tau_n \le \sigma_n + \delta \le t$,*

$$\sup_{n \ge n_0} \mathbf{P}(d(X^n_{\sigma_n}, X^n_{\tau_n}) \ge \eta) \le \epsilon.$$

[A$_3$] *For each $t > 0$*

$$\lim_{\delta \to 0} \limsup_{n \to \infty} \sup_{\tau \le t} \sup_{h \in [0,\delta]} \mathbf{E}[d(X^n_\tau, X^n_{\tau+h}) \wedge 1] = 0,$$

where the first sup *is over all $\mathcal{F}^n$-stopping times $\tau \le t$.*

[A$_4$] *For each $t > 0$*

$$\lim_{\delta \to 0} \limsup_{n \to \infty} \sup_{\sigma,\tau} \mathbf{E}[d(X^n_\sigma, X^n_\tau) \wedge 1] = 0,$$

where sup *is over all $\mathcal{F}^n$-stopping times τ, σ such that $\sigma \le \tau \le \sigma + \delta \le t$.*

Proof. (i) [A] $\iff$ [A$_1$] Condition [A] means that for each $t, \epsilon, \eta > 0$, a sequence of stopping times $\tau_n \le t$ and a sequence of positive numbers h_n tending to zero as $n \to \infty$, there exists m_0 such that

$$\sup_{n \ge m_0} \mathbf{P}(d(X^n_{\tau_n}, X^n_{\tau_n + h_n}) \ge \eta) \le \epsilon. \tag{4.83}$$

If [A$_1$] holds, one can satisfy (4.83) by choosing $m_0 = n_0 \wedge m$, where m is such that $h_n \le \delta$ for $n > m$. Conversely, suppose [A$_1$] does not hold. Then there exist $t, \epsilon, \eta > 0$ such that for any δ and n_0 there exists a sequence of stopping times $\tau_n \le t$ with

$$\sup_{n \ge n_0} \mathbf{P}(d(X^n_{\tau_n}, X^n_{\tau_n + \delta}) \ge \eta) > \epsilon.$$

Consequently, given a sequence of positive numbers $h_1, h_2, \dots$ converging to zero, we can construct a sequence of stopping times $\tau_n \le t$ such that for all $n = 1, 2, \dots$

$$\mathbf{P}(d(X^n_{\tau_n}, X^n_{\tau_n + h_n}) \ge \eta) > \epsilon,$$

which contradicts (4.83).

(ii) [A$_1$] $\iff$ [A$_3$]. Condition [A$_1$] means that for each $t, \eta > 0$

$$\lim_{\delta \to 0} \limsup_{n \to \infty} \sup_{\tau \le t} \sup_{h \in [0,\delta]} \mathbf{P}(d(X^n_\tau, X^n_{\tau+h}) \ge \eta) = 0,$$

and this is equivalent to [A$_3$] by the same argument as in Proposition 1.1.1.

(iii) [A$_2$] $\iff$ [A$_4$]. The same as (ii).

(iv) [A$_4$] $\Longrightarrow$ [A$_3$], because for any $\mathcal{F}_n$-stopping time τ

$$\sup_{h \in [0,\delta]} \mathbf{E}[d(X^n_\tau, X^n_{\tau+h})] \le \sup_{\sigma \in [\tau, \tau+h]} \mathbf{E}[d(X^n_\sigma, X^n_\tau)],$$

where σ are also $\mathcal{F}^n$-stopping times. The inverse implication $[A_3] \Longrightarrow [A_4]$ follows from the inequality

$$\sup_{\sigma,\tau} \mathbf{E}[d(X_\sigma, X_\tau) \wedge 1] \leq 3 \sup_\tau \sup_{h\in[0,2\delta]} \mathbf{E}[d(X_\tau, X_{\tau+h}) \wedge 1] \tag{4.84}$$

To obtain (4.84) integrate the triangle inequality

$$d(X_\sigma, X_\tau) \leq d(X_\sigma, X_{\tau+h}) + d(X_\tau, X_{\tau+h}),$$

yielding

$$\delta d(X_\sigma, X_\tau) \leq \int_0^\delta [d(X_\sigma, X_{\tau+h}) + d(X_\tau, X_{\tau+h})]\, dh.$$

Since $[\tau, \tau+\delta] \subset [\sigma, \sigma+2\delta]$ for $0 \leq \tau - \sigma \leq \delta$, this implies

$$\delta d(X_\sigma, X_\tau) \leq \int_0^{2\delta} [d(X_\sigma, X_{\sigma+h}) + d(X_\tau, X_{\tau+h})]\, dh,$$

and (4.84) follows. □

Theorem 4.8.2 (Aldous criterion for tightness). *Condition* [A] *implies the basic tightness condition* (ii) *of Theorem* 4.8.1.

Proof. We follow Kallenberg [154]. By changing, if necessary, the metric d to the equivalent metric $d \wedge 1$, we can and will consider d to be uniformly bounded by 1. For any natural n and $\eta > 0$ let us define recursively the stopping times

$$\sigma_{k+1}^n = \inf\{s > \sigma_k^n : d(X_{\sigma_k^n}^n, X_s^n) > \eta\}, \quad n = 0, 1, \ldots,$$

starting with $\sigma_0^n = 0$. Given $h, t > 0$, the subsequence $\{\sigma_{k_l}^n\}$, $k_l = 1, 2, \ldots$, defined by the prescription

$$\sigma_{k_{l+1}}^n = \inf_m \{\sigma_{k_l+m}^n : \sigma_{k_l+m}^n - \sigma_{k_l+m-1}^n \geq h\},$$

specifies a partition of $[0, t]$ with the minimal increment not less than h. Hence

$$\tilde{w}(X^n, t, h) \leq \sup_{k_l : \sigma_{k_l}^n \leq t} \sup_{\sigma_{k_l} \leq r \leq s < \sigma_{k_l+m}^n} d(X_r^n, X_s^n). \tag{4.85}$$

Consequently, for any m,

$$\tilde{w}(X^n, t, h) \leq 2\eta + \sum_{k<m} \mathbf{1}_{\{\sigma_{k+1}^n - \sigma_k^n < h, \sigma_k^n < t\}} + \mathbf{1}_{\{\sigma_m^n < t\}}. \tag{4.86}$$

By construction $d(X_{\sigma_{k+1}^n}^n, X_{\sigma_k^n}^n) > \eta$. Hence by Markov's inequality

$$\mathbf{P}(\sigma_{k+1}^n - \sigma_k^n < h, \sigma_k^n < t) \leq \eta^{-1} \nu_n(t+h, h),$$

where

$$\nu_n(t+h,h) = \sup_{\sigma \le \tau \le \sigma+h \le t} \mathbf{E}\, d(X^n_\sigma, X^n_\tau).$$

From [A$_4$] and (4.86) we deduce

$$\lim_{h\to 0} \limsup_{n\to\infty} \mathbf{E}\tilde{w}(X^n, t, h) \le 2\eta + \limsup_{n\to\infty} \mathbf{P}(\sigma^n_m < t). \tag{4.87}$$

Finally, by Markov's inequality

$$\mathbf{P}(\sigma^n_m < t) \le e^t \mathbf{E}(e^{-\sigma^n_m} \mathbf{1}_{\sigma^n_m < t}).$$

By the elementary inequality (4.88), proved in the lemma below, this does not exceed

$$e^t [e^{-mc} + \eta^{-1} \nu_n(t+c,c)].$$

By [A$_4$], we can choose $c > 0$ so that the second term is arbitrarily small uniformly for all $n \ge n_0$ with some n_0. Then taking m large enough we see that the second term in (4.87) vanishes in the limit $m \to \infty$. As η was arbitrary this implies that

$$\lim_{h\to 0} \limsup_{n\to\infty} \mathbf{E}\tilde{w}(X^n, t, h) = 0,$$

which implies (4.82). □

Lemma 4.8.1. *For any $c > 0$ and any collection of non-negative random variables $\xi_1, \dots, \xi_n$,*

$$\mathbf{E}\exp\{-(\xi_1 + \cdots + \xi_n)\} \le e^{-nc} + \max_{k \le n} \mathbf{P}(\xi_k < c), \tag{4.88}$$

Proof. By Hölder's inequality the l.h.s. does not exceed

$$\prod_k (\mathbf{E}e^{-n\xi_k})^{1/n}.$$

Using for each term the evident estimate

$$\mathbf{E}e^{-n\xi_k} \le e^{-nc} + \mathbf{P}(\xi_k < c),$$

yields (4.87). □

The Aldous condition is not necessary for tightness. To see this, it is enough to take a single deterministic indicator function $\mathbf{1}_A$ for any interval A on $[0,1]$. This remark anticipates the following result.

Proposition 4.8.2. *If a process X is the limit in distribution of processes X^n satisfying Aldous criterion, then X is stochastically (or in probability) continuous.*

Proof. Suppose X is not stochastically continuous. Then there exist $\eta, \epsilon, t > 0$ such that for any h there exists s with $|s| < h$ such that

$$\mathbf{P}(|X_{t+s} - X_t| > \eta) > \epsilon.$$

But this contradicts [A_1] with $\tau_n = t$. □

As an easy consequence of the Aldous criterion one can get the following crucial link between the *convergence of semigroups* and *weak convergence for Feller processes*.

Theorem 4.8.3. *Let S be locally compact and $X, X^1, X^2, \dots$ be S-valued Feller processes with the corresponding Feller semigroups $T_t, T_t^1, T_t^2, \dots$. If the compact containment condition holds and the semigroups T^n converge to T_t strongly and uniformly for bounded times, then the sequence $\{X^n\}$ is tight. In particular, if additionally the initial distributions of X^n converge weakly to the initial distribution of X, the distributions of X^n converge to the distribution of X.*

Proof. By Theorems 4.8.1, 4.8.2 and the strong Markov property of Feller processes, one needs only show that $d(X_0^n, X_{h_n}^n) \to 0$ in probability as $n \to \infty$ for any initial distributions μ_n that may arise from optional stopping of X^n and any positive constants $h_n \to 0$. By the compact containment condition (and Prohorov's criterion for tightness) we may assume μ_n converge weakly to a certain μ. Let X_0 be a random variable with law μ. By the assumed uniform-in-time semigroup convergence, $T_{h_n}^n g \to g$ for any $g \in C_\infty(S)$, implying

$$\mathbf{E}[f(X_0^n)g(X_{h_n}^n)] = \mathbf{E}(f T_{h_n}^n g)(X_0^n) \to \mathbf{E}(fg)(X_0)$$

for $f, g \in C_\infty(S)$. Consequently $(X_0^n, X_{h_n}^n) \to (X_0, X_0)$ in distribution. Then $d(X_0^n, X_{h_n}^n) \to d(X_0, X_0) = 0$ in distribution and hence also in probability. □

Similarly, one proves the following discrete analogue of Theorem 4.8.3.

Theorem 4.8.4. *Let X_t be a Feller process in a locally compact space S, specified by the semigroup T_t and generator A with a core D. Let Y^n, $n \in \mathbb{N}$, be a sequence of discrete-time Markov chains in S with transition operators U_n and h_n a sequence of positive numbers converging to zero. If the operators $T_t^n = U_n^{[t/h_n]}$ converge to T_t strongly and uniformly on compact time intervals, the compact containment condition holds for Y^n, and the initial distributions of Y_n converge to a distribution of a random variable X_0, then the processes $Y^{[t/h_n]}$ converge in distribution to the process X_t with the initial distribution given by X_0.*

4.9 Weak convergence of martingales

Here the general theory of tightness developed above is applied to the analysis of convergence of martingales and solutions to martingale problems.

Lemma 4.9.1. (i) *Let Y and X be two càdlàg non-negative processes on a filtered probability space such that Y is non-decreasing and $\mathbf{E}(X_\tau) \le \mathbf{E}(Y_\tau)$ for every finite stopping time. Then*

$$\mathbf{P}(\sup_{s\le\tau} X_s > \epsilon) \le \mathbf{E}(Y_\tau)/\epsilon$$

for every finite stopping time τ and $\epsilon > 0$.

(ii) *If additionally Y is continuous, then for every $\eta > 0$*

$$\mathbf{P}(\sup_{s\le\tau} X_s > \epsilon) \le \mathbf{P}(Y_\tau \ge \eta) + \mathbf{E}(Y_\tau \wedge \eta)/\epsilon \le \mathbf{P}(Y_\tau \ge \eta) + \eta/\epsilon. \tag{4.89}$$

Proof. (i) Let $S = \inf\{t : X_t > \epsilon\} \wedge \tau$. Then

$$\epsilon\mathbf{P}(\sup_{s\le\tau} X_s > \epsilon) \le \epsilon\mathbf{P}(X_S \ge \epsilon) \le \mathbf{E}(X_S) \le \mathbf{E}(Y_S) \le \mathbf{E}(Y_\tau).$$

(ii) We have

$$\mathbf{P}(\sup_{s\le\tau} X_s > \epsilon) \le \mathbf{P}(Y_\tau \ge \eta) + \mathbf{P}(\sup_{s\le\tau} X_s > \epsilon, Y_\tau < \eta)$$

Let $S = \inf\{t > 0 : Y_\tau \ge \eta\}$. As Y is continuous, $Y_S = \eta$ whenever $Y_0 \le \eta$. Moreover, the events $\{Y_\tau < \eta\}$ and $\{\tau < S\}$ coincide. Hence

$$\begin{aligned}\mathbf{P}(\sup_{s\le\tau} X_s > \epsilon, Y_\tau < \eta) &= \mathbf{P}(\sup_{s\le\tau} X_s > \epsilon, \tau < S) \\ &\le \mathbf{P}(\sup_{s\le\tau\wedge S} X_s > \epsilon, Y_\tau < \eta) \le \mathbf{E}(Y_{\tau\wedge S})/\epsilon = \mathbf{E}(Y_\tau \wedge \eta)/\epsilon.\end{aligned}$$ □

The following result is usually referred to as the *Rebolledo criterion* for tightness. We present it here only for martingales with a continuous predictable quadratic variation.

Theorem 4.9.1. *Let M^n be a family of $\mathbb{R}^d$-valued càdlàg square-integrable martingales, which have continuous predictable quadratic variations $\langle M^n\rangle_t$. Then if $\langle M^n\rangle_t$ enjoy the Aldous condition* [A], *then so do also the sequences of the martingales M^n and of the quadratic variations $[M^n]_t$.*

Proof. For any square-integrable martingale M_t one has $\mathbf{E}\|M_s\|^2 = \mathbf{E}[M,M]_t = \mathbf{E}\langle M\rangle_t$. Hence, if $\langle M\rangle_t$ is continuous, one can apply Lemma 4.9.1 (ii) to $X_t = \|M_t\|^2$ or $X_t = [M]_t$ and $Y = \langle M\rangle_t$. Choosing a bounded stopping time T let us apply (4.89) to the martingale $L_t^T = M_t - M_{T\wedge t}$ and its predictable variation $\langle L^T\rangle_t$. Thus for every $b, a, \delta > 0$

$$\mathbf{P}(\sup_{T\le s\le T+\delta} \|M_s - M_T\| \ge b) \le \mathbf{P}((\langle M\rangle_{T+\delta} - \langle M\rangle_T) \ge a) + a/b^2.$$

Set $a = b^2\epsilon/2$. Consequently, if there exists $\delta > 0$ such that

$$\mathbf{P}((\langle M\rangle_{T+\delta} - \langle M\rangle_T) \ge b^2\epsilon/2) \le \epsilon/2,$$

then

$$\mathbf{P}(\sup_{T\le s\le T+\delta} \|M_s - M_T\| \ge b) \le \epsilon,$$

implying that if [A] holds for $\langle M^n\rangle$, then it holds for M^n. The statement about $[M^n]_t$ is obtained similarly. □

The following result is a direct corollary of the Rebolledo criterion.

Proposition 4.9.1. *Let X_t^n be a family of processes*

$$X_t^n = M_t^n + V_t^n,$$

where V_t^n are càdlàg processes of finite variation and M_t^n are square-integrable martingales with continuous Meyer increasing processes $\langle M^n\rangle_t$. Then the family X_t^n satisfies the Aldous condition [A] *whenever V_t^n and $\langle M^n\rangle_t$ satisfy this condition.*

Theorem 4.9.2. *Let X^n solve the martingale problem in $\mathbb{R}^d$ for operators A^n with the common domain $D \subset C(\mathbb{R}^d)$ that forms an algebra and is dense in $C(\mathbb{R}^d)$ in the topology of uniform convergence on compact sets (typically $D = C_c^2(\mathbb{R}^d)$). Suppose the family X^n satisfies the compact containment condition. Then the family X^n is tight. If additionally $\|(A_n - A)f\| \to 0$, as $n \to \infty$ for any $f \in D$, then any convergent subsequence converges to a solution of the martingale problem for A on D.*

Proof. As X^n solves the martingale problem,

$$f(X_t^n) = M_t^{f,n} + \int_0^t Af(X_s^n)\,ds,$$

and thus we are in the setting of Proposition 4.9.1. Since by Proposition 4.7.11

$$\langle f(X_t^n), g(X_t^n)\rangle_t = \int_0^t (A^n(fg) - fA^n g - gA^n f)(X_s^n)\,ds$$

for $f, g \in D$, and the integrands are uniformly bounded, it follows straightforwardly that $f(X_t^n)$ satisfy the Aldous condition [A]. Taking into account the compact containment condition and Aldous criterion we conclude that the family $f(X_t^n)$ is tight for any $f \in D$. Finally, for any $\epsilon > 0$, there exists a compact set K such that the probability that any X^n has values outside K does not exceed ϵ, and there exists $f \in D$ such that $\sup_{x \in K} |f(x) - x| \leq \epsilon$. This clearly implies the Aldous condition for X_t^n. By dominated convergence it is clear that any limiting point of X_t^n solves the martingale problem for A. □

Remark 36. As a direct consequence of this theorem we can deduce a new proof of Theorem 4.8.3.

4.10 Martingale problems and Markov processes

Roughly speaking, this section is devoted to the connection between uniqueness and Markovianity of the solutions to a martingale problem.

We begin by observing that being a solution to a martingale problem is a property of finite-dimensional distributions. Indeed, X_t solves the (L, D)-martingale problem if and only if $\mathbf{E}\eta(X) = 0$ for any partition $0 \leq t_1 < t_2 < \cdots < t_{n+1}$ of $\mathbb{R}_+$, $f \in D$ and $h_1, \ldots, h_n \in C(S)$, where

$$\eta(X) = \left(f(X_{t_{n+1}}) - f(X_{t_n}) - \int_{t_n}^{t_{n+1}} Lf(X_s)\, ds \right) \prod_{k=1}^{n} h_k(X_{t_k}). \tag{4.90}$$

The following crucial result shows that uniqueness of solutions to a martingale problem implies the Markov property.

Theorem 4.10.1. *Let L be a linear operator $L : D \mapsto B(S)$, $D \in C(S)$. Suppose that for any $\mu \in \mathcal{P}(S)$ any two solutions X and Y of the (L, D)-martingale problem have the same one-dimensional distributions, i.e.*

$$\mathbf{P}(X_t \in B) = \mathbf{P}(Y_t \in B), \quad B \in \mathcal{B}(S). \tag{4.91}$$

Then any solution of the (L, D)-martingale problem with respect to a filtration $\mathcal{F}_t$ has the Markov property with respect to $\mathcal{F}_t$:

$$\mathbf{E}[f(X_{r+t})|\mathcal{F}_r] = \mathbf{E}[f(X_{r+t})|X_r]), \quad f \in B(S),\ r, t \geq 0, \tag{4.92}$$

and even the strong Markov property:

$$\mathbf{E}[f(X_{\tau+t})|\mathcal{F}_\tau] = \mathbf{E}[f(X_{\tau+t})|X_\tau]), \quad f \in B(S),\ t \geq 0, \tag{4.93}$$

for any finite $\mathcal{F}_t$-stopping time τ.

Proof. Let X be a solution of (L, D)-martingale problem with respect to a filtration $\mathcal{F}_t$. Equation (4.92) is equivalent to

$$\int_F \mathbf{E}[f(X_{r+t})|\mathcal{F}_r]\mathbf{P}(d\omega) = \int_F \mathbf{E}[f(X_{r+t})|X_r])\mathbf{P}(d\omega), \quad F \in \mathcal{F}_r. \tag{4.94}$$

It suffices to establish this fact for sets F of positive measure. In order to achieve this, it is enough to show that the measures $\mathbf{P}_1, \mathbf{P}_2$ on $(\Omega, \mathcal{F})$ coincide, where

$$\mathbf{P}_1(B) = (\mathbf{P}(F))^{-1} \int_F \mathbf{E}[\mathbf{1}_B|\mathcal{F}_r]\mathbf{P}(d\omega),$$
$$\mathbf{P}_2(B) = (\mathbf{P}(F))^{-1} \int_F \mathbf{E}[\mathbf{1}_B|X_r]\mathbf{P}(d\omega),$$

or equivalently, in terms of expectations,

$$\mathbf{E}_1(h) = (\mathbf{P}(F))^{-1} \int_F \mathbf{E}[h(\omega)|\mathcal{F}_r]\mathbf{P}(d\omega),$$
$$\mathbf{E}_2(h) = (\mathbf{P}(F))^{-1} \int_F \mathbf{E}[h(\omega)|X_r]\mathbf{P}(d\omega).$$

Note that if $B \in \sigma(X_r)$, then

$$\mathbf{P}_1(B) = \mathbf{P}_2(B) = \mathbf{P}(B \cap F)/\mathbf{P}(F) = \mathbf{P}(B|F).$$

Set $Y_t = X_{r+t}$. Since X solves the martingale problem, $\mathbf{E}[\eta(X_{r+\cdot})|\mathcal{F}_r] = 0$, where η is given by (4.90). But then $\mathbf{E}[\eta(X_{r+\cdot})|X_r] = 0$, implying that $\mathbf{E}_1(\eta(Y)) = \mathbf{E}_2(\eta(Y)) = 0$. Hence Y is a solution of (L, D)-martingale problem on $(\Omega, \mathcal{F}, \mathbf{P}_1)$ and $(\Omega, \mathcal{F}, \mathbf{P}_2)$. Consequently by (4.91), $\mathbf{E}_1 f(Y_t) = \mathbf{E}_2 f(Y_t)$ for all $t \geq 0$ and $f \in B(S)$, which is equivalent to (4.94).

By the optional sampling theorem $\mathbf{E}[\eta(X_{\tau+\cdot})|\mathcal{F}_\tau] = 0$ for a finite stopping time τ. Consequently the same proof is applied to obtain the strong Markov property. □

The next result shows that the uniqueness of one-dimensional distributions of the solutions to a martingale problem implies uniqueness. We shall not use this fact, but give it here for the sake of completeness. It also presents another illustration of the idea used in the proof of the previous theorem.

Proposition 4.10.1. *Under the assumptions of Theorem* 4.10.1, *any two solutions of the (L, D)-martingale problem have the same finite-dimensional distributions.*

Proof. Let X^i, $i = 1, 2$, be solutions of (L, D)-martingale problem defined on probability spaces $(\Omega^i, \mathcal{F}^i, \mathbf{P}^i)$. We need to show that

$$\mathbf{E}^1 \prod_{k=1}^m h_k(X^1_{t_k}) = \mathbf{E}^2 \prod_{k=1}^m h_k(X^2_{t_k}) \tag{4.95}$$

for all $0 \le t_1 < t_2 < \cdots < t_m$ and $h_1, \dots, h_m \in C(S)$. It is sufficient to consider only positive h_k. We shall use induction in m. For $m = 1$ this holds by (4.91). Suppose it holds for $m \le n$. Let us pick a finite sequence $0 \le t_1 < t_2 < \cdots < t_n$ and let probabilities $\tilde{\mathbf{P}}^i$, $i = 1, 2$, be defined via the expectations

$$\tilde{\mathbf{E}}^i(h) = \frac{\mathbf{E}^i[h(\omega_i)\prod_{k=1}^n h_k(X^i_{t_k})]}{\mathbf{E}^i[\prod_{k=1}^n h_k(X^i_{t_k})]}, \quad \omega_i \in \Omega^i,$$

i.e. $\tilde{\mathbf{P}}^i(B) = \tilde{\mathbf{E}}^i(\mathbf{1}_B)$. Set $\tilde{X}^i_t = X^i_{t_n+t}$. As X^i solves the (L, D)-martingale problem on the probability spaces $(\Omega^i, \mathcal{F}^i, \mathbf{P}^i)$, this implies that $\tilde{E}^i\eta(\tilde{X}^i) = 0$, where η is defined by (4.90). Hence $\tilde{X}^i_t$ solves the (L, D)-martingale problem on $(\Omega^i, \mathcal{F}^i, \tilde{\mathbf{P}}^i)$. By (4.95) with $m = n$ we conclude that

$$\tilde{\mathbf{E}}^1[f(\tilde{X}^1_0)] = \tilde{\mathbf{E}}^2[f(\tilde{X}^2_0)],$$

so that $\tilde{X}^1$ and $\tilde{X}^2$ have the same initial distributions. Consequently, by (4.91),

$$\tilde{\mathbf{E}}^1[f(\tilde{X}^1_t)] = \tilde{\mathbf{E}}^2[f(\tilde{X}^2_t)],$$

which is precisely (4.95) with $m = n + 1$, $f_{n+1} = f$ and $t_{n+1} = t_n + t$. □

By Theorem 4.10.1, well-posedness of a martingale problem implies Markovianity of the solutions. It turns out that it usually also implies continuous dependence on the initial data. To formulate this result we need to introduce two important notions. Namely, let us say that a family of solutions X^x_t (x denotes the initial point) of the (L, D)-martingale problem on a locally compact metric space S satisfies the *compact containment condition for compact initial data*, if for any $\eta, T > 0$ and a compact set $K \subset S$, there exists a compact set $\Gamma_{\eta,T,K} \subset S$ such that

$$\inf_{x \in K} \mathbf{P}(X^x(t) \in \Gamma_{\eta,T,K} \ \forall\, t \in [0, T]) \ge 1 - \eta. \tag{4.96}$$

The family X^x_t is said to be *uniformly stochastically continuous* if for any compact set K

$$\lim_{t \to 0} \sup_{x \in K} \mathbf{P}(\sup_{s \le t} \|X^x_t - x\| > r) = 0. \tag{4.97}$$

Theorem 4.10.2. (i) *If (L, D)-martingale problem on a locally compact metric space S is well posed and the compact containment condition for compact initial data holds, then the distribution of $D([0, T], S)$-valued processes X^x_t depends continuously on x and the corresponding sub-Markov semigroup preserves constants. In particular, our martingale problem is measurably well posed.*

(ii) *If additionally, the processes X^x_t are uniformly stochastic continuous, then the corresponding Markov semigroup preserves the space $C(S)$ and consequently is C-Feller.*

Proof. (i) Let K be any compact subset of S. By Theorem 4.9.2 with $A_n = A$, the set $\{X^x|_{x\in K}\}$ of the solutions of the (L, D)-martingale problem with initial values in K is compact in the weak topology. Hence the projection $X^x \mapsto x$, $x \in K$, is a continuous bijection of compact sets, implying that the inverse mapping $x \mapsto X^x$ is also continuous. Consequently the function $\mathbf{E} f(X_t^x)$ depends continuously on x for any continuous f.

In order to see that the dynamics of averages preserves constants, one has to show that $\lim_{n\to\infty} \mathbf{E}\chi[d(X_t^x, x)/n)] = 1$ for any $\chi \in C_\infty(\mathbb{R}_+)$ that equals one in a neighborhood of the origin, where d is the distance in S. Clearly the limit exists and does not exceed one. But by (4.96), it is not less than $1-\epsilon$ for any $\epsilon > 0$.

(ii) Notice that the continuous dependence of the distribution X_t^x on x does not allow us to conclude that the semigroup preserves the space $C(S)$, because the evaluation map $t \mapsto X_t$ is not continuous in the Skorohod topology. But the uniform stochastic continuity implies that for any t all processes X^x have no jumps at time t a.s. And at a continuity point of a càdlàg path the evaluation map becomes continuous. This implies the required claim. □

Finally we prove here the following fact.

Theorem 4.10.3. *Suppose X_t is a Feller process with the generator L given on its core D. Then the (L, D)-martingale problem is measurably well posed.*

Proof. Process X_t solves the (L, D)-martingale problem with the initial condition x and its distribution depends measurably (even continuously) on x. So only uniqueness needs to be proved.

Suppose Y_t solves the (L, D)-martingale problem. From Proposition 3.9.3 it follows that for $\lambda > 0$ and $f \in D$

$$f(Y_t)e^{-\lambda t} = \mathbf{E}\left[f(Y_{t+s})e^{-\lambda(t+s)} + \int_t^{t+s} e^{-\lambda\tau}(\lambda - L)f(Y_\tau)\,d\tau \middle| \mathcal{F}_t \right].$$

Multiplying this equation by $e^{\lambda t}$, passing to the limit $s \to \infty$ and shifting the variable of integration yields

$$f(Y_t) = \mathbf{E}\left[\int_0^\infty e^{-\lambda\tau}(\lambda - L)f(Y_{t+\tau})\,d\tau \middle| \mathcal{F}_t \right]. \tag{4.98}$$

Since L generates a Feller semigroup, the resolvent operator $R_\lambda = (\lambda - L)^{-1}$ is a well-defined bounded operator. Consequently, applying (4.98) to the function $f = R_\lambda h$ yields

$$R_\lambda h(Y_t) = \mathbf{E}\left[\int_0^\infty e^{-\lambda\tau} h(Y_{t+\tau})\,d\tau \middle| \mathcal{F}_t \right]. \tag{4.99}$$

In particular, for any two solutions Y_t^1, Y_t^2 of the (L, D)-martingale problem with the initial point x one has

$$\int_0^\infty e^{-\lambda\tau}\mathbf{E}h(Y_\tau^1)\,d\tau = \int_0^\infty e^{-\lambda\tau}\mathbf{E}h(Y_\tau^1)\,d\tau$$

for any $\lambda > 0$. By the uniqueness of the Laplace transform, this implies that $\mathbf{E}h(Y_\tau^1) = \mathbf{E}h(Y_\tau^2)$ for any h and $\tau > 0$, so that the one-dimensional distributions of Y_t^1, Y_t^2 coincide. This implies uniqueness by Proposition 4.10.1. □

Let us stress the importance of the assumption that D is a core in the above theorem. In particular, if X_t is a Feller process with generator L having a domain containing $C_c^2(\mathbb{R}^d)$, but we do not know whether this space is a core, we can conclude neither that a Feller process with such a property is unique, nor that the $(L, C_c^2(\mathbb{R}^d))$-martingale problem is well posed.

4.11 Stopping and localization

In this section (S, d) denotes a complete separable metric space.

For an adapted process X_t and a stopping time τ on a filtered probability space $(\Omega, \mathcal{F}, \mathcal{F}_t, \mathbf{P})$ the *stopped process* is defined as $X_t^\tau = X_{t\wedge\tau}$. By Proposition 3.10.3 the random variable X_τ is $\mathcal{F}_\tau$-measurable. Hence X_t^τ is adapted, because

$$(X_t^\tau \subset B) = [(X_t \subset B) \cap (\tau > t)] \cup [(X_t \subset B) \cap (\tau \leq t)].$$

Proposition 4.11.1. *If X_t is a càdlàg Markov (or strong Markov) process in S and a stopping time τ is such that $(\tau \leq t) \in \sigma(X_t)$ for all t, then the stopped process X_t^τ is also Markov (or strong Markov respectively).*

Proof. Let $s < t$. Then

$$\begin{aligned}\mathbf{E}[f(X_{t\wedge\tau})|\mathcal{F}_s] &= \mathbf{E}[f(X_s)\mathbf{1}_{\tau\leq s}|\mathcal{F}_s] + \mathbf{E}[f(X_{t\wedge\tau})\mathbf{1}_{\tau>s}|\mathcal{F}_s]\\ &= \mathbf{1}_{\tau\leq s}f(X_s) + \mathbf{1}_{\tau>s}\mathbf{E}[f(X_{t\wedge\tau})|X_s] = \mathbf{E}[f(X_{t\wedge\tau})|X_s].\end{aligned}$$

The case of a stopping time s is analyzed similarly in case of a strong Markov process X_t. □

Let X_t be a process with sample paths in $D([0, \infty), S)$ and initial distribution μ. For an open subset $U \subset S$, define the exit time from U as

$$\tau_U = \inf\{t \geq 0 : X_t \notin U\}. \tag{4.100}$$

As paths of X_t are right-continuous, one has

$$\tau_U = \min\{t \geq 0 : X_t \notin U\}. \tag{4.101}$$

We shall write τ_U^x when stressing the initial point.

Proposition 4.11.2. *Let X_t be a càdlàg Markov (or strong Markov) process in S, and let U be an open set. Then the stopped process $X_t^{\tau_U}$ is also Markov (or strong Markov respectively).*

Proof. It follows from Proposition 4.11.1. □

Remark 37. Some authors (see e.g. Ethier and Kurtz [110]) prefer to work with exit times defined by

$$\tilde{\tau}_U = \inf\{t \geq 0 : X_t \notin U \text{ or } X(t-) \notin U\} \tag{4.102}$$

rather than (4.100). The following is worth noting.

(i) The process $X_t^{\tilde{\tau}_U}$ may not be Markov for a Markov X_t.

(ii) $\tilde{\tau}_U$ respects the limits of approximations: if $U = \bigcup_k U_k$ with $U_1 \subset U_2 \subset \dots$ and the boundaries of U_k approach the boundary of U, then

$$\tilde{\tau}_U = \lim_{k\to\infty} \tau_k = \lim_{k\to\infty} \tilde{\tau}_k.$$

(iii) The results below on the stopped martingale problem (and their proofs) remain valid for $\tilde{\tau}_U$, though the stopped processes $X_t^{\tau_U}$ and $X_t^{\tilde{\tau}_U}$ may differ.

(iv) Both definitions coincide for processes with continuous paths, or at least for exits from transmission admissible domains (see Section 6.1 for the latter).

Let L be an operator in $C(S)$ with the domain D. One says that the process Z_t solves the *stopped (L, D)-martingale problem* in U starting with $\mu \in \mathcal{P}(S)$ if $Z_t = Z_{t\wedge\tau_U}$ a.s. and

$$f(Z_t) - \int_0^{t\wedge\tau_U} Lf(Z_s)\,ds$$

is a martingale for any $f \in D$. We say that this stopped martingale problem is well posed (respectively measurably well posed), if for any initial law $\mu \in \mathcal{P}(S)$ there exists a unique solution (respectively when additionally it depends measurably on μ). Notice that this definition does not require Z to be obtained by stopping a solution to the corresponding martingale problem in the whole space. Nevertheless, the following result shows that this is often the case.

Theorem 4.11.1. *Let L be a operator in $C(S)$ with the domain D. Let the (L, D)-martingale problem be measurably well posed in S and U be an open subset of S. Its solution starting at $x \in S$ will be denoted X_t^x. Then for any solution Z of the stopped (L, D)-martingale problem in U there exists a solution Y of the (L, D)-martingale problem in S such that the processes Z_t and $Y_{t\wedge\tau_U}$ have the same distribution.*

Proof. The idea is clear. After $\tau = \tau_U$ the process should be defined by the solutions of the martingale problem started at Z_τ. In other words, we define

$$Y_t = \begin{cases} Z_t, & t \leq \tau, \\ X_t^{Z_\tau}, & t > \tau. \end{cases}$$

To have this process well defined on a probability space we use the randomization Lemma 1.1.1, which is applicable, because the distribution of X^x is assumed to depend measurably on x. To show that this process is indeed a solution to the (L, D)-martingale problem in S, we need to prove that $\mathbf{E}\eta(Y) = 0$, where η is given by (4.90).

Because

$$g(t) - g(s) = [g(t \wedge \tau) - g(s \wedge \tau)] + [g(t \vee \tau) - g(s \vee \tau)]$$

for $t > s$ and any function g, we may write

$$\eta(Y) = \eta_1(Y) + \eta_2(Y),$$

with

$$\eta_1(Y) = \left(f(Y_{t_{n+1}\wedge\tau_U}) - f(Y_{t_n\wedge\tau_U}) - \int_{t_n\wedge\tau_U}^{t_{n+1}\wedge\tau_U} Lf(Y_s)\,ds \right) \prod_{k=1}^n h_k(Y_{t_k}),$$

$$\eta_2(Y) = \left(f(Y_{t_{n+1}\vee\tau_U}) - f(Y_{t_n\vee\tau_U}) - \int_{t_n\vee\tau_U}^{t_{n+1}\vee\tau_U} Lf(Y_s)\,ds \right) \prod_{k=1}^n h_k(Y_{t_k}).$$

Since $\eta_1(Y)$ vanishes for $\tau_U \le t_n$, its value will not be changed, if we write $h_k(Y_{t_k\wedge\tau_U})$ instead of $h_k(Y_{t_k})$ in the expression for $\eta_1(Y)$. Hence $\mathbf{E}\eta_1(Y) = \mathbf{E}\eta(Z) = 0$. Next, by continuity of functions f, Lf, h_k, in order to prove $\mathbf{E}\eta_2(Y) = 0$ it suffices to prove that $\mathbf{E}\eta_2^d(Y) = 0$, where

$$\eta_2^d(Y) = \left(f(Y_{t_{n+1}\vee\tau_U^d}) - f(Y_{t_n\vee\tau_U^d}) - \int_{t_n\vee\tau_U^d}^{t_{n+1}\vee\tau_U^d} Lf(Y_s)\,ds \right) \prod_{k=1}^n h_k(Y_{t_k}),$$

and τ_U^d is a discrete approximation to τ with values in a finite set R. Consequently it suffices to show that

$$\mathbf{E}[\eta_2^d(Y)\mathbf{1}_{\tau_U^d=r}] = 0$$

for any $r \in R$. For definiteness, assume $r \in (t_l, t_{l+1}]$ with $l = 1, \dots, n-1$. Then

$$\mathbf{E}[\eta_2^d(Y)\mathbf{1}_{\tau_U^d=r}] = \mathbf{E}\prod_{k=1}^l h_k(Z_{t_k})$$
$$\times \mathbf{E}\left[\left(f(Y_{t_{n+1}}) - f(Y_{t_n}) - \int_{t_n}^{t_{n+1}} Lf(Y_s)\,ds \right) \prod_{k=l+1}^n h_k(Y_{t_k})\mathbf{1}_{\tau_U^d=r} \,\middle|\, \mathcal{F}_r\right].$$

This expression equals zero as the internal expectation vanishes (because $\tau_U^d = r$ implies $X_r \in S \setminus U$ and $Y_t = X_{t-r}^{Z_r}$ for $t \ge r$). □

As a straightforward corollary we get the following main result of this section.

Theorem 4.11.2. *Suppose the (L, D) martingale problem is measurably well posed in S. Then for any open set U the corresponding stopped martingale problem is also measurably well posed.*

The following result is needed when constructing a solution to a martingale problem by gluing the localized solutions.

Theorem 4.11.3. *Let U and V be two open subsets of a S. If the stopped (L, D) martingale problem are measurably well posed in U and V, then it is also measurably well posed in $U \cup V$.*

Proof. It is enough to consider the Dirac initial conditions δ_x only. Assume $x \in U \cup V$. Define the stopping times τ_j and the time $\tau_{U \cup V}$ by the following recursive rule. First let $\tau_0 = 0$ and $\tau_1 = \min\{t \geq 0 : X_t \notin U\}$. If $X_{\tau_1} \notin V$, set $\tau_{U \cup V} = \tau_1$. Otherwise define $\tau_2 = \min\{t \geq \tau_1 : X_t \notin V\}$. If $X_{\tau_2} \notin U$, set $\tau_{U \cup V} = \tau_2$. And so on. For $t \in [\tau_k, \tau_{k+1}]$ the process is of course defined as the solution to the stopped (L, D) martingale problem in U or V. By right continuity, $\tau_{k+1} > \tau_k$ for any k except possibly for $k = 0$. One shows as in Theorem 4.11.1 that the process so defined is a solution to the stopped (L, D) martingale problem in $U \cup V$. Conversely, if X_t^x is a solution to the stopped (L, D) martingale problem in $U \cup V$, one can define the corresponding stopping times τ_j. By the uniqueness in each interval $[\tau_k, \tau_{k+1}]$, one gets uniqueness for X_t^x. □

Theorem 4.11.4. *Suppose L and L_k, $k = 1, 2, \dots$, are operators in $C(S)$ with a common domain $D = C_c^2(\mathbb{R}^d)$, and $U_1 \subset U_2 \subset \dots$ is an open covering of S such that $L_k f(x) = Lf(x)$ for $x \in U_k$, $f \in D$. Assume that the martingale problem for each L_k is measurably well posed and for any $\mu \in \mathcal{P}(S)$ either*

(i) *there exists a solution to the (L, D)- martingale problem starting from μ, or*

(ii) *the family of solutions to the (L_k, D)- martingale problem starting from μ and stopped in U_k is tight.*

Then the martingale problem for L is also measurably well posed.

Proof. Firstly, by Theorem 4.9.2 condition (ii) implies condition (i). Secondly, for any solution to the (L, D)-martingale problem, we define stopping times τ_k by recursive equations $\tau_{k+1} = \min\{t \geq \tau_k : X_t \notin U_k\}$ and observe that on each interval $[\tau_k, \tau_{k+1}]$ the process is uniquely specified, as it solves the stopped (L_k, D)-martingale problem. □

Practically, local solutions are often constructed on a covering (U_k) which is not ordered by inclusion. However, Theorem 4.11.3 allows one to convert such a covering into an ordered one; see Section 5.7 for an example.

Part II

Markov processes and beyond

Chapter 5
Processes in Euclidean spaces

In Section 4.6 we described the class of Markov processes that can be constructed via the standard stochastic calculus. This chapter is devoted to some other methods of the reconstruction of a Markov process in $\mathbb{R}^d$ from its given pre-generator of the Lévy (or Lévy–Khintchine) type. Namely, starting with the direct methods, we introduce the methods based on limits of Lie–Trotter formula type and T-products, stochastic monotonicity, martingale problem and ΨDE in the Sobolev spaces, Lyapunov functions, and stochastic integration driven by non-linear Lévy noise. In the last sections we touch upon some qualitative and quantitative analysis of Markov processes, sketching the theory of stochastic monotonicity and stochastic scattering, and introducing nonlinear Markov processes as dynamic LLN limits for Markov models of interacting particles. Other methods, e.g. semiclassical asymptotics, are mentioned in the Comments. The chapter is written in such a way that all these methods can be read about practically independent of each other.

Let us stress that just building a semigroup from a pre-generator, though important as a first step, does not lead directly to any practical applications. What one needs for this is some continuity (or even better differentiability) of the semigroup with respect to natural parameters (which often referred to as the sensitivity analysis and is crucial for the purposes of the calibration, or statistical estimation, of these parameters) as well as to initial data (which can never be specified precisely). The latter question is closely related to the problem of identifying an invariant core of differentiable functions for the generator of a semigroup together with appropriate bounds for the derivatives evolving in time. Therefore, paying attention to these problems (as we shall do) is caused by clear practical reasons.

Of special interest for a semigroup T_t or a propagator $U^{t,s}$ is the possibility of having an estimate of the form $\|T_t\| \le e^{Kt}$ or $\|U_{t,s}\| \le e^{K(t-s)}$ for its growth, where K is a constant. We call such semigroups or propagators *regular*. Of course contraction semigroups enjoy this property with $K = 0$. For a propagator in $C(\mathbb{R}^d)$ (even without the assumption of positivity), this property allows one to obtain straightforward path-integral representation (or probabilistic interpretation) for the corresponding evolution, see Chapter 9. On the other hand, the contraction propagators in $C(\mathbb{R}^d)$, which have this kind of growth when projected to the space of smooth functions $C^k(\mathbb{R}^d)$ can be easily combined via the Lie–Trotter limit formulas, see Section 5.3.

5.1 Direct analysis of regularity and well-posedness

In this section we introduce the most straightforward approach to analyzing the regularity as well as well-posedness of Markov semigroup, which is based on direct analysis of the corresponding evolution equations for the derivatives of a Markov evolution. The method is elementary in that it does not exploit any advanced theory. We shall consider three examples of application of this idea.

We start with the processes generated by integro-differential (or pseudo-differential) operators *of order at most one*, i.e. by the operators

$$Lf(x) = (b(x), \nabla f(x)) + \int_{\mathbb{R}^d \setminus \{0\}} (f(x+y) - f(x))\nu(x, dy) \tag{5.1}$$

with Lévy measures $\nu(x, .)$ having finite first moment $\int_{B_1} |y|\nu(x, dy)$.

Theorem 5.1.1. *Assume that $b \in C^1(\mathbb{R}^d)$ and $\nabla\nu(x, dy)$, gradient of the Lévy kernel with respect to x, exists in the weak sense as a signed measure and depends weakly continuously on x. Moreover, assume*

$$\sup_x \int \min(1, |y|)\nu(x, dy) < \infty, \quad \sup_x \int \min(1, |y|)|\nabla\nu(x, dy)| < \infty, \tag{5.2}$$

and for any $\epsilon > 0$ there exists a $K > 0$ such that

$$\sup_x \int_{\mathbb{R}^d \setminus B_K} \nu(x, dy) < \epsilon, \quad \sup_x \int_{\mathbb{R}^d \setminus B_K} |\nabla\nu(x, dy)| < \epsilon, \tag{5.3}$$

$$\sup_x \int_{B_{1/K}} |y|\nu(x, dy) < \epsilon. \tag{5.4}$$

Then L generates a conservative Feller semigroup T_t in $C_\infty(\mathbb{R}^d)$ with invariant core $C^1_\infty(\mathbb{R}^d)$. Moreover T_t reduced to $C^1_\infty(\mathbb{R}^d)$ is also a strongly continuous semigroup in the Banach space $C^1_\infty(\mathbb{R}^d)$, where it is regular in the sense that

$$\|T_t\|_{C^1_\infty(\mathbb{R}^d)} \le e^{Kt} \tag{5.5}$$

with a constant K.

Proof. Notice first that (5.2) implies that for any $\epsilon > 0$

$$\sup_x \int_{\mathbb{R}^d \setminus B_\epsilon} \nu(x, dy) < \infty, \quad \sup_x \int_{\mathbb{R}^d \setminus B_\epsilon} |\nabla\nu(x, dy)| < \infty. \tag{5.6}$$

Next, since the operator

$$\int_{\mathbb{R}^d \setminus B_1} (f(x+y) - f(x))\nu(x, dy) \tag{5.7}$$

is bounded in the Banach spaces $C(\mathbb{R}^d)$ and $C^1(\mathbb{R}^d)$ (by (5.2)) and also in the Banach spaces $C_\infty(\mathbb{R}^d)$ and $C^1_\infty(\mathbb{R}^d)$ (by (5.3)), by the standard perturbation argument (see Theorem 1.9.2) we can reduce the situation to the case when all $\nu(x, dy)$ have support in B_1, which we shall assume from now on.

Let us introduce the approximation

$$L_h f(x) = (b(x), \nabla f(x)) + \int_{\mathbb{R}^d \setminus B_h} (f(x+y) - f(x))\nu(x, dy). \tag{5.8}$$

For any $h > 0$ this operator generates a conservative Feller semigroup T_t^h in $C_\infty(\mathbb{R}^d)$ with invariant core $C^1_\infty(\mathbb{R}^d)$, because so does the first term in (5.8) and the second term is a bounded operator in the Banach spaces $C_\infty(\mathbb{R}^d)$ and $C^1_\infty(\mathbb{R}^d)$ (by (5.6)), so that perturbation theory (Theorem 1.9.2) applies (conservativity also follows from the perturbation series representation).

Differentiating the equation $\dot{f}(x) = L_h f(x)$ with respect to x yields the equation

$$\begin{aligned}\frac{d}{dt}\nabla_k f(x) &= L_h \nabla_k f(x) + (\nabla_k b(x), \nabla f(x)) \\ &\quad + \int_{B_1 \setminus B_h} (f(x+y) - f(x))\nabla_k \nu(x, dy).\end{aligned} \tag{5.9}$$

Considering this as an evolution equation for $g = \nabla f$ in the Banach space $C_\infty(\mathbb{R}^d \times \{1, \dots, d\}) = C_\infty(\mathbb{R}^d) \times \dots \times C_\infty(\mathbb{R}^d)$, observe that the r.h.s. is represented as the sum of the diagonal operator that generates a Feller semigroup and of the two bounded (uniformly in h by (5.2)) operators. Hence this evolution is well posed.

To show that the derivative of $f(x)$ is actually given by the semigroup generated by (5.9), we first approximate b, ν by a sequence of the twice continuously differentiable objects b_n, ν_n, $n \to \infty$. The corresponding approximating generators of type (5.9) have an invariant core $C^1_\infty(\mathbb{R}^d)$, hence the uniqueness of the solutions to the corresponding evolution equation holds (by Theorem 1.9.4) implying that this solution coincides with the derivative of the corresponding $(T_t^h)_n f$. Letting $n \to \infty$ completes the argument.

Hence $\nabla_k T_t^h f$ is uniformly bounded for all $h \in (0, 1]$ and t from any compact interval whenever $f \in C^1_\infty(\mathbb{R}^d)$. Therefore, writing

$$(T_t^{h_1} - T_t^{h_2})f = \int_0^t T_{t-s}^{h_2}(L_{h_1} - L_{h_2})T_s^{h_1}\, ds$$

for arbitrary $h_1 > h_2$ and estimating

$$\begin{aligned}|(L_{h_1} - L_{h_2})T_s^{h_1} f(x)| &\le \int_{B_{h_1} \setminus B_{h_2}} |(T_s^{h_1} f)(x+y) - (T_s^{h_1} f)(x)|\nu(x, dy) \\ &\le \int_{B_{h_1}} \|\nabla T_s^{h_1} f\| |y| \nu(x, dy) = o(1)\|f\|_{C^1_\infty}, \quad h_1 \to 0,\end{aligned}$$

by (5.4), yields

$$\|(T_t^{h_1} - T_t^{h_2})f\| = o(1)t\|f\|_{C^1_\infty}, \quad h_1 \to 0. \tag{5.10}$$

Therefore the family $T_t^h f$ converges to a family $T_t^h f$, as $h \to 0$. Clearly the limiting family T_t specifies a strongly continuous semigroup in $C_\infty(\mathbb{R}^d)$. Writing

$$\frac{T_t - f}{t} = \frac{T_t - T_t^h f}{t} + \frac{T_t^h - f}{t}$$

and noting that by (5.10) the first term is of order $o(1)\|f\|_{C^1_\infty}$ as $h \to 0$ allows one to conclude that $C^1_\infty(\mathbb{R}^d)$ belongs to the domain of the generator of the semigroup T_t in $C_\infty(\mathbb{R}^d)$ and that it is given there by (5.1).

Applying to T_t the procedure applied above to T_t^h (differentiating the evolution equation with respect to x) shows that T_t defines also a strongly continuous semigroup in $C^1_\infty(\mathbb{R}^d)$, as its generator differs from the diagonal operator with all entries on the diagonal being L by a bounded additive term. The perturbation series representation also implies estimate (5.5). □

As another example we analyze the simplest version of a decomposable generator, containing just one term. However, unlike the discussion in Section 5.7 no assumption whatsoever will be made on the underlying Lévy measure.

Theorem 5.1.2. *Let*

$$Lf(x) = a(x) \int [f(x+y) - f(x) - (y, \nabla f(x))]\nu(dy)$$

where ν is a Lévy measure with finite second moment, i.e. $\int |y|^2 \nu(dy) < \infty$, and the nonnegative function $a(x)$ is from $C^2(\mathbb{R}^d)$ and such that

$$\sup_{x,j} \left| \frac{\partial a}{\partial x_j} \right| \frac{1}{a} = \kappa < \infty, \quad \sup_{x,j} \left| \frac{\partial^2 a}{\partial x_j \partial x_i} \right| \frac{1}{a} = \tilde{\kappa} < \infty.$$

Then L generates a Feller semigroup T_t in $C_\infty(\mathbb{R}^d)$ with a core containing $C^2_\infty(\mathbb{R}^d)$. Moreover, if $f \in C^4(\mathbb{R}^d)$, then $T_t f \in C^2(\mathbb{R}^d)$ for all t and yields a classical solution to the equation $\dot{f} = Lf$.

Proof. As in the previous theorem, let $\nu_h(dy) = \mathbf{1}_{|y|>h}(y)\nu(dy)$ and

$$L_h f(x) = a(x) \int [f(x+y) - f(x) - (y, \nabla f(x))]\nu_h(dy).$$

Clearly each operator L_h, $h > 0$ generates a Feller semigroup T_h^t in $C_\infty(\mathbb{R}^d)$ with invariant cores C^j_∞, $j = 1, 2, 3$. Since $L_h T_t^h = T_t^h L_h$, it follows that

$$\|L_h T_t^h f_0\| \le \|L_h f_0\|$$

for any $f_0 \in C^1_\infty(\mathbb{R}^d)$. If $f_0 \in C^2_\infty(\mathbb{R}^d)$, then $L_h f_0 \in C^1_\infty(\mathbb{R}^d)$, and hence $L_h T^h_t f_0 \in C^1_\infty(\mathbb{R}^d)$ and

$$\|L_h T^h_t f_0\| \le \|f_0''\| \|a\| \int |y|^2 \nu(dy).$$

Let $f^h_t = T^h_t f_0$ and

$$g_j = \frac{\partial f_0}{\partial x_j}, \quad g^h_{j,t} = \frac{\partial T^h_t f_0}{\partial x_j},$$

which is well defined and belong to $C^1_\infty(\mathbb{R}^d)$ for any $f_0 \in C^2_\infty(\mathbb{R}^d)$. Differentiating the equation for f_h gives

$$\dot g^h_{j,t} = L_h g^h_{j,t} + \frac{\partial a}{\partial x_j}\frac{1}{a} L_h f^h_t.$$

Hence

$$\|g^h_{j,t}\| \le \|g_j\| + t\kappa \|L_h f_0\|,$$

implying that the first derivatives of $T^h_t f_0$ remains bounded uniformly in h.

Applying the above estimates for the derivatives to $L_h f_0$ yields

$$\left\| \frac{\partial}{\partial x_j} L_h f^h_t \right\| \le \left\| \frac{\partial L_h f_0}{\partial x_j} \right\| + t\kappa \|L_h L_h f_0\|.$$

Since

$$\frac{\partial}{\partial x_j} L_h f^h_t = L_h g^h_{j,t} + \frac{\partial a}{\partial x_j}\frac{1}{a} L_h f^h_t,$$

it follows that

$$\|L_h g^h_{j,t}\| \le \kappa \|L_h f_0\| + \left\| \frac{\partial L_h f_0}{\partial x_j} \right\| + t\kappa \|L_h L_h f_0\|.$$

Differentiating the equation for f again (and taking into account a straightforward cancellation) yields

$$\frac{d}{dt}\frac{\partial^2 f^h_t}{\partial x_j \partial x_i} = L_h \frac{\partial^2 f^h_t}{\partial x_j \partial x_i} + \frac{\partial a}{\partial x_i}\frac{1}{a} L_h g^t_{j,h} + \frac{\partial a}{\partial x_j}\frac{1}{a} L_h g^t_{i,h} + \frac{\partial^2 a}{\partial x_i \partial x_j}\frac{1}{a} L_h f^h_t,$$

implying

$$\left\| \frac{\partial^2 f^h_t}{\partial x_j \partial x_i} \right\| \le \left\| \frac{\partial^2 f_0}{\partial x_j \partial x_i} \right\| + 2t\kappa \left(\kappa \|L_h f_0\| + \left\| \frac{\partial L_h f_0}{\partial x_j} \right\| + t\kappa \|L_h L_h f_0\| \right) + t\kappa\tilde\kappa \|L_h f_0\|.$$

As in the previous theorem, we shall look for a limit of T^h as $h \to 0$. Assuming that $f_0 \in C^4(\mathbb{R}^d)$ we conclude that the first and second derivatives of f^h_t are uniformly bounded, which allows us to conclude (as in the proof of Theorem 5.1.1) that the approximations $T_t^h f_0$ converge as $h \to 0$ to a function that we denote by $T_t f_0$. As all T_t^h are contraction semigroups, we can extend this convergence to all $f \in C_\infty(\mathbb{R}^d)$, and conclude that the limiting family of operators T_t also forms a contraction semigroup. Again as in the proof of Theorem 5.1.1, differentiating the evolution equation for $T_t f$ with respect to x shows that if $f_0 \in C^4(\mathbb{R}^d)$ then the first two derivatives remain continuous and bounded and that f_0 belongs to the domain of the generator of T_t. As this generator is closed, one deduces that any $f \in C_\infty^2(\mathbb{R}^d)$ also belongs to the domain. □

Notice the crucial difference between the above two theorems. In the latter case we did not get any invariant core (to get boundedness in time of the second spatial derivatives we need the fourth derivative of the initial function). If $\ln a \in C^\infty(\mathbb{R}^d)$ it is not difficult to show by sequential differentiation that the space $C_\infty^\infty(\mathbb{R}^d)$ becomes an invariant core, but this still does not imply any nice estimates for the growth of these derivatives.

Our final example is one-dimensional, where the absolutely continuous part of the Lévy kernel dominates in some sense its derivatives. We shall denote the derivatives with respect to the space variable x by primes.

Theorem 5.1.3. *Let*

$$Lf(x) = \frac{1}{2}G(x)f''(x) + b(x)f'(x) + \int (f(x+y) - f(x) - f'(x)y)\nu(x,dy), \tag{5.11}$$

with

$$\sup_x \int (|y| \wedge |y|^2)\nu(x,dy) < \infty, \tag{5.12}$$

where $G, b \in C^2(\mathbb{R})$ and the first two derivatives $\nu'(x,dy)$ and $\nu''(x,dy)$ of ν with respect to x exist weakly and define continuous signed Lévy kernels such that

$$\sup_x \int (|y| \wedge |y|^2)|\nu'(x,dy)| < \infty, \quad \sup_x \int (|y| \wedge |y|^2)|\nu''(x,dy)| < \infty. \tag{5.13}$$

Let the density $\nu_{\text{abs}}(x,y)$ of the absolutely continuous part of $\nu(x,dy)$ satisfy the following conditions.

(i) *There exist $h_+, h_- > 0$ such that*

$$-2\int_w^{h_+} \nu'(x,dy) \le \nu_{\text{abs}}(x,w), \quad 2\int_{-h_-}^{-w} \nu'(x,dy) \le \nu_{\text{abs}}(x,w) \tag{5.14}$$

for all positive $w < h_+$ and $w < h_-$ respectively.

(ii) *There exists a positive sequence* $y_j \to 0$ *as* $j \to \infty$ *such that*

$$y_j^3 \sup_x \nu_{\mathrm{abs}}(x, \pm y_j) \to 0 \tag{5.15}$$

as $j \to \infty$.

Then L *generates a unique Feller semigroup with generator given by* (5.11) *on the subspace* $C^2_\infty(\mathbb{R})$ *and the latter space is its invariant core, where the semigroup is regular in the sense that*

$$\|T_t\|_{C^2_\infty(\mathbb{R}^d)} \le e^{Kt} \tag{5.16}$$

with a constant K.

Remark 38. Condition (ii) is a very mild regularity assumption, as for any fixed x it holds automatically due to the boundedness of $\int y^2 \nu_{\mathrm{abs}}(x, y)\, dy$.

Proof. We shall use the following Taylor formulas:

$$f(x+y) - f(x) - f'(x)y = \int_0^y (f'(x+z) - f'(x))dz = \int_0^y (y-z) f''(x+z)dz,$$

where of course $\int_0^y = -\int_y^0$ for $y < 0$.

Let us introduce the approximating operator L_h, $h \in [0, h_+ \wedge h_-]$, as

$$L_h f(x) = \frac{1}{2} G(x) f''(x) + b(x) f'(x) + \int (f(x+y) - f(x) - f'(x)y)\nu_h(x, dy),$$

where

$$\nu_h(x, dy) = \mathbf{1}_{|y| \ge h} \nu(x, dy) + \mathbf{1}_{|y|<h} \sup_z \nu_{\mathrm{abs}}(z, h) dy.$$

We shall use it actually only for $h = y_j$, implying in particular that the corresponding sup is finite. For $h > 0$, L_h is the sum of a diffusion operator and a bounded operator in $C_\infty(\mathbb{R})$. Hence (by the standard theory of one-dimensional diffusion, see e.g. Theorem 4.6.1[1] and the perturbation theory) it generates a conservative Feller semigroup T_t^h. Let $f_t^h = T_t^h f$ for $f \in C_\infty(\mathbb{R})$.

Differentiating the equation $\dot{f}^h = L_h f^h$ with respect to the spatial variable x yields the following equation for $g^h = (f^h)'$:

$$\frac{d}{dt} g^h(x) = (L_h^0 + K_h^0 + L^1 + K^1) g^h(x) + \frac{1}{2} G'(x)(g^h)'(x) + b'(x) g^h(x), \tag{5.17}$$

[1] One can avoid referring to this theory by working with an approximation to $f''(x)$ by an appropriate integral operator, say $3h^{-3} \int_{-h}^h (f(x+y) - f(x))dy$.

where

$$L^1 g(x) = \frac{1}{2} G(x) g''(x) + b(x) g'(x)$$
$$+ \left(\int_{h_+}^{\infty} + \int_{-\infty}^{-h_-} \right) (g(x+y) - g(x) - g'(x)y) \nu_h(x, dy),$$
$$L_h^0 g(x) = \left(\int_0^{h_+} + \int_{-h_-}^{0} \right) (g(x+y) - g(x) - g'(x)y) \nu_h(x, dy),$$
$$K^1 g(x) = \left(\int_{h_+}^{\infty} + \int_{-\infty}^{-h_-} \right) (f(x+y) - f(x) - f'(x)y) \nu'(x, dy),$$
$$K_h^0 g(x) = \left(\int_0^{h_+} + \int_{-h_-}^{0} \right) (f(x+y) - f(x) - f'(x)y) \nu_h'(x, dy).$$

Using the above Taylor formula, one can write

$$K_h^0 g(x) = \int_0^{h_+} \left(\int_0^{y} (g(x+z) - g(x)) dz \right) \nu_h'(x, dy)$$
$$- \int_{-h_-}^{0} \left(\int_y^{0} (g(x+z) - g(x)) dz \right) \nu_h'(x, dy),$$

and consequently

$$(L_h^0 + K_h^0) g(x) = \int_0^{h_+} (g(x+z) - g(x)) \left[\nu_h(x, dz) + \left(\int_z^{h_+} \nu_h'(x, dy) \right) dz \right]$$
$$+ \int_{-h_-}^{0} (g(x+z) - g(x)) \left[\nu_h(x, dz) - \left(\int_{-h_-}^{z} \nu_h'(x, dy) \right) dz \right]$$
$$- \int_{-h_-}^{h_+} y \nu_h(x, dy) g'(x),$$

and similarly for K^1.

Our key observation is the following. Due to (5.14), and since

$$\int_z^{h_+} \nu_h'(x, dy) = \int_h^{h_+} \nu_h'(x, dy), \quad z < h,$$

the operator $L_h^0 + K_h^0$ is bounded, conditionally positive and satisfies the positive maximum principle for any $h > 0$, because it has the standard Lévy–Khintchine form.

Consequently, it generates a bounded positivity-preserving contraction semigroup in $C_\infty(\mathbb{R}^d)$. Therefore we can conclude (by perturbation theory) that equation (5.17) generates a bounded semigroup in $C_\infty(\mathbb{R}^d)$ uniformly in h. Hence the first derivatives in x of the function $T_t^h f(x)$ are uniformly bounded in h and $t \leq t_0$ for any t_0 and $f \in C^1(\mathbb{R})$.

Differentiating the equation for f^h once more we get for $v^h = (g^h)'$ the equation

$$\begin{aligned}\frac{d}{dt} v^h(x) &= (L_h^0 + 2K_h^0 + L^1 + 2K^1)v(x)\\ &\quad + G'(x)(v^h)'(x) + 2b'(x)v^h(x) + \frac{1}{2}G''(x)(v^h)(x) + b''(x)g^h(x)\\ &\quad + \int (f(x+y) - f(x) - f'(x)y)\nu_h''(x,dy).\end{aligned}$$

Since the last integral equals

$$\int \nu_h''(x,dy) \int_0^y (y-z)v(x+z)dz$$

and represents a bounded operator of v uniformly for h, we conclude as above (here the coefficient 2 in (5.14) is needed) that for any bounded continuous g this equation generates a family of strongly continuous operators in $C_\infty(\mathbb{R})$ (giving the solution to its Cauchy problem), uniformly bounded in h, implying that $T_t^h f(x) \in C^2(\mathbb{R})$ uniformly in h and finite t whenever $f \in C^2(\mathbb{R})$.

Let us choose the sequence of approximations L_{y_j}, which we briefly denote L_j, where y_j are from condition (ii). Therefore, writing

$$(T_t^{y_k} - T_t^{y_j})f = \int_0^t T_{t-s}^{y_j}(L_j - L_k)T_s^{y_k}\, ds$$

for arbitrary $j < k$, $y_j < h^+ \wedge h_-$, and estimating

$$\begin{aligned}&|(L_k - L_j)T_s^{y_k} f(x)|\\ &\quad \leq \|T_s^{y_k} f\|_{C_2(\mathbb{R})} \left[2y_j^3 \sup_z(\nu_{\text{abs}}(z,y_j) + \nu_{\text{abs}}(z,y_k)) + \int_{-y_j}^{y_j} y^2 \nu(x,dy)\right]\\ &\quad = o(1)\|f\|_{C^2(\mathbb{R})}, \quad j,k \to \infty,\end{aligned}$$

yields

$$\|(T_t^{y_k} - T_t^{y_j})f\| = o(1)t\|f\|_{C^2(\mathbb{R})}, \quad j \to \infty. \tag{5.18}$$

Therefore the family $T_t^{y_j} f$ converges to a family $T_t f$, as $j \to \infty$. Clearly the limiting family T_t specifies a strongly continuous semigroup in $C_\infty(\mathbb{R})$.

Applying to T_t the same procedure, as was applied above to T_t^ϵ (differentiating the evolution equation with respect to x), shows that T_t defines also a contraction semigroup in $C_\infty(\mathbb{R}) \cap C^1(\mathbb{R})$, preserving the spaces $C_\infty^1(\mathbb{R})$ and $C_\infty^2(\mathbb{R})$. The proof is completed as in Theorem 5.1.1. □

5.2 Introduction to sensitivity analysis

As we already mentioned, for using Markov processes as a modeling tool in concrete problems it is important to be able to assess how sensitive is the behavior of the process to changing the key parameters of the model. Ideally one would like to have some kind of smooth dependence of the evolution on these parameters. Here we shall touch on this problem for the examples of the processes discussed above.

Theorem 5.2.1. *Under the assumptions of Theorem* 5.1.1, *suppose that b and ν depend on a real parameter α such that all estimates in the condition of this theorem are uniform with respect to this parameter. Let T_t^α denote the corresponding semigroups. Suppose the initial function f_0^α is continuously differentiable with respect to α and belongs to $C^1(\mathbb{R}^d)$ as a function of x uniformly in α. Then the function $T_t^\alpha f_0^\alpha$ is continuously differentiable with respect to α for any $t \geq 0$, and*

$$\sup_\alpha \left\| \frac{\partial}{\partial \alpha} T_t f_0^\alpha \right\| \leq \sup_\alpha \left\| \frac{\partial}{\partial \alpha} f_0^\alpha \right\| + O(t) \sup_\alpha \| f_0^\alpha \|_{C^1_\infty(\mathbb{R}^d)} \tag{5.19}$$

(where the norms refer to f^α as functions of the position x).

Proof. Differentiating the equation $\dot{f} = Lf$ with respect to α yields the equation

$$\frac{d}{dt}\frac{\partial f}{\partial \alpha} = L\frac{\partial f}{\partial \alpha} + \left(\frac{\partial b}{\partial \alpha}, \frac{\partial f}{\partial x}\right) + \int_{\mathbb{R}^d} (f(x+y) - f(x)) \frac{\partial \nu}{\partial \alpha}(x, dy). \tag{5.20}$$

Since the last two terms are bounded by $\sup_\alpha \| f_0^\alpha \|_{C^1_\infty(\mathbb{R}^d)}$ (because $C^1_\infty(\mathbb{R}^d)$ is an invariant core for all T_t^α), it follows that the derivative of f with respect to α satisfies the same equation as f itself up to a bounded non-homogeneous additive term, which implies (5.19). □

Similar result can be obtained on the basis of Theorem 5.1.3 and many other results given below, where a regular enough invariant core for the semigroup is identified, say for Proposition 4.6.2.

5.3 The Lie–Trotter type limits and T-products

The formula

$$e^{L_1+L_2} = \lim_{n\to\infty} (e^{L_1/n} e^{L_2/n})^n \tag{5.21}$$

(Lie–Trotter Daletski–Chernoff) was established and widely applied under various assumptions on the linear operators L_1, L_2. However usually it is obtained under the condition of the existence of the semigroup $\exp\{t(L_1 + L_2)\}$. We are going to discuss here a situation where the semigroups generated by L_1, L_2 are regular enough to allow one to deduce also the existence of the semigroup generated by $L_1 + L_2$, to identify its invariant core and to get the precise rates of convergence. Further we extend this result to several generators and to a time-nonhomogeneous case.

For given operators L_1, L_2 generating bounded semigroups e^{tL_1} and e^{tL_2} in a Banach space and a given $\tau > 0$ let us define the family of bounded operators U_t^τ, $t > 0$, in the following way. For a natural number k, let

$$U_t^\tau = e^{(t-2k\tau)L_1}(e^{\tau L_2}e^{\tau L_1})^k, \qquad 2k\tau \le t \le (2k+1)\tau, \tag{5.22}$$

$$U_t^\tau = e^{(t-(2k+1)\tau)L_2}e^{\tau L_1}(e^{\tau L_2}e^{\tau L_1})^k, \qquad (2k+1)\tau \le t \le (2k+2)\tau. \tag{5.23}$$

We shall work with three Banach spaces B_0, B_1, B_2 with the norms denoted by $\|.\|_i$, $i = 0, 1, 2$, such that $B_0 \subset B_1 \subset B_2$, B_0 is dense in B_1, B_1 is dense in B_2 and $\|.\|_0 \ge \|.\|_1 \ge \|.\|_2$.

Theorem 5.3.1. *Suppose the linear operators L_1, L_2 in B_2 generate strongly continuous semigroups e^{tL_1} and e^{tL_2} in B_2 with B_1 being their common invariant core. Suppose additionally that*

(i) *L_1, L_2 are bounded operators $B_0 \to B_1$ and $B_1 \to B_2$,*

(ii) *B_0 is also invariant under both e^{tL_1} and e^{tL_2} and these operators are bounded as operators in B_0, B_1, B_2 with norms not exceeding e^{Kt} with K constant (the same for all B_j and L_i).*

Then

(i) *for any $T > 0$ and $f \in B_2$ the curves $U_t^{2^{-k}} f$ converge in $C([0,T], B_2)$ to a curve $U_t f$, and for $f \in B_1$ this convergence holds in $C([0,T], B_1)$ and $U_t f \in C([0,T], B_1)$;*

(ii) *the norms $\|U_t\|_{B_2}$ and $\|U_t\|_{B_1}$ are bounded by e^{Kt};*

(iii) *for $f \in B_0$ and $2^{-k} \le t/2$*

$$\|(U_t^{2^{-k}} - U_t)f\|_2 = \|f\|_0 O(t) 2^{-k}. \tag{5.24}$$

(iv) *the operators U_t form a strongly continuous semigroup in B_2 with the generator $(L_1 + L_2)/2$ having B_1 as its invariant core;*

Proof. Let an arbitrary $T > 0$ be chosen. By t we shall denote positive numbers not exceeding T. First note that condition (ii) implies that

$$\|U_t^\tau\|_{B_j} \le e^{Kt} \quad \forall \tau \le 1,\ j = 0, 1, 2. \tag{5.25}$$

Next, for any $f \in B_1$ and $i = 1, 2$

$$e^{tL_i} f - f = \int_0^t L_i e^{sL_i} f \, ds, \tag{5.26}$$

implying firstly that

$$\|e^{tL_i} f - f\|_2 = O(t)\|L_i\|_{B_1\to B_2} e^{Kt}\|f\|_1, \tag{5.27}$$

and secondly that in case $f \in B_0$

$$\|e^{tL_i} f - f\|_1 = O(t)\|L_i\|_{B_0\to B_1} e^{Kt}\|f\|_0, \tag{5.28}$$

and moreover

$$\begin{aligned} e^{tL_i} f &= f + tL_i f + \int_0^t L_i(e^{sL_i} f - f)\, ds \\ &= f + t(L_i f + \|L_i\|_{B_1\to B_2}\|L_i\|_{B_0\to B_1}\|f\|_0[O(t)]_2), \end{aligned} \tag{5.29}$$

where by $[O(t)]_2$ we denoted a vector with a B_2-norm of order $O(t)$. Consequently,

$$\begin{aligned} e^{tL_1}e^{tL_2} f &= e^{tL_1} f + te^{tL_1}(L_2 f + \|f\|_0[O(t)]_2) \\ &= f + tL_1 f + t\|f\|_0[O(t)]_2 + te^{tL_1}L_2 f \\ &= f + t(L_1 + L_2) f + t\|f\|_0[O(t)]_2, \end{aligned} \tag{5.30}$$

where the last equation comes from estimate (5.27) applied to $L_2 f$ instead of f. Consequently

$$\|(e^{tL_1}e^{tL_2} - e^{tL_2}e^{tL_1}) f\|_2 = O(t^2)\|f\|_0,$$

and therefore

$$\begin{aligned} \|(e^{2tL_1}e^{2tL_2} - e^{tL_1}e^{tL_2}e^{tL_1}e^{tL_2}) f\|_2 &= \|e^{tL_1}(e^{tL_1}e^{tL_2} - e^{tL_2}e^{tL_1})e^{tL_2} f\|_2 \\ &= O(t^2)\|f\|_0. \end{aligned}$$

Writing now

$$\begin{aligned} &(e^{\tau L_2}e^{\tau L_1})^k - (e^{\tau L_2/2}e^{\tau L_1/2})^{2k} \\ &\quad = \sum_{l=1}^{k} (e^{\tau L_2}e^{\tau L_1})^{k-l}[e^{\tau L_2}e^{\tau L_1} - (e^{\tau L_2/2}e^{\tau L_1/2})^2](e^{\tau L_2/2}e^{\tau L_1/2})^{2l-2}, \end{aligned}$$

we can conclude that

$$\|((e^{\tau L_2}e^{\tau L_1})^k - (e^{\tau L_2/2}e^{\tau L_1/2})^{2k}) f\|_2 = k\|f\|_0 O(\tau^2),$$

so that

$$\|(U_t^\tau - U_t^{\tau/2}) f\|_2 = \|f\|_0 O(t\tau). \tag{5.31}$$

Consequently for a natural number k

$$\|(U_t^\tau - U_t^{\tau 2^{-k}}) f\|_2 = \|f\|_0 O(t\tau)(1 + \cdots + 2^{-k+1}) = \|f\|_0 O(t\tau),$$

implying that $U_t^{2^{-k}}$ converges in $C([0,T],B_2)$ as $k\to\infty$ to a curve, which we denote by $U_t f$, and that estimate (5.24) holds. Since U_t are uniformly bounded, we deduce by the usual density argument that $U_t^{2^{-k}} f$ converge for any $f\in B_2$, and that the limiting set of operators U_t forms a bounded semigroup in B_2.

Next let us observe that the family $U_t^\tau f$ is relatively compact in $C([0,T],B_1)$ by Arzelà–Ascoli theorem for any $f\in B_0$, because

$$\prod_{i=1}^k e^{\tau\tilde L_i} f - f = \sum_{i=1}^k \int_{\tau(i-1)}^{\tau i} \tilde L_i e^{(s-\tau(i-1))\tilde L_i} \prod_{l=1}^{i-1} e^{\tau \tilde L_l} f\, ds,$$

where each $\tilde L_i$ is any of the operators L_1, L_2, so that

$$\left\| \prod_{i=1}^k e^{\tau\tilde L_i} f - f \right\|_1 = O(k\tau)\|f\|_0.$$

Hence the sequence $U_t^{2^{-k}} f$ contains a convergent subsequence in $C([0,T],B_1)$. But the limit of such a subsequence is uniquely defined (it is its limit in $C([0,T],B_2)$), implying that the whole sequence $U_t^{2^{-k}} f$ converges in $C([0,T],B_1)$, as $k\to\infty$. Again by the density argument we conclude that this convergence still holds for any $f\in B_1$, implying that U_t form a bounded semigroup in B_1.

It remains to prove statement (iv) of the theorem. Denote $\tau_k = 2^{-k}$. Let first $f\in B_0$ and let t be a binary rational. Then $t/2\tau_k\in\mathbb{N}$ for k large enough so that

$$U_t^{\tau_k} f = (e^{\tau_k L_2} e^{\tau_k L_1})^{t/2\tau_k}$$

and

$$U_t^{\tau_k} f - f = \sum_{l=0}^{(t/2\tau_k)-1} (e^{\tau_k L_2} e^{\tau_k L_1} - 1)(e^{\tau_k L_2} e^{\tau_k L_1})^l f.$$

Therefore, by (5.30),

$$U_t^{\tau_k} f - f = \tau_k \sum_{l=0}^{(t/2\tau_k)-1} (L_1+L_2)(e^{\tau_k L_2} e^{\tau_k L_1})^l f + \|f\|_0[O(\tau_k)]_2.$$

This can be rewritten as

$$U_t^{\tau_k} f - f = \frac{1}{2}(2\tau_k) \sum_{l=0}^{(t/2\tau_k)-1} (L_1+L_2) U_{2l\tau_k} f$$
$$+ \tau_k \sum_{l=0}^{(t/2\tau_k)-1} (L_1+L_2)[U_{2l\tau_k}^{\tau_k} - U_{2l\tau_k}] f + \|f\|_0[O(\tau_k)]_2.$$

Passing to the limit as $k \to \infty$ in the topology of B_2 yields

$$U_t f - f = \frac{1}{2} \int_0^t (L_1 + L_2) U_s f \, ds. \tag{5.32}$$

By the density argument the same formula holds for any $f \in B_1$. Using now the continuity of $U_t f$ in B_1 it follows from (5.32) that

$$\frac{d}{dt} U_t f = \frac{1}{2}(L_1 + L_2) f$$

for any $f \in B_1$, where the derivative is defined in the topology of B_2. □

Let us extend the result to several generators. Let the operators L_i, $i = 1, \dots, n$, generating bounded semigroups e^{tL_i} in a Banach space and a $\tau > 0$ be given. For natural k and $l \in \{0, \dots, n-1\}$ let

$$U_t^\tau = e^{(t-(nk+l)\tau)L_{l+1}} e^{\tau L_l} \cdots e^{\tau L_1} (e^{\tau L_n} \cdots e^{\tau L_1})^k,$$
$$(nk + l)\tau \le t \le (nk + l + 1)\tau. \tag{5.33}$$

Theorem 5.3.2. *Suppose that*

(i) *the linear operators L_i, $i = 1, \dots, n$, in B_2 generate strongly continuous semigroups e^{tL_i} in B_2, with B_1 their common invariant core,*

(ii) *all L_i are bounded as operators $B_0 \to B_1$ and $B_1 \to B_2$,*

(iii) *B_0 is invariant under all e^{tL_i} and the latter operators are bounded as operators in B_0, B_1, B_2 with the norms not exceeding e^{Kt} with a constant K (the same for all B_j and L_i).*

Then statements (i)–(iii) *of Theorem* 5.3.1 *hold for the family* (5.33) *and the operators U_t form a strongly continuous semigroup in B_2 with the generator $(L_1 + \cdots + L_n)/n$ having B_1 as its invariant core.*

Proof. This is the same as above with obvious modifications. Say, instead of (5.30) one gets

$$e^{tL_n} \cdots e^{tL_1} f = f + t(L_1 + \cdots + L_n) f + t \|f\|_0 [O(t)]_2, \tag{5.34}$$

and

$$(e^{tL_n} \cdots e^{tL_1})^k - (e^{tL_n/2} \cdots e^{tL_1/2})^{2k}$$
$$= \sum_{l=1}^k (e^{tL_n} \cdots e^{tL_1})^{k-l} [e^{tL_n} \cdots e^{tL_1} - (e^{tL_n/2} \cdots e^{tL_1/2})^2] (e^{tL_n/2} \cdots e^{tL_1/2})^{2l-2},$$

yielding again (5.31). □

Let us extend further to the time non-homogeneous situation. For families L_i^s, $i = 1, \dots, n$, $s \geq 0$, of linear operators in a Banach space such that each L_i^s generates a bounded semigroup, and for a given $\tau > 0$, let us define the family U_t^τ in the following way. For natural k and $m < l \in \{0, \dots, n-1\}$ let

$$U_{t,s}^\tau = \exp\{(t-s)L_{l+1}^{(nk+l)\tau}\}, \quad (nk+l)\tau \leq s \leq t \leq (nk+l+1), \tag{5.35}$$

and

$$U_{t,s}^\tau = \exp\{(t-(nk+l)\tau)L_{l+1}^{(nk+l)\tau}\} \times \prod_{j=m+2}^{l} \exp\{\tau L_j^{(nk+j-1)\tau}\} \exp\{((nk+m+1)\tau - s)\tau L_{m+1}^{(nk+m)\tau}\} \tag{5.36}$$

for

$$(nk+m)\tau \leq s \leq (nk+m+1)\tau \leq (nk+l)\tau \leq t \leq (nk+l+1)\tau.$$

For other $s \leq t$, these operators are defined by gluing together to form a propagator. In particular, if $s = kn\tau$, $t = (k+m)n\tau$, $m, k \in \mathbb{N}$, then

$$U_{t,s}^\tau = (\exp\{\tau L_n^{[(k+m)n-1]\tau}\} \cdots \exp\{\tau L_1^{(k+m-1)n\tau}\}) \cdots \times (\exp\{\tau L_n^{(kn+n-1)\tau}\} \cdots \exp\{\tau L_2^{(kn+1)\tau}\} \exp\{\tau L_1^{kn\tau}\}).$$

As above we denote $\tau_k = 2^{-k}$.

Theorem 5.3.3. *Suppose that*

(i) *linear operators L_i^s, $i = 1, \dots, n$, $s \in [0, T]$, generate strongly continuous semigroups $e^{tL_i^s}$ in B_2 with common invariant core B_1,*

(ii) *all L_i^s are bounded as operators $B_0 \to B_1$ and $B_1 \to B_2$ with a common bound,*

(iii) *B_0 is invariant under all $e^{tL_i^s}$ and the latter operators are bounded as operators in B_0, B_1, B_2 with the norms not exceeding e^{Kt} with a constant K (the same for all B_j and L_i^s),*

(iv) *L_j^t depend Lipschitz continuous on t in the following sense:*

$$\|(L_j^{t+\tau} - L_j^t)f\|_2 = O(\tau)\|f\|_0 \tag{5.37}$$

uniformly for finite t, τ.

Then

(i) *the propagators* $U_{t,s}^{\tau_k}$ *converge in* $C([0,T],B_2)$ *and in* $C([0,T],B_1)$ *to a regular propagator* $U_{t,s}$, *i.e., such that*

$$\|U_{t,s}\|_{B_2} \le e^{K(t-s)}, \quad \|U_{t,s}\|_{B_1} \le e^{K(t-s)}$$

for some constant K*;*

(ii) *for* $f \in B_0$ *and* $\tau_k \le t/n$,

$$\|(U_{t,s}^{\tau_k} - U_{t,s})f\|_2 = \|f\|_0 O(t-s)\tau_k;$$

(iii) *the propagator* $U_{t,s}$ *is generated by the family* $(L_1^t + \cdots + L_n^t)/n$ *(in the sense of the definition given before Theorem* 1.9.3*).*

Proof. This follows the same lines as above. Instead of (5.34) one gets

$$\exp\{tL_{i_k}^{t_k}\}\cdots\exp\{tL_{i_1}^{t_1}\}f = f + t(L_{i_1}^{t_1} + \cdots + L_{i_k}^{t_k})f + t\|f\|_0[O(t)]_2, \quad (5.38)$$

for any finite collection of the operators $L_{i_l}^{t_l}$ with $[O(t)]_2$ being uniform for bounded k, t, t_j. This implies

$$\begin{aligned}
&(U_{(k+1)n\tau,kn\tau}^{\tau} - U_{(k+1)n\tau,kn\tau}^{\tau/2})f \\
&\quad = \frac{1}{2}\tau(2L_n^{(k+1)n-1)\tau} - L_n^{(2kn+2n-1)n-1)\tau/2} - L_n^{(kn+n-1)\tau/2} + \cdots \\
&\qquad + 2L_1^{kn\tau} - L_1^{(2kn+n)\tau/2} - L_1^{2kn\tau/2})f + \tau\|f\|_0[O(\iota)]_2 = \tau\|f\|_0[O(\tau)]_2
\end{aligned}$$

by (5.37), yielding the crucial estimate for subdivisions. □

The case $n = 1$ in Theorem 5.3.3 describes the construction of a propagator as a T-product. But in this case simpler assumptions are sufficient, and we shall prove the corresponding result independently.

Let $L_t : B_1 \mapsto B_2$, $t \ge 0$, be a family of uniformly (in t) bounded operators such that the closure in B_2 of each L_t is the generator of a strongly continuous semigroups of bounded operators in B_2. For a partition $\Delta = \{0 = t_0 < t_1 < \cdots < t_N = t\}$ of an interval $[0,t]$ let us define a family of operators $U_\Delta(\tau,s)$, $0 \le s \le \tau \le t$, by the rules

$$\begin{aligned}
U_\Delta(\tau,s) &= \exp\{(\tau - s)L_{t_j}\}, \quad t_j \le s \le \tau \le t_{j+1}, \\
U_\Delta(\tau,r) &= U_\Delta(\tau,s)U_\Delta(s,r), \quad 0 \le r \le s \le \tau \le t.
\end{aligned}$$

Let $\Delta t_j = t_{j+1} - t_j$ and $\delta(\Delta) = \max_j \Delta t_j$. If the limit

$$U(s,r)f = \lim_{\delta(\Delta)\to 0} U_\Delta(s,r)f \quad (5.39)$$

exists for some f and all $0 \leq r \leq s \leq t$ (in the norm of B_2), it is called the *T-product* (or *chronological exponent* of L_t) and is denoted by $T\exp\{\int_r^s L_\tau\, d\tau\} f$. Intuitively, one expects the T-product to give a solution to the Cauchy problem

$$\frac{d}{dt}\phi = L_t\phi, \quad \phi_0 = f, \tag{5.40}$$

in B_2 with the initial conditions f from B_1.

Theorem 5.3.4. *Let a family $L_t f$, $t \geq 0$, of linear operators in B_2 be given such that*

(i) *each L_s generates a strongly continuous semigroup e^{tL_s} in B_2 with invariant core B_1,*
(ii) *L_s are uniformly bounded operators $B_0 \to B_1$ and $B_1 \to B_2$,*
(iii) *B_0 is also invariant under all e^{sL_t} and these operators are uniformly bounded as operators in B_0, B_1,*
(iv) *L_t, as a function $t \mapsto B_2$, depends continuously on t locally uniformly in f (i.e. for f from bounded domains of B_1).*

Then

(i) *the T-product $T\exp\{\int_0^s L_\tau\, d\tau\} f$ exists for all $f \in B_2$, and the convergence in (5.39) is uniform in f on any bounded subset of B_1;*
(ii) *if $f \in B_0$, then the approximations $U_\Delta(s,r)$ converge also in B_1;*
(iii) *this T-product defines a strongly continuous (in t, s) family of uniformly bounded operators in both B_1 and B_2,*
(iv) *this T-product $T\exp\{\int_0^s L_\tau\, d\tau\} f$ is a solution of problem (5.40) for any $f \in B_1$.*

Proof. (i) Since B_1 is dense in B_2, the existence of the T-product for all B_2 follows from its existence for $f \in B_1$, in which case it follows from the formula

$$U_\Delta(s,r) - U_{\Delta'}(s,r) = U_{\Delta'}(s,\tau)U_\Delta(\tau,r)|_{\tau=r}^{\tau=s} = \int_r^s \frac{d}{d\tau} U_{\Delta'}(s,\tau)U_\Delta(\tau,r)\, d\tau$$

$$= \int_r^s U_{\Delta'}(s,\tau)(L_{[\tau]_\Delta} - L_{[\tau]_{\Delta'}})U_\Delta(\tau,r)\, d\tau$$

(where we denoted $[s]_\Delta = t_j$ for $t_j \leq s < t_{j+1}$) and the uniform continuity of L_t.

(ii) If $f \in B_0$, then it follows from the approximate equations

$$U_\Delta(s,r) = \int_r^s L_{[\tau]_\Delta} U_\Delta(\tau,r)\, d\tau,$$

that the family $U_\Delta(s,t)$ is uniformly Lipschitz continuous in B_1 as a function of t. Hence one can choose a subsequence converging in $C([0,T], B_1)$ subsequence. But the limit is unique (it is the limit in B_2), implying the convergence of the whole family $U_\Delta(s,t)$.

(iii) It follows from (ii) that the limiting propagator is bounded. Strong continuity in B_1 is deduced first for $f \in B_0$ and then for all $f = inB_1$ by the density argument.

(iv) Since B_0 is dense in B_1, it is enough to prove the claim for $f \in B_0$. To this end, one can pass to the limit in the above approximate equations. □

An application of T-products will be given in Section 5.7 below.

To conclude the section we present an example of the application of Theorem 5.3.3 (in the corresponding time homogeneous case Theorem 5.3.2 is of course sufficient). Namely we shall construct a nonhomogeneous diffusions combined with mixtures of possibly degenerate *stable-like process* and processes generated by the operators of order at most one. Namely, let

$$L_t f(x) = \frac{1}{2}\operatorname{tr}(\sigma_t(x)\sigma_t^T(x)\nabla^2 f(x)) \tag{5.41}$$
$$+ (b_t(x), \nabla f(x)) + \int (f(x+y) - f(x))\nu_t(x, dy)$$
$$+ \int_P (dp) \int_0^K d|y| \int_{S^{d-1}} a_{p,t}(x,s) \frac{f(x+y) - f(x) - (y, \nabla f(x))}{|y|^{\alpha_{p,t}(x,s)+1}} d|y|\omega_{p,t}(ds),$$

where $s = y/|y|$, $K > 0$ and (P, dp) is a Borel space with a finite measure dp and $\omega_{p,t}$ are certain finite Borel measures on S^{d-1}.

Proposition 5.3.1. *Let the functions σ, b, a, α and the finite measure $|y|\nu(x, dy)$ be of smoothness class C^5 with respect to all variables (the measure is smooth in the weak sense), and a_p, α_p take values in compact subintervals of $(0, \infty)$ and $(0, 2)$ respectively. Then the family of operators L_t of form* (4.64) *generates a propagator $U_{t,s}$ on the invariant domain $C^2_\infty(\mathbb{R}^d)$ (in the sense of the definition given before Theorem* 1.9.3*), and hence a unique Markov process.*

Proof. This follows from Theorems 5.3.3, 5.1.1 and Proposition 4.6.2. Notice that since ν is not supposed to be bounded the perturbation theory argument combined with (a nonhomogeneous version of) Proposition 4.6.2 would not suffice. □

5.4 Martingale problems for Lévy type generators: existence

Here we prove a rather general existence result for the martingale problem corresponding to a pseudo-differential (or integro-differential) operator of the form

$$Lu(x) = \frac{1}{2}(G(x)\nabla, \nabla)u(x) + (b(x), \nabla u(x))$$
$$+ \int [u(x+y) - u(x) - (y, \nabla u(x))\mathbf{1}_{B_1}(y)]\nu(x, dy), \tag{5.42}$$

where $\nu(x, .)$ is a Lévy measure for all x.

But let us start with the basic criterion of stochastic continuity. It can be used to verify the conditions of Theorem 4.10.2, and will be essential for the applications to processes with boundary.

Theorem 5.4.1. *Let a family of the solutions $X_t^{x,\alpha}$ of the martingale problems for Lévy-type generators L_α with the common domain $C_c^2(\mathbb{R}^d)$ of form (5.42) with coefficients G_α, b_α, ν_α and with the common domain $C_c^2(\mathbb{R}^d)$ be given (α from an arbitrary index set, x denotes the starting point) such that the coefficients are uniformly bounded in bounded domains, i.e.*

$$\sup_{x\in K,\alpha}\left(\|G_\alpha(x)\| + \|b_\alpha(x)\| + \int (1\wedge |y|^2)\nu_\alpha(x,dy)\right) < \infty \tag{5.43}$$

and the compact containment condition for compact initial data holds uniformly in α, i.e., for any $\eta, T > 0$ and compact set $K \subset S$, there exists a compact set $\Gamma_{\eta,T,K} \subset \mathbb{R}^d$ such that

$$\inf_{x\in K,\alpha} \mathbf{P}\{X_t^{x,\alpha} \in \Gamma_{\eta,T,K}\ \forall\, t\in[0,T]\} \geq 1-\eta. \tag{5.44}$$

Then the family of the processes $X_t^{x,\alpha}$ is uniformly stochastically continuous, in the sense that for any compact set K

$$\lim_{t\to 0}\sup_{x\in K,\alpha} \mathbf{P}\{\sup_{s\leq t}\|X_s^{x,\alpha}-x\| > r\} = 0. \tag{5.45}$$

Proof. Let ρ denote an even non-negative function from $C_c^2(\mathbb{R})$ such that $\rho(r) = r^2$ for r from a neighborhood of the origin. Let $f_{n,x}(y) = \rho(|y-x|/n)$. By (5.43),

$$\sup_{x\in K,\alpha} |Lf_{n,x}| \leq C(K,n).$$

Hence, applying the Dynkin martingale built on the function $f_{n,x}$ yields

$$\sup_{x\in K,\alpha} \mathbf{E}\rho(|X_{x,\alpha}^t - x|/n) \leq tC(K,n),$$

and consequently by Doob's inequality

$$\sup_{x\in K,\alpha} \mathbf{P}\{\sup_{s\leq t}\rho(|X_{x,\alpha}^s - x|/n) > r^2\} \leq C(K,n)\frac{t}{r^2}. \tag{5.46}$$

Finally, choosing η and T we can find n large enough so that $\rho(|y-x|/n) = (y-x)^2$ for $y\in\Gamma_{\eta,T,K}$. By (5.44) and (5.46) we have

$$\sup_{x\in K,\alpha} \mathbf{P}\{\sup_{s\leq t}\|X_t^{x,\alpha}-x\| > r\} \leq 2\eta$$

for small enough t. □

The following is the basic existence result for martingale problems in $\mathbb{R}^d$.

Theorem 5.4.2. *Suppose the symbol*

$$p(x,\xi) = \frac{1}{2}(G(x)\xi,\xi) - i(b(x),\xi) + \int (1 - e^{i\xi y} + i\mathbf{1}_{|y|\leq 1}(y)(\xi,y))\nu(x,dy) \quad (5.47)$$

of the pseudo-differential operator $(-L)$ *is continuous,*

$$\sup_x \left(\frac{\|G(x)\|}{1+|x|^2} + \frac{|b(x)|}{1+|x|} + \frac{\int_{B_1} |y|^2 \nu(x,dy)}{1+|x|^2} + \int_{\{|y|>1\}} \nu(x,dy) \right) < \infty \quad (5.48)$$

and

$$\sup_x \int_{|y|>1} \ln^+(|y|)\nu(x,dy) < \infty. \quad (5.49)$$

Then the martingale problem corresponding to L *has a solution* P_μ *for any initial probability distribution* μ. *Moreover, one can construct these solution in such a way that the compact containment condition for compact initial data (introduced before* Theorem 4.10.2*) holds.*

Remark 39. The solutions still exist without assumption (5.49), because the operators with these two conditions differ by a bounded operator and one can apply the general perturbation theory for martingale problems (see e.g. Proposition 4.10.2 from [110]). Condition (5.49) is not very restrictive, but it supplies a handy explicit growth rate estimate for the corresponding processes.

Proof. Suppose first that the coefficients of the generator L are bounded, i.e.

$$\sup_x \left\{ |G(x)| + |b(x)| + \int_{|y|\leq 1} y^2 \nu(x,dy) \right\} < \infty \quad (5.50)$$

holds instead of (5.48). Then the natural discrete approximations of L, given by

$$L_h u(x) = \frac{1}{2}\Delta_h u(x) + \frac{1}{h}\sum_{i=1}^d b^i(x)(u(x+he_i) - u(x)) \quad (5.51)$$
$$+ \int_{|y|>h} [u(x+y) - u(x) - (y, \nabla u(x))\mathbf{1}_{B_1}(y)]\nu(x,dy), \quad h > 0,$$

where Δ_h is a discrete approximation of $(G(x)\nabla, \nabla)$ on $C_c^2(\mathbb{R}^d)$, are bounded operators. Hence the corresponding martingale problems are measurably well posed. Let us denote by $X_t^{x,h}$ their solutions starting from $x \in \mathbb{R}^d$.

Choosing a positive increasing smooth function $f_{\ln}$ on $\mathbb{R}_+$ such that $f_{\ln}(y) = \ln y$ for $y \geq 2$, we claim that the process

$$M_t^h = \phi(X_t^{x,h}) - \int_0^t L\phi(X_s^{x,h})\, ds$$

is a martingale for $\phi(y) = f_{\ln}(|y|)$. Indeed, approximating ϕ by the increasing sequence of positive functions $\phi_n(y) = f_{\ln}(|y|)\chi(|y|/n)$ from $C_c^2(\mathbb{R}^d)$, $n = 1, 2, \dots$, where χ is a smooth function $[0, \infty) \mapsto [0, 1]$, which has compact support and equals 1 in a neighborhood of the origin, observe that $|L_h \phi_n(x)|$ is a uniformly bounded function of $x \in R^d$, $h \in (0, 1]$ and $n \in \mathbb{N}$. Consequently, by dominated convergence, we establish the martingale property of M_t^h. Hence $E f_{\ln}(|X_t^{x,h}|) \leq f_{\ln}(|x|) + ct$ with some constant $c > 0$ independent of h, n, x. From Doob's maximum inequality we conclude that

$$\mathbf{P}(\sup_{0 \leq s \leq t} f_{\ln}(|X_s^{x,h}|) \geq r) \leq \frac{C(t + f_{\ln}(|x|))}{r} \tag{5.52}$$

for all $r > 0$ and some $C > 0$. This implies the compact containment condition for the family of processes $X_t^{x,h}$, $h \in (0, 1]$. Hence Theorem 4.9.2 yields the existence of solutions to the martingale problem for L on the domain $C_c^2(\mathbb{R}^d)$. Passing to the limit as $h \to 0$ in (5.52) yields

$$\mathbf{P}(\sup_{0 \leq s \leq t} f_{\ln}(|X_s^x|) \geq r) \leq \frac{C(t + f_{\ln}(|x|))}{r}, \tag{5.53}$$

implying the compact containment condition for the compact initial data for processes X_t^x.

In the general case, we approximate $G(x), \beta(x)$ and $\nu(x, .)$ by a (uniformly on compact sets) convergent sequence of bounded G_m, β_m, ν_m such that all estimates required in (i) are uniform for all m and all operators L_m obtained from L by changing G, β, ν by G_m, β_m, ν_m respectively. It follows that $|L_m \phi(x)|$ is a uniformly bounded function of m and x for $\phi(y) = f_{\ln}(|y|)$. Hence the proof is completed as above by passing to the limit $m \to \infty$, implying (5.53) and the compact containment condition for compact initial data for processes X_t^x. □

Proposition 5.4.1. *Under the assumptions of Theorem* 5.4.2*, suppose additionally that the martingale problem is well posed. Then the corresponding process is a conservative Feller and C-Feller process.*

Proof. By Theorems 4.10.2, 5.4.1 and 5.4.2 the corresponding semigroup is conservative and C-Feller. It remains to show that the space $C_\infty(\mathbb{R}^d)$ is preserved, as strong continuity is then evident (it holds for $f \in C_c^2(\mathbb{R}^d)$ by the martingale property, and hence for other functions by straightforward approximation arguments).

To this end, suppose first that the coefficients are bounded, i.e. (5.50) holds. Then the function $(L\phi_x)(y)$ is uniformly bounded as a function of two variables, where $\phi_x(y) = \phi(y-x) = f_{\ln}(|y-x|)$. Using the corresponding Dynkin's martingale yields the estimate $E|X_t^x - x| \leq ce^{ct}$ for some $c > 0$ uniformly for all x. Hence $\mathbf{P}(\sup_{0\leq s\leq t}|X_s^x - x| > r)$ tends to zero as $r \to \infty$ uniformly for all x. Consequently, for $f \in C_\infty(\mathbb{R}^d)$, one has $Ef(X_t^x) \to 0$ as $x \to \infty$. Returning to the general case, first observe that by standard perturbation theory (if A generates a Feller semigroup and B is bounded in $C_\infty(\mathbb{R}^d)$ and satisfies the positive maximum principle, then $A+B$ generates a Feller semigroup), it is enough to prove the statement under the additional assumption that all measures $\nu(x,.)$ have support in the unit ball.

In this case, the proof is easily reduced to the case of bounded coefficients by a change of variable. Namely, let Ω be a diffeomorphism of $\mathbb{R}^d$ on itself and $\tilde\Omega$ be the corresponding linear contraction in $C(\mathbb{R}^d)$ defined as $\tilde\Omega f(x) = f(\Omega(x))$. Then the generator of the martingale problem obtained by transforming a solution to the (L,D)-martingale problem by Ω has the generator $\tilde L = \tilde\Omega^{-1}L\tilde\Omega$. If L is given by (5.42), then

$$L\tilde\Omega f(x) = L(f(\Omega))(x) = \left((\nabla f)(\Omega(x)), \frac{\partial\Omega}{\partial x}b(x)\right) \tag{5.54}$$
$$+\frac12\sum_{k,l}G^{kl}(x)\left[\sum_{i,j}\frac{\partial\Omega_i}{\partial x_k}\frac{\partial\Omega_j}{\partial x_l}\frac{\partial^2 f}{\partial z_i\partial z_j}(\Omega(x)) + \sum_j\frac{\partial^2\Omega_j}{\partial x_k\partial x_l}\frac{\partial f}{\partial z_j}(\Omega(x))\right]$$
$$+\int\left[f(\Omega(x+y)) - f(\Omega(x)) - \left((\nabla f)(\Omega(x)), \frac{\partial\Omega}{\partial x}y\right)\mathbf{1}_{B_1}(y)\right]\nu(x,dy),$$

and hence

$$\tilde L f(z) = L(f(\Omega))(\Omega^{-1}(z))$$
$$= \left((\nabla f)(z), \frac{\partial\Omega}{\partial x}(\Omega^{-1}(z))b(\Omega^{-1}(z))\right)$$
$$+\frac12\sum_{k,l,j}G^{kl}(\Omega^{-1}(z))\left[\frac{\partial^2\Omega_j}{\partial x_k\partial x_l}(\Omega^{-1}(z))\frac{\partial f}{\partial z_j}(z)\right.$$
$$\left.+\sum_i\left(\frac{\partial\Omega_i}{\partial x_k}\frac{\partial\Omega_j}{\partial x_l}\right)(\Omega^{-1}(z))\frac{\partial^2 f}{\partial z_i\partial z_j}(z)\right]$$
$$+\int\Big[f(\Omega(\Omega^{-1}(z)+y)) - f(z)$$
$$-\left((\nabla f)(z), \frac{\partial\Omega}{\partial x}(\Omega^{-1}(z))y\right)\mathbf{1}_{B_1}(y)\Big]\nu(\Omega^{-1}(z),dy). \tag{5.55}$$

Let the diffeomorphism $\Omega(x) = \tilde{x}$ be such that $\tilde{x} = x$ for $|x| \le 1$, $\tilde{x}/|\tilde{x}| = x/|x|$ for all x, and $|\tilde{x}| = \ln|x|$ for $|x| \ge 3$. Then it is straightforward to see that $\tilde{L}$ has bounded coefficients. □

5.5 Martingale problems for Lévy type generators: moments

The aim of this section is to show that strengthening condition (5.49) on the moments of Lévy measure allows one to strengthen the moment estimates for the solutions to the corresponding martingale problem.

Theorem 5.5.1. *Under the assumptions of Theorem* 5.4.2, *assume additionally that the moment condition*

$$\sup_x (1+|x|)^{-p} \int_{\{|y|>1\}} |y|^p \nu(x,dy) < \infty \tag{5.56}$$

holds for $p \in (0,2]$. *Let* X_t *be a solution to the martingale problem for* L *with domain* $C_c^2(\mathbb{R}^d)$, *which exists according to Theorem* 5.4.2. *Then*

$$\mathbf{E} \min(|X_t^x - x|^2, |X_t^x - x|^p) \le (e^{ct} - 1)(1 + |x|^2) \tag{5.57}$$

for all t, *with a constant* c. *Moreover, for any* $T > 0$ *and a compact set* $K \subset \mathbb{R}^d$,

$$\mathbf{P}(\sup_{s \le t} |X_s^x - x| > r) \le \frac{t}{r^p} C(T, K) \tag{5.58}$$

for all $t \le T$, $x \in K$ *and large enough* r *with some constant* $C(T, K)$.

Proof. Notice first that from the Cauchy inequality

$$\int_{\{|y|>1\}} |y|^q \nu(x,dy) \le \left(\int_{\{|y|>1\}} |y|^p \nu(x,dy) \right)^{q/p} \left(\int_{\{|y|>1\}} \nu(x,dy) \right)^{(p-q)/p},$$

it follows that (5.56) together with (5.48) imply

$$\sup_x (1+|x|)^{-q} \int_{\{|y|>1\}} |y|^q \nu(x,dy) < \infty$$

for all $q \in (0,p]$. Now let $f_p(r)$ be an increasing smooth function on $\mathbb{R}_+$ that equals r^2 in a neighborhood of the origin, r^p for $r > 1$ and is not less than r^2 for $r < 1$. For instance, we can take $f(r) = r^2$ when $p = 2$. Also let $\chi_q(r)$ be a smooth non-increasing function $[0,\infty) \mapsto [0,1]$ that equals 1 for $r \in [0,1]$ and r^{-q} for $r > 2$. To get a bound for the average of the function $f_p^x(y) = f_p(\|y - x\|)$ we approximate it by the increasing sequence of functions $g_n(y) = f_p^x(y)\chi_q(|y-x|/n), n = 1, 2, \dots,$ $q > p$. The main observation is that

$$|Lg_n(y)| \leq c(g_n(y) + x^2 + 1) \tag{5.59}$$

for some constant c uniformly for x, y and n. To see this we analyze separately the action of all terms in the expression for L. For instance,

$$\left|\operatorname{tr}\left(G(y)\frac{\partial^2}{\partial y^2}g_n(y)\right)\right| \leq c(1+|y|^2)[\min(1, |y-x|^{p-2})\chi_q(|y-x|/n)$$
$$+ f_p(|y-x|)\chi_q''(|y-x|/n)/n^2 + f_p'(|y-x|)\chi_q'(|y-x|/n)/n].$$

Taking into account the obvious estimate

$$\chi_q^{(k)}(z) \leq c_k(1+|z|^k)^{-1}\chi_q(z)$$

(which holds for any k, though we need only $k = 1, 2$) and using $|y|^2 \leq 2(y-x)^2 + 2x^2$ yields

$$\left|\operatorname{tr}\left(G(y)\frac{\partial^2}{\partial y^2}g_n(y)\right)\right| \leq c(|g_n(y)| + x^2 + 1),$$

as required. Also, as $g_n(x) = 0$,

$$\int_{\{|y|>1\}}(g_n(x+y) - g_n(x))\nu(x, dy) = \int_{\{|y|>1\}} f_p(|y|)\chi_q(|y|/n)\nu(x, dy)$$
$$\leq \int_{\{|y|>1\}} |y|^p\nu(x, dy) \leq c(1+|x|)^p) \leq c(1+|x|^2),$$

and so on.

Next, as $q > p$ the function $g_n(y)$ belongs to $C_\infty(\mathbb{R}^d)$ and we can establish, by an obvious approximation, that the process

$$M_{g_n}(t) = g_n(X_t^x) - \int_0^t Lg_n(X_s^x)\,ds$$

is a martingale. Using now the dominated and monotone convergence theorems when passing to the limit $n \to \infty$ in the equation $\mathbf{E}M_{g_n}(t) = g_n(x) = 0$ (representing the martingale property of M_{g_n}) yields the inequality

$$\mathbf{E}f_p(\|X_t^x - x\|) \leq c\int_0^t [\mathbf{E}f_p(\|X_s^x - x\|) + x^2 + 1]\,ds.$$

This implies

$$\mathbf{E}f_p(|X_t^x - x|) \leq (e^{ct} - 1)(1+|x|^2)$$

by Gronwall's lemma, and (5.57) follows.

Once the upper bound for $Ef_p(|X_t^x - x|)$ is obtained it is straightforward to show, by the same approximation as above, that M_f is a martingale for $f = f_p^x$. Moreover, passing to the limit in (5.59) we obtain

$$|Lf_p^x(y)| \le c(f_p^x(y) + x^2 + 1). \tag{5.60}$$

Applying Doob's maximal inequality yields

$$\mathbf{P}\left(\sup_{s\le t} |f_p^x(X_s^x) - \int_0^s Lf_p^x(X_\tau^x)\,d\tau| \ge r\right) \le \frac{1}{r}tc(T)(1+|x|^2) \le \frac{1}{r}tc(T,K)).$$

Hence with a probability not less than $1 - tc(T,K)/r$

$$\sup_{s\le t}\left| f_p^x(X_s^x) - \int_0^s Lf_p^x(X_\tau^x)\,d\tau\right| \le r,$$

implying by Gronwall's lemma and (5.60)

$$\sup_{t\le T} f_p^x(X_t^x) \le c(T)(r + x^2 + 1) \le 2C(T)r$$

for $x^2 + 1 \le r$. This in turn implies (with a different constant $C(T,K)$)

$$\mathbf{P}(\sup_{s\le t} f_p(|X_s^x - x|) > r) \le \frac{t}{r}C(T,K).$$

Since $|X_s^x - x| > r$ if and only if $f_p(|X_s^x - x|) > r^p$, estimate (5.58) follows. □

Exercise 5.5.1. Show that if the coefficients of L are bounded, i.e. (5.50) holds, then

$$\mathbf{E}\min(|X_t^x - x|^2, |X_t^x - x|^p) \le (e^{ct} - 1) \tag{5.61}$$

uniformly for all x, and also that (5.58) holds for all x with $C(T,K)$ not depending on K.

5.6 Martingale problems for Lévy type generators: unbounded coefficients

To clarify the semigroup structure of the processes with unbounded coefficients, it is convenient to work with weighted spaces of continuous function. Recall that if $f(x)$ is a continuous positive function on a locally compact space S tending to infinity as $x \to \infty$, we denote by $C_f(S)$ (resp. $C_{f,\infty}(S)$) the space of continuous functions g on S such that $g/f \in C(S)$ (resp. $g/f \in C_\infty(S)$) with the norm $\|g\|_{C_f} = \|g/f\|$.

Similarly we define $C_f^k(S)$ (resp. $C_{f,\infty}^k(S)$) as the space of k times continuously differentiable functions such that $g^{(l)}/f \in C(S)$ (resp. $g^{(l)}/f \in C_\infty(S)$) for all $l \leq k$.

The next result establishes the existence of solutions to the martingale problem in case of unbounded coefficients, whenever an appropriate Lyapunov function is available.

Theorem 5.6.1. *Let an operator L be defined in $C_c^2(\mathbb{R}^d)$ by (5.42) and*

$$\int_{\{|y|\geq 1\}} |y|^p \nu(x, dy) < \infty$$

for a $p \leq 2$ and any x. Assume a positive function $f_L \in C^2_{1+|x|^p}$ is given (the subscript L stands either for the operator L or for the Lyapunov function) such that $f_L(x) \to \infty$ as $x \to \infty$ and

$$L f_L \leq c(f_L + 1) \tag{5.62}$$

for some constant c. Then the martingale problem for L in $C_c^2(\mathbb{R}^d)$ has a solution such that

$$\mathbf{E} f_L(X_t^x) \leq e^{ct}(f_L(x) + c), \tag{5.63}$$

and

$$\mathbf{P}(\sup_{0\leq s\leq t} f_L(X_s^x) > r) \leq \frac{c(t, f_L(x))}{r}. \tag{5.64}$$

Proof. Set

$$s_L^p(x) = \|G(x)\| + |b(x)| + \int \min(|y|^2, |y|^p)\nu(x, dy).$$

For a given $q > 1$ the martingale problem for the 'normalized' operators $L_n = \chi_q(s_L^p(x)/n)L$, $n = 1, 2, \ldots$, with bounded coefficients has a solution in $C_c^2(\mathbb{R}^d)$. Let $X_{t,m}$ denote the solutions to the corresponding martingale problems. Approximating f_L by $f_L(y)\chi_p(y/n)$ as in the above proof of Theorem 5.5.1, it is straightforward to conclude that the processes

$$M_m(t) = f_L(X_{t,m}^x) - \int_0^t L_m f_L(X_{s,m}^x)\, ds$$

are martingales for all m. Moreover, since $\chi_p \leq 1$, it follows from our assumptions that $L_m f_L \leq c(f_L + 1)$ for all m, implying again by Gronwall's lemma that

$$\mathbf{E} f_L(X_{t,m}^x) \leq e^{ct}(f_L(x) + c). \tag{5.65}$$

Since by (5.62) and (5.65) the expectation of the negative part of the martingale $M_m(t)$ is uniformly (for $t \le T$) bounded by $c(T)(f_L(x)+1)$, we conclude that the expectation of its magnitude is also bounded by $c(T)(f_L(x)+1)$ (in fact, for any martingale $M(t)$ one has $M(0) = \mathbf{E}M(t) = \mathbf{E}M^+(t) - \mathbf{E}M^-(t)$, where $M^{\pm}(t)$ are the positive and negative parts of $M(t)$, implying $\mathbf{E}M^+(t) = \mathbf{E}M^-(t) + M(0)$). Hence, by the same argument as in the proof of (5.58) above, one deduces from Doob's inequality for martingales that

$$\sup_m \mathbf{P}(\sup_{0\le s\le t} f_L(X^x_{s,m}) > r) \le \frac{c(t, f_L(x))}{r}$$

uniformly for $t \le T$ with arbitrary T. Since $f_L(x) \to \infty$ as $x \to \infty$, this implies the *compact containment condition* for $X_{t,m}$:

$$\lim_{r\to\infty} \sup_m \mathbf{P}(\sup_{0\le s\le t} |X^x_{s,m}| > r) = 0$$

uniformly for x on compacts and $t \le T$ with arbitrary T. Hence the existence of a solution to the martingale problem of L follows from Theorem 4.9.2. Estimates (5.63) and (5.64) are of course established by passing to the limit $m \to \infty$ from the corresponding estimates for $X^x_{s,m}$. □

Theorem 5.6.2. *Under the assumptions of Theorem* 5.6.1 *assume that for a given* $q > 1$ *the martingale problem for each 'normalized' operator* $L_n = \chi_q(s^p_L(x)/n)L$, $n = 1, 2, \ldots,$ *with bounded coefficients, introduced in the proof of the previous theorem, is well posed in* $C^2_c(\mathbb{R}^d)$ *and the corresponding process is a conservative Feller process. Then the martingale problem for* L *in* $C^2_c(\mathbb{R}^d)$ *is also well posed, the corresponding process* X_t *is strong Markov and its contraction semigroup preserves* $C(\mathbb{R}^d)$ *and extends from* $C(\mathbb{R}^d)$ *to a strongly continuous semigroup in* $C_{f_L,\infty}(\mathbb{R}^d)$ *with a domain containing* $C^2_c(\mathbb{R}^d)$*. Moreover, the semigroup in* $C_{f_L,\infty}(\mathbb{R}^d)$ *is a contraction whenever* $c = 0$ *in* (5.62).

Proof. Let us estimate the difference between the Feller semigroups of $X_{s,n}, X_{s,m}$. By the compact containment condition, for any $\epsilon > 0$ there exists $r > 0$ such that for $f \in C(\mathbb{R}^d)$

$$|\mathbf{E}f(X^x_{t,m}) - \mathbf{E}f(X^x_{t,n})| \le |\mathbf{E}[f(X^x_{s,m})\mathbf{1}_{t<\tau^m_r}] - \mathbf{E}[f(X^x_{s,n})\mathbf{1}_{t<\tau^n_r}]| + \epsilon\|f\|,$$

where τ^m_r is the exit time of $X^x_{t,m}$ from the ball B_r (i.e. is given by (4.100) with $U = B_r$). Note that for large enough n, m the generators of $X^x_{t,m}$ and $X^x_{t,n}$ coincide in B_r and hence by Theorem 4.11.2 the first term on the r.h.s. of the above inequality vanishes. Consequently,

$$|\mathbf{E}f(X^x_{t,m}) - \mathbf{E}f(X^x_{t,n})| \to 0$$

as $n, m \to \infty$ uniformly for x from any compact set. And this fact clearly implies that the limit

$$T_t f(x) = \lim_{n\to\infty} \mathbf{E} f(X_{t,n}^x)$$

exists and that T_t is a Markov semigroup preserving $C(\mathbb{R}^d)$ (i.e. is a C-Feller semigroup) and continuous in the topology of uniform convergence on compact sets, i.e. such that $T_t f(x)$ converges to $f(x)$ as $t \to 0$ uniformly for x on compacts. Clearly the compact containment implies the tightness of the family of the transition probabilities for the Markov processes $X_{t,m}^x$, leading to the conclusion that the limiting semigroup T_t has form (3.38) for certain transitions p_t and hence specifies a Markov process, which therefore solves the required martingale problem. Uniqueness follows by localization, i.e. by Theorem 4.11.4. It remains to observe that (5.65) implies (5.63), and this in turn implies (5.64) by the same argument as for the approximations $X_{t,m}$ above. Consequently T_t extends by monotonicity to a semigroup on $C_f(\mathbb{R}^d)$. Since the space $C(\mathbb{R}^d) \subset C_f(\mathbb{R}^d)$ is invariant and T_t is continuous there in the topology of uniform convergence on compact sets it follows that $T_t f$ converges to f as $t \to 0$ in the topology of $C_f(\mathbb{R}^d)$ for any $f \in C(\mathbb{R}^d)$ and hence (by a standard approximation argument) also for any $f \in C_{f,\infty}(\mathbb{R}^d)$, implying the required strong continuity. □

5.7 Decomposable generators

The construction and analysis of martingale problems and corresponding Markov semigroups can be essentially fertilized by using advanced functional analytic techniques, especially Fourier analysis. As these methods work more effectively in Hilbert spaces and the original Feller semigroups act in the Banach space $C_\infty(\mathbb{R}^d)$, one looks for auxiliary Hilbert spaces where the existence of a semigroup can be shown as a preliminary step. As these auxiliary spaces it is natural to use the *Sobolev spaces* $H^s(\mathbb{R}^d)$ defined as the completions of the Schwartz space $S(\mathbb{R}^d)$ with respect to the Hilbert norm

$$\|f\|_s = \|f\|_{H^s}^2 = \int f(x)(1-\Delta)^s f(x)\, dx.$$

In particular, H^0 coincides with the usual L^2. Hence, in this section we shall denote by $\|f\|_0$ the norm of f in $L_2(\mathbb{R}^d)$. Recall that $\|f\|$ for a function on $\mathbb{R}^d$ always denotes its supnorm.

The celebrated *Sobolev embedding lemma* states that H^s is continuously embedded in $(C_\infty \cap C^l)(\mathbb{R}^d)$ whenever $s > l + d/2$. Consequently, if we can show the existence of a semigroup in H^s, it supplies automatically an invariant dense domain (and hence a core) for its extension to C_∞. For a detailed discussion of Fourier analysis and Sobolev spaces in the context of Markov processes we refer to Jacob [142].

As an example of the application of the techniques mentioned above, we shall discuss the Markov semigroups with the so-called decomposable generators.

Let ψ_n, $n = 1, \dots, N$, be a finite family of generators of Lévy processes in $\mathbb{R}^d$, i.e. for each n

$$\psi_n f(x) = \frac{1}{2}(G^n \nabla, \nabla) f(x) + \left(b^n, \frac{\partial}{\partial x}\right) f(x) + \int (f(x+y) - f(x) - \nabla f(x) y) \nu^n(dy) + \int (f(x+y) - f(x)) \mu^n(dy), \tag{5.66}$$

where $G^n = (G^n_{ij})$ is a non-negative symmetric $d \times d$-matrix, $b^n \in \mathbb{R}^d$, ν^n and μ^n are Radon measures on the ball $\{|y| \le 1\}$ and on $\mathbb{R}^d$ respectively (Lévy measures) such that

$$\int |y|^2 \nu^n(dy) < \infty, \quad \int \min(1, |y|) \mu^n(dy) < \infty, \quad \nu^n(\{0\}) = \mu^n(\{0\}) = 0.$$

Such a decomposition of the Lévy measure in two parts makes our further assumptions on this measure more transparent. The function

$$p_n(\xi) = \frac{1}{2}(G^n \xi, \xi) - i(b^n, \xi) + \int (1 - e^{i\xi y} + i\xi y) \nu^n(dy) + \int (1 - e^{i\xi y}) \mu^n(dy) \tag{5.67}$$

is the symbol of the operator $-\psi_n$. We shall denote by p_n^ν, p_n^μ the corresponding integral terms in (5.67), i.e. $p_n^\mu(\xi) = \int (1 - e^{i\xi y}) \mu^n(dy)$. We also denote $p_0 = \sum_{n=1}^N p_n$.

Let a_n be a family of positive continuous functions on $\mathbb{R}^d$. With some abuse of notation we shall denote also by a_n the operator of multiplication by a_n. Let us say that the Lévy–Khintchine type operator is *decomposable*, if it is represented in the form $\sum_{n=1}^N a_n \psi_n$.

We shall prove the existence and uniqueness of the Markov process with generator $\sum_{n=1}^N a_n \psi_n$ under the following assumptions on the symbols p_n: there exists $c > 0$ and constants $\alpha_n > 0$, $\beta_n < \alpha_n$ such that for each $n = 1, \dots, N$

(A1) $|\mathrm{Im}\, p_n^\mu(\xi) + \mathrm{Im}\, p_n^\nu(\xi)| \le c |p_0(\xi)|$,

(A2) $\mathrm{Re}\, p_n^\nu(\xi) \ge c^{-1} |\mathrm{pr}_{\nu^n}(\xi)|^{\alpha_n}$ and $|\nabla(p_n^\nu)(\xi)| \le c |\mathrm{pr}_{\nu^n}(\xi)|^{\beta_n}$, where pr_{ν^n} is the orthogonal projection on the minimal subspace containing the support of the measure ν^n.

Remark 40. Condition (A2) is not very restrictive in practice. It allows, in particular, any α-stable measures ν (perhaps degenerate) with $\alpha \ge 1$ (the case $\alpha < 1$ can be included in μ_n). Moreover, if $\int |\xi|^{1+\beta_n} \nu_n(d\xi) < \infty$, then the second condition in

(A2) holds, because $|e^{ixy}-1| \le c|xy|^\beta$ for any $\beta \le 1$ and some $c > 0$. In particular, the second inequality in (A2) always holds with $\beta_n = 1$. Hence, in order that (A2) holds it is enough to have the first inequality in (A2) with $\alpha_n > 1$.

For the multiplication operators we shall assume the following bounds:

(A3) $a_n(x) = O(|x|^2)$ as $x \to \infty$ for those n where $G^n \neq 0$ or $\nu^n \neq 0$, $a_n(x) = O(|x|)$ as $x \to \infty$ for those n where $\beta^n \neq 0$, and $a_n(x)$ is bounded whenever $\mu^n \neq 0$.

Clearly these bounds ensure that estimate (5.48) holds.

The aim of this section is to prove the following.

Theorem 5.7.1. *Suppose* (A1), (A2) *hold for the family of operators ψ_n, and suppose that all a_n are positive functions from $C^s(\mathbb{R}^d)$ for $s > 2 + d/2$ such that* (A3) *holds. Then there exists a unique extension of the operator $\sum_{n=1}^N a_n \psi_n$ (with the initial domain being $C_c^2(\mathbb{R}^d)$) that generates a Feller semigroup in $C_\infty(\mathbb{R}^d)$.*

We are aiming at using the localization procedure based on Theorem 4.11.2. Therefore we shall start with a local version of Theorem 5.7.1 under the additional assumption on the symbols p_n:

(A1′) $|\mathrm{Im}\, p_n(\xi)| \le c|p_0(\xi)|$ and hence $|p_n(\xi)| \le (1+c)|p_0(\xi)|$.

Clearly (A1′) is a version of (A1) for the whole symbol, which thus combines (A1) with some restriction on the drift.

Let a_n and ψ_n be as in Theorem 5.7.1. Set $L_0 = \sum_{n=1}^N \psi_n$ and

$$L = L_0 + \sum_{n=1}^N a_n \psi_n$$

(the pseudo-differential operator with the symbol $-\sum_{n=1}^N (1 + a_n(x)) p_n(\xi)$).

Proposition 5.7.1. *Suppose* (A1′) *and* (A2) *hold for the family of operators ψ_n, all $a_n \in C_b^s(\mathbb{R}^d)$ for $s > 2 + d/2$ and*

$$2(c+1) \sum_{n-1}^N \|a_n\| < 1, \tag{5.68}$$

where the constant c is taken from condition (A1′) *(let us stress that $\|.\|$ always denotes the usual sup-norm of a function). Then the closure of $\sum_{n=1}^N a_n \psi_n$ (with the initial domain $C_c^2(\mathbb{R}^d)$) generates a Feller semigroup in $C_\infty(\mathbb{R}^d)$ and the (strongly) continuous semigroups in all Sobolev spaces $H^{s'}$, $s' \le s$, including $H^0 = L^2$.*

The proof will be preceded by the sequence of lemmas. We start by defining an equivalent family of norms on H^s. Namely, let $b = \{b_I\}$ be any family of (strictly) positive numbers parametrized by multi-indices $I = \{i_1, \dots, i_d\}$ such that $0 < |I| = i_1 + \cdots + i_d \leq s$ and $i_j \geq 0$ for all j. Then the norm $\|.\|_{s,b}$ defined by

$$\|f\|_{s,b} = \|f\|_0 + \sum_{0<|I|\leq s} b_I \left\| \frac{\partial^{|I|}}{\partial x_I} f \right\|_0$$

$$= \sqrt{\int |\hat{f}(\xi)|^2 \, d\xi} + \sum_{0<|I|\leq s} b_I \sqrt{\int |\xi|^{2I} |\hat{f}(\xi)|^2 \, d\xi},$$

where $|\xi|^{2I} = |\xi_1|^{2i_1} \cdots |\xi_d|^{2i_d}$ for $I = \{i_1, \dots, i_d\}$, is a norm in $S(\mathbb{R}^d)$ which is obviously equivalent to norm $\|.\|_s$. We shall denote by $H^{s,b}$ the corresponding completion of $S(\mathbb{R}^d)$ which coincides with H^s as a topological vector space.

Lemma 5.7.1. *Let $a(x) \in C_b^s(\mathbb{R}^d)$. Then for an arbitrary $\epsilon > 0$ there exists a collection $b = \{b_I\}$, $0 < |I| \leq s$, of positive numbers such that the operator A of multiplication by $a(x)$ is bounded in $H^{s,b}$ with norm not exceeding $\|a\| + \epsilon$ (i.e. the bounds on the derivatives of $a(x)$ are essentially irrelevant for the norm of A).*

Proof. To simplify the formulas, we shall give a proof for the case $s = 2, d = 1$. In this case we have

$$\|f\|_{2,b} = \|f\|_0 + b_1\|f'\|_0 + b_2\|f''\|_0$$

and

$$\|Af\|_{2,b} \leq (\|a\| + b_1\|a'\| + b_2\|a''\|)\|f\|_0 + (b_1\|a\| + 2b_2\|a'\|)\|f'\|_0 + b_2\|a\| \, \|f''\|_0.$$

Clearly by choosing b_1, b_2 small enough we can ensure that the coefficient of $\|f\|_0$ is arbitrarily close to $\|a\|$, and then by decreasing (if necessary) b_2 we can make the coefficient at $\|f'\|_0$ arbitrarily close to $b_1\|a\|$. The proof is complete. □

Next we shall prove a simple auxiliary result on semigroups in H^s generated by ΨDO. Let us called a subset M of H^s *spectrally localized* if for any $\epsilon > 0$ there exists a compact set $K \subset \mathbb{R}^d$ such that $\int_{\mathbb{R}^d \setminus K} (1 + |\xi|^{2s}) |\hat{f}(\xi)|^2 \, d\xi < \epsilon$ for all $f \in M$. Clearly, bounded subsets of $H^{s'}$ with any $s' > s$ are localized subsets in H^s.

Lemma 5.7.2. *Let $p(\xi)$ be an arbitrary continuous function $\mathbb{R}^d \to \mathbb{C}$ with everywhere non-negative real part. Then the ΨDO $-p(i^{-1}\nabla)$ with the symbol $-p$ generates a strongly continuous semigroup of contractions in every H^s. Moreover,*

$$\|(e^{-tp(i^{-1}\nabla)} - 1)f\|_s \to 0$$

as $t \to 0$ uniformly on spectrally localized subsets of H^s; in particular, on bounded subsets of $H^{s'}$ with any $s' > s$. Finally, if $|p(\xi)| \le c(1+|\xi|^2)$ with a positive constant c, then H^{s+2} is a core for the semigroup in H^s for any $s > 0$.

Proof. The operator $p(i^{-1}\nabla)$ multiplies the Fourier transform of a function by $p(\xi)$. This directly implies the required contraction property. The operator $e^{-tp(i^{-1}\nabla)} - 1$ multiplies the Fourier transform of a function by $e^{-tp(\xi)} - 1$. This function is uniformly bounded and in any bounded region of ξ it tends to 0 uniformly as $t \to 0$. Hence, uniformly for f from a localized set M,

$$\int (1 + |\xi|^{2s})|(e^{-tp(\xi)} - 1)\hat{f}(\xi)|^2 \, d\xi \to 0,$$

as $t \to 0$. The last statement is clear, as the domain of the generator of our semigroup in H^s consists precisely of those functions f, for which

$$\int (1 + |\xi|^{2s})|p(\xi)\hat{f}(\xi)|^2 \, d\xi < \infty. \qquad \square$$

We will now construct a semigroup in L^2 and H^s with generator L which is considered as a perturbation of L_0. To this end, for a family of functions ϕ_s, $s \in [0,t]$, on $\mathbb{R}^d$ let us define a family of functions $\mathcal{F}_s(\phi)$, $s \in [0,t]$, on $\mathbb{R}^d$ by

$$\mathcal{F}_s(\phi) = \int_0^s e^{(s-\tau)L_0}(L - L_0)\phi_\tau \, d\tau. \tag{5.69}$$

From perturbation theory one knows that formally the solution to the Cauchy problem

$$\dot{\phi} = L\phi, \quad \phi(0) = f \tag{5.70}$$

is given by the perturbation series

$$\phi = (1 + \mathcal{F} + \mathcal{F}^2 + \cdots)\phi^0, \quad \phi_s^0 = e^{sL_0} f. \tag{5.71}$$

In order to carry out a rigorous proof on the basis of this formula, we shall study carefully the properties of the operator $\mathcal{F}$. We shall start with the family of operators F_t on the Schwartz space $S(\mathbb{R}^d)$ defined as

$$F_t(f) = \int_0^t e^{sL_0}(L - L_0)f \, ds = \int_0^t e^{sL_0} \sum_{n=1}^N a_n \psi_n f \, ds.$$

We denote by $\|F\|_s$ the norm of an operator F in the Sobolev space H_s.

Lemma 5.7.3. *F_t is a bounded operator in $L^2(\mathbb{R}^d)$ for all $t > 0$. Moreover, for an arbitrary $\epsilon > 0$, there exists $t_0 > 0$ such that for all $t \le t_0$*

$$\|F_t\|_0 \le 2(c+1)\sum_n \|a_n\| + \epsilon$$

and hence $\|F_t\|_0 < 1$ for small enough ϵ.

Proof. We have

$$\begin{aligned}
\psi_n(a_n f) &= a_n(x)(b^n, \nabla f(x)) + f(x)(b^n, \nabla a_n(x)) + \frac{1}{2}a_n(x)(G^n\nabla,\nabla)f(x)\\
&\quad + \frac{1}{2}f(x)(G^n\nabla,\nabla)a_n(x) + (G^n\nabla a_n, \nabla f)(x)\\
&\quad + \int [(a_n f)(x+y) - (a_n f)(x) - (\nabla(a_n f)(x), y)]\nu^n(dy)\\
&\quad + \int [(a_n f)(x+y) - (a_n f)(x)]\mu^n(dy)\\
&= [a_n\psi_n(f) + f\psi_n(a_n)](x) + (G^n\nabla a_n, \nabla f)(x)\\
&\quad + \int (a_n(x+y) - a_n(x))(f(x+y) - f(x))(\nu^n(dy) + \mu^n(dy)).
\end{aligned}$$

Consequently,

$$\begin{aligned}
[a_n\psi_n(f)](x) &= [\psi_n(a_n f) - f\psi_n(a_n)](x) - (G^n\nabla a_n, \nabla f)(x)\\
&\quad - \int (a_n(x+y) - a_n(x))(f(x+y) - f(x))(\nu^n(dy) + \mu^n(dy)).
\end{aligned}$$

The idea is to represent the last two terms as a sum of an operator bounded in $L^2(\mathbb{R}^d)$ and a pseudo-differential operator with differentiation acting at the end (i.e. staying at the beginning of the pseudo-differential expression). The latter can be then combined with the regularizing term e^{sL_0} in the expression for F_t. Since $e^{(y,\nabla)}g(x) = g(x+y)$, we can write

$$\begin{aligned}
\nabla a_n(x)(f(x+y) - f(x)) &= f(x+y)\nabla a_n(x+y) - f(x)\nabla a_n(x)\\
&\quad - f(x+y)(\nabla a_n(x+y) - \nabla a_n(x))\\
&= (e^{(y,\nabla)} - 1)[f\nabla a_n](x) - f(x+y)(e^{(y,\nabla)} - 1)\nabla a_n(x).
\end{aligned}$$

And consequently

$$a_n\psi_n(f) = R_n^1 f + R_n^2 f,$$

where

$$R_n^1 f(x) = \psi_n(a_n f)(x) - \sum_{k,l} \nabla_k (G_{kl}^n (\nabla_l a_n) f)(x)$$
$$- \int (y, (e^{(y,\nabla)} - 1)[f \nabla a_n](x)) \nu^n(dy),$$
$$R_n^2 f(x) = \Big(\sum_{k,l} \nabla_k (G_{kl}^n \nabla_l a_n) - \psi_n(a_n)\Big)(x) f(x)$$
$$- \int (a_n(x+y) - a_n(x) - (\nabla a_n(x), y))(f(x+y) - f(x)) \nu^n(dy)$$
$$+ \int [f(x+y)(\nabla a_n(x+y) - \nabla a_n(x), y) \nu^n(dy)$$
$$- \int (a_n(x+y) - a_n(x))(f(x+y) - f(x)) \mu^n(dy).$$

Since an operator of the form $f \mapsto \int f(x+y)\eta(dy)$ is a contraction in $L^2(\mathbb{R}^d)$ for any probability measure η, we conclude that R_n^2 is a bounded operator in $L^2(\mathbb{R}^d)$ with norm not exceeding

$$\|\nabla^2 a_n\| \left(\|G^n\| + 2 \int |y|^2 \nu^n(dy) \right) + \|\psi_n(a_n)\| + \|\nabla a_n\| \int (1 \wedge |y|) \mu^n(dy).$$

Hence

$$F_t(f) = \sum_{n=1}^N \int_0^t e^{sL_0} \psi_n(a_n f)\, ds - \sum_{n=1}^N \int_0^t e^{sL_0} \sum_{k,l} \nabla_k (G_{kl}^n (\nabla_l a_n) f)\, ds$$
$$- \sum_{n=1}^N \int_0^t e^{sL_0}\, ds \int (y, (e^{(y,\nabla)} - 1)[f \nabla a_n]) \nu^n(dy) + O(t)\|f\|_0.$$

As the operator $\int_0^t e^{sL_0} \psi_n\, ds$ multiplies the Fourier transform of a function by the function $\int_0^t e^{-sp_0(\xi)} p_n(\xi)\, ds$, we can estimate the first term by

$$\left\| \int_0^t e^{sL_0} \psi_n a_n\, ds \right\|_0 \le \|a_n\| \left\| \int_0^t e^{-sp_0(\xi)} |p_n(\xi)|\, ds \right\|$$
$$\le \|a_n\| \left\| \frac{p_n}{p_0}(1 - e^{-tp_0}) \right\| \le 2(1+c)\|a_n\| \tag{5.72}$$

(the last inequality is due to condition (A1′)), the second term as

$$\left\| \int_0^t e^{sL_0} \nabla_k G_{kl}^n (\nabla_l a_n)\, ds \right\|_0 \le \|\nabla a_n\| \sup_\xi \left(\int_0^t e^{-s(G^n \xi, \xi)} \|G^n(\xi)\| ds \right)$$
$$= O(t^{1/2}) \|\nabla a_n\| \, \|\sqrt{G}\|,$$

and the last term using

$$\left\| \int_0^t e^{sL_0}\, ds \int (e^{i(y,\nabla)} - 1) y \nu^n(dy) \right\|_0 \leq \sup_\xi \left(\int_0^t e^{-s \operatorname{Re} p_0} \|\nabla p_n^\nu(\xi)\| ds \right)$$
$$\leq \sup_\omega \left(\int_0^t e^{-s\omega^\alpha/c} c\omega^\beta \, ds \right) = O(1) \int_0^t s^{-\beta/\alpha}\, ds = O(t^{1-\beta/\alpha})$$

(which holds due to (A2)). □

It turns out that the same holds in H^s.

Lemma 5.7.4. *For any $\epsilon > 0$ there exists $t_0 > 0$ and a family of positive numbers $b = \{b_I\}$, $0 < |I| \leq s$ such that for all $t \leq t_0$ and $s' \leq s$*

$$\|F_t\|_{s',b} \leq 2(c+1) \sum_n \|a^n\| + \epsilon.$$

Proof. Follows by the same arguments as the proof of Lemma 5.7.3 with the use of Lemma 5.7.1 and the definition of the norm $\|.\|_{s,b}$. In fact, the norms of the pseudo-differential operators with symbols of the form $p(\xi)$ do not exceed $\|p\| = \sup_\xi |p(\xi)|$ if considered as operators in every $H^{s,b}$, and the operators of the form $f \mapsto \int f(x+y)\eta(dy)$ are contractions in every $H^{s,b}$ for any probability measure η. □

We can now deduce the necessary properties of the operator $\mathcal{F}$.

For a Banach space B of functions on $\mathbb{R}^d$, let us denote by $C([0,t], B)$ the Banach space of continuous functions ϕ_s from $[0,t]$ to B with the usual sup-norm $\sup_{s\in[0,t]} \|\phi_s\|_B$. We shall identify B with a closed subspace of functions from $C([0,t], B)$ which do not depend on $s \in [0,t]$.

Lemma 5.7.5. *Under conditions of Proposition* 5.7.1, *the operator $\mathcal{F}$ defined by* (5.69) *is a continuous operator in $C([0,t], H^{s',b})$ for any $s' < s$ and $\|\mathcal{F}\| < 1$ for small enough t.*

Proof. The statement about the norm follows from Lemma 5.7.4 and holds for all $s' \leq s$. Let us show that $\mathcal{F}(\phi) \in C([0,t], H^{s,b})$ whenever $\phi \in C([0,t], H^{s',b})$ with $s' < s$. By the standard density argument it is sufficient to assume that $\phi \in C([0,t], H^{s,b})$. One has

$$\mathcal{F}_{t+\tau}(\phi) - \mathcal{F}_t(\phi) = \int_0^t (e^{\tau L_0} - 1) e^{(t-s)L_0} (L - L_0)\phi_s \, ds$$
$$+ \int_t^{t+\tau} e^{(t+\tau-s)L_0}(L - L_0)\phi_s \, ds. \tag{5.73}$$

Manipulating with these integrals as in the proof of Lemma 5.7.3 reduces the problem to the case when instead of $L - L_0$ one plugs in the operator ψ_n. And for this case, the estimate (5.72) amounts to the application of Lemma 5.7.2. □

As a direct consequence we get the following result.

Lemma 5.7.6. *Under the conditions of Proposition* 5.7.1, *there exists* t_0 *such that the series* (5.71) *converges in* $C([0,t], H^{s',b})$ *for all* $s' \leq s$ *and* $t \leq t_0$. *Moreover, the r.h.s. of* (5.71) *defines a strongly continuous family of bounded operators* $f \mapsto T_t f$ *in all* $H^{s'}$, $s' < s$.

We can prove now Proposition 5.7.1.

By the Sobolev lemma, H^s can be continuously embedded in $C_\infty \cap C^l$ whenever $s > l + d/2$. As $s > 2 + d/2$, $\mathcal{F}_t(\phi)$ is differentiable in t for any $\phi \in C([0,t], H^{s,b})$ and, as follows from equation (5.73) and Lemma 5.7.2,

$$\frac{d}{dt}\mathcal{F}_t(\phi) = \lim_{\tau \to 0} \frac{1}{\tau}(\mathcal{F}_{t+\tau}(\phi) - \mathcal{F}_t(\phi)) = L_0 \mathcal{F}_t + (L - L_0)\phi_t, \tag{5.74}$$

where the limit is understood in the norm of H^{s-2}. Therefore, one can differentiate the series (5.71) to show that for $f \in H^{s'}$ with $s' < s$, the function $T_t f$ gives a (classical) solution to the Cauchy problem (5.70). Since a classical solution in C_∞ for such a Cauchy problem is positivity-preserving and unique, because L is an operator with the PMP property (see Theorem 4.1.1 and its corollary), we conclude that T_t defines a positivity-preserving semigroup in each $H^{s'}$, $s' \leq s$. In fact, using the semigroup property one can extend T_t to all finite $t > 0$, thus removing the restriction $t \leq t_0$. Moreover, again by Theorem 4.1.1, the operators T_t do not increase the sup-norm. Hence, by continuity, they extend to contractions in $C_\infty(\mathbb{R}^d)$ defining a Feller semigroup.

Finally, we shall prove Theorem 5.7.1 in two steps. First we shall remove the restriction (A1′) using T-products, and then complete the proof via localization. Thus first let us show that the statement of Proposition 5.7.1 still holds if we assume (A1) instead of (A1′). The difference between these conditions concerns only the drift terms of L. So, our statement will be proved, if we can show that if L is as in Proposition 5.7.1 and γ is an arbitrary vector field of the class $C_b^s(\mathbb{R}^d)$, then the statements of Proposition still holds for the generator $L + (\gamma(x), \nabla)$. Let S_t be the family of diffeomorphisms of $\mathbb{R}^d$ defined by the equation $\dot{x} = -\gamma(x)$ in $\mathbb{R}^d$. With some abuse of notation we shall denote by S_t also the corresponding action on functions, i.e. $S_t f(x) = f(S_t(x))$. In the interaction representation (with respect to the group S_t), the equation

$$\dot{\phi} = (L + (\gamma(x), \nabla))\phi, \quad \phi(0) = f, \tag{5.75}$$

has the form

$$\dot{g} = L_t g = (S_t^{-1} L S_t) g, \quad g(0) = f, \tag{5.76}$$

i.e. equations (5.75) and (5.76) are equivalent for g and $\phi = S_t g$. We now apply Theorem 5.3.4 to the operators L_t from (5.76) using the Banach spaces $B_0 = H^s$, $B_1 = H^{s'}$ and $B_2 = H^{s'-2}$, where $s > s' > 2 + d/2$. The extension of the semigroup in H^s to the semigroup in $C_\infty(\mathbb{R}^d)$ is then carried out using the PMP, as above in the proof of Proposition 5.7.1. Finally, the proof of the theorem is completed via localization, i.e. by the application of Theorems 5.4.2, 4.11.3, 4.11.4 and Proposition 5.4.1.

Combining Theorems 5.7.1 and 5.6.2, we directly obtain the following well-posedness result for decomposable operators with unbounded coefficients.

Theorem 5.7.2. *Suppose* (A1), (A2) *hold for the family of operators* ψ_n *and all* a_n *are positive functions from* $C^s(\mathbb{R}^d)$ *for* $s > 2 + d/2$. *Instead of* (A3) *assume now that there exists a positive function* $f_L \in C^2_{1+|x|^p}$, $p > 0$, *such that* $f_L(x) \to \infty$ *as* $x \to \infty$ *and, moreover,* $\int |y|^p \mu^n(dy) < \infty$ *and* $a_n \psi_n f_L(x) \le c$ *for some constant* c *for all* n. *Then there exists a unique solution to the martingale problem of the operator* $\sum_{n=1}^N a_n \psi_n$ *with the domain* $C_c^2(\mathbb{R}^d)$ *defining a strong Markov* C*-Feller process.*

5.8 SDEs driven by nonlinear Lévy noise

In Section 4.6 we have shown how classical SDE allows one to construct Markov processes in the case when the Lévy kernel $\nu(x, dy)$ in the generator can be expressed as a push forward of a certain fixed measure $\nu(dy)$ along a family of regular enough transformations $F_x : \mathbb{R}^d \to \mathbb{R}^d$. Here we develop a method of weak SDE driven by nonlinear noise, which can be used to construct Markov processes for kernels $\nu(x, dy)$ that depend Lipschitz continuously on x in the Wasserstein–Kantorovich metric W_p.

We start with the analysis of processes generated by a time-dependent family of Lévy–Khintchine operators

$$L_t f(x) = \frac{1}{2}(G_t \nabla, \nabla) f(x) + (b_t, \nabla f)(x) + \int [f(x+y) - f(x) - (y, \nabla f(x)) \mathbf{1}_{B_1}(y)] \nu_t(dy), \tag{5.77}$$

where for any t, G_t is a non-negative symmetric $d \times d$-matrix, $b_t \in \mathbb{R}^d$ and ν_t is a Lévy measure. The set of Lévy measures is equipped with the weak topology, where the continuous dependence of the family ν_t on t means that $\int f(y) \nu_t(dy)$ depends continuously on t for any continuous f on $\mathbb{R}^d$ with $|f(y)| \le c \min(|y|^2, 1)$.

Proposition 5.8.1. *For a given family* $\{L_t\}$ *of form* (5.77) *with bounded coefficients* G_t, b_t, ν_t, *i.e.*

$$\sup_t \left(\|G_t\| + \|b_t\| + \int (1 \wedge y^2) \nu_t(dy) \right) < \infty,$$

that depend continuously on t a.s., i.e. outside a fixed zero-measure subset $S \subset \mathbb{R}$, there exists a unique family $\{\Phi^{s,t}\}$ of positive linear contractions in $C_\infty(\mathbb{R}^d)$ depending strongly continuously on $s \le t$ such that for any $f \in C^2_\infty(\mathbb{R}^d)$ the functions $f_s = \Phi^{s,t} f$ belong to $C^2_\infty(\mathbb{R}^d)$ and solve a.s. (i.e. for s outside a null set) the inverse-time Cauchy problem

$$\dot{f}_s = -L_s f_s, \quad s \le t, \quad f_t = f \tag{5.78}$$

(where the derivative is taken in the Banach topology of $C(\mathbb{R}^d)$).

Proof. Let f belong to the Schwartz space $S(\mathbb{R}^d)$. Then its Fourier transform

$$g(p) = (Ff)(p) = \int_{\mathbb{R}^d} e^{-ipx} f(x)\,dx$$

also belongs to $S(\mathbb{R}^d)$. As the Fourier transform of equation (5.78) has the form

$$\dot{g}_s(p) = -\left[-\frac{1}{2}(G_s p, p) + i(b_s, p) + \int (e^{ipy} - 1 - ipy\mathbf{1}_{B_1})\nu_s(dy)\right] g_s(p),$$

it has the obvious unique solution

$$g_s(p) = \exp\left\{\int_s^t \left[-\frac{1}{2}(G_\tau p, p) + i(b_\tau, p) \right.\right. \\ \left.\left. + \int (e^{ipy} - 1 - ipy\mathbf{1}_{B_1})\nu_\tau(dy)\right] d\tau\right\} g(p) \tag{5.79}$$

(the integral is defined in both the Lebesgue and the Riemann sense, as the discontinuity set of the integrand has measure zero), which belongs to $L^1(\mathbb{R}^d)$, so that $f_s = F^{-1} g_s = \Phi^{s,t} f$ belongs to $C_\infty(\mathbb{R}^d)$. As for any fixed s, t the operator $\Phi^{s,t}$ coincides with an operator from the semigroup of a certain homogeneous Lévy process, each $\Phi^{s,t}$ is a positivity-preserving contraction in $C_\infty(\mathbb{R}^d)$ preserving the spaces $(C_\infty \cap C^2)(\mathbb{R}^d)$ and $C^2_\infty(\mathbb{R}^d)$. Strong continuity is then obtained first for $f \in (C^2 \cap C_\infty)(\mathbb{R}^d)$ and then for general f by the density argument. Finally, uniqueness follows from Theorem 1.9.4. □

Let us define a time-nonhomogeneous Lévy process generated by the family $\{L_t\}$ as a time-nonhomogeneous càdlàg Markov process X_t such that

$$\mathbf{E}(f(X_t)|X_s = x) = (\Phi^{s,t} f)(x), \quad f \in C(\mathbb{R}^d),$$

where $\Phi^{s,t}$ is the propagator of positive linear contractions in $C_\infty(\mathbb{R}^d)$ from Proposition 5.8.1 (notice that a Markov process is defined uniquely up to a modification by its transition probabilities). Processes of this kind are called sometimes *additive processes* (they are stochastically continuous and have independent increments), see e.g. [289], pp. 51–68. We use the term 'nonhomogeneous Lévy' stressing their translation invariance and the analytic properties of their propagators, which represent the most straightforward time-nonhomogeneous extensions of the semigroups of Lévy processes.

We wish to interpret the nonhomogeneous Lévy processes as weak stochastic integrals. For this purpose, it will be notational more convenient to work with generator families depending on time via a multidimensional parameter. Namely, let L_η be a family of operators of form (5.77) with coefficients G_η, b_η, ν_η depending continuously on a parameter $\eta \in \mathbb{R}^n$ (ν_η is continuous as usual in the above specified weak sense). Let ξ_t be a curve in $\mathbb{R}^n$ with at most countably many discontinuities and with left and right limits existing everywhere. Then the family of operators L_{ξ_t} satisfies the assumptions of Proposition 5.8.1. Clearly, the resulting propagator $\{\Phi^{s,t}\}$ does not depend on the values of ξ_t at the points of discontinuity.

Applying Lemma 1.1.1 to the distributions of the family of the Lévy processes $Y_t(\eta)$ (corresponding to the generators L_η, we can define them on a single probability space (actually on the standard Lebesgue space) in such a way that they depend measurably on the parameter η.

Let ξ_s, α_s be piecewise constant left-continuous functions (deterministic, to begin with) with values in $\mathbb{R}^n$ and $d \times d$-matrices respectively, that is

$$\xi_s = \sum_{j=0}^{n} \xi^j \mathbf{1}_{(t_j,t_{j+1}]}(s), \quad \alpha_s = \sum_{j=0}^{n} \alpha^j \mathbf{1}_{(t_j,t_{j+1}]}(s), \tag{5.80}$$

where $0 = t_0 < t_1 < \cdots < t_{n+1}$. Then it is natural to define the stochastic integral with respect to the nonlinear Lévy noise $Y_s(\xi_s)$ by the formula

$$\int_0^t \alpha_s d Y_s(\xi_s) = \sum_{j=0}^{n} \alpha^j Y^j_{t\wedge t_{j+1}-t_j}(\xi^j) \mathbf{1}_{t_j<t}, \tag{5.81}$$

where $Y_t^j(\eta)$ are independent copies of the families of $Y_t(\eta)$ defined above via the randomization lemma. It is clear that the process so defined $\int_0^t \alpha_s d Y_s(\xi_s)$ is a nonhomogeneous Lévy process constructed by Proposition 5.8.1 from the generator family

$$\begin{aligned} L_t^{\alpha,\xi} f(x) &= \frac{1}{2}((\alpha_t G_{\xi_t} \alpha_t')\nabla, \nabla) f(x) + (\alpha_t b_{\xi_t}, \nabla f)(x) \\ &\quad + \int [f(x+\alpha_t y) - f(x) - (\alpha_t y, \nabla f(x)) \mathbf{1}_{B_1}(y)] \nu_{\xi_t}(dy), \end{aligned} \tag{5.82}$$

which coincides with L_{ξ_t} for $\alpha_t = 1$. Next, if ξ_t and α_t are arbitrary càdlàg function, let us define its natural piecewise constant approximation as

$$\xi_t^\tau = \sum_{\tau j<t}^{n} \xi_{\tau j} \mathbf{1}_{(\tau j, \tau(j+1)]}, \quad \alpha_t^\tau = \sum_{\tau j<t}^{n} \alpha_{\tau j} \mathbf{1}_{(\tau j, \tau(j+1)]},$$

As usual the integral $\int_0^t \alpha_s dY_s(\xi_s)$ should be defined as a limit (if it exists in some sense) of the integrals over its approximations $\int_0^t \alpha_s^\tau dY_s(\xi_s^\tau)$.

Theorem 5.8.1. *The distribution of the process of integrals $x + \int_0^t \alpha_s dY_s(\xi_s)$ is well defined as the weak limit, as $\tau \to 0$, of the distributions on the Skorohod space $D([0,T], \mathbb{R}^d)$ of the approximating simple integrals $x + \int_0^t \alpha_s^\tau dY_s(\xi_s^\tau)$, and is the distribution of the Lévy process started at x and generated by the family* (5.82). *This limit also holds in the sense of the convergence of the propagators of the corresponding nonhomogeneous Lévy processes.*

Proof. The right-continuous versions ξ_{s+}^τ converge to ξ_t in the sense of the Skorohod topology. Hence, by Theorem 1.9.5 the corresponding processes $\int_0^t dY_s(\xi_s^\tau)$ converge to the nonhomogeneous Lévy process generated by the family (5.82) in the sense of convergence of propagators. By a straightforward time-nonhomogeneous extension of either Theorem 4.8.3 or Theorem 4.9.2, this implies the weak convergence as distributions on the Skorohod space. □

In particular, we have constructed the probability kernel on the space $D([0,T], \mathbb{R}^d)$ of càdlàg paths that takes a curve ξ_t to the distribution of the integral $x + \int_0^t dY_s(\xi_s)$. The main point is that the invariant measure for this kernel defines a weak solution to the stochastic equation $\xi_t = \xi_0 + \int_0^t dY_s(\xi_s)$, and its uniqueness is closely linked with the Markovianity of the corresponding process, as we shall see shortly.

Now let us describe a couple of more specific situations. We shall use the following elementary inequalities.

Proposition 5.8.2. (i) *For any $p \in [1,2]$ and $x \geq -1$ one has*

$$0 \leq (1+x)^p - 1 - px \leq (p-1)x^2, \tag{5.83}$$

(ii) *For any $p \in [1,2]$, $d \in \mathbb{N}$ and $A, B \in \mathbb{R}^d$*

$$0 \leq |A+B|^p - |A|^p - p(A,B)|A|^{p-2} \leq c_p |B|^p, \tag{5.84}$$

where

$$c_p = \max_{x \in [0,1/2]} [(1-x)^p - x^p + px^{p-1}]. \tag{5.85}$$

The next statement describes jump-type processes when L_p-estimates are available.

Proposition 5.8.3. *Suppose $Y_s(\eta)$ is a family of Lévy processes in $\mathbb{R}^d$ with càdlàg paths, depending on a parameter $\eta \in \mathbb{R}^n$ and specified by their generators*

$$L_\eta f(x) = \int [f(x+y) - f(x) - (y, \nabla) f(x)] \nu_\eta(dy), \tag{5.86}$$

where

$$\nu_\eta(\{0\}) = 0, \quad \sup_\eta \int |y|^p \nu_\eta(dy) = \kappa < \infty \tag{5.87}$$

for some $p \in [1, 2]$. *Then the process* $\int_0^t \alpha_s dY_s(\xi_s)$ *is a martingale, and*

$$\mathbf{E}\left|\int_0^t \alpha_s dY_s(\xi_s)\right|^p \le 2c_p\kappa_1 \int |\alpha_s|^p \, ds. \tag{5.88}$$

Proof. In view of Theorem 5.8.1 it is enough to prove the statement for the approximations $\int_0^t \alpha_s^\tau dY_s(\xi_s^\tau)$. Then the martingale property follows from the property of Lévy processes. Finally, by (5.84) one has

$$\mathbf{E}\left|\int_0^t \alpha_s^\tau dY_s(\xi_s^\tau)\right|^p \le c_p \sum_{j=0}^{[t/\tau]} \|(\alpha_{j\tau}\|^p \mathbf{E}|Y^j_{t\wedge\tau(j+1)-\tau j}(\xi_{\tau j})|^p,$$

implying (5.88). Alternatively, one can get this estimate from the martingale problem associated with the propagator of the corresponding non-homogeneous Lévy process, but the proof given extends straightforwardly to the case of random ξ. □

As another situation of interest, let us consider the case of $Y_s(\eta)$ that are compound Poisson processes, i.e. they are generated by the family

$$L_\eta f(x) = \int [f(x+y) - f(x)]\nu_\eta(dy) \tag{5.89}$$

with uniformly bounded measures ν_η.

Proposition 5.8.4. *For any càdlàg curve* ξ_t, *the process* $x + \int_0^t dY_s(\xi_s)$ *has the following probabilistic description. Starting from the initial point* x, *it waits there a (non-homogeneous exponential) random time* τ_1 *with* $\mathbf{P}(\tau_1 > t) = \exp\{-\int_0^t \|\nu_{\xi_s}\| \, ds\}$ *and then jumps to a point* y_1 *distributed according to the law* $\nu_{\xi_{\tau_1}}/\|\nu_{\xi_{\tau_1}}\|$. *Then it waits a random time* τ_2 *with* $\mathbf{P}(\tau_2 > t) = \exp\{-\int_{\tau_1}^{t+\tau_1} \|\nu_{\xi_s}\| \, ds\}$ *and jumps to a point* $y_1 + y_2$ *with* y_2 *distributed according to the law* $\nu_{\xi_{\tau_2}}/\|\nu_{\xi_{\tau_2}}\|$, *etc.*

Proof. This follows from Theorem 5.8.1 and the basic series expansion of the propagator of a process generated by family (5.89); see Section 3.7. □

In particular, to link with the usual stochastic integration over a Poisson measure, let us note that if $N(dsdx)$ is the Poisson measure of a Levy process Z_t in $\mathbb{R}^d$ and $g(s, x)$ is a continuous bounded function on $\mathbb{R}_+ \times \mathbb{R}_d$, then

$$\int_0^t \int_A g(s, x) N(dsdx)$$

is a process of the type described in Proposition 5.8.4, specified by the family

$$L_t f(x) = \int [f(x + g(t, y)) - f(x)]\nu(dy).$$

Remark 41. Similarly, one can define the integral $\int_0^t g_s(dY_s(\xi_s))$ for a càdlàg family of nonlinear mappings $g_s : \mathbb{R}^n \to \mathbb{R}^d$ of the class $C^2(\mathbb{R}^n)$, as the limit of the approximating sums

$$\sum_{j=0}^{n} g_{j\tau}(Y^j_{t\wedge(j+1)\tau-j\tau}(\xi_{j\tau}))\mathbf{1}_{j\tau<t},$$

which converge to the non-homogeneous Lévy process generated by the family of operators

$$\begin{aligned} L_t^{g,\xi} f(x) &= \frac{1}{2}\left(\left[\frac{\partial g_t}{\partial y}(0)G_{\xi_t}\left(\frac{\partial g_t}{\partial y}\right)'(0)\right]\nabla, \nabla\right) f(x) \\ &\quad + \left(\frac{1}{2}G_{\xi_t}\frac{\partial^2 g_t}{\partial y^2}(0) + \frac{\partial g_t}{\partial y}(0)b_{\xi_t}, \nabla f\right)(x) \\ &\quad + \int \left[f(x + g_t(y)) - f(x) - \left(\frac{\partial g_t}{\partial y}(0)y, \nabla f(x)\right)\mathbf{1}_{B_1}(y)\right]\nu_{\xi_t}(dy). \end{aligned} \tag{5.90}$$

In order to solve the equation

$$X_t = x + \int_0^t dY_s(X_s), \tag{5.91}$$

with x a random variable with given law μ, it is convenient to use Euler-type approximations. For $\tau > 0$, let the process $X_t^{\mu,\tau}$ be defined by the recursive equation

$$X_t^{\mu,\tau} = X_{l\tau}^{\mu,\tau} + Y^l_{t-l\tau}(X_{l\tau}^{\mu,\tau}), \quad \mathcal{L}(X_0^{\mu,\tau}) = \mu, \tag{5.92}$$

for $l\tau < t \le (l+1)\tau$, where $\mathcal{L}(X)$ is the law of X. Clearly these approximation processes are càdlàg. According to the above definition of the stochastic integral, this process satisfies the equation

$$X_t^{\mu,\tau} = X_0 + \int_0^t dY_s(X_{[s/\tau]}^{\mu,\tau}), \quad \mathcal{L}(X_0) = \mu. \tag{5.93}$$

By conditioning, one deduces that $X_t^{\mu,\tau}$ solves the following martingale problem. For any $f \in C_c^2(\mathbb{R}^d)$, the process

$$M_\tau(t) = f(X_t^{\mu,\tau}) - f(X_0) - \int_0^t L[X_{[s/\tau]\tau}^{\mu,\tau}]f(X_s^{\mu,\tau})\,ds, \quad \mu = \mathcal{L}(X_0), \tag{5.94}$$

is a martingale, where we transferred the lower subscript to square brackets, i.e. $L[X] = L_X$ in the previous notations. Due to the basic convergence criteria for martingale problem solutions (see Section 4.9), it follows that the family of processes $X_t^{\mu,\tau}$, $\tau > 0$, is tight (as was noted by many authors, see e.g. Stroock [302] or Böttcher and Schilling [65]), and hence relatively compact, and moreover, any limiting process X_t^{μ} solves the martingale problem for the family L_η: for any $f \in C_c^2(\mathbb{R}^d)$, the process

$$M_\tau(t) = f(X_t^{\mu}) - f(x) - \int_0^t L[X_s^{\mu}]f(X_s^{\mu})\, ds, \quad \mu = \mathcal{L}(x), \tag{5.95}$$

is a martingale. Moreover, using the Skorohod coupling (Theorem 1.1.3), one can choose a probability space where all theses approximations are simultaneously defined and converge a.s. Consequently, by Theorem 5.8.1, X_t^x also solves the stochastic equation (5.91). In particular, this equation has a solution.

Theorem 5.8.2. *Suppose for any x the probability kernel on the subset of $D([0,t],\mathbb{R}^d)$ of paths starting at x given by the integral constructed above in Theorem* 5.8.1 *can have at most one invariant measure, i.e. equation* (5.91) *has at most one solution. Then the approximations* (5.92) *converge weakly to the unique solution X_t^x of the equation* (5.91)*, which is a Markov process solving the martingale problem* (5.95).

Proof. Convergence follows from the uniqueness of the limit. Passing to the limit $\tau \to 0$ in the Markov property for the approximations

$$\mathbf{E}(f(X_t^{\mu,\tau}) \mid \sigma(X_u^{\mu,\tau})|_{u\le j\tau}) = \mathbf{E}(f(X_t^{\mu,\tau}) \mid X_{j\tau}^{\mu,\tau})$$

yields the Markov property for the limit X_t^{μ}. □

In order to apply these results, we should be able to compare the Lévy measures. To this end, we introduce an extension of the Wasserstein–Kantorovich distance to unbounded measures. Namely, let $\mathcal{M}_p(\mathbb{R}^d)$ denote the class of Borel measures μ on $\mathbb{R}^d \setminus \{0\}$ (not necessarily finite) with finite pth moment (i.e. such that $\int |y|^p \mu(dy) < \infty$). For a pair of measures ν_1, ν_2 in $\mathcal{M}_p(\mathbb{R}^d)$ we define the distance $W_p(\nu_1,\nu_2)$ by

$$W_p(\nu_1,\nu_2) = \left(\inf_\nu \int |y_1 - y_2|^p \nu(dy_1 dy_2)\right)^{1/p}, \tag{5.96}$$

where inf is taken over all $\nu \in \mathcal{M}_p(\mathbb{R}^{2d})$ such that condition

$$\int_{\mathbb{R}^{2d}} (\phi_1(x) + \phi_2(y))\nu(dxdy) = (\phi_1,\nu_1) + (\phi_2,\nu_2)$$

holds for all ϕ_1, ϕ_2 satisfying $\phi_i(.)/|.|^p \in C(\mathbb{R}^d)$. It is easy to see that for finite measures this definition coincides with the usual one.

Moreover, by the same argument as for finite measures (see Rachev and Rüschendorf [270] or Villani [314]) we can show that whenever the distance $W_p(\nu_1, \nu_2)$ is finite, the infimum in (5.96) is achieved; i.e. there exists a measure $\nu \in \mathcal{M}_p(\mathbb{R}^{2d})$ such that

$$W_p(\mu_1, \mu_2) = \left(\int |y_1 - y_2|^p \nu(dy_1 dy_2) \right)^{1/p}. \tag{5.97}$$

To compare the distributions on Skorohod spaces we shall use the corresponding Wasserstein–Kantorovich distances. These distances depend on the choice of distances between individual paths. The most natural choice of these distances is uniform, leading to the distance on the distributions given by (1.53), that is

$$W_{p,T}(X^1, X^2) = \inf(\mathbf{E} \sup_{t \le T} |X_t^1 - X_t^2|^p)^{1/p}, \tag{5.98}$$

where inf is taken over all couplings of the distributions of the random paths X_1, X_2.

Let us now approach the *stochastic differential equation* (*SDE*) driven by *nonlinear Lévy noise* of the form

$$X_t = x + \int_0^t dY_s(g(X_{s-})) + \int_0^t b(X_{s-})\, ds + \int_0^t G(X_{s-}) dW_s, \tag{5.99}$$

where W_s is the standard Wiener process in $\mathbb{R}^d$ and $Y_t(\eta)$ is a family of pure-jump Lévy processes from Proposition 5.8.3.

Theorem 5.8.3. *Let* (5.87), (5.100) *hold for the family* (5.86) *with a* $p \in (1, 2]$. *Let* b, G *be bounded functions from* $\mathbb{R}^d$ *to* $\mathbb{R}^d$ *and to* $d \times d$*-matrices respectively, and* b, G, ν *be Lipschitz continuous with a common Lipschitz constant* κ_2*, where* $\nu(x, .)$ *are equipped with* W_p*-metric, i.e.*

$$W_p(\nu(x_1, .), \nu(x_2, .)) \le \kappa_2 \|x_1 - x_2\|. \tag{5.100}$$

Finally let x *be a random variable independent of all* $Y_s(z)$. *Then the solution to* (5.99) *exists in the sense of distribution (that is, the distributions of the l.h.s. and r.h.s. coincide) and is unique. Moreover, the solutions to this equation specify a Feller process in* $\mathbb{R}^d$*, whose generator contains the set* $C_c^2(\mathbb{R}^d)$*, where it is given by the formula*

$$\begin{aligned} Lf(x) &= \frac{1}{2}(G(x)\nabla, \nabla) f(x) + (b(x), \nabla f)(x) \\ &\quad + \int [f(x+y) - f(x) - (y, \nabla f(x))] \nu(x, dy). \end{aligned} \tag{5.101}$$

Proof. This is based on the contraction principle in the complete metric space $M_p(t)$ of distributions on the Skorohod space of càdlàg paths $\xi \in D([0,t],\mathbb{R}^d)$ with finite pth moment $W_{p,t}(\xi,0) < \infty$ and with metric $W_{p,t}$. For any $\xi \in M_p(t)$, let $\Phi(\xi) = \Phi^1(\xi) + \Phi^2(\xi)$ with

$$\Phi^1(\xi)_t = x + \int_0^t dY_s(\xi_{s-}), \quad \Phi^2(\xi)_t = \int_0^t b(\xi_{s-})\,ds + \int_0^t G(X_{s-})dW_s.$$

One has

$$W_{p,t}^p(\Phi(\xi^1),\Phi(\xi^2)) = \inf_{\xi^1,\xi^2} W_{p,t,\mathrm{cond}}^p(\Phi(\xi^1),\Phi(\xi^2)),$$

where the first infimum is over all couplings of ξ^1,ξ^2 and $W_{p,t,\mathrm{cond}}$ denotes the distance (1.53) conditioned on the given values of ξ^1,ξ^2. Hence

$$W_{p,t}^p(\Phi(\xi^1),\Phi(\xi^2)) \le 2W_{p,t}^p(\Phi^1(\xi^1),\Phi^1(\xi^2)) + 2W_{p,t}^p(\Phi^2(\xi^1),\Phi^2(\xi^2)).$$

By Proposition 5.8.2, applied recursively to the increments of the discrete approximations of our stochastic integrals, one obtains

$$\sup_{s\le t}\mathbf{E}|\Phi^1(\xi_s^1) - \Phi^1(\xi_s^2)|^p \le 2t\kappa_2 c_p \sup_{s\le t}|\xi_s^1 - \xi_s^2|^p$$

for the coupling of Φ^1,Φ^2 given by Proposition 5.8.5 proved below. Using now Doob's maximum inequality for martingales yields

$$W_{p,t,\mathrm{cond}}^p(\Phi^1(\xi^1),\Phi^1(\xi^2)) \le 4\kappa_2 t c_p \sup_{s\le t}|\xi_s^1 - \xi_s^2|^p,$$

implying

$$W_{p,t}^p(\Phi^1(\xi^1),\Phi^1(\xi^2)) \le 4\kappa_2 t c_p W_{p,t}^p(\xi^1,\xi^2).$$

On the other hand, by standard properties of Brownian motion,

$$\mathbf{E}|\Phi^2(\xi^1) - \Phi^2(\xi^2)|^2 \le t\kappa_2 \sup_{s\le t}|\xi_s^1 - \xi_s^2|^2.$$

implying

$$W_{2,t}^2(\Phi^2(\xi^1),\Phi^2(\xi^2)) \le 4\kappa_2 t W_{2,t}^2(\xi^1,\xi^2),$$

and consequently

$$W_{p,t}^p(\Phi^2(\xi^1),\Phi^2(\xi^2)) \le (4\kappa_2 t)^{p/2} W_{p,t}^p(\xi^1,\xi^2).$$

Thus finally

$$W_{p,t}^p(\Phi(\xi^1),\Phi(\xi^2)) \le c(t,\kappa_2) W_{p,t}^p(\xi^1,\xi^2).$$

Hence the mapping $\xi \mapsto \Phi(\xi)$ is a contraction in $M_p(t)$ for small enough t. This implies the existence and uniqueness of a fixed point and hence of the solution to (5.99) for this t. For large t this construction is extended by iterations.

Consequently the main condition of Theorem 5.8.2 is satisfied. The Feller property is easy to check (see details in [196]) and the form of the generator reads out from the martingale problem formulation discussed in the previous section. □

Theorem 5.8.3 reduces the problem of constructing a Feller process from a given pre-generator to a Monge–Kantorovich mass transportation (or optimal coupling) problem, where essential progress was made recently, see e.g. the book Rachev and Rüschendorf [270]. The usual approach of stochastic analysis works in the case when all measures $\nu(x, dy)$ can be expressed as images of a sufficiently regular family of mappings F_x of a certain given Lévy measure ν, see Section 4.6. To find such a family an optimal solution (or its approximation) to the Monge problem is required. Our extension allows one to use instead the solutions of its more tractable extension, called the Kantorovich problem (whose introduction in the last century signified a major breakthrough in dealing with mass transportation problems). It is well known (and easy to see in examples using Dirac measures) that the optimal coupling of probability measures (Kantorovich problem) can not always be realized via a mass transportation (a solution to the Monge problem), thus leading to examples when the construction of the process via standard stochastic calculus would not work.

Let us note that the main assumption on ν is satisfied if one can decompose the Lévy measures $\nu(x;.)$ as countable sums $\nu(x;.) = \sum_{n=1}^{\infty} \nu_n(x;.)$ of probability measures so that $W_p(\nu_i(x;.), \nu_i(z;.)) \le a_i|x - z|$ and the series $\sum a_i^p$ converges.

We complete this section by proving an easy coupling result used above.

Proposition 5.8.5. *Let Y_s^i, $i = 1, 2$, be two Lévy processes in $\mathbb{R}^d$ specified by their generators*

$$L_i f(x) = \int (f(x + y) - f(x) - (\nabla f(x), y))\nu_i(dy) \tag{5.102}$$

with $\nu_i \in \mathcal{M}_p(\mathbb{R}^d)$, $p \subset [1, 2]$. Let $\nu \in \mathcal{M}_p(\mathbb{R}^{2d})$ be a coupling of ν_1, ν_2, i.e.

$$\iint (\phi_1(y_1) + \phi_2(y_2))\nu(dy_1 dy_2) = (\phi_1, \nu_1) + (\phi_2, \nu_2) \tag{5.103}$$

holds for all ϕ_1, ϕ_2 satisfying $\phi_i(.)/|.|^p \in C(\mathbb{R}^d)$. Then the operator

$$\begin{aligned} Lf(x_1, x_2) = \int & [f(x_1 + y_1, x_2 + y_2) - f(x_1, x_2) \\ & - ((y_1, \nabla_1) + (y_2, \nabla_2)) f(x_1, x_2)]\nu(dy_1 dy_2) \end{aligned} \tag{5.104}$$

(where ∇_i means the gradient with respect to x_i) specifies a Lévy process Y_s in $\mathbb{R}^{2d}$ with characteristic exponent

$$\eta_{x_1,x_2}(p_1, p_2) = \int (e^{iy_1 p_1 + iy_2 p_2} - 1 - i(y_1 p_1 + y_2 p_2))\nu(dy_1 dy_2),$$

that is, a coupling of Y_s^1, Y_s^2 in the sense that the components of Y_s have the distribution of Y_s^1 and Y_s^2 respectively. Moreover, if $f(x_1, x_2) = h(x_1 - x_2)$ with a function

$h \in C^2(\mathbb{R}^d)$, then

$$Lf(x_1,x_2) = \int [h(x_1-x_2+y_1-y_2)-h(x_1-x_2)-(y_1-y_2,\nabla h(x_1-x_2)]\nu(dy_1dy_2). \tag{5.105}$$

Finally,

$$\mathbf{E}|\xi + Y_t^1 - Y_t^2|^p \le |\xi|^p + tc_p \iint |y_1 - y_2|^p \nu(dy_1dy_2). \tag{5.106}$$

Proof. It is straightforward to see from the definition of coupling that if f depends only on x_1 (resp. x_2), then the operator (5.104) coincides with L_1 (resp. L_2). Similarly one sees that the characteristic exponent of Y_s coincides with the characteristic exponent of Y^1 (resp. Y^2) for $p_2 = 0$ (resp. $p_1 = 0$). Formula (5.105) is a consequence of (5.104). To get (5.106) one uses Dynkin's formula for the function $f(x_1,x_2) = |\xi + x_1 - x_2|^p$, observing that by (5.105) and (5.84) one has

$$|Lf(x_1,x_2)| \le c_p \iint |y_1 - y_2|^p \nu(dy_1dy_2). \qquad \square$$

5.9 Stochastic monotonicity and duality

A Markov process X_t in $\mathbb{R}$ is called *stochastically monotone* if the function $\mathbf{P}(X_t^x \ge y)$ (as usual, x stands for the initial point here) is nondecreasing in x for any $y \in \mathbb{R}$, $t \in \mathbb{R}_+$, or, equivalently (which one sees by linearity and approximation), if the corresponding Markov semigroup preserves the set of non-decreasing functions.

Remark 42. This notion of monotonicity is closely linked with the notion of *stochastic dominance* widely used in economics. Namely, a random variable X with distribution function F_X is said to *stochastically dominate* Y whenever $F_X(x) \le F_Y(x)$ for all x. One says that a Markov process X_t^x stochastically dominates a Markov process Y_t^y if $X_t^{z_1}$ dominates $Y_t^{z_2}$ for all $t > 0$ and $z_1 \ge z_2$. Hence, a Markov process X_t^x is stochastically monotone if it stochastically dominates itself or, equivalently, if the mapping $x \mapsto X_t^x$ is non-decreasing, with random variables ordered by stochastic dominance.

A Markov process Y_t in $\mathbb{R}$ is called *dual* to X_t if

$$\mathbf{P}(Y_t^y \le x) = \mathbf{P}(X_t^x \ge y) \tag{5.107}$$

for all $t > 0$, $x, y \in \mathbb{R}$. If a dual Markov process exists it is obviously unique. Existence is settled in the following result.

Theorem 5.9.1. *Let X be a stochastically monotone Feller process s.t., for any y, the function $P(X_t^x \ge y)$ depends continuously on x and tends to 1 and 0 for x tending to $+\infty$ and $-\infty$ respectively. Then the dual Markov process exists.*

Proof. The one-dimensional margins of the dual process Y_t^y are defined by (5.107). Hence its transition operators should be given by

$$(\Phi_t g)(y) = \int g(z) dF_t^y(z),$$

where

$$F_t^y(x) = \mathbf{P}(Y_t^y \le x) = \mathbf{P}(X_t^x \ge y).$$

By stochastic monotonicity of X this function is in fact a distribution function. So the only thing to check is that the operators Φ form a semigroup, or equivalently, that the Chapman-Kolmogorov equation holds. Thus we need to show that

$$\int_{\mathbb{R}^2} g(v) dF_s^z(v) dF_t^y(z) = \int_{\mathbb{R}} g(z) dF_{t+s}^y(z). \tag{5.108}$$

And it suffices to show that for $g = \mathbf{1}_{(-\infty,b]}$, in which case (5.108) reduces to

$$\int \mathbf{P}(X_s^b \ge z) dF_t^y(z) = \mathbf{P}(X_{t+s}^b \ge y). \tag{5.109}$$

one way to see that this equation holds is via discrete approximations. Assume that the integral on l.h.s. exists as a Riemann integral. Then this l.h.s. rewrites as

$$\begin{aligned}
&\lim_{h\to 0} \sum_j \mathbf{P}(X_s^b \ge jh)[\mathbf{P}(X_t^{(j+1)h} \ge y) - \mathbf{P}(X_t^{jh} \ge y)] \\
&= \lim_{h\to 0} \sum_j [\mathbf{P}(X_s^b \ge (j-1)h) - \mathbf{P}(X_s^b \ge jh)] \mathbf{P}(X_t^{jh} \ge y) \\
&= \int \mathbf{P}(X_t^z \ge y) dG_s^b(z)
\end{aligned}$$

(where G_s^x is the distribution function of X_s^x), which clearly coincides with the r.h.s. of (5.109). □

Remark 43. 1. In the general formulation of duality widely used in the theory of martingale problems and super-processes (see e.g. Etheridge [109] or Ethier and Kurtz [110]) the two families of process X_t^x, Y_t^y in a metric space S, parametrized by their initial conditions, are called dual with respect to a function f on $S \times S$ if $\mathbf{E} f(X_t^x, y) = \mathbf{E} f(x, Y_t^y)$. Our duality given by (5.107) fits to the above general notion for a function $f(x, y) = \mathbf{1}_{\{x \le y\}}$.

2. There exists another notion of duality, also of extreme importance for the theory of Markov processes, where two Markov processes in S, with the semigroups T_t and Φ_t, are called dual with respect to a measure μ if the operators T_t and Φ_t are dual as operators in $L^2(S, \mu)$.

Theorem 5.9.2. *Let L be given by* (5.11) *with* $G, b \in C^2(\mathbb{R})$ *and a continuously differentiable in x Lévy kernel* ν *such that* (5.12) *and* (5.13) *hold. If for any* $a > 0$ *the functions*

$$\int_a^\infty \nu(x, dy), \quad \int_{-\infty}^{-a} \nu(x, dy) \tag{5.110}$$

are non-decreasing and non-increasing respectively, then L generates a unique Feller semigroup with the generator given by (5.11) *on the subspace* $C^2_\infty(\mathbb{R})$, *which is an invariant core for it. The corresponding process is stochastically monotone.*

Proof. This is a direct corollary of Theorem 5.1.3 (and its proof). In fact, under the given assumptions the inequalities (5.14) are satisfied automatically with $h_\pm = \infty$, and as approximations for ν one can take just $\nu_h(x, dy) = \mathbf{1}_{|y|>h}\nu(x, dy)$. Moreover, from the monotonicity of (5.110) it follows that the generator of the equation for g^h is conditionally positive, implying that the evolution of the derivative of f preserves positivity, giving the stochastic monotonicity property. □

Proposition 5.9.1. *Under the assumptions of Theorem* 5.9.2 *suppose that the Lévy measures are supported on* $\mathbb{R}_+$ *(for measures on* $\mathbb{R}_-$ *the formulas are of course similar) and* $\nu(x, dy) = \nu(x, y)M(dy)$ *with* $\nu(x, y)$ *differentiable in x and a certain Borel measure M. Then the generator of the dual Markov process acts by*

$$\begin{aligned} Lf(x) &= \frac{1}{2}G(x)f''(x) - \left[\frac{1}{2}G'(x) + b(x)\right]f'(x) \\ &+ \int_0^\infty [f(x-y) - f(x) + f'(x)\mathbf{1}_{B_1}(y)]\tilde{\nu}(x, dy) \\ &+ f'(x)\int_0^1 y(\nu - \tilde{\nu})(x, dy) \end{aligned} \tag{5.111}$$

on $C^2_c(\mathbb{R})$, *where*

$$\tilde{\nu}(x, dy) = \nu(x-y, y)M(dy) + \frac{\partial}{\partial x}\left(\int_y^\infty \nu(x-y, z)M(dz)\right) dy.$$

Proof. This can be obtained from the explicit form of the transition probabilities. We omit the details that can be found in Kolokoltsov [197]. □

Let us discuss a more general situation including unbounded coefficients, where we include a separate term in the generator to handle in a unified way a simpler situation of Lévy measures with a finite first moment. Let $C_{\infty,|.|}(\mathbb{R})$ denote the Banach space of continuous functions g on $\mathbb{R}$ such that $\lim_{x\to\infty} g(x)/|x| = 0$, equipped with the norm $\|g\|_{C_{\infty,|.|}} = \sup_x(|g(x)|/(1+|x|))$.

Theorem 5.9.3. *For the operator*

$$Lf(x) = \frac{1}{2}G(x)f''(x) + b(x)f'(x) \\ + \int (f(x+y) - f(x) - f'(x)y)\nu(x,dy) \\ + \int (f(x+y) - f(x))\mu(x,dy), \tag{5.112}$$

let the following conditions hold:

(i) *The functions $G(x)$ and $b(x)$ are twice continuously differentiable, G is non-negative, and the first two derivatives of ν and μ with respect to x exist weakly as signed Borel measures and are continuous in the sense that the integral*

$$\int f(y)(\nu(x,dy) + |\nu'(x,dy)| + |\nu''(x,dy)|)$$

is bounded and depends continuously on x for any continuous $f(y) \le |y| \wedge |y|^2$, and the integral

$$\int f(y)(\mu(x,dy) + |\mu'(x,dy)| + |\mu''(x,dy)|)$$

is bounded and depends continuously on x for any continuous $f(y) \le |y|$.

(ii) *For any $a > 0$ the functions* (5.110) *are non-decreasing and non-increasing respectively.*

(iii) *For some constant $c > 0$*

$$b(x) + \int |y|(\mu(x,dy) + \int_{-\infty}^{-x} |y+x|\nu(dy) \le c(1+x), \quad x > 1,$$
$$-b(x) + \int |y|(\mu(x,dy) + \int_{-x}^{\infty} |y+x|\nu(dy) \le c(1+|x|), \quad x < -1. \tag{5.113}$$

Then the martingale problem for L in $C_c^2(\mathbb{R})$ is well posed, the corresponding process X_t^x is strong Markov with

$$\mathbf{E}|X_t^x| \le e^{ct}(|x| + c),$$

its contraction Markov semigroup preserves $C(\mathbb{R})$ and extends from $C(\mathbb{R})$ to a strongly continuous semigroup in $C_{\infty,|.|}(\mathbb{R})$ with domain containing $C_c^2(\mathbb{R})$. If additionally, for each $a > 0$ the functions

$$\int_a^{\infty} \mu(x,dy), \quad \int_{-\infty}^{-a} \mu(x,dy)$$

are non-decreasing and non-increasing respectively, then the process X_t^x is stochastically monotone.

Proof. By condition (5.113) and the method of Lyapunov functions (see Section 5.6) with the Lyapunov function f_L being a regularized absolute value, i.e. $f_L(x)$ is a twice continuously differentiable positive convex function coinciding with $|x|$ for $|x| > 1$, the theorem is reduced to the case of bounded coefficients, that is to Theorem 5.9.2. □

Let us extend the stochastic monotonicity property to a finite-dimensional setting. Suppose that the space $\mathbb{R}^d$ is equipped with its natural partial order, that is $x \le y$ if this relation holds for each co-ordinate of x and y. A real function on $\mathbb{R}^d$ will be called monotone if $x \le y \Longrightarrow f(x) \le f(y)$. Let us say that a Markov process in $\mathbb{R}^d$ is *stochastically monotone* if the set of monotone functions is preserved by its Markov semigroup.

Theorem 5.9.4. *Let*

$$Lf(x) = \frac{1}{2}(G(x)\nabla,\nabla)f(x) + (b(x),\nabla f(x)) + \int_{\mathbb{R}_d^+} [f(x+y) - f(x)]\nu(x,dy).$$

Assume that

(i) *ν is a Lévy measure with support on $\mathbb{R}_+^d$ and with a finite first moment*

$$\sup_x \int |y|\nu(x,dy) < \infty;$$

(ii) *$G,b \in C^2(\mathbb{R}^d)$ and ν is twice continuously differentiable in x with*

$$\sup_{x,j} \int |y| \frac{\partial}{\partial x_j}\nu(x,dy) < \infty, \quad \sup_x \int |y|\nabla^2\nu(x,dy) < \infty;$$

(iii) *the matrix elements G_{ij} depend only on x_i, x_j, and finally*

$$\frac{\partial b_j}{\partial x_i} \ge 0, \quad i \ne j,$$

and $\frac{\partial \nu}{\partial x_i}(x,dy)$ are non-negative measures on $\mathbb{R}_+^d$.

Then L generates a uniquely defined Feller process, which is stochastically monotone. Moreover, the spaces $C_\infty^1(\mathbb{R}^d)$ and $C_\infty^2(\mathbb{R}^d)$ are invariant under the semigroup T_t of this process, and T_t is regular there in the sense that

$$\|T_t\|_{C_\infty^1(\mathbb{R}^d)} \le e^{Kt}, \quad \|T_t\|_{C_\infty^2(\mathbb{R}^d)} \le e^{Kt}$$

for some constant K. Finally, $C_\infty^2(\mathbb{R}^d)$ is an invariant core for T_t.

Proof. Well-posedness follows from Proposition 4.6.2. However, a direct proof via the same line of arguments but taking monotonicity into account is straightforward. Differentiating equation $\dot{f} = Lf$ (and using the assumption that G_{ij} depend only on x_i, x_j) yields the equation

$$\frac{d}{dt}\frac{\partial f}{\partial x_k} = L\frac{\partial f}{\partial x_k} + \frac{1}{2}\frac{\partial G_{kk}}{\partial x_k}\frac{\partial}{\partial x_k}\frac{\partial f}{\partial x_k} + \sum_{i\neq k}\frac{\partial G_{ki}}{\partial x_k}\frac{\partial}{\partial x_i}\frac{\partial f}{\partial x_k} + \frac{\partial b_k}{\partial x_k}\frac{\partial f}{\partial x_k}$$

$$+ \sum_{i\neq k}\frac{\partial b_i}{\partial x_k}\frac{\partial f}{\partial x_i} + \int_{\mathbb{R}^d_+}\int_0^1 (\nabla f(x+\theta y), y)d\theta \frac{\partial \nu}{\partial x_k}(x, dy).$$

We see that the generator of the evolution of the gradient ∇f decomposes into the sum of a conditionally positive operator and a bounded positive operator. Hence the proof can be completed as in Theorem 5.1.1 by first introducing the approximations L_h by changing ν to the cut-off measure $\mathbf{1}_{|y|>h}\nu(x, dy)$, where well-posedness becomes obvious (diffusion process perturbed by a bounded operator) and where the evolution of the derivatives preserves positivity and is bounded uniformly in h. □

5.10 Stochastic scattering

We make a short introduction to *stochastic scattering*, referring to original papers for generalizations and detailed proofs.

This story represents an aspect of the general question of describing the long-time behavior of a stochastic evolution. Two basic classes of processes from this point of view are *recurrent* and *transient*. Roughly speaking, *recurrent processes* in a locally compact metric space are those that a.s. return to any open bounded domain infinitely many times, like, say one dimensional BM. On the other hand, *transient processes* have trajectories that a.s. go to infinity. It turns out that for a rather general class of Markov processes the so called *transience-recurrence dichotomy* can be proved that states that the processes of this class are either recurrent or transient simultaneously for all initial conditions.

If a process is recurrent, the natural way to analyze its large-time behavior is by looking for an invariant measure. Once such a measure is found, one can look at ergodicity, that is at convergence to an invariant measure as time goes to infinity in various senses. If there are several invariant measures, one can try to classify the initial states by their limiting distribution, which eventually describe the long-time behavior pretty well.

But how one can describe the long-time behavior of a transient process? Well, it goes to infinity, but how? Here the natural idea is to compare its long-time behavior with some simple process, whose behavior is well understood. More precisely, one can ask whether the large-time behavior of a given process is the same as that of a simple process, but with possibly shifted initial condition. This is precisely the idea of scattering, and this shift of initial data is often called the *wave operator* or the *scattering operator*. We shall explain this in some detail on the most trivial meaningful example of the stochastic evolution given by SDE

$$dX_t = dB_t + K(X_t)dt, \tag{5.114}$$

or in integral form

$$X_t = x + \int_0^t K(X_s)ds + B_t,$$

where B_t is standard d-dimensional BM. In other words, this is the Markov process discussed in Section 4.4.

It is evident that for a random process $Y_t = X_t - B_t$ the corresponding SDE reduces to the ODE with random coefficients

$$\dot{Y}_t = K(Y_t + B_t).$$

Theorem 5.10.1 (Simplest stochastic scattering). *Let $d \geq 3$ and let K be a uniformly bounded and Lipschitz continuous function. Suppose there exist positive constants α and κ such that*

$$\frac{1}{\alpha} < \frac{1}{2} - \frac{1}{d}$$

and

$$|K(x)| \leq \kappa |x|^{-\alpha}, \quad x \in \mathbb{R}^d, \tag{5.115}$$

$$|K(x_1) - K(x_2)| \leq \kappa |r|^{-\alpha} |x_1 - x_2|, \quad |x_1|, |x_2| > r. \tag{5.116}$$

Then for any $x_\infty \in \mathbb{R}^d$ there exists a unique solution X_t of equation (5.114) *such that a.s.*

$$\lim_{t\to\infty} |X_t - x_\infty - B_t| = 0. \tag{5.117}$$

Proof. Let X_t be the required solution. Then $u_t = X_t - x_\infty - B_t$ satisfies the equation $\dot{u}_t = K(u_t + x_\infty + B_t)$. By the condition on α there exists an $\epsilon > 0$ such that

$$\left(\frac{1}{2} - \frac{1}{d} - \epsilon\right)\alpha > 1.$$

Hence by (5.115) and Theorem 2.2.3, u_t satisfies the integral equation

$$u_t = \int_t^\infty K(u_s + B_s + x_\infty)\, ds \tag{5.118}$$

(in particular, the integral is convergent).

Let $C([T,\infty), \mathbb{R}^d)$ denote as usual the Banach space of $\mathbb{R}^d$-valued bounded continuous functions on the unbounded interval $[T,\infty)$. The above argument shows that if a function u_t belongs to the unit ball B_1^T of the space $C([T,\infty),\mathbb{R}^d)$, then $X_t = u_t + x_\infty + B_t$ satisfies the requirements of the theorem if and only if u_t satisfies (5.118), that is, it is a fixed point of a mapping

$$u \mapsto \mathcal{F}(u)_t = \int_t^\infty K(u_s + B_s + x_\infty)\, ds.$$

Finally, condition (5.115) and Theorem 2.2.3 ensure that $\mathcal{F}$ takes B_1^T to itself for large enough T, and condition (5.116) insures that $\mathcal{F}$ is a contraction. Hence the contraction principle implies the existence of a unique fixed point. □

Theorem 5.10.1 allows one to define the *random wave operator* as the mapping $\Omega : \mathbb{R}^d \to \mathbb{R}^d$ that takes x_∞ into the initial condition x_0 of the solution X_t satisfying (5.117). This operator yields of course a rather precise description of the process X_t at large times. Of interest are the questions of regularity of the mapping Ω, and most importantly its invertibility, that is the existence of Ω^{-1}. In the general scattering theory this latter property is called the *completeness* of the wave operator Ω.

Similarly, using Theorem 2.2.4 instead of Theorem 2.2.3 one can analyze scattering for a Newtonian particle driven by a white noise force, that is for the system (4.34):

$$\dot{x} = y, \quad dy = \frac{\partial V}{\partial x} + dB_t. \tag{5.119}$$

This stochastic Newtonian system formally describes the dynamics of particle in the (formal) potential field $V(x) - x\dot{B}_t$ (see Section 5.12 for references including quantum extensions).

5.11 Nonlinear Markov chains, interacting particles and deterministic processes

As a final topic in this chapter we briefly introduce the quickly developing field of nonlinear Markov processes, showing how they arise as the dynamic LLN limit for interacting particles. We shall touch upon the simplest case of a finite number of types (discrete state space setting), referring for a full story to the book [196].

We first recall the basic fact about first-order PDE linking it with deterministic Markov processes. Consider a PDE of the form

$$\frac{\partial S}{\partial t} = \left(b(x), \frac{\partial S}{\partial x}\right) = \sum_{j=1}^d b^j(x)\frac{\partial S}{\partial x_j}, \quad x \in \mathbb{R}^d,\ t \geq 0, \tag{5.120}$$

where $b \in C^1(\mathbb{R}^d)$. The solutions to the ODE $\dot{x} = b(x)$ are called the *characteristics* of the linear first-order PDE (5.120). It is well known that for $S_0 \in C^1\mathbb{R}^d)$ there exists a unique solution to the Cauchy problem for equation (5.120), given by the formula $S(t,x) = S_0(X_t^x)$, where X_t^x denotes the characteristics starting at x, that is $X_0^x = x$. Let us show how this fact can be deduced from the theory of semigroups.

We shall be interested in the slightly more general setting where state space Ω is a polyhedron in $\mathbb{R}^d$, that is a compact set with an non-empty interior that can be represented as the intersection of a finite number of half-spaces (though more general Ω can be easily fitted to the argument).

Proposition 5.11.1 (PDE and deterministic Markov processes). *Let Ω be a polyhedron in $\mathbb{R}^d$ and $b(x)$ a vector-valued function on Ω of the class $C^1(\Omega)$ (i.e. it has bounded continuous partial derivatives up to the boundary of Ω). Assume that for any $x \in \Omega$ there exists a unique solution to the ODE $\dot{X}_t^x = b(X_t^x)$ with the initial condition x that stays in Ω for all times. Then the operators $T_t f(x) = f(X_t^x)$ form a Feller semigroup in $C(\Omega)$ such that the space $C^1(\Omega)$ is an invariant core for its generator. Finally, if $b \in C^2(\Omega)$, then the space $C^2(\Omega)$ is also an invariant core.*

Proof. It is clear that T_t is Feller and that the spaces $C^1(\Omega)$ and $C^2(\Omega)$ are invariant. It remains to show that any $f \in C^1(\Omega)$ belongs to the generator. But for such an f we have

$$T_t f(x) - f(x) = (\nabla f(x), X_t^x - x) + o(t) = (\nabla f(x), b(x)t + o(t)) + o(t),$$

so that

$$\lim_{t\to 0} \frac{1}{t}(T_t f(x) - f(x)) = (\nabla f(x), b(x)). \qquad \square$$

In particular, we have shown that, for $f \in C^1(\Omega)$, the function $f(X_t^x)$ represents a classical solution to equation (5.120), as expected.

Let us move to interacting particles. Suppose our initial state space is a finite collection $\{1,\dots,d\}$, which can be interpreted as the types of a particle (say, possible opinions of individuals on a certain subject). Let $Q(\mu) = (Q_{ij})(\mu)$ be a family of Q-matrices depending on a vector μ from the simplex

$$\Sigma_d = \Big\{\mu = (\mu_1,\dots,\mu_d) \in \mathbb{R}_+^d : \sum_{j=1}^d = 1\Big\},$$

as on a parameter. Each such matrix specifies a Markov chain on the state space $\{1,\dots,d\}$ with the generator described by the formulas (3.60) and with the intensity of jumps being

$$|Q_{ii}| = -Q_{ii}(\mu) = \sum_{j\neq i} Q_{ij}(\mu).$$

Suppose we have a large number of particles distributed arbitrary among the types $\{1,\dots,d\}$. More precisely our state space S consists of all sequences of d non-negative integers $N = (n_1,\dots,n_d)$, where each n_i specifies the number of particles in the state i. Let $|N|$ denote the total number of particles in state N: $|N| = n_1 + \cdots + n_d$. For $i \neq j$ and a state N with $n_i > 0$ denote by N^{ij} the state obtained from N by removing one particle of type i and adding a particle of type j, that is n_i and n_j are changed to $n_i - 1$ and $n_j + 1$ respectively. We are interested in the Markov process on S specified by the following generator:

$$Lf(N) = \sum_{i=1}^{d} n_i Q_{ij}(N/|N|)[f(N^{ij}) - f(N)]. \tag{5.121}$$

Recalling the probabilistic description of a pure jump Markov process given in Theorem 3.7.3, we see that operator (5.121) generates the following process. Starting from any time and current state N one attaches to each particle a $|Q_{ii}|(N/|N|)$-exponential random waiting time (where i is the type of this particle). If the shortest of the waiting times τ that occur is attached to a particle of the type i, this particle jumps to a state j according to the distribution $(Q_{ij}/|Q_{ii}|)(N/|N|)$. Briefly, with this distribution and at rate $|Q_{ii}|(N/|N|)$, any particle of type i can turn to a type j. After any such transition the process starts again from the new state N^{ij}.

Such processes are usually called *mean-field interacting* Markov chains (as their transitions depend on the empirical measure $N/|N|$ or the mean field). Notice that since the number of particles $|N|$ is preserved by any jump, this process is in fact a Markov chain with a finite state space. To shorten the formulas, we shall denote the inverse number of particles by h, that is $h = 1/|N|$.

Normalizing the states to $N/|N| \in \Sigma_d$ leads to a generator of the form

$$L_h f(N/|N|) = \sum_{i=1}^{d}\sum_{j=1}^{d} \frac{n_i}{|N|}|N| Q_{ij}(N/|N|)[f(N^{ij}/|N|) - f(N/|N|)], \tag{5.122}$$

or equivalently

$$L_h f(x) = \sum_{i=1}^{d}\sum_{j=1}^{d} x_i Q_{ij}(x)\frac{1}{h}[f(x - he_i + he_j) - f(x)], \quad x \in h\mathbb{Z}_+^d, \tag{5.123}$$

where $e_1, \dots, e_d$ denotes the standard basis in $\mathbb{R}^d$. We shall be interested in the asymptotics as $h \to 0$, that is in the dynamic LLN for a Markov chain specified by generator (5.123).

If $f \in C^1(\Sigma_d)$, then

$$\lim_{|N|\to\infty,\, N/|N|\to x} |N|[f(N^{ij}/|N|) - f(N/|N|)] = \frac{\partial f}{\partial x_j}(x) - \frac{\partial f}{\partial x_i}(x),$$

where $x = N/|N|$, so that

$$\lim_{|N|\to\infty,\, N/|N|\to x} L_h f(N/|N|) = \Lambda f(x),$$

where

$$\Lambda f(x) = \sum_{i=1}^{d} \sum_{j \neq i} x_i Q_{ij}(x) \left[\frac{\partial f}{\partial x_j} - \frac{\partial f}{\partial x_i} \right](x)$$

$$= \sum_{k=1}^{d} \sum_{i \neq k} [x_i Q_{ik}(x) - x_k Q_{ki}(x)] \frac{\partial f}{\partial x_k}(x). \tag{5.124}$$

The limiting operator Λf is a first-order PDO with characteristics equation

$$\dot{x}_k = \sum_{i \neq k} [x_i Q_{ik}(x) - x_k Q_{ki}(x)] = \sum_{i=1}^{k} x_i Q_{ik}(x), \quad k = 1, \dots, d, \tag{5.125}$$

called the *kinetic equations* for the process of interaction described above. The characteristics specify the dynamics of the deterministic Markov Feller process in Σ_d defined via the generator Λ (and Proposition 5.11.1).

By general results on convergence of semigroups, see e.g. Maslov [231] or Reed and Simon [274], the convergence of the generators L_h to the generator Λ on all functions from a core $C^1(\Sigma_d)$ of the latter implies convergence of the corresponding semigroups, and consequently (by Theorem 4.8.3) convergence of the distributions of the corresponding processes. However, let us formulate a result with a stronger regularity assumptions that also give rates of convergence.

Proposition 5.11.2. *Let all the elements $Q_{ij}(\mu)$ (of a given family of Q-matrices) belong to $C^2(\Sigma)$. Then the semigroups T_t^h, $h = 1/|N|$, generated by L_h converge to the semigroup of the deterministic process X_t^x given by the solutions to the kinetic equation* (5.125)*. The corresponding processes hN_t converge in distribution to the corresponding (deterministic) characteristics. Finally, if $f \in C^2(\Sigma_d)$, then*

$$\sup_{s \leq t} \sup_{N \in \mathbb{Z}_+^d : |N| = 1/h} [T_t^h f(hN) - f(X_t^{hN})] = O(h) \|f\|_{C^2(\Sigma_d)}.$$

Proof. Convergence of semigroups and the required estimate follow from Theorem 8.1.1 proved in Section 8.1. Convergence of the distributions follows from Theorem 4.8.3, as was already noted. □

A semigroup of measurable transformations of probability measures is called a *nonlinear Markov semigroup* . It is not difficult to show (see Stroock [302] or Kolokoltsov [196]) that, under mild regularity assumption, each nonlinear Markov semigroup on Σ_d (the set of probability laws on $\{1, \dots, d\}$) arises from equations of the form (5.125). We have shown that the solution to these equations describe the *dynamic law of large numbers* (*LLN*) (as the limit $|N| \to \infty$) of the mean-field interacting Markov chains specified by the generator (5.122).

Once a dynamic LLN is found, it is natural to describe the fluctuations, where one expects the corresponding dynamic CLT to hold. In our simple finite-dimensional example this is easily obtained. Namely, consider the process of fluctuations

$$Z_t = \frac{hN_t^N - X_t^x}{\sqrt{h}} \tag{5.126}$$

when $hN \to x$ and $(hN - x)/\sqrt{h}$ tend to a finite limit, as $h = 1/|N| \to 0$. Using Exercise 3.6.3, in particular formula (3.50) we conclude that the process Z_t is a time non-homogeneous Markov with the propagator generated by the family of operators

$$\begin{aligned} A_t f(y) &= -\frac{1}{\sqrt{h}}(\dot{X}_t, \nabla f(y)) \\ &+ \sum_{i,j=1}^d (\sqrt{h}y_i + X_{t,i})Q_{ij}(\sqrt{y} + X_t)\frac{1}{h}[f(y - he_i + he_j) - f(y)]. \end{aligned} \tag{5.127}$$

Using Taylor approximation for small h, in particular

$$\frac{1}{h}[f(y-he_i+he_j)-f(y)] = \frac{1}{\sqrt{h}}\left(\frac{\partial f}{\partial y_j} - \frac{\partial f}{\partial y_i}\right) + \frac{1}{2}\left(\frac{\partial^2 f}{\partial y_j^2} - 2\frac{\partial^2 f}{\partial y_j \partial y_i} + \frac{\partial^2 f}{\partial y_i^2}\right),$$

one observe that, due to the kinetic equations for X_t, the terms of order $h^{-1/2}$ in (5.127) cancel (as should be expected) yielding

$$A_t f = O_t f + O(\sqrt{h}),$$

where

$$\begin{aligned} O_t f(y) &= \frac{1}{2}\sum_{i,j=1}^d X_{t,i} Q_{ij}(X_t)\left(\frac{\partial^2 f}{\partial y_j^2} - 2\frac{\partial^2 f}{\partial y_j \partial y_i} + \frac{\partial^2 f}{\partial y_i^2}\right) \\ &+ \sum_{i,j=1}^d [y_i Q_{ij}(X_t) + X_{t,i}(\nabla Q_{ij}(X_t), y)]\left(\frac{\partial f}{\partial y_j} - \frac{\partial f}{\partial y_i}\right) \end{aligned} \tag{5.128}$$

and $O(\sqrt{h})$ depend on the bounds for the second derivatives for Q and third derivatives for f. Operator O_t is a second-order PDO with a linear drift and position-independent diffusion coefficients. Hence it generates a Gaussian diffusion process (a kind of time non-homogeneous OU process). By the same argument as in Proposition 5.11.2, we arrive at the following.

Proposition 5.11.3 (Dynamic CLT for simplest mean-field interactions). *Let all the elements $Q_{ij}(\mu)$ (of a given family of Q-matrices) belong to $C^2(\Sigma)$. Then the process of fluctuations* (5.126) *converge (both in the sense of convergence of propagators and the distributions on paths) to the Gaussian diffusion generated by the second order* PDO (5.128).

The examples of Markov chains of type (5.121) are numerous. For instance, in modeling a pool of voters, the transition $N \to N^{ij}$ is interpreted as the change of opinion.

In model (5.121) a mean-field interaction was assumed. Similarly one can model binary, ternary or generally *k-th order interaction*, where change may occur after an interaction of a group of particles of size k. Say, assuming that any two particles i, j of different type (binary interaction) can be transformed to a pair of type k, l (say, two agents i, j communicated and i changed her opinion to j) with rates Q_{ij}^{kl}, yields, instead of (5.121), the generator

$$Lf(N) = \sum_{i \neq j} \sum_{k,l=1}^{d} n_i n_j Q_{ij}^{kl}(N/|N|)[f(N^{ij,kl}) - f(N)],$$

where $N^{ij,kl}$ is obtained from N by changing two particles of type i, j to two particles of type k, l. Appropriate scaling leads, instead of (5.124), to the limiting operator

$$\Lambda f = \sum_{i \neq j} \sum_{k,l=1}^{d} x_i x_j Q_{ij}^{kl} \left[\frac{\partial f}{\partial x_k} + \frac{\partial f}{\partial x_l} - \frac{\partial f}{\partial x_j} - \frac{\partial f}{\partial x_i} \right](x),$$

and the corresponding kinetic equations (characteristics) of binary interaction, with the r.h.s. depending quadratically on the position.

Let us note finally that dynamic LLN may not be deterministic as above. Stochastic dynamic LLN are addressed later in Section 6.9.

5.12 Comments

Theorem 5.1.1 is taken from [196], where one can find extensions to cover C-Feller semigroups, as well as time-nonhomogeneous and even nonlinear generalizations.

The results of Section 5.4 represent more or less straightforward consequences of the general theory applied to Lévy–Khintchine type generators. Section 5.6 is an application of the method of Lyapunov functions as barriers in the context of martingale problems. The exposition is based on Kolokoltsov [187] and [196].

In the extensive literature on Feller processes with pseudo-differential generators, decomposable generators play an essential role, because analytically they are simpler to deal with, but at the same time their properties capture the major qualitative features of the general case. On the other hand, decomposable generators appear naturally in connection with the interacting particle systems, see Chapter 6. In the context of interacting particle systems, the corresponding functions a_n are usually unbounded but smooth. Theorems 5.7.1 and 5.7.2 and their proofs are taken from [187]. Previously, using another method (resolvents and Hille–Yosida theorem), the well-posedness for decomposable generators was proved by Hoh [132] under the assumptions of the reality of symbols (all $p_n(\xi)$ are real) and strong non-degeneracy: $\sum_{n=1}^N p_n(\xi) \geq c|\xi|^\alpha$ with some positive c, α. These conditions are rather restrictive, as they do not include even standard degenerate diffusions.

For a more complete discussion of the pseudo-differential operator technique applied to Markov processes (including extensive bibliography) we refer to the fundamental three-volume work of N. Jacob [142]. Some interesting advances in the analysis of the martingale problem can be found in Bass and Perkins [32]. Another useful approach for the construction of Markov processes, that we do not discuss here, is based on the study of *Dirichlet forms*, which for an operator A of Lévy type are defined as the corresponding quadratic form (f, Af) (pairing in the L^2-sense); see Fukushima, Oshima and Takeda [118] or Ma and Röckner [225].

We did not of course exhaust all the approaches to the construction of Markov semigroups. Several special methods can be used for particular classes. For instance, the case of pure integral generators (corresponding to pure jump processes, as discussed in Section 3.7) can be dealt with analytically for rather general class of unbounded kernels, see e.g. Kolokoltsov [194], [196]. This treatment is based on the extension to continuous state space of the classical construction of the solutions to Kolmogorov's equation for Markov chains, as detailed e.g. in [18] and [86]. Another useful class is the *affine processes* (which can be roughly characterized by having the coefficients of the generator depending linearly on the position), developed in detail by Duffie, Filipovic and Schachermayer [104].

An interesting application of probabilistic methods combined with differential geometry leads to the analysis of Markov processes on curvilinear domains, or more generally on manifolds. We refer to the books Stroock [301], Elworthy, Le Jan and Xue-Mei Li [107], Kolokoltsov [179] and [196], Bismut and Lebeau [59] and references therein for various "curvilinear" directions of research. In particular, in [196] a *curvilinear Ornstein–Uhlenbeck process* on a cotangent bundle T^*M to a compact Riemannian manifold driven by a (nonhomogeneous) Lévy-type noise, was analyzed. Such a process is specified via a generator of the form

$$\begin{aligned} Lf(x,p) = {} & \frac{\partial H}{\partial p}\frac{\partial f}{\partial x} - \frac{\partial H}{\partial x}\frac{\partial f}{\partial p} - \left(\alpha(x)p, \frac{\partial f}{\partial p}\right) + \frac{1}{2}\mathrm{tr}\left(A(x)\frac{\partial^2 f(x,p)}{\partial p^2}\right) \\ & + \int \left[f(x,p+q) - f(x,p) - \frac{\partial f(x,p)}{\partial p} q \right] \frac{(\det G(x))^{1/2} dp}{\omega(x,p)}, \end{aligned} \tag{5.129}$$

where $(x,p) \in T^*M$, $G(x)$ is the inverse matrix of the Riemannian metric, $H(x,p) = (G(x)p, p)$, and A, α, ω are certain tensors. If $\alpha = 0$, this process can be also called a *stochastic geodesic flow* driven by a Lévy-type noise (for instance a stable one).

Of interest is an application of the Feller property to semigroups in C^*-algebras that appear in quantum mechanics, see Carbone and Fagnola [72], [73], and for nonlinear extensions Section 11.3 of [196].

Stochastic monotonicity for Markov chains is well studied and applied for the analysis of many practical models, see e.g. Anderson [18], Conlisk [88], Dardoni [92], Maasoumi [226]. Stochastic monotonicity and the related duality are also well studied for diffusions (see Kallenberg [154], Chen and Wang [80]) and jump-type Markov processes (see Chen and Zhang [78], Chen [79]) and Zhang [325]). In Samorodnitski and Taqqu [287] monotonicity for stable processes was analyzed. For general Markov processes the analysis of stochastic monotonicity and related duality was initiated in author's paper [183] devoted to the case of one-dimensional processes with polynomial coefficients (note some nasty typos in the expression of the dual generator in [183]). This was related to interacting particle models (see also Lang [218]), and related Markov models in financial mathematics. In Jie-Ming Wang [316] the theory of monotonicity was extended to multidimensional processes of Lévy–Khintchine type with Lévy measures having a finite first internal moment, i.e. with $\int_{|y|<1} \nu(x, dy) < \infty$ (slightly more generally in fact). In section 5.9 we follow mostly Kolokoltsov [197], that extends the approach from [183]. Unlike [316], we are not aiming at necessary and sufficient conditions for monotonicity under the assumption of the well-posedness of a semigroup, but rather at the simplest conditions that ensure both stochastic monotonicity and well-posedness. We also pay attention to the construction of the dual process and its generator.

An accessible introduction to non-stochastic scattering is provided by Reed and Simon [275].

For the development of stochastic scattering for diffusion processes we refer to the papers Albeverio, Hilbert and Kolokoltsov [8], [9] and Kolokoltsov [171], [173], [174]. In particular, there one can find the discussion of scattering for the *Newtonian particle driven by a random force*. Namely, the system of SDE $dx = v\,dt,\ dv = dB_t$, called sometimes the *Langevin equation*, describes the behavior of a Newtonian particle with the white noise force and velocity being a Brownian motion. The process (x, v) is often called a *physical Brownian motion* or *Kolmogorov's diffusion*. Allowing for an additional deterministic force leads to system (4.34). For potentials V decreasing at infinity quickly enough, the scattering theory can be developed. Some infinite-dimensional extensions are presented in Albeverio and Kolokoltsov [13].

Natural extension to the stable case is described by a process solving the corresponding *stable noise driven Langevin equation* $dx = v\,dt,\ dv = dW_t^\alpha$, where W^α is an α-stable Lévy motion. Including a usual linear friction and a potential force leads to the system

$$\begin{cases} dx = v\,dt \\ dv = -\beta v\,dt - \frac{\partial V}{\partial x}\,dt + dW_t^\alpha, \end{cases}$$

describing a Newtonian particle driven by a stable random force. In Kolokoltsov and Tyukov [204] the scattering theory for such a system is initiated. In particular, the random wave operators are proved to exist. As a starting point for this theory one needs an estimate for the *rate of escape of the integrated stable Lévy motions.* For instance, the following is established in [204].

Theorem 5.12.1. *Let $d \geq 2$, $\alpha \in (0,2)$ and $f(t)$ be an increasing positive function on $\mathbb{R}_+$ such that $f(t) = o(t^{1+1/\alpha})$ and $t/f(t) = o(1)$ for $t \to \infty$ and*

$$\int_1^\infty (f(t)t^{-(1+1/\alpha)})^d \, dt < \infty.$$

Then a.s.

$$\lim_{t\to\infty} \frac{1}{f(t)} \left| \int_0^t W_s^\alpha \, ds \right| = \infty.$$

Other result on the rate of escape of various processes can be found in Khoshnevisan [159], [161], Alparslan [17], Barchielli and Paganoni [26], Lachal [216] and references therein.

The study of classical mechanics problems with random forcing provides an interesting application for the methods of stochastic analysis with many directions and open problems. Seemingly even the following simple problem is still open: find necessary and sufficient conditions on a smooth function V, bounded below, ensuring that the process specified by (5.119) is transient in dimension $d = 1, 2$ (for $d \geq 3$ the answer is fully settled by Exercise 4.4.2; in $d = 1$ only a necessary (but not a sufficient) condition for transience is known).

The stochastic Schrödinger equation obtained by quantization of the evolution of a Newtonian particle with random force has the form

$$ih\, d\psi = \left(-\frac{h^2}{2}\Delta + V(x)\right)\psi \, dt - x\psi \, d_S B, \tag{5.130}$$

where d_S is the so called Stratonovich differential connected with the more familiar Itô's differential via the formula $\psi d_S B = \psi dB + \frac{1}{2} d\psi dB$. This leads to the Itô's form of the *stochastic Schrödinger equation*:

$$ih\, d\psi = \left(-\frac{h^2}{2}\Delta + V(x)\right)\psi \, dt - \frac{i}{2h} x^2 \psi dt - x\psi \, dB. \tag{5.131}$$

the corresponding "free" stochastic evolution is obtain for $V = 0$, i.e. it is

$$ih\, d\psi = -\frac{h^2}{2}\Delta\psi \, dt - \frac{i}{2h} x^2 \psi dt - x\psi \, dB. \tag{5.132}$$

The following result is obtained in [173].

Theorem 5.12.2. *Let the potential V in* (5.131) *belong to the class $L_r(\mathbb{R}^d)$ for some $r \in [2, d)$ and let the dimension $d > 2$. Then for each solution of* (5.132) *(defined by an initial function $\psi_0 \in L_2(\mathbb{R}^d)$) there exists with probability one a solution ϕ of* (5.131) *such that, in $L_2(\mathbb{R}^d)$,*

$$\lim_{t \to \infty} (\psi(t) - \phi(t)) = 0.$$

This theorem allows one to define the *quantum random wave operator* as the mapping $\psi_0 \mapsto \phi_0$.

Much more nontrivial object than (5.132) is the *Belavkin's quantum filtering equation* for normalized states (see also Section 9.4), which represents a *nonlinear stochastic Schrödinger equation* of the form

$$d\phi = \frac{1}{2}(i\Delta\phi - (x - \langle x \rangle_\phi)^2 \phi)\, dt + (x - \langle x \rangle_\phi)\phi\, dB_t, \tag{5.133}$$

where

$$\langle x \rangle_\psi = \int x|\psi|^2(x)\, dx \|\psi\|^{-2}$$

denotes the mean position of the state ψ.

It is easy to see that for a Gaussian initial condition the solution to this equation localizes around the path of classical Newtonian particle driven by a white noise force, i.e. around a solution to (5.119). In particular, they do not spread out over the space, as is the case with the standard free Schrödinger equation. In physics, this phenomenon is often referred to as the *watch-dog effect*. This fact was observed by Diosi [101] and Belavkin [37]. For general initial data it was proved by the author in [171] (with some cavalier estimates of [171] detailed in [175]).[2] Let us state this result precisely.

Theorem 5.12.3. *Let ϕ_t be the solution of the Cauchy problem for equation* (5.133) *with an arbitrary initial function $\phi_0 \in L_2$, $\|\phi_0\| = 1$. Then for a.a. trajectories of the Brownian motion B_t,*

$$\|\phi_t - \pi^{1/4} g^{1-i}_{q(t),p(t)}\| = O(e^{-\gamma t})$$

as $t \to \infty$, for arbitrary $\gamma \in (0, 1)$, where

$$q(t) = q + pt + B_t + \int_0^t B_s\, ds + O(e^{-\gamma t})$$

and $p(t) = p + B_t + O(e^{-\gamma t})$ for some random constants q, p.

[2]Paper [175] suggests also a second proof of this theorem, which is slightly flawed by a non-correct use of the spectral theory of non-selfadjoint operators. However, it is not difficult to correct it by applying the modern theory of discrete spectra from Davies [95].

It follows in particular that the mean position of the solution behaves like the integral of Brownian motion, which provides one of the motivations for the study of this process.

Finite-dimensional analogues of this watch-dog effect and their applications are discussed e.g. in Kolokoltsov [172], [176], Yuriev [152], [153], Barchielli and Paganoni [26].

Section 5.11 presents simplified version of the results from Kolokoltsov [188] (related discussions can be found in Darling and Norris [93] or Benaïm and Le Boudec [42]). This is an introduction to the analysis of *nonlinear Markov processes* that are described by nonlinear evolution equations of the weak form

$$\frac{d}{dt}(f, \mu_t) = (A_{\mu_t} f, \mu_t), \quad f \in C(\mathbb{R}^d), \tag{5.134}$$

where A_t is a family of Lévy–Khintchine type operators with coefficients (diffusion, drift and Lévy measure) depending on a measure as a parameter (see [196], where an extensive bibliography can be also found). For instance, a *nonlinear Lévy process* is obtained, when each operator A_μ in (5.134) generates a Lévy process. A popular particular setting of nonlinear processes represent conservation laws, see e.g. Jourdain, Méléard and Woyczynski [316] or Biler, Karch and Woyczyn'ski [54]. Apart from interacting particles, other natural sources for nonlinear problems in stochastic analysis are filtering theory, for which we refer to the monograph Bain and Crisan [25] and references therein, and optimal control theory, which is addressed briefly in Section 7.8, and where non-smooth coefficients become imminent.

A powerful method, that we do not discuss here, for constructing tractable approximations to Markov semigroups and specially to their Green functions (or heat kernels) is based on *semiclassical* or *quasi-classical* asymptotics, which is often referred to as the *small-diffusion* asymptotics, when applied to diffusion processes. This method looks for the asymptotics of the solutions to parabolic equations of the type

$$\frac{\partial u}{\partial t} = \left(b(x), \frac{\partial u}{\partial x}\right) + \frac{h}{2}\mathrm{tr}\left(G(x)\frac{\partial^2}{\partial x^2}\right)u$$

for $h \to 0$. Approximations for $h \to 0$ are often consistent with the approximations $t \to 0$. Thus the small-time asymptotics (which are of importance for many applications) can be usually obtained as a by-product of the small-diffusion asymptotics.

The method of small diffusion asymptotics for degenerate diffusions was developed in Kolokoltsov [179]. In particular, a remarkable class of degenerate diffusions is singled out there, for which the phase and amplitude of small time and small h asymptotics for the Green function can be expanded in asymptotic power series in time and small distances. That is, the Green function $G_h(t, x, x_0)$ for these diffusions has the form

$$C(h)(1 + O(h))\phi(t, x) \exp\left\{-\frac{1}{h}S(t, x)\right\},$$

where $C(h)$ is some normalizing constant, and both ϕ and S are represented as asymptotic power expansion in $(t, x - x_0)$, multiplied by t^{-M} with some positive M (possibly different for ϕ and S). It turns out that this class of diffusions is also characterized by the property that the main term of the small-time and small-distance asymptotics of their Green functions $G(t, x, x_0)$ in a neighborhood of x_0 coincides with the Green function of the *Gaussian diffusion approximation*

$$\frac{\partial u}{\partial t} = \frac{h}{2}\mathrm{tr}\left(G(x_0)\frac{\partial^2 u}{\partial x^2}\right) + \left(b(x_0) + \frac{\partial b}{\partial x}(x_0)(x - x_0), \frac{\partial u}{\partial x}\right).$$

The Green functions of Gaussian diffusions can be of course written explicitly. It turns out that their forms can be classified in terms of the Young schemes or Young tableaux.

For other results on the asymptotics and Gaussian bounds for degenerate diffusions we refer e.g. to Ben Arous [41] and Léandre [219] and references therein.

The method of quasi-classical asymptotics can be also applied to nonlinear parabolic equations, yielding in particular the *fast dying out asymptotics* for superprocesses. Namely, small h asymptotics (quasi-classics) for the solutions to the equation

$$\frac{\partial u}{\partial t} = \frac{h}{2}\left(G(x)\frac{\partial}{\partial x}, \frac{\partial}{\partial x}\right)u - \left(A(x), \frac{\partial u}{\partial x}\right) - \frac{1}{h}V(x)u - \frac{1}{h}a(x)u^{1+h} \tag{5.135}$$

describes the small-diffusion and fast dying approximation for the (spatially nonhomogeneous) *Dawson–Watanabe superprocess*. In particular, the standard (spatially homogeneous) *Dawson–Watanabe super-process* corresponds to the equation

$$\frac{\partial u}{\partial t} = \frac{h}{2}\Delta u - \frac{1}{h}u^{1+h},$$

see e.g. Etheridge [109]. It is shown in Kolokoltsov [180] that the small h limit of equation (5.135) is described in terms of the first-order PDE, a *generalized Hamilton–Jacobi–Bellman equation* (*HJB equation*)

$$\frac{\partial S}{\partial t} + H\left(x, \frac{\partial S}{\partial x}, S\right) = 0$$

with the generalized Hamiltonian function

$$H(x, p, s) = \frac{1}{2}(G(x)p, p) - (A(x), p) - V(x) - a(x)e^{-s},$$

that depends not only on the derivatives of an unknown function S, but also on S itself. In its turn, solutions to this HJB equation are obtained via non-Hamiltonian (or generalized Hamiltonian) characteristics, that solve the ODE system

$$\begin{cases} \dot{x} = \frac{\partial H}{\partial p}(x, p, s) \\ \dot{p} = -\frac{\partial H}{\partial x}(x, p, s) - \frac{\partial H}{\partial s}(x, p, s)p \\ \dot{s} = p\frac{\partial H}{\partial p}(x, p, s) - H(x, p, s) \end{cases}$$

For the corresponding super-process $Z_h(t)$ this method allows one to calculate the *doubly logarithmic limit*

$$\lim_{h\to 0} h \log(-\log \mathbf{E}_x \exp\{-\theta Z_h^x(t, x_0)\}).$$

Finally, another advanced method of representing Markov semigroups that we did not touch on here is based on quantum stochastic differential equations. This method allows one eventually to express all stochastic integrals in terms of the usual Riemann integral (using certain nontrivial representation) and an arbitrary stochastic evolution in terms of the boundary-value problem for an appropriate Dirac type operator, see e.g. Belavkin and Kolokoltsov [38], [39], and Chebotarev [76].

Chapter 6
Processes in domains with a boundary

Analysis of boundary-value problems (as well as mixed initial and boundary problems) is one of the central themes in the classical theory of partial differential equations. This chapter is devoted to the basics of the corresponding theory for general Lévy–Khintchine type ΨDO. As an initial domain we shall mostly use the space $C_c^2(\mathbb{R}^d)$. If U is an open set in $\mathbb{R}^d$ and X_t^x solves the $(L, C_c^2(\mathbb{R}^d))$-martingale problem in $\mathbb{R}^d$, one expects (at least from the experience gained from diffusion processes) the function $u(x) = \mathbf{E}\psi(X_{\tau_U}^x)$ to solve the Dirichlet problem: $Lu = 0$ in U, $u|_{\partial U} = \psi$; and the function $u(t,x) = \mathbf{E}\psi(X_{t\wedge\tau_U}^x)$ to solve the mixed problem: $\frac{\partial u}{\partial t} = Lu$ in U and u coincides with ψ on the boundary ∂U and for $t = 0$. If these functions do not solve the problem classically, one can expect them to represent some generalized solutions. The main point to check is then continuity, as otherwise taking given boundary values does not make much sense. The basic results are given in Section 6.2. The rest of the chapter is devoted to various criteria used to check the conditions of the main theorems and to their more concrete performances and applications.

6.1 Stopped processes and boundary points

For an open set D in $\mathbb{R}^d$ we denote by ∂D and $\bar{D}$ its boundary and closure respectively. By the ϵ-neighborhood of a set D we mean as usual the open set $\{y : \exists x \in D : \|x - y\| < \epsilon\}$.

Let U be an open subset of $\mathbb{R}^d$ and $\{U_k\}$, $k \in \mathbb{N}$, be a family of open sets such that $\bar{U}_k \subset U_{k+1}$ for any k, $U = \bigcup_k U_k$ and the distance ρ_k between U_k and U defined as

$$\rho_k = \sup_{x\in\partial U_k} \inf_{y\in\partial U} \|x - y\|$$

tends to zero as $k \to \infty$. Let L be a Lévy type operator in U with continuous coefficients, i.e. it is given by

$$Lu(x) = \frac{1}{2}(G(x)\nabla, \nabla)u(x) + (b(x), \nabla u(x))$$
$$+ \int [u(x+y) - u(x) - (y, \nabla u(x))\mathbf{1}_{B_1}(y)]\nu(x, dy), \qquad (6.1)$$

with $G(x)$, $b(x)$, $\nu(x,.)$ being continuous functions on the closure $\bar{U}$ (continuity of ν means the weak continuity of the family of finite measures $(1 \wedge |y|^2)\nu(x,dy)$). From now on, we shall assume the following, concerning the triple $(U, \{U_k\}, L)$:

(U) the operators L_m obtained from L by the restriction to U_m can be extended to operators on $\mathbb{R}^d$ of the form (6.1) (which we again denote by L_m) with continuous coefficients which are uniformly bounded on compact sets, i.e.

$$\sup_{x \in K, m} \left(\|G_m(x)\| + \|b_m(x)\| + \int (1 \wedge |y|^2) \nu_m(x, dy) \right) < \infty$$

for any compact set K, and the corresponding martingale problems are well posed in $C_c^2(\mathbb{R}^d)$ and specify a family of strong Markov processes $X_{t,m}^x$ satisfying the *uniform compact containment with compact initial data condition*: for any $\epsilon > 0$, $T > 0$ and a compact set $K \subset \mathbb{R}^d$, there exists a compact set $\Gamma_{\epsilon,T,K} \subset \mathbb{R}^d$ such that

$$\sup_{x \in K, m \in \mathbb{N}} \mathbf{P}(X_{t,m}^x \in \Gamma_{\epsilon,T,K} \ \forall\, t \in [0,T]) \geq 1 - \epsilon.$$

By Theorems 4.10.2 and 5.4.1, this condition implies that the distributions of the processes $X_{t,m}^x$ depend weakly continuous on x, these processes are C-Feller, and the solutions of the martingale problem with an arbitrary initial probability law η can be obtained by the integration with respect to η of the corresponding solutions with the Dirac initial conditions.

If the operator L can be extended to an operator with a continuous symbol on the whole of $\mathbb{R}^d$ of form (6.1) in such a way that the corresponding martingale problem is well-posed and specifies a process X_t, the stopped process $X_{\min(t,\tau_U)}$ (where τ_U is the exit time defined in (4.100)) is well defined and solves the corresponding stopped martingale problem (see Theorem 4.11.2). But in many reasonable situations this is not the case. The simplest example is given by the operator $x^\beta \frac{d^2}{dx^2}$ on $\mathbb{R}_+$ with a small positive β that can not be extended to $\mathbb{R}$ at least as a diffusion operator with Lipschitz continuous coefficients. Our first objective is to define the stopped process properly in the case of non-extendable L.

For an open $D \subset U_m$ with some m (including $D = U_m$) the *exist time* τ_D from D is defined by formula (4.100), i.e.

$$\tau_D = \inf\{t \geq 0 : X_t \notin D\}.$$

If $D \subset U$, it will be defined as

$$\tau_D = \lim_{k \to \infty} \tau_{D \cap U_k} = \sup_k \tau_{D \cap U_k},$$

which clearly agrees with the first definition if $D \subset U_m$. As usual, we shall write τ_D^x when stressing the initial point x. Sometimes we shall denote τ_{U_m} briefly by τ_m.

Remark 44. Notice that in case of L extendable beyond ∂U thus defined τ_U is given by $\tilde{\tau}_U$ from Remark 37, i.e. it equals

$$\inf\{t \geq 0 : X_t \notin U \text{ or } X_{t-} \notin U\}.$$

Proposition 6.1.1. (i) *Under condition* (U), *if* $\tau_U = \lim_{m\to\infty} \tau_{U_m} < \infty$ *a.s., then* $\lim_{m\to\infty} X_{\tau_m,m}$ *exists a.s. and belongs to* $\mathbb{R}^d \setminus U$. *This allows us to define the stopped process* X_t^{stop} *in* $\mathbb{R}^d$ *by the requirements* $X_t^{\text{stop}} = X_{t,m}$ *for* $t \leq \tau_{U_m}$ *with any* m *and* $X_t^{\text{stop}} = \lim_{m\to\infty} X_{\tau_m,m}$ *for* $t \geq \tau_U$.

(ii) *This stopped process is strong Markov. It can be characterized as the unique solution of the stopped martingale problem in* $\bar{U}$, *i.e. for any initial probability measure* η *on* $\mathbb{R}^d$ *it defines a unique measure* P_η^{stop} *on* $D([0,\infty), \mathbb{R}^d)$ *such that* X_0 *is distributed according to* η, $X_t = X_{t\wedge\tau_U}$ *and*

$$\phi(X_t) - \phi(X_0) - \int_0^{\min(t,\tau_U)} L\phi(X_s)\, ds$$

is a P_η^{stop}*-martingale for any* $\phi \in C_c^2(\mathbb{R}^d)$.

(iii) *If for any* t

$$\lim_{m\to\infty} \mathbf{P}(\tau_m \leq t) = 0 \tag{6.2}$$

(in other words, if $\tau_m \to \infty$ *in probability) for any initial probability measure supported on* U, *then for any initial probability law* η *on* U *the distribution* P_η^{stop} *from* (ii) *lives on* $D([0,\infty), U)$ *and is such that*

$$\phi(X_t) - \phi(X_0) - \int_0^t L\phi(X_s)\, ds$$

is a P_η^{stop}*-martingale for any* $\phi \in C_c^2(\mathbb{R}^d)$.

Proof. (i) By the uniform compact containment condition there exists, almost surely, a finite limit point of the sequence $X_{\tau_{U_m},m}$, which does not belongs to U. By uniform stochastic continuity of $X_{t,m}$ (see Theorem 5.4.1) this limit point is unique, because if one supposes that there are two different limit points, y_1 and y_2, say, the process must perform infinitely many transitions from any fixed neighborhood of y_1 to any fixed neighborhood of y_2 and back in a finite time, which is impossible by the Borel–Cantelli lemma and equation (5.45).

(ii) Theorem 4.11.1 implies that the stopped (at U_m) processes $X_{t,m}^{\text{stop}}$ give unique solutions to the corresponding stopped martingale problem in U_m. By the dominated convergence theorem we get from (i) that X_t^{stop} is a solution to the stopped martingale problem in U. Uniqueness is clear, because the (uniquely defined) stopped processes $X_{t,m}^{\text{stop}}$ defines X_t^{stop} uniquely for $t < \tau_U$, and hence up to τ_U inclusive (due to (i)). After τ_U the behavior of the process is fixed by the definition.

(iii) Condition (6.2) implies $\tau_U = \infty$ and hence $\min(t, \tau_U) = t$ a.s. □

We shall say that the process X_t *leaves a domain* $D \subset U$ *a.s. (respectively with a finite expectation)* if $\mathbf{P}(\tau_D^x < \infty) = 1$ for all x (respectively if $\mathbf{E}\tau_D^x < \infty$ for all $x \in D$). Furthermore, we shall say that

(i) a boundary point $x_0 \in \partial U$ is *t-regular* if $\tau_U^x \to 0$ in probability as $x \to x_0$;

(ii) a point $x_0 \in \partial U$ is *normally regular*, if there exists a neighborhood V of x_0 such that $\mathbf{E}\tau_{U\cap V}^x \leq c\|x - x_0\|$ with a constant c;

(iii) a point $x_0 \in \partial U$ is an *entrance boundary* if for any positive t and ϵ there exist an integer m and a neighborhood V of x_0 such that $\mathbf{P}(\tau_{V\cap(U\setminus U_m)}^x > t) < \epsilon$ and $\mathbf{P}(\tau_{V\cap(U\setminus U_m)}^x = \tau_U^x) < \epsilon$ for all $x \in V \cap U$;

(iv) a point $x_0 \in \partial U$ is a *natural boundary*, if for any m and positive t, ϵ there exists a neighborhood V of x_0 such that $\mathbf{P}(\tau_{U\setminus U_m}^x < t) < \epsilon$ for all $x \in V \cap U$;

(v) a subset $\Gamma \subset \mathbb{R}^d \setminus U$ is *inaccessible from an open set* V if $\mathbf{P}_x(X_{\tau_{U\cap V}} \in \Gamma, \tau_{U\cap V} < \infty) = 0$ for all $x \in V$; it is *inaccessible* if it inaccessible from the whole U.

In particular, condition (6.2) implies that the whole set $\mathbb{R}^d \setminus U$ is inaccessible.

Exercise 6.1.1. Convince yourself that the notions introduced above are properties of U and L and do not depend on the choice of the approximating family of domains $\{U_k\}$ (as long as the required conditions are satisfied).

Exercise 6.1.2. A simple illustrative example can be given by the generator $L = -\frac{\partial}{\partial x}$ in the square domain $U = \{(x, y) \in \mathbb{R}^2 : x \in (0, 1), y \in (0, 1)\}$. Show that the closed left side $\{(x, y) \in \partial U : x = 0\}$ consists of regular points, the open right side $\{(x, y) \in \partial U : x = 1, y \in (0, 1)\}$ consists of entrance points and other points of the boundary are natural. Pay special attention to the corner points.

We shall denote by ∂U_{treg} the set of t-regular points of ∂U (with respect to some given process).

The notion of t-regularity is crucial for the analysis of the continuity of stopped semigroups (see the next sections) and the corresponding boundary-value problems. The normal regularity of a point is required if one is interested in the regularity of the solutions to a boundary-value problem beyond the simple continuity (see e.g. [117] for the case of degenerate diffusions with extendable L).

Finally let us note important simplifications that occur in case of L with the *transmission property*. For an open U let U_{ext} be defined as

$$U_{\text{ext}} = \Big\{ \bigcup_{x\in U} \operatorname{supp} \nu(x, .) \Big\} \cup U$$

We shall say that U is *transmission-admissible* with respect to L, or L satisfies the *transmission property* in U, if $U = U_{\text{ext}}$. The following statement is straightforward.

Proposition 6.1.2. *Under condition* (U), *if U is transmission-admissible, then*

$$\tau_U = \inf\{t : X_t^x \in \partial U\}, \tag{6.3}$$

the trajectories of X_t^x are almost surely continuous at $t = \tau_U$, they live in $D([0,\infty), \bar{U})$ (and not $D([0,\infty), \mathbb{R}^d)$ as in general case), and $\tau_U > \tau_{U_m}$ a.s. for any m. Moreover, if a subset $\Gamma \subset \partial U$ is inaccessible from an open neighborhood V of Γ, then Γ is inaccessible.

For transparency, we shall mostly work with transmission-admissible domains.

6.2 Dirichlet problem and mixed initial-boundary problem

The Markov semigroup T_t^{stop} of the process stopped on the boundary is defined as

$$(T_t^{\text{stop}} u)(x) = Eu(X_{t\wedge\tau_U}^x) \tag{6.4}$$

on the space of bounded measurable functions on $\bar{U}$.

An important question is whether this semigroup is Feller or C-Feller.

To begin with, let us observe that the exit time τ_U is not a continuous functional on the space of càdlàg paths. However, the exit time from the closure of any domain D defined as

$$\tau_{\bar{D}} = \inf\{t \geq 0 : X_t \notin \bar{D}\} \tag{6.5}$$

and the corresponding exit point $X_{\tau_{\bar{D}}}$ are continuous functionals on the space of càdlàg paths equipped with Skorohod topology, because if $X_{t,n} \to X_t$ in Skorohod topology and $X_{t_0} \notin \bar{D}$, then for any $\epsilon > 0$ and large enough n there exists t such that $|t - t_0| < \epsilon$ and $|X_t^n - X_{t_0}| < \epsilon$.

Theorem 6.2.1. *Under condition* (U) *suppose U is transmission-admissible, X_t leaves U almost surely, and $\partial U \setminus \partial U_{t\text{reg}}$ is an inaccessible set. Then*

(i) *for any $h \in C_b(\partial U_{t\text{reg}})$, the function $\mathbf{E}h(X_{\tau_U}^x)$ is continuous in $U \cup \partial U_{t\text{reg}}$;*

(ii) *the set $C_b(U \cup \partial U_{t\text{reg}})$ of bounded continuous functions on $U \cup \partial U_{t\text{reg}}$ is preserved by the semigroup T_t^{stop}; in particular, if $\partial U = \partial U_{t\text{reg}}$, the semigroup T_t^{stop} is C- Feller in $\bar{U}$;*

(iii) *for any $u \in C_b(U \cup \partial U_{t\text{reg}})$ and $x \in U$ there exists a limit*

$$\lim_{t\to\infty} T_t^{\text{stop}} u(x) = \mathbf{E}u(X_{\tau_U}^x); \tag{6.6}$$

(iv) *the functions $\mathbf{E}h(X_{\tau_U}^x)$ are invariant under the action of T_t^{stop}.*

Proof. (i) Let $x \in U$ and $x_n \to x$ as $n \to \infty$. By the compact containment for compact initial data, for any neighborhood V_x of $x \in U$ and any given $\epsilon > 0$ there exists a compact set K such that the trajectories starting from V_x remain in K, with probability not less than $1 - \epsilon$, until their exit from U. By the uniform stochastic continuity, for any $\eta > 0$ there exists τ such that

$$\sup_{x \in K, m} \mathbf{P}(\sup_{s \le \tau} \|X^x_{s,m} - x\| > \eta) < \epsilon. \tag{6.7}$$

By definition of t-regularity and compactness of K, there exists $\delta_1 > 0$ such that

$$\sup_{x \in U^{\delta_1}_{t\mathrm{reg}} \cap K} \mathbf{P}(\tau^x_{\bar{U}} > \tau) < \epsilon. \tag{6.8}$$

Next, as $U_{\mathrm{inac}} = U \setminus U_{t\mathrm{reg}}$ is inaccessible, there exists a.s. a (random) $r > 0$ such that the trajectory X^x_t does not visit the r-neighborhood U^r_{inac} of U_{inac}. Hence, by σ-additivity of probability measures, there exists δ_2 such that with probability not less than $1 - \epsilon$ the trajectories of X^x_t do not visit the open set $U^{\delta_2}_{\mathrm{inac}}$. Next, by the convergence of the boundaries of U_m to the boundary of U it follows that there exists M such that $\partial U_m \cap K \subset U^\delta \cap K$ for all $m \ge M$, where $\delta = \delta_1 \wedge \delta_2$. Summarizing: with probability arbitrarily close to one, the trajectories started from x and x_n remain in K, do not enter U^δ_{inac}, and at time $\tau_{\bar{U}_M}$ they turn out to be in $U^\delta_{t\mathrm{reg}}$; moreover, after the moment $\tau_{\bar{U}_M}$ they exit U in time not exceeding τ and do not deviate from their positions by a distance exceeding η.

By the continuous dependence of the solution to the martingale problems for any m and by the continuity of the exit times $\tau_{\bar{U}_m}$ and the corresponding exit points, we conclude that for any m, $\tau^{x_n}_{\bar{U}_m} \to \tau^x_{\bar{U}_m}$ and $X^{x_n}_{\tau_{\bar{U}_m}} \to X^x_{\tau_{\bar{U}_m}}$ weakly as $n \to \infty$. Using the Skorohod coupling (Theorem 1.1.3) we can make this convergence a.s., and hence in probability. Therefore, with probability not less than $1 - \epsilon$,

$$\|X^{x_n}_{\tau_{\bar{U}_M}} - X^x_{\tau_{\bar{U}_M}}\| < \eta, \quad |\tau^{x_n}_{\bar{U}_M} - \tau^x_{\bar{U}_M}| < \tau$$

for large enough n. On the other hand, again up to an arbitrarily small probability, these points belong to $U^\delta_{t\mathrm{reg}}$ and exit U faster than time τ. By (6.7),

$$\|X^{x_n}_{\tau_{\bar{U}}} - X^x_{\tau_{\bar{U}}}\| < 3\eta, \quad |\tau^{x_n}_{\bar{U}} - \tau^x_{\bar{U}}| < 2\tau,$$

implying that $\mathbf{E}h(X^{x_n}_{\tau_{\bar{U}}}) \to \mathbf{E}h(X^x_{\tau_{\bar{U}}})$ as $n \to \infty$ for any $h \in C_b(\partial U_{t\mathrm{reg}})$. For $x \in \partial U_{t\mathrm{reg}}$, this convergence is obvious, since then $\mathbf{E}h(X^x_{\tau_{\bar{U}}}) = h(x)$.

(ii) Pick any $t > 0$. With probability one the trajectories X^x_t and $X^{x_n}_t$ are continuous at t. Next, if $t > \tau^x_{\bar{U}_M}$, then, with probability arbitrarily close to one, $t > \tau^{x_n}_{\bar{U}_M}$ for $n > N$ with large enough N, and hence

$$\|X^{x_n}_{t \wedge \tau_{\bar{U}}} - X^x_{t \wedge \tau_{\bar{U}}}\| < 3\eta.$$

Since

$$f(X^{x_n}_{t\wedge\tau_U}) - f(X^x_{t\wedge\tau_U}) = [f(X^{x_n}_{t\wedge\tau_U}) - f(X^x_{t\wedge\tau_U})]\mathbf{1}_{t>\tau^x_{\tilde U_M}} + [f(X^{x_n}_t) - f(X^x_t)]\mathbf{1}_{t\le\tau^x_{\tilde U_M}},$$

this implies that $\mathbf{E} f(X^{x_n}_{t\wedge\tau_U}) \to \mathbf{E} f(X^x_{t\wedge\tau_U})$ as $n\to\infty$ for a continuous f.

(iii) To prove (6.6), let us write

$$T^{\mathrm{stop}}_t u(x) = \mathbf{E}(u(X^x_t)\mathbf{1}_{\tau^x_U\ge t}) + \mathbf{E}(u(X^x_{\tau_U})\mathbf{1}_{\tau^x_U<t}). \tag{6.9}$$

The first term here tends to zero, because we assumed that the process leaves the domain almost surely in finite time, and the second term tends to the r.h.s. of (6.6) by the dominated convergence theorem.

(iv) For $g(x) = \mathbf{E}h(X^x_{\tau_U})$, let us write

$$T^{\mathrm{stop}}_t g(x) = \mathbf{E}(g(X^x_t)\mathbf{1}_{\tau^x_U\ge t}) + \mathbf{E}(h(X^x_{\tau_U})\mathbf{1}_{\tau^x_U<t}), \tag{6.10}$$
$$g(x) = \mathbf{E}(h(X^x_{\tau_U})\mathbf{1}_{\tau^x_U\ge t}) + \mathbf{E}(h(X^x_{\tau_U})\mathbf{1}_{\tau^x_U<t}).$$

Hence one has to show that the first terms in these expressions coincide. But they both equal to

$$\int_t^\infty \mathbf{E}(h(X^x_s)|X^x_t)p_x(ds),$$

where p_x denotes the distribution of the random variable τ^x_U. □

As we mentioned, we shall work mostly with transmission-admissible domains. However, in many cases, the adjustments needed to remove this restriction are not very essential. Let us formulate the corresponding analogue of the above theorem.

Theorem 6.2.2. *Under condition* (U) *suppose X_t leaves U almost surely, and $\partial U\setminus\partial U_{t\mathrm{reg}}$ belongs to an inaccessible set A. Then*

(i) *for any $h\in C_b((\mathbb{R}^d\setminus U)\setminus A)$, the function $\mathbf{E}h(X^x_{\tau_U})$ is continuous in $\mathbb{R}^d\setminus A$;*

(ii) *the set $C_b(\mathbb{R}^d\setminus A)$ is preserved by the semigroup T^{stop}_t; in particular, if $\partial U = \partial U_{t\mathrm{reg}}$, the semigroup T^{stop}_t is C-Feller in $\mathbb{R}^d$;*

(iii) *for any $u\in C_b(\mathbb{R}^d\setminus A)$ and $x\in U$ there exists a limit*

$$\lim_{t\to\infty} T^{\mathrm{stop}}_t u(x) = \mathbf{E}u(X^x_{\tau_U}).$$

Proof. As for Theorem 6.2.1. □

Remark 45. In some situations, in particular in some problems from insurance mathematics (see Section 6.11), the assumption that the process leaves the domain a.s. is not satisfied. Moreover, one is often interested in estimating the probability of exit (called *ruin probability* in insurance mathematics). Hence it is worth noting that the assumption that X_t leaves U a.s. is not essential in Theorem 6.2.1, but is made for simplicity. Without this assumptions the results of Theorems 6.2.1 or 6.2.2 remain the same, but the function $\mathbf{E}[h(X^x_{\tau_U})\mathbf{1}_{\tau_U<\infty}]$ should be taken instead of $\mathbf{E}h(X^x_{\tau_U})$.

A natural application of Theorem 6.2.1 is in the study of the Dirichlet problem. Let $h \in C_b(\partial U_{t\mathrm{reg}})$. A function $u \in C_b(U \cup \partial U_{t\mathrm{reg}})$ is called *a generalized solution of the Dirichlet problem* for L in U if u coincides with h on $\partial U_{t\mathrm{reg}}$ and $T^{\mathrm{stop}}u = u$.

Remark 46. Of course, the last condition implies that $L^{\mathrm{stop}}u = 0$, where L^{stop} is the generalized generator of the semigroup T^{stop}.

In potential analysis, solutions to the Dirichlet problem for L in U are called *harmonic functions*.

To show that this definition of generalized solution is reasonable, one should prove that any classical solution (i.e. a function $u \in C_b(U \cup \partial U_{t\mathrm{reg}})$ which satisfies the boundary condition, is twice continuously differentiable and satisfies $Lu = 0$ in U), is also a generalized solution. This question, as well as the well-posedness of the problem, is addressed in the following theorem.

Theorem 6.2.3. *Suppose the assumptions of Theorem* 6.2.1 *hold. Then*

(i) *a generalized solution exists, is unique, and is given by the formula*

$$u(x) = E_x h(X_{\tau_U})$$

for any $h \in C_b(\partial U_{t\mathrm{reg}})$*;*

(ii) *any classical solution is a generalized solution.*

Proof. (i) By Theorem 6.2.1, this function u is a solution. To show uniqueness, suppose u is a solution vanishing at $\partial U_{t\mathrm{reg}}$. Hence $T_t^{\mathrm{stop}}u = u$ and from (6.6) it follows that $u = \lim_{t\to\infty} T_t^{\mathrm{stop}}u = 0$.

(ii) If $u \in C^2(\mathbb{R}^d) \cap C_b(\mathbb{R}^d)$, $Lu = 0$, then $T_t^{\mathrm{stop}}u = u$, because $u(X_t^{x,\mathrm{stop}})$ is a martingale. If $u \in C^2(U)$ only, consider a sequence of functions $u_m \in C^2 \cap C_b$, $Lu \in C_\infty$ such that u_m coincide with u in U_m and vanishes outside U_{m+1}. Hence

$$E_x u(X_{\min(t,\tau_m)}) - u(x) = E_x \int_0^{\min(t,\tau_m)} Lu(X_s)\,ds = 0,$$

where τ_m denote the exit times from U_m, and by the dominated convergence theorem $T_t^{\mathrm{stop}}u = u$. □

An important further question is whether the semigroup of the stopped process preserves the set of functions which are continuous up to the boundary of U. Let us start with the following result in this direction.

Theorem 6.2.4. *Let the assumptions of Theorem* 6.2.1 *hold, and let the inaccessible set $\partial U \setminus \partial U_{treg}$ consists of natural boundary points only. Then the semigroup of the stopped process T_t^{stop} preserves the subspace $C_0(\bar{U})$ of $C_b(\bar{U})$ consisting of functions vanishing at ∂U.*

Proof. If $x \in \partial U_{treg}$, then $T_t^{\text{stop}} f(x) = f(x)$ for all $t > 0$. Next, the definition of natural boundary implies that for any given time t, if the initial point of the process tends to a natural boundary point, the process is obliged to stay near thc boundary the whole time t. Consequently $T_t^{\text{stop}} f(x) = E_x f(X_t^{\text{stop}})$ tends to zero as x approaches the boundary ∂U whenever $f(x)$ vanishes on the boundary. □

It is not difficult to give an example when T_t^{stop} does not preserve the whole space $C_b(U \cup \partial U)$. The following simple example is instructive. Let $U = (0,1) \times (0,1) \in \mathbb{R}^2$ and $Lf(x,y) = -\partial f/\partial x$. It is straightforward to see that $\partial U_{treg} = \{(x,y) : x = 0\}$ and $T_t^{\text{stop}} f(x,y) = f((x-t) \vee 0, y)$. In particular, assuming $f(x,y) = 0$ if and only if $x = 0$, then $T_t^{\text{stop}} f(x,y) = 0$ for $t \geq 1$ and all $(x,y) \in U$. Since $T_t^{\text{stop}} f(x,y) = f(x,y)$ for any $(x,y) \in \partial U$, the resulting function $T_t^{\text{stop}} f$ is not continuous in $\bar{U}$.

This example suggests the following modification of the notion of the stopped process. Suppose for each $x \in \partial U$, the family of distributions of $X^{y,\text{stop}}$ converges weakly to a limit, as $y \to x$, $y \in U$. Let us then define the process $\tilde{X}^{x,\text{stop}}$ as $X^{x,\text{stop}}$ for $x \in U$ and as the weak limit of the processes $X^{y,\text{stop}}$ as $y \to x$ for $x \in \partial U$ (unlike $X_t^{x,\text{stop}} = x$ for $x \in \partial U$).

For instance, in the above example, $\tilde{X}^{x,\text{stop}}$ exists and the corresponding semigroup acts as $\tilde{T}_t^{\text{stop}} f(x,y) = f((x-t) \vee 0, y)$ on $C(\bar{U})$ and preserves this space. This example is of course trivial in the sense that L is extendable beyond U and $\tilde{X}_t^{x,\text{stop}}$ can be defined simply as the process X_t^x (defined by L in $\mathbb{R}^2$) restricted to $\bar{U}$ and stopped at ∂U_{treg}.

Theorem 6.2.5. *Let the assumptions of Theorem* 6.2.1 *hold, and let the inaccessible set $\partial U \setminus \partial U_{treg}$ consists of entrance boundary points only. Then the process $\tilde{X}^{x,\text{stop}}$ exists in $\bar{U}$ and the space $C_b(\bar{U})$ (not only $C_b(U \cup \partial U_{treg})$ as in Theorem* 6.2.1*) is preserved by the corresponding semigroup $\tilde{T}_t^{\text{stop}}$. Moreover, for any continuous bounded function h on ∂U_{treg}, the function $E_x h(X_{\tilde{\tau}_U})$ is continuous in $\bar{U}$ and for any $u \in C_b(\bar{U})$ there exists a limit*

$$\lim_{t \to \infty} \tilde{T}_t^{\text{stop}} u(x) = E_x u(X_{\tilde{\tau}_U}).$$

Proof. As in Theorem 4.9.2 we show that any sequence of the solutions $X^{y_n,\text{stop}}$ to the stopped martingale problem with $y_n \to x \in \partial U$, $y_n \in U$, is tight, and the limit of any convergent subsequence specifies a process with càdlàg paths in $\bar{U}$ solving the martingale problem stopped at $\partial U_{t\text{reg}}$. One needs only show that this limit is uniquely defined and depends continuously on $x \in \partial U$. If $x \in \partial U_{t\text{reg}}$, then such a limit is clearly the constant process $X_t^x = x$. Hence only the case of x being an entrance boundary needs to be analyzed. But in this case, by the definition of entrance, $\inf\{t : X_t^x \in U\} = 0$ a.s. Consequently (by the uniqueness of the stopped martingale problem solution with stating point in U), the distribution of this process is uniquely specified after any given time $t > 0$. Hence it is uniquely defined, because a.s. $X_0 = \lim_{t\to 0_+} X_t$. The rest of the proof is the same as in Theorem 6.2.1. □

It is worth noting that, as in the simple example above, if L is extendable beyond ∂U (or at least beyond $\partial U \setminus \partial U_{t\text{reg}}$) the process $\tilde{X}^{x,\text{stop}}$ constructed in Theorem 6.2.5 can be defined simply as the process X_t^x restricted to $\bar{U}$ and stopped at $\partial U_{t\text{reg}}$.

Remark 47. The results of Theorems 6.2.4 and 6.2.5 makes our classification of boundary points consistent with the well-known classification of the boundary points for one-dimensional diffusions, see e.g. [228].

Exercise 6.2.1. The aim here is to show that continuity of the stopped process on the initial point (and consequently the preservation of continuous functions by the corresponding Markov semigroup) proved in above results is not an obvious fact generally. As an example to illustrate this point, let us consider the operator $-\frac{\partial}{\partial x}$ on the non-convex two-dimensional domain

$$U = \{(x, y) : -1 < x < 1,\ -1 < y < |x|\}.$$

(i) Show that the point $(0, 0)$ is neither entrance, nor natural, nor regular, but it is accessible in the sense that there exist starting points x from which one can reach this boundary point with a positive probability (in fact with probability one in this case).

(ii) The distributions $P_{x,y}$ of the stopped process starting at (x, y) are discontinuous on the interval $0 < x < 1$, $y = 0$. In particular, the limit of $P_{x,y}$ as $(x, y) \to (0, 0)$ depends on the way the points (x, y) are approaching the origin.

(iii) Show that the situation change drastically if one adds a diffusion term, i.e. if one works in this domain with the operator $-\frac{\partial}{\partial x} + \frac{\partial^2}{\partial y^2}$. In this case the open right side of U, where $x = 1$, $y \in (-1, 1)$, consists of entrance points, and all other boundary points are regular.

Exercise 6.2.2. Under the assumptions of Theorem 6.2.1, suppose that $\mathbf{P}(\tau_U^x > t) \to 0$ uniformly in x (in particular, if $\sup_x \mathbf{E}\tau_U^x < \infty$). Then the limit in (6.6) is uniform, that is, it is a limit in the topology of $C_b(U \cup \partial U_{t\text{reg}})$. In particular, in this case, the last statement of Theorem 6.2.1 becomes straightforward from the contraction property of

T_t^{stop}. Hint: In this case, the first term (6.9) tends to zero uniformly as well as the second one

$$E_x\left(u(X_{\tau_U})\mathbf{1}_{\tau_U<t}\right) - E_x(u(X_{\tau_U})) = E_x\left(u(X_{\tau_U})\mathbf{1}_{\tau_U\geq t}\right).$$

Exercise 6.2.3. The semigroup of the *process killed on the boundary* is defined by

$$(T_t^{\text{kil}}u)(x) = E_x(u(X_t)\chi_{t<\tau_U}) \tag{6.11}$$

on the space of bounded measurable functions on $\bar{U}$. Show that under the assumptions of Theorem 6.2.1 the subset of $C_b(U \cup \partial U_{t\text{reg}})$ consisting of functions vanishing at $\partial U_{t\text{reg}}$ is preserved by T_t^{stop}. The restriction of T_t^{stop} coincides with the semigroup T_t^{kil}. Hence T_t^{stop} is a conservative extension of T_t^{kil}. However, unless $\partial U_{t\text{reg}}$ is empty, the semigroup T_t^{kil} itself is not conservative. All these facts are easily visualized on the one-dimensional example of the operator $-\nabla$ on $[0, 1]$.

6.3 The method of Lyapunov functions

We assume everywhere that the condition (U) holds and U is transmission-admissible. Let us say that $\phi \in C^2(\mathbb{R}^d)$ satisfies the *martingale condition* if

$$f(X_{t,m}^x) - \int_0^t L_m f(X_{s,m}^x)\, ds$$

is a martingale for all x and L_m. By condition (U), all $\phi \in C_c^2(\mathbb{R}^d)$ satisfy the martingale condition.

For the analysis of the exit times from a domain and for the classification of the boundary points, the major role is played by the method of *Lyapunov (or barrier) functions*. The main idea can be seen from the following simple statement.

Proposition 6.3.1 (Method of Lyapunov functions). *Let $f \in C^2(\mathbb{R}^d)$ satisfy the martingale condition, and be non-negative in D_{ext} for $D \subset U$.*

(i) *If $Lf(x) \leq 0$ for $x \in D$, then for all $t > 0$ and $x \in D$*

$$f(x) \geq \mathbf{E}_x f(X_{t\wedge\tau_D}). \tag{6.12}$$

If, moreover, the process leaves D almost surely, then also

$$f(x) \geq \mathbf{E}_x(f(X_{\tau_D})). \tag{6.13}$$

(ii) *If $Lf(x) \leq -c$ for $x \in D$ with some $c > 0$, then for all $t > 0$*

$$f(x) \geq c\mathbf{E}_x(t \wedge \tau_D). \tag{6.14}$$

In particular, the process leaves D a.s. and with a finite expectation

$$\mathbf{E}_x(\tau_D) \leq f(x)/c. \tag{6.15}$$

Proof. Let $D_m = D \cap U_m$. As f satisfies the martingale condition,

$$\mathbf{E}_x f(X_{\min(t,\tau_{D_m})}) = f(x) + \mathbf{E}_x \int_0^{\min(t,\tau_{D_m})} Lf(X_s)\, ds.$$

Consequently

$$f(x) \geq \mathbf{E}_x f(X_{t\wedge\tau_{D_m}})$$

in case (i) and

$$f(x) \geq c\mathbf{E}_x(t \wedge \tau_{D_m})$$

in case (ii). Passing to the limit $m \to \infty$, using Fatou's lemma and the continuity of the processes at exit times (see Proposition 6.1.2) yields (6.12) in case (i) and (6.14) in case (ii). Passing to the limit $t \to \infty$ completes the proof. □

We shall show how Proposition 6.3.1 works by deducing from it some criteria of t-regularity, inaccessibility and entrance.

Proposition 6.3.2 (Criteria for regularity). *Suppose*

(i) $x_0 \in \partial U$,

(ii) *there exist* $f \in C^2(U)$ *such that* $f(x_0) = 0$, *and* $f(x) > 0$ *for all* $x \in \bar{U} \setminus \{x_0\}$,

(iii) *the restrictions of* f *to* U_m *can be extended to* $C^2(\mathbb{R}^d)$ *functions that satisfy the martingale conditions for* L_m,

(iv) *there exists a neighborhood* V *of* x_0 *such that* $Lf(x) \leq -c$ *for* $x \in U \cap V$ *with some* $c > 0$.

Then x_0 *is a* t*-regular point, and*

$$\mathbf{E}_x(\tau_{V\cap U}) \leq f(x)/c \tag{6.16}$$

for $x \in V \cap U$. *In particular, if* $f(x) \leq |x - x_0|$ *for* $x \in V \cap U$, *then* x_0 *is normally regular.*

Proof. Proposition 6.3.1 (ii) implies (6.16). Hence $E_x(\tau_{V\cap U}) \to 0$ as $x \to x_0$. And consequently $\mathbf{P}_x(\tau_{V\cap U} > t) \to 0$ as $x \to x_0$ for any $t > 0$. By Proposition 6.3.1 (i), for $x \in U \cap V$

$$\begin{aligned} f(x) &\geq \mathbf{E}_x f(X_{\tau_{V\cap U}}) = \mathbf{E}_x(f(X_{\tau_U})\mathbf{1}_{\tau_{V\cap U}=\tau_U}) + \mathbf{E}_x(f(X_{\tau_{V\cap U}})\mathbf{1}_{\tau_{V\cap U}<\tau_U}) \\ &\geq \min_{y\in U\setminus V} f(y)\mathbf{P}_x(\tau_{V\cap U} < \tau_U). \end{aligned}$$

Hence $\mathbf{P}_x(\tau_{V\cap U} < \tau_U) \to 0$ as $x \to x_0$. Consequently,

$$\mathbf{P}_x(\tau_{V\cap U} < t \text{ and } \tau_{V\cap U} = \tau_U) \to 1$$

as $x \to x_0$, $x \in V \cap U$, and so does

$$\mathbf{P}_x(\tau_U < t) > \mathbf{P}_x(\tau_U < t, \tau_{V\cap U} = \tau_U) = \mathbf{P}_x(\tau_{V\cap U} < t, \tau_{V\cap U} = \tau_U). \qquad \square$$

Proposition 6.3.3 (Criteria for inaccessibility). *Let Γ be a subset of the boundary ∂U. Suppose there is a neighborhood V of Γ and a twice continuously differentiable non-negative function f on U such that f vanishes outside a compact subset of $\mathbb{R}^d$, $Lf(x) \le 0$ for $x \in V \cap U$, and $f(x) \to \infty$ as $x \to \Gamma$, $x \in V \cap U$. Then Γ is inaccessible from V.*

Proof. For any m let us choose a function $f_m \in C_c^2(\mathbb{R}^d)$ that coincides with f in U_m. For any $r > 0$ there exists a neighborhood V_r of Γ such that $\bar{V}_r \subset V$ and $\inf\{f(y) : y \in V_r \cap U\} \ge r$. By Proposition 6.3.1, for $x \in U_m \cap V$,

$$f(x) = f_m(x) \ge \mathbf{E}_x f_m(X_{\min(t,\tau_{U_m \cap V})})$$
$$\ge \min\{f_m(y) : y \in V_r \cap \partial U_m\} \mathbf{P}_x(\tau_{(U_m \cap V)} \le t,\ X_{\tau_{U_m \cap V}} \in V_r).$$

Hence

$$\mathbf{P}_x(\tau_{(U_m \cap V)} \le t,\ X_{\tau_{U_m \cap V}} \in V_r) \le f(x)/r$$

for all t, and consequently

$$\mathbf{P}_x(\tau_{(U \cap V)} \le t,\ X_{\tau_{U \cap V}} \in V_r \cap \partial U) \le f(x)/r.$$

Hence

$$\mathbf{P}_x(\tau_{(U \cap V)} \le \infty,\ X_{\tau_{U \cap V}} \in V_r \cap \partial U) \le f(x)/r.$$

Since $\bigcap_{r=1}^{\infty} V_r \supset \Gamma$, the proof is complete. □

Proposition 6.3.4 (Criteria for entrance). *Suppose $x_0 \in \partial U$ and V is a neighborhood of x_0 such that the set $V \cap \partial U$ is inaccessible. Suppose for any $\delta > 0$, there exist a positive integer m and a non-negative function $f \in C_c^2(\mathbb{R}^d)$ such that $f(x) \in [0, \delta]$ and $Lf(x) \le -1$ for $x \in V \cap (U \setminus U_m)$. Then x_0 is an entrance boundary.*

Proof. Since $V \cap \partial U$ is inaccessible, the probability of leaving $V \cap (U \setminus U_m)$ via ∂U vanishes. To check another condition in the definition of an entrance boundary, observe that by Chebyshev's inequality and Proposition 6.3.1

$$P(\tau^x_{V \cap (U \setminus U_m)} > t) \le \frac{1}{t} E_x(\tau_{V \cap (U \setminus U_m)}) \le \frac{1}{t} f(x) \le \frac{\delta}{t}$$

for $x \in V \cap (U \setminus U_m)$, which can be made arbitrary small because δ is arbitrary small. □

6.4 Local criteria for boundary points

Here we shall give concrete criteria in terms of the coefficients of the generator L that we write in a form, where the Lévy measure is decomposed in two parts (with singularities at infinity and at the origin), that is

$$Lu(x) = \frac{1}{2}\text{tr}\left(G(x)\frac{\partial^2}{\partial x^2}\right)u(x) + (b(x), \nabla)u(x)$$
$$+ \int_{B_1}(u(x+y) - u(x) - (y, \nabla)u(x))\nu(x, dy)$$
$$+ \int (u(x+y) - u(x))\mu(x, dy), \tag{6.17}$$

assuming

$$\int_{B_1} y^2\nu(x, dy) < \infty, \quad \int |y|\mu(x, dy) < \infty$$

uniformly on x from compact sets. This representation for L allows us to include naturally the criteria arising when $\nu = 0$.

We concentrate on local criteria for points lying on smooth parts of the boundary (they can be used also for piecewise smooth boundaries). Since locally all these parts look like hyper-spaces (can be reduced to them by an appropriate change of the variables), we take U here to be the half-space

$$U = \mathbb{R}_+ \times \mathbb{R}^{d-1} = \{(z, v) \in \mathbb{R}^d : z > 0, v \in \mathbb{R}^{d-1}\},$$

and we denote by b_z and b_v the corresponding components of the vector field b and by $G_{zz}(x)$ the first entry of the matrix $G(x)$. We assume that the supports of all $\nu(x, .)$ and $\mu(x, .)$ belong to U and that condition (U) holds with $U_m = \{z > 1/m\}$. Let us pick up positive numbers a and r, and for any $\epsilon > 0$ let

$$V_\epsilon = \{z \in (0, a), |v| \leq r + \epsilon\}.$$

Proposition 6.4.1. *If*

$$|\min(b_z(x), 0)| = O(z), \quad G_{zz}(x) = O(z^2), \quad \int_{B_1} \tilde{z}^2\nu(x, d\tilde{x}) = O(z^2) \tag{6.18}$$

in V_ϵ *where* $\tilde{x} = (\tilde{z}, \tilde{v}), x = (z, v)$, *then the ball* $\{(0, v) : |v| \leq r\}$ *belongs to the inaccessible part of the boundary* ∂U.

Proof. A direct application of Proposition 6.3.3 is not enough here, but a proof given below is in the same spirit. Let a non-negative $f \in C^2(U)$ be such that it is decreasing in z, equals $1/z$ in V_ϵ and vanishes for large v or z. By f_m we denote a function $f \in C_c^2(\mathbb{R}^d)$ that coincides with f in U_m. Let τ_m denote the exit time from $V \cap U_m$. Condition (6.18) implies that $Lf(x) \leq cf(x)$ for all $x \in V$ and some constant $c \geq 0$.

Hence, considering the stopped martingale problem in $V \cap U_m$ and taking as a test function f_m, one obtains

$$\mathbf{E}_x f(X_{\min(t,\tau_m)}) - f(x) = \mathbf{E}_x \int_0^{\min(t,\tau_m)} Lf(X_s)\,ds$$
$$\leq c\mathbf{E}_x \int_0^{\min(t,\tau_m)} f(X_s)\,ds \leq c\mathbf{E}_x \int_0^t f(X_{\min(s,\tau_m)})\,ds.$$

Consequently, applying Gronwall's lemma yields the estimate

$$\mathbf{E}_x f(X_{\min(t,\tau_m)}) \leq f(x)e^{ct}.$$

Hence

$$P_x(\tau_m \leq t,\ X_{\tau_m} \in \partial U_m \cap V_\epsilon) \leq \frac{1}{m} f(x)e^{ct},$$

which implies that (a neighborhood of) Γ is inaccessible by taking the limit as $m \to \infty$. □

Remark 48. (i) The measure μ does not enter this condition, as the possibility of jumping away from the boundary cannot spoil the property of being inaccessible.

(ii) This criterion can be used also for piecewise smooth boundaries. For example, let $\tilde{U} = U \cap \{v : v^1 > 0\}$ and condition (6.18) holds in $V_\epsilon \cap \tilde{U}$. Then the same proof as below shows that $\{|v| \leq r\} \cap \{v : v^1 > 0\}$ is inaccessible. The same remark concerns other propositions below.

Exercise 6.4.1. Suppose there exist constants $0 \leq \delta_1 < \delta_2 \leq 1$ such that

$$G_{zz}(x) = O(z^{1+\delta_2}), \qquad \int \tilde{z}^2 \nu(x, d\tilde{x}) = O(z^{1+\delta_2})$$

in V_ϵ (as above $\tilde{x} = (\tilde{z}, \tilde{v}), x = (z, v)$), and also either

$$b_z(x) \geq \omega z^{\delta_1}$$

or $b_z(x) \geq 0$ and

$$\int_{V_\epsilon \cap \{z \leq \tilde{z} \leq 3z/2\}} \tilde{z}\mu(x, d\tilde{x}) \geq \omega z^{\delta_1}$$

in V_ϵ with some $\omega > 0$. Then the ball $\{(0, v) : |v| \leq r\}$ belongs to the inaccessible part of the boundary.

Hint. Use Proposition 6.3.3 with function f from Proposition 6.4.1. Observe that under the given conditions the diffusion term and the integral term depending on ν in L are both of order $O(z^{\delta_2 - 2})$ and either the drift term is negative of order $z^{\delta_1 - 2}$ and the integral term depending on μ is negative (because f is decreasing in z) or the drift term is negative and the integral term depending on μ is negative of order $z^{\delta_1 - 2}$.

Proposition 6.4.2. *Suppose that for* $|v| \leq r + \epsilon$ *either* (i) $G_{zz}(0, v) \neq 0$, *or* (ii) $b_z(0, v) < 0$, *or*

$$\text{(iii)} \qquad \frac{1}{\beta} \int_{V_\epsilon \cap \{\tilde{z} \leq \beta\}} \tilde{z}^2 \nu((0, v), d\tilde{x}) \geq \omega$$

with some $\omega > 0$ *and all sufficiently small* β. *Then the origin* 0 *belongs to* $\partial U_{t\text{reg}}$.

Proof. Let f be defined as

$$f(x) = cv^2 + \min\left(\frac{\beta}{4}, z - \frac{z^2}{\beta}\right) = \begin{cases} cv^2 + z - z^2/\beta, & z < \beta/2 \\ cv^2 + \beta/4, & z \geq \frac{\beta}{2} \end{cases} \tag{6.19}$$

in V_ϵ, belong to C^2 and be bounded from below and above by some positive constants. Then

$$Lf(0, v) = b_z(0, v) - \frac{2}{\beta} G_{zz}(0, v) + \int \min\left(\tilde{z} - \frac{\tilde{z}^2}{\beta}, \frac{\beta}{4}\right) \mu((0, v), d\tilde{x})$$
$$+ \int \min\left(-\frac{\tilde{z}^2}{\beta}, \frac{\beta}{4} - \tilde{z}\right) \nu((0, v), d\tilde{x}) + O(c).$$

Clearly the integral term depending on μ tends to zero as $\beta \to 0$, and the integral depending on ν over the subset $\tilde{z} \geq 1/2\beta$ is negative. Hence

$$Lf(0, v) \leq b_z(0, v) - \frac{2}{\beta} G_{zz}(0, v) - \frac{1}{\beta} \int_{V_\epsilon \cap \{\tilde{z} \leq \beta/2\}} \tilde{z}^2 \nu((0, v), d\tilde{x}) + o(1),$$

where $o(1)$ tends to zero if $c \to 0$ and $\beta \to 0$. The assumptions of the proposition clearly imply that this expression becomes negative for small c and β. By continuity, $Lf(z, v)$ will be also negative for small enough z. Consequently, application of Proposition 6.3.2 completes the proof. □

Proposition 6.4.3. *Suppose* $G_{zz}(z, v) = \kappa(v) z(1 + o(1))$ *as* $z \to 0$ *in* V_ϵ.

(i) *If* $\kappa(v) > b_z(z, v)$ *for* $|v| \leq r + \epsilon$ *and small* z, *then the ball* $\{|v| \leq r\}$ *belongs to* $\partial U_{t\text{reg}}$.

(ii) *If* $\kappa(v) < b_z(z, v)$ *for* $|v| \leq r + \epsilon$ *and small* z, *then the ball* $\{|v| \leq r\}$ *belongs to the inaccessible part of the boundary.*

Proof. (i) As a barrier function, let us take f as $z^\gamma + cv^2$ with a $\gamma \in (0, 1)$ in V_ϵ and smooth and bounded from below and above by positive constants outside. Then near the boundary the sum of the drift and the diffusion terms of Lf is

$$\gamma z^{\gamma - 1}[b_z(z, v) - (1 - \gamma)\kappa(v) + o(1)] + O(1),$$

which can be made negative by choosing small enough γ. The integral term depending on ν is negative and the integral term depending on μ can be made smaller than $\epsilon z^{\gamma-1}$ with arbitrary ϵ by changing f outside an arbitrarily small neighborhood of the boundary. Then the origin belongs to $\partial U_{t\text{reg}}$ by Proposition 6.3.2, and similarly one deals with other points of $\{|v| \leq r\}$.

(ii) This follows from Proposition 6.3.3 if one uses the same barrier function as in Proposition 6.4.1 and Exercise 6.4.1. □

The set where $\kappa(v) = b_z(0, v)$ is a nasty set for the classification even in the case of diffusions (see e.g. [117], [309]. The following result is intended to show what kind of barrier function can be used to deal with this situation.

Exercise 6.4.2. Let the boundary of the open set $\Gamma = \{v : b_z(0, v) > 0\}$ in ∂U be smooth, the vector field $b(x)$ on $\partial\Gamma$ have a positive component in the direction of outer normal η to $\partial\Gamma$, and the diffusion term and the integral terms vanish in a neighborhood of Γ in $\bar{U}$. Then the closed subset $\bar{\Gamma} = \Gamma \cup \partial\Gamma$ of the boundary ∂U is inaccessible.

Hint. To apply Proposition 6.3.3, consider the barrier function $f = (z^2 + \rho_\Gamma(v)^2)^{-1}$, where ρ_Γ denotes the distance to Γ. Then

$$Lf \leq -2\frac{zb_z(x)}{z^2 + (\rho_\Gamma(v))^2} - 2\rho_\Gamma(v)\frac{(b_v(x), \eta)}{z^2 + (\rho_\Gamma(v))^2},$$

and the second term dominates in a neighborhood of the boundary of Γ, because $b_z(x)$ is of order $\rho(v)$.

Exercise 6.4.3. Let the ball $\{(0, v) : |v| \leq r + \epsilon\}$ be inaccessible. Suppose $b_z(x) \geq c > 0$ and $\int \tilde{z}^2 \nu(x, d\tilde{x}) = O(z)$ in V_ϵ. Then all points from the ball $\{(0, v) : |v| \leq r\}$ are entrance boundaries.

Hint. It is enough to prove the statement for the origin. Suppose for brevity that $\nu(x, .)$ vanishes in V_ϵ (the modifications required in the general case are as above). Then use Proposition 6.3.3 (ii) with Lyapunov function $f(x)$ that equals $\delta - z/c$ for $z < c\delta/2$ and is non-negative and decreasing in z. Then $f(x) \in [\delta/2, \delta]$ for $z \leq c\delta/2$ and $Lf(x) \leq -1$ for these z, because the contributions from the diffusion part of L and the integral part depending on μ are clearly negative.

6.5 Decomposable generators in $\mathbb{R}^d_+$

There is a variety of situations when the state space of a stochastic model is parametrized by positive numbers only. This happens, for instance, if one is interested in the evolution of the number (or the density) of particles (or species or prices) of different kinds. In this case, the state space of a system is $\mathbb{R}^d_+$. Consequently, one of the most natural application of the results discussed above concerns the situation when

$U = \mathbb{R}_+^d$. By U_j we shall denote the internal part of the boundary with vanishing j-th coordinate: $U_j = \{x \in \mathbb{R}_+^d : x_j = 0, x_{k \neq j} > 0\}$.

This section is devoted to some examples of the applications of the general theory developed above to processes in $U = \mathbb{R}_+^d$. For definiteness, we shall work with decomposable ΨDO studied in Section 5.7, that is with operators $L = \sum_{n=1}^N a_n(x)\psi_n$, where ψ_n are given by (5.66), i.e.

$$\psi_n f(x) = \frac{1}{2}(G^n\nabla, \nabla)f(x) + \left(b^n, \frac{\partial}{\partial x}\right)f(x)$$
$$+ \int (f(x+y) - f(x) - \nabla f(x)y)\nu^n(dy)$$
$$+ \int (f(x+y) - f(x))\mu^n(dy),$$

where ν^n and μ^n are Radon measures on the ball $\{|y| \leq 1\}$ and on $\mathbb{R}^d$ respectively such that

$$\int |y|^2\nu^n(dy) < \infty, \quad \int \min(1, |y|)\mu^n(dy) < \infty, \quad \nu^n(\{0\}) = \mu^n(\{0\}) = 0.$$

If $l \in \mathbb{R}_+^d$, we shall say that L is *l-subcritical* (respectively, *l-critical*) if $\psi_n f_l \leq 0$ (respectively, $\psi_n f_l = 0$) for all $x \in \mathbb{R}_+^d$ and n, where $f_l(x) = (l, x)$. Notice that

$$\psi_n f_l = (b^n, l) + \int (l, y)\mu^n(dy).$$

We say that l-subcritical L is *strictly subcritical* if there is n such that $\psi_n f_l < 0$.

We shall study the continuity property (Feller property) of the stopped semigroups for decomposable generators under the following condition:

(B1) $a_n \in C(\bar{U})$ for all n and they are (strictly) positive and of class $C^s(U)$ in U with $s > 2 + d/2$;

(B2) for any n, the support of the measure $\mu^n + \nu^n$ is contained in the closure $\bar{U}$ of $\mathbb{R}^d$ for all n (this condition ensures that U is transmission-admissible) and $\int |y|\mu^n(dy) < \infty$;

(B3) there exists $l \in \mathbb{R}_+^d$ such that L is l-subcritical.

The following simple result is a starting point for the analysis.

Proposition 6.5.1. *Suppose the conditions* (A1), (A2) *(introduced in Section* 5.7*), as well as* (B1)–(B3) *hold for L. Then the domain U is transmission-admissible and the condition* (U) *from Section* 6.1 *holds for the triple $(U, \{U_k\}, L)$, where $U_k = U + e/k$, $e = (1, \dots, 1) \in \mathbb{R}^d$, implying the applicability of Proposition* 6.1.1 *and hence the well-posedness of the stopped $(L, C_c^2(\mathbb{R}^d))$-martingale problem in $\mathbb{R}_+^d$.*

Proof. Assume first that a_n are extendable beyond the boundary ∂R^d_+ as positive smooth functions. Let us show how to choose an appropriate extension of a_n and f_l in a way to meet the requirements of Theorem 5.7.2. First let us choose an extension of a_n in such a way that they are everywhere smooth and positive, and uniformly bounded outside $U - e$. Let ρ_ϵ be a smooth function $\mathbb{R} \to (-2\epsilon, \infty)$ such that $\rho(y) = y$ for $y \geq -\epsilon$ and $\rho(y) = |y|$ for $y \leq -2\epsilon$ and define

$$f_L = \sum_{k=1}^{d} l_k(\rho_\epsilon(x_k) + 2\epsilon).$$

Clearly this f_L is smooth, positive, has uniformly bounded first and second derivatives and $f_L(x) \to \infty$ as $x \to \infty$. Hence, $Lf_L \leq c$ outside $U - e$, because a_n are bounded there (and taking into account the last condition in (B2)). On the other hand, $\psi_n f_L \leq 0$ in $U - e$, because there f_L coincides with f_l up to a constant. Consequently $Lf_L \leq c$ everywhere, and the conditions of Theorem 5.7.2 are met implying the well-posedness of the $(L, C_c^2(\mathbb{R}^d))$-martingale problem in $\mathbb{R}^d$.

Applying this procedure to the restrictions of a_n to U_m, we prove the validity of the condition (U). □

Theorem 6.5.1. (i) *Suppose the assumptions of Proposition* 6.5.1 *hold for a decomposable pseudo-differential operator L in $U = \mathbb{R}^d_+$. For any $j = 1, \dots, d$ and $n = 1, \dots, N$, let $a_n(x) = O((x^j)^2)$ in a neighborhood of $\bar{U}_j$ uniformly on compact sets whenever $G^n_{jj} \neq 0$ or $\int (x^j)^2 \nu^n(dx) \neq 0$, and $a_n(x) = O(x^j)$ uniformly on compact sets whenever $\beta^n_j < 0$. Then the whole boundary ∂U is inaccessible, and there exists a unique solution to the martingale problem for L in U, which is a Markov process with a C-Feller semigroup T_t, i.e. this semigroup preserves the space $C_b(U)$.*

(ii) *Suppose additionally that $a_n(x) = O(x^j)$ uniformly on compact sets whenever either $\beta^n_j \neq 0$ or $\int x^j \mu^n(dx) \neq 0$. Then all points of the boundary are natural and T_t preserves the subspace of $C_b(\bar{U})$ of functions vanishing on the boundary.*

(iii) *If additionally conditions* (A3) *(introduced before Theorem* 5.7.1*) on the growth of a_n hold, then T_t is a strongly continuous Feller semigroup on the Banach space $C_\infty(U)$ of continuous functions on U vanishing when x approaches infinity or the boundary of U.*

Proof. (i) This follows directly from Theorem 6.2.1 and Propositions 6.5.1, 6.4.1.

(ii) Taking into account Theorem 6.2.4, we only need to prove that for an arbitrary $j \in \{1, \dots, d\}$ any point $x_0 \in U_j$ is a natural boundary. To this end, let us show that

$$\lim_{R \to \infty, x \to x_0, x \in U} P_x(\tau^R_\epsilon < t) = 0, \tag{6.20}$$

$t > 0$, $\epsilon > 0$, uniformly for x_0 from an arbitrary compact subset K of U_j, where $B_R = \{x_0 \in U_j : |x_0| \le R\}$ and τ_ϵ^R denotes the exit time from $V_\epsilon^R = \{x : d(x, B_R) < \epsilon\}$ (d denotes the usual distance, of course). By the uniform compact containment, by taking R large enough one can ensure that the process does not leave the domain $\{x \in U : |x| \le R\}$ up to any given time with probability arbitrary close to 1 when started in $x \in K$. Let $f(x)$ be a function from $C^2(\{x : x^j > 0\})$ such that $f(x) = (x^j)^\gamma$ with some $\gamma \in (0,1)$ in a neighborhood of V_ϵ^R and f vanishes outside some compact set. Then $|Lf(x)| \le cf$ in a neighborhood of V_ϵ^R with some constant c and as in the proof of Proposition 6.4.1 (and taking into account that the whole boundary ∂U is inaccessible) one shows that

$$\mathbf{E} f(X^x_{\min(t,\tau_\epsilon^R)}) \le f(x) e^{ct}$$

for x from V_ϵ^R. As (up to an arbitrary small probability which allows the process to leave the domain $\{x : |x| \le R\}$) the l.h.s. of this inequality can be estimated from below by

$$P_x(\tau_\epsilon < t) \min\{f(x) : x^j = \epsilon\} = P_x(\tau_\epsilon < t)\epsilon^\gamma,$$

the limiting formula (6.20) follows.

(iii) this follows from Theorem 5.7.1 (or directly via Proposition 5.4.1). □

We shall conclude this section with simple estimates for exit times from U that can be used to verify assumptions for the theorems given above. Let us assume for the rest of this section that $U = \mathbb{R}_+^d$ and the assumptions of Proposition 6.5.1 hold.

Proposition 6.5.2. *Let the assumptions of Proposition* 6.5.1 *hold. The process* X_t *leaves the domain* $U = \mathbb{R}_+^d$ *almost surely if there exists* n *such that* $a_n(x) \ge a > 0$ *and either*

(i) $\psi_n f_l = -c < 0$, *or*

(ii) *there exist* j *such that* $\beta_j^n + \int y_j \mu^n(dy) = -c < 0$, *or*

(iii) $G^n \ne 0$, *or* (iv) $\int_{B_1} |y|^2 \nu^n(dy) \ne 0$.

Moreover, in cases (i), (ii), the estimates

$$\mathbf{E}\tau_U^x \le \frac{(l,x)}{ac}, \quad \mathbf{E}\tau_U^x \le \frac{x_j}{ac}$$

for the expectation of the exit time hold respectively.

Proof. In case (i), (ii), the statement together with the required estimates follow directly from Proposition 6.3.1 (ii) using the barrier functions $f_l(x) = (l,x)$ and $f(x) = x_j$ respectively. Assume now the conditions (iii) or (iv). A key argument in the proof is based on the observation that (by subcriticality), the process $Z_t = f_l(X_{\min(t,\tau_U)})$ is a positive supermartingale, and hence it has a finite limit as

$t \to \infty$ almost surely (see Theorem 3.9.2). Hence there exists a.s. a compact subset $U^b = \{x : (l, x) \leq b\}$ in $\bar{U}$, which the process does not exit. Consequently, up to arbitrary small probability, the process remains in U^b for all time. Consequently, to prove the statement, one only needs to show that the process leaves the interior of any subset U^b almost surely. Under assumption (ii) or (iv), there exists j such that either $G^n_{jj} \neq 0$ or $\int_{B_1} y_j^2 \nu^n(dy) \neq 0$. It remains to use Proposition 6.3.1 (ii) with a barrier function that equals $f(x) = \lambda^{-1}(e^{\lambda R} - e^{\lambda x_j})$ in U^b, where R and λ are large enough. □

If a_n is not positive up to the boundary, one can combine the above exit criteria applied U_m with local criteria around the boundary ∂U.

6.6 Gluing boundary

Here we study a natural class of processes which have possibly accessible boundary but which do not stop on it, but just stick to it as soon as they reach it. Unlike the previous section, we shall work with general Lévy type ΨDO of form (6.1), but for simplicity will restrict our attention to the domain $U = \mathbb{R}^d_+$. If the coordinates x_j are interpreted as number (or the density of) particles of type j, the gluing property of the part U_j of the boundary is naturally interpreted as non-revivability of the corresponding particles (see Section 6.9).

Let us say that the face U_j of the boundary is *gluing* for L if for any $f \in C^2_C(\mathbb{R}^d)$, $Lf(x)|_{x \in U_j}$ does not depend on the values of f outside U_j. In terms of the coefficients of L this is clearly equivalent to the requirement that

$$G_{ik}(x)|_{x_j=0} = 0 \quad \forall k = 1, \dots, d, \quad b_i(x)|_{x_j=0} = 0,$$

and

$$\nu(x, dy)|_{x_j=0} = \delta(y_j)\tilde{\nu}(x, dy_1 \cdots dy_{j-1} dy_{j+1} \cdots dy_d)$$

with a certain Lévy measure $\tilde{\nu}$. Yet another way to express this property is by the equation

$$\frac{\partial}{\partial \xi_j} p(x, \xi)|_{x_j=0} = 0,$$

where $p(x, \xi)$ denotes the symbol of the ΨDO L. More generally, for a subset I of the set of indices $\{1, \dots, d\}$, let us say that the face $U_I = \bigcap_{j \in I} U_j$ of the boundary is *gluing* if for all $j \in I$ and all ξ

$$\frac{\partial}{\partial \xi_j} p(x, \xi)|_{x \in U_I} = 0,$$

or equivalently, if the values of $Lf(x)$ for $x \in U_I$ do not depend on the behavior of f outside U_I. This is the key property of the gluing boundary that allows the process (with generator L) to live on it without leaving it. In the Theorem below, we shall call U_j accessible if it is not inaccessible.

Theorem 6.6.1. *Let the condition* (U) *from Section* 6.1 *hold for* $u = \mathbb{R}^d_+$ *and a* $\Psi DO\ L$.

(i) *Suppose that for any* j*, the boundary* U_j *is either inaccessible or gluing and the same hold for the restrictions of L to any accessible* U_j*, i.e. for the process on* U_j *defined by the restriction of L to* U_j *(well defined due to the gluing property) each of its boundaries* U_{ji}*,* $i \neq j$ *is either inaccessible or gluing, and the same holds for the restriction of L to each accessible* U_{ji} *and so on. Then there exists a unique Markov process* $X_t^{x,\text{glue}}$ *in* $\bar{U}$ *with sample paths in* $D_{\bar{U}}[0,\infty)$ *such that*

$$\phi(X_t^{x,\text{glue}}) - \phi(x) - \int_0^t L\phi(X_s^{x,\text{glue}})\, ds$$

is a martingale for any $x \in U$ *and any* $\phi \in C_c^2(\mathbb{R}^d)$ *and such that* $X_t^{x,\text{glue}} \in U_j$ *for all* $t \geq s$ *almost surely whenever* $X_s^{x,\text{glue}} \in U_j$*. This process can be characterized alternatively as the process* $X_t^{x,\text{glue}}$ *that represents a (unique) solution to the stopped martingale problem in U up to the time* τ_1 *when it reaches the boundary at some point* $y \in U_{j_1}$ *with some* j_1 *such that* U_{j_1} *is not inaccessible and hence gluing. Starting from y it evolves like a unique solution to the stopped martingale problem in* U_{j_1} *(with the same generator L) till it reaches a boundary point at* $U_{j_1} \cap U_{j_2}$ *with some* j_2*, hence it evolves as the unique solution of the stopped martingale problem in* $U_{j_1} \cap U_{j_2}$ *and so on, so that it either stops at the origin or ends at some* U_I *with an inaccessible boundary.*

(ii) *If additionally* $\partial U \setminus \partial U_{t\text{reg}}$ *is an inaccessible set (for all restrictions of L to all accessible boundary spaces), then the corresponding semigroup preserves the set of functions* $C_b(U \cup \partial U_{t\text{reg}})$*. In particular, if either* $\partial U = \partial U_{t\text{reg}}$ *or* $\partial U \setminus \partial U_{t\text{reg}}$ *consists of entrance boundaries only, then the space* $C_b(\bar{U})$ *is preserved.*

Proof. (i) The construction (and the proof if uniqueness) of the solutions of the martingale problem for L in $\bar{U}$ by gluing the solutions on various U_I is done as by gluing the stopped solutions in different domains in Theorems 4.11.3 and 4.11.2. (ii) This follows from Theorems 6.2.1 and 6.2.5. □

For conclusion let us give a simple extinction result.

Proposition 6.6.1. *Suppose the assumptions of Proposition* 6.5.1 *hold for a decomposable pseudo-differential operator L in* $U = \mathbb{R}^d_+$*, and the assumptions of Theorem* 6.6.1 (i) *are valid so that the process* $X_t^{x,\text{glue}}$ *is well defined. Assume finally that there exists n such that* $a_n(x) \geq a > 0$ *and* $\psi_n f_l = -c < 0$*. Then a.s.* $X_t^{x,\text{glue}} \to 0$ *as* $t \to \infty$*. More precisely,* $X_t^{x,\text{glue}}$ *becomes zero in a finite time not exceeding* $(l,x)/(ac)$.

Proof. Straightforward, because the process $Z_t = f_l(X_t^{x,\text{glue}})$ is a positive supermartingale such that $\mathbf{E}Z_t \le Z_0 - act$. □

6.7 Processes on the half-line

Of course, many problems become simpler for one-dimensional problems. As an example, let us formulate a well-posedness result for stochastically monotone processes on the half-line.

Theorem 6.7.1. *Let an operator L be given by* (5.112) *for* $x > 0$ *and the following conditions hold:*

(i) *The supports of measures* $\nu(x,.)$ *and* $\mu(x,.)$ *belong to* $\mathbb{R}_+ = \{x > 0\}$,

$$\sup_{x\in(0,1]}\left[|b(x)| + G(x) + \int |y|\mu(x,dy) + \int (y \wedge y^2)\nu(x,dy)\right] < \infty,$$

and the condition (i) *of Theorem* 5.9.3 *holds for* $x > 0$.

(ii) *For any* $a > 0$ *the functions*

$$\int_a^\infty \nu(x,dy), \quad \int_a^\infty \mu(x,dy)$$

are non-decreasing in x.

(iii) *For a constant* $c > 0$

$$b(x) + \int y\mu(x,dy) \le c(1+x), \quad x > 1.$$

Then the stopped martingale problem for L in $C_c^2(\mathbb{R})$ *is well-posed and specifies a stochastically monotone Markov process* X_t^x *in* $\bar{\mathbb{R}}_+ = \{x \ge 0\}$.

Proof. It follows from Theorem 5.9.3 and the localization procedure for martingale problems, see Theorem 4.11.4. □

Applying the local criteria for boundary points, developed above, one can deduce various regularity properties of the stopped process and its semigroup (extending the results from [183] obtained there under restrictive technical assumptions). For example, one obtains the following.

Corollary 17. *Let the assumptions of Theorem* 6.7.1 *hold.*

(i) *Suppose*

$$G(x) = O(x^2), \quad \int_0^1 z^2 \nu(x, dz) = O(x^2), \quad |b(x) \wedge 0| = O(x),$$

for $x \to 0$. *Then the point* 0 *is inaccessible for* X_t^x *and its semigroup preserves the space* $C(\mathbb{R}_+)$ *of bounded continuous functions on* $\mathbb{R}_+$.

(ii) *Suppose the* $\lim_{x \to 0} b(x)$ *exists and* $G(x) = \alpha x(1 + o(1))$ *as* $x \to 0$ *with a constant* $\alpha > 0$. *If* $\alpha < b(0)$, *then again the point* 0 *is inaccessible for* X_t^x *and its semigroup preserves the space* $C(\mathbb{R}_+)$. *If* $\alpha > b(0)$, *then the boundary point* 0 *is* t*-regular for* X_t^x *and its semigroup preserves the space* $C(\bar{\mathbb{R}}_+)$ *of bounded continuous functions on* $\bar{\mathbb{R}}_+$.

6.8 Generators of reflected processes

In this chapter we are working mostly with the stopped or killed processes that yield the simplest reduction of a Markov dynamics to a domain with a boundary. Another well-studied reduction is given by processes reflected at the boundary. We shall only touch upon this subject here by deducing some more or less straightforward corollaries from the results of Section 3.8. The notations introduced in that section will be used here without further reminder. We shall describe the natural cores of the generators of the reflected processes in the simplest case of the reflection of processes defined in the whole space (not discussing a boundary such that the original process is not extendable beyond it, when the classification of the boundary points should be taken into consideration as in the case of stopped processes).

Theorem 6.8.1. *Let* X_t^x *be a Feller process in* $\mathbb{R}^d$ *with semigroup of transition operators* Φ_t *and with generator* L *having the space* $C_\infty^k(\mathbb{R}^d)$ *with some* $k \geq 0$ *as an invariant core. Suppose* L *commutes with the reflection* R_i, *that is*

$$L\tilde{R}_i f = \tilde{R}_i L f \tag{6.21}$$

for $f \in C_\infty^k(\mathbb{R}^d)$. *Then the reflected process* $Y_t^y = \bar{R}_i(X_t^x)$, $y = \bar{R}_i(x)$, *is a Feller process in* $\bar{\mathbb{R}}_i^d$ *with generator having the space* $C_{\infty,i}^k(\mathbb{R}^d)$ *of the functions from* $C_\infty^k(\mathbb{R}^d)$ *that are invariant under* R_i *as an invariant core.*

Proof. Condition (6.21) implies that Φ_t commute with $\tilde{R}_i$ (because $\Phi_t \tilde{R}_i f$ and $\tilde{R}_i \Phi_t f$ satisfy the same Cucahy problem for the equation $\dot{f}_t = L\tilde{R}_i f_t$). Hence by Proposition 3.8.2 the reflected process is Markov with semigroup Φ_t^i acting as

$$\Phi_t^i f(x) = \Phi_t f(x), \quad x \in \bar{R}_i^d, \ f \in C_\infty(\bar{R}_i^d),$$

where f on the r.h.s. denotes (with some abuse of notations) the extension of f to a function on $\mathbb{R}^d$ invariant under $\tilde{R}_i$. Clearly Φ_t^i specify a Feller semigroup. Moreover, since the space $C^k_{\infty,i}(\mathbb{R}^d)$ is invariant and belongs to the domain of the generator, it specifies a core. (More precisely, this core is given by the space of functions from $C_\infty(\bar{\mathbb{R}}^d_i)$ that can be obtained as the restriction to $\bar{\mathbb{R}}^d_i$ of a function from $C^k_{\infty,i}(\mathbb{R}^d)$.) □

Exercise 6.8.1. Let L have the standard Lévy–Khintchine form

$$Lu(x) = \frac{1}{2}(G(x)\nabla, \nabla)u(x) + (b(x), \nabla u(x)) + \int [u(x+y) - u(x) - (y, \nabla u(x))\mathbf{1}_{B_1}(y)]\nu(x, dy).$$

Show that condition (6.21) holds for this L whenever

$$G(R_i x) = G(x), \quad b(R_i x) = -b(x), \quad \nu(x, d(R_i y)) = \nu(R_i x, dy), \tag{6.22}$$

and in this case the generator of Φ^i_t is given by the formula

$$Lu(x) = \frac{1}{2}(G(x)\nabla, \nabla)u(x) + (b(x), \nabla u(x)) + \int [u(\bar{R}_i(x+y)) - u(x) - (y, \nabla u(x))\mathbf{1}_{B_1}(y)]\nu(x, dy).$$

Notice that if $f \in C^k_{\infty,i}(\mathbb{R}^d)$ with $k \geq 1$, then

$$\left.\frac{\partial f}{\partial x_i}\right|_{x_i=0} = 0, \tag{6.23}$$

so that the function $f_t = \Phi^i_t f = \Phi_t f$ yields a solution to the *mixed initial-boundary-value problem* in $\mathbb{R}^d_i$ with the *Neumann boundary condition* (6.23).

Similarly to Theorem 6.8.1 one shows the following.

Theorem 6.8.2. *Let X^x_t be a Feller process in $\mathbb{R}^d$ with semigroup of transition operators Φ_t and with generator L having the space $C^k_\infty(\mathbb{R}^d)$ with some $k \geq 0$ as an invariant core. Suppose L commutes with all reflections R_i, $i = 1, \ldots, d$. Then the process Y^y_t obtained by reflection of X^x_t with respect to all coordinate hyperplanes via Proposition* 3.8.3 *(with G being the group generated by all these reflections) is a Feller process in $\bar{\mathbb{R}}^d_+$ with generator having the space $C^k_{\infty,\mathrm{even}}(\mathbb{R}^d)$ of the functions from $C^k_\infty(\mathbb{R}^d)$ such that*

$$f(\pm x^1, \ldots, \pm x^d) = f(x^1, \ldots, x^d)$$

(better to say the space of functions from $C^k_\infty(\bar{\mathbb{R}}^d_+)$ that can be obtained as the restrictions of the functions from $C^k_{\infty,\mathrm{even}}(\mathbb{R}^d)$) as an invariant core.

As in the previous case one sees that the corresponding semigroup yields a solution to the *mixed initial-boundary-value problem* in $\mathbb{R}^d_+$ with the *Neumann boundary condition*, that is satisfying (6.23) for all $i = 1, \ldots, d$.

6.9 Application to interacting particles: stochastic LLN

Here we show that Markov chains in $\mathbb{Z}_+^d$ describing binary or more generally k-ary interacting particles of d different types approximate (in the continuous-state limit) Markov processes on $\mathbb{R}_+^d$ having pseudo-differential generators $p(x, i\frac{\partial}{\partial x})$ with symbols $p(x,\xi)$ depending polynomially (degree k) on x. The case when the limiting Markov process was deterministic was already discussed in Section 5.11.

Our general scheme of continuous-state (or finite-dimensional measure-valued) limits to processes of k-nary interaction yields a unified description of these limits for a large variety of models that are intensively studied in different domains of natural science from interacting particles in statistical mechanics (e.g. coagulation-fragmentation processes) to evolutionary games and multidimensional birth and death processes from biology and social sciences.

Let $\mathbb{Z}^d$ denote the integer lattice in $\mathbb{R}^d$ and let $\mathbb{Z}_+^d$ be its positive cone (which consists of vectors with non-negative coordinates). We equip $\mathbb{Z}^d$ with the usual partial order saying that $N \le M$ iff $M - N \in \mathbb{Z}_+^d$. A state $N = \{n_1, \dots, n_d\} \in \mathbb{Z}_+^d$ will designate a system consisting of n_1 particles of the first type, n_2 particles of the second type, etc. For such a state we shall denote by $\mathrm{supp}(N) = \{j : n_j \neq 0\}$ the support of N (considered as a measure on $\{1, \dots, d\}$). We shall say that N has *full support* if $\mathrm{supp}(N)$ coincides with the whole set $\{1, \dots, d\}$. We shall write $|N|$ for $n_1 + \cdots + n_d$.

Roughly speaking, *k-ary (or k-th order) interaction* means that any group of k particles (chosen randomly from a given state N) are allowed to have an act of interaction, with the result that some of these particles (maybe all or none of them) may die, producing a random number of offspring of different types. More precisely, each sort of k-ary interaction is specified by:

(i) a vector $\Psi = \{\psi_1, \dots, \psi_d\} \in \mathbb{Z}_+^d$, which we shall call the *profile of the interaction*, with $|\Psi| = \psi_1 + \cdots + \psi_d = k$, so that this sort of interaction is allowed to occur only if $N \ge \Psi$ (i.e. ψ_j denotes the number of particles of type j which take part in this act of interaction);

(ii) a family of non-negative numbers $g_\Psi(M)$ for $M \in \mathbb{Z}^d$, $M \neq 0$, vanishing whenever $M \ge -\Psi$ does not hold.

The generator of a Markov process (with the state space $\mathbb{Z}_+^d$) describing *k-nary interacting particles of types* $\{1, \dots, d\}$ is then an operator on $B(\mathbb{Z}_+^d)$ defined as

$$(G_k f)(N) = \sum_{\Psi \le N, |\Psi| = k} C_{n_1}^{\psi_1} \cdots C_{n_d}^{\psi_d} \sum_M g_\Psi(M)(f(N+M) - f(N)), \quad (6.24)$$

where C_n^k denote the usual binomial coefficients. Notice that each $C_{n_j}^{\psi_j}$ in (6.24) appears from the possibility of choosing randomly (with the uniform distribution) any ψ_j particles of type j from a given group of n_j particles. Consequently, the generator

$\sum_{k=0}^{|K|} G_k$ of the interactions of order not exceeding K can be written as

$$(G_K f)(N) = \sum_{\Psi \le K} C_{n_1}^{\psi_1} \cdots C_{n_d}^{\psi_d} \sum_{M \in \mathbb{Z}^d} g_\Psi(M)(f(N+M) - f(N)), \quad (6.25)$$

where we used the usual convention that $C_n^k = 0$ for $k > n$. The term with $\Psi = 0$ corresponds to the external input of particles.

We shall show that the continuous state limits of the Markov chains with generators (6.25) are given by Markov processes on $\mathbb{R}_+^d$ having decomposable pseudo-differential generators with polynomially growing symbols (studied in Section 6.5). This limiting procedure can be used to prove the existence and non-explosion of such Markov processes on $\mathbb{R}_+^d$, as well for numeric calculations of their basic characteristics (say, exit times).

To this end, instead of Markov chains on $\mathbb{Z}_+^d$ we shall consider the corresponding scaled Markov chains on $h\mathbb{Z}_+^d$, h being a positive parameter, with generators of type

$$(G_K^h f)(hN) = \sum_{\Psi \le K} h^{|\Psi|} C_{n_1}^{\psi_1} \cdots C_{n_d}^{\psi_d} \sum_{M \in \mathbb{Z}^d} g_\Psi(M)(f(Nh + Mh) - f(Nh)), \quad (6.26)$$

which clearly can be considered as the restriction on $B(h\mathbb{Z}_+^d)$ of an operator on $B(\mathbb{R}_+^d)$ (which we shall again denote by G_K^h with some abuse of notations) defined as

$$(G_K^h f)(x) = \sum_{\Psi \le K} C_\Psi^h(x) \sum_{M \in \mathbb{Z}^d} g_\Psi(M)(f(x + Mh) - f(x)), \quad (6.27)$$

where we introduced a function C_Ψ^h on $\mathbb{R}_+^d$ defined as

$$C_\Psi^h(x) = \frac{x_1(x_1 - h) \cdots (x_1 - (\psi_1 - 1)h)}{\psi_1 !} \cdots \frac{x_d(x_d - h) \cdots (x_d - (\psi_d - 1)h)}{\psi_d !}$$

in case $x_j \ge (\psi_j - 1)h$ for all j and $C_\Psi^h(x)$ vanishes otherwise.

As

$$\lim_{h \to 0} C_\Psi(x) = \frac{x^\Psi}{\Psi !} = \prod_{j=1}^d \frac{x_j^{\psi_j}}{\psi_j},$$

one can expect that (with an appropriate choice of $g_\Psi(M)$, possibly depending on h) the operators G_K^h will tend to the generator of a stochastic process on $\mathbb{R}_+^d$ which has the form of a polynomial in x with “coefficients” being generators of spatially homogeneous processes with i.i.d. increments (i.e. Lévy processes) on $\mathbb{R}_+^d$, which are given therefore by the Lévy–Khintchine formula with the Lévy measures having support in $\mathbb{R}_+^d$.

Clearly, as is usual in the theory of superprocesses and interacting superprocesses, we can consider points on $\mathbb{Z}_+^d$ as integer-valued measures on $\{1, \dots, d\}$ (empirical measures). The limit $Nh \to x$, $h \to 0$, describes the limit of empirical measures as the number of particles tend to infinity but the "mass" of each particle is re-scaled in such a way that the whole mass tend to x.

Of course, the continuous-state limit obtained depends on the scaling of the coefficients $g_\Psi(M)$. Roughly speaking, if one accelerates some short range interactions (say, with $|M| = 1$ in (6.27)), one gets a second-order parabolic operator as part of a limiting generator, and if one slows down the long-range interactions (large M in (6.27)), one gets non-local (Lévy-type) terms.

Let us proceed to more detail. By $Z_t(G_K)$ (respectively $Z_t(G_K^h)$ we shall denote the minimal Markov chain on $\mathbb{Z}_+^d$ (respectively on $h\mathbb{Z}_+^d$) specified by the generator of type (6.25) (respectively (6.26)). For a given $L \in \mathbb{Z}_+^d$, we shall say that $Z_t(G_K)$ and the generators G_K, G_K^h are *L-subcritical* (respectively *L-critical*) if

$$\sum_{M \neq 0} g_\Psi(M)(L, M) \le 0 \tag{6.28}$$

for all $\Psi \le K$ (respectively, if the equality holds in (6.28)), where (L, M) denotes the usual scalar product in $\mathbb{R}^d$. Putting for convenience $g_\Psi(0) = -\sum_{M \neq 0} g_\Psi(M)$, we conclude from (6.28) that the Q-matrix Q^K of the chain $Z_t(G_K)$ defined as

$$Q_{NJ}^K = \sum_{\Psi \le K} C_{n_1}^{\psi_1} \cdots C_{n_d}^{\psi_d} g_\Psi(J - N) \tag{6.29}$$

satisfies the condition $\sum_J Q_{NJ}^K (L, J - N) \le 0$ for all $N = \{n_1, \dots, n_d\}$.

Proposition 6.9.1. *If G_K is L-subcritical with some L having full support, then*

(i) *$Z_t(G_K)$ is a unique Markov chain with the Q-matrix* (6.29),

(ii) *$Z_t(G_K)$ is a regular jump process (i.e. it is non-explosive),*

(iii) *$(L, Z_t(G_K))$ is a non-negative supermartingale, which is a martingale iff G_K is L-critical.*

Proof. This is a direct consequence of (6.28) and the standard theory of continuous-time Markov chains, see Section 3.7. For example, statement (iii) follows from Dynkin's formula. □

Let us describe now precisely the generators of limiting processes on $\mathbb{R}_+^d$ and the approximating chains in $\mathbb{Z}_+^d$. Suppose that to each $\Psi \le K$ there correspond

(i) a non-negative symmetric $d \times d$-matrix $G(\Psi) = G_{ij}(\Psi)$ such that $G_{ij}(\Psi) = 0$ whenever i or j does not belong to $\mathrm{supp}(\Psi)$,

(ii) vectors $\beta(\Psi) \in \mathbb{R}_+^d$, $\gamma(\Psi) \in \mathbb{R}_+^d$ such that $\gamma_j(\Psi) = 0$ whenever $j \notin \mathrm{supp}(\Psi)$,

(iii) Radon measures ν_Ψ and μ_Ψ on $\{|y| \leq 1\} \subset \mathbb{R}^d$ and on $\mathbb{R}^d_+ \setminus \{0\}$ respectively (Lévy measures) such that

$$\int |\xi|^2 \nu_\Psi(d\xi) < \infty, \int |\xi| \mu_\Psi(d\xi) < \infty, \quad \mu(\{0\}) = \nu(\{0\}) = 0$$

and supp ν_Ψ belongs to the subspace in $\mathbb{R}^d$ spanned by the unit vectors e_j with $j \in$ supp(Ψ).

These objects specify an operator in $C(\mathbb{R}^d_+)$ by the formula

$$(\Lambda_K f)(x) = -\sum_{\Psi \leq K} \frac{x^\Psi}{\Psi!} p_\Psi(-i\nabla), \tag{6.30}$$

where

$$\begin{aligned} -p_\Psi(-i\nabla) = \operatorname{tr}\left(G(\Psi)\frac{\partial^2}{\partial x^2}\right) f &+ \sum_{j=1}^d (\beta_j(\Psi) - \gamma_j(\Psi)) \frac{\partial f}{\partial x_j} \\ &+ \int (f(x+y) - f(x) - f'(x)y)\nu_\Psi(dy) \\ &+ \int (f(x+y) - f(x))\mu_\Psi(dy) \end{aligned} \tag{6.31}$$

is the pseudo-differential operator with the symbol $-p_\Psi(\xi)$, where

$$\begin{aligned} p_\Psi(\xi) &= (\xi, G(\Psi)\xi) - i(\beta - \gamma, \xi) \\ &+ \int (1 - e^{iy\xi} + iy\xi)\nu_\Psi(dy) + \int (1 - e^{iy\xi})\mu_\Psi(dy) \end{aligned}$$

and where as usual

$$\operatorname{tr}\left(G(\Psi)\frac{\partial^2}{\partial x^2}\right) f = \sum_{i,j=1}^d G_{ij}(\Psi) \frac{\partial^2 f}{\partial x_i \partial x_j}.$$

Remark 49. Conditions in (i) and (iii) concerning the supp(Ψ) mean simply that a particle of type i can not kill a particle of type j without an interaction. Condition (iii) highlights the fact that in the framework of interacting particles, it is natural to write the generators of a Lévy process in the form (6.31) with two measures ν and μ. It corresponds to Lévy–Itô decomposition of Lévy processes into parts with large and small jumps.

We shall say that operator (6.31) is *L-subcritical* (respectively *L-critical*) for an $L \in \mathbb{Z}^d_+$ if for all Ψ

$$\left(\beta(\Psi) - \gamma(\Psi) + \int y\mu_\Psi(dy), L\right) \leq 0 \tag{6.32}$$

(respectively, if the equality holds).

Next, let $\Delta_h(\Psi, G)$ be a finite-difference operator of the form

$$(\Delta_h(\Psi, G) f)(x) = \frac{1}{h^2} \sum_{i \in \mathrm{supp}(\Psi)} \omega_i(\Psi)(f(x+he_i) + f(x-he_i) - 2f(x))$$
$$+ \frac{1}{h^2} \sum_{i \neq j : i,j \in \mathrm{supp}(\Psi)} [\omega_{ij}(\Psi)(f(x+he_i+he_j) + f(x-he_i-he_j) - 2f(x))$$
$$+ \tilde{\omega}_{ij}(\Psi)(f(x+he_i-he_j) + f(x-he_i+he_j) - 2f(x))] \qquad (6.33)$$

with some constants $\omega_i, \omega_{ij}, \tilde{\omega}_{ij}$ (where e_j are the vectors of the standard basis in $\mathbb{R}^d$) that approximate $\mathrm{tr}\,(G(\Psi)\frac{\partial^2}{\partial x^2})$ in the sense that

$$\left\| \mathrm{tr}\left(G(\Psi) \frac{\partial^2}{\partial x^2} \right) - \Delta_h(\Psi, G)) f \right\| = O(h) \| f''' \| \qquad (6.34)$$

for $f \in C^3(\mathbb{R}^d)$. If $f \in C^4(\mathbb{R}^d)$, then the l.h.s. of (6.34) can be better estimated by $O(h^2)\| f^{(4)} \|$.

Putting $B_h = \{x \in \mathbb{R}^d_+ : 0 \leq x_j < h \,\forall j\}$ and choosing an arbitrary $\omega \in (0,1)$, we can now define an operator as

$$\Lambda_K^h = \sum_{\Psi \leq K} C_\Psi^h(x) \pi_\Psi^h,$$

with

$$\pi_\Psi^h = (\Delta_h(\Psi, G) f)(x)$$
$$+ \frac{1}{h} \sum_j (\beta_j(\Psi)(f(x+he_j) - f(x)) + \gamma_j(\Psi)(f(x-he_j) - f(x)))$$
$$+ \sum_{M : M_j \geq h^{-\omega} \forall j} \Big(f(x+Mh) - f(x) + \sum_j M_j (f(x-he_j) - f(x)) \Big) v(M,h)$$
$$+ \sum_{M : M_j \geq h^{-\omega} \forall j} (f(x+Mh) - f(x)) \mu_\Psi(B_h + Mh), \qquad (6.35)$$

where

$$v(M,h) = \frac{1}{h^2 M^2} \tilde{\nu}(B_h + Mh), \quad \tilde{\nu}(dy) = y^2 \nu(dy).$$

Proposition 6.9.2. *Operator* (6.35) *is L-subcritical, if and only if*

$$(\beta(\Psi) - \gamma(\Psi) + \sum_M M \mu_\Psi(B_h + Mh), L) \leq 0.$$

In particular, if Λ_K is L-subcritical or critical, then the same holds for its approximation Λ_K^h.

Proof. It follows from a simple observation that operator (6.33) and the operator given by the sum in (6.35) that depends on the measure ν are always L-critical for any L, i.e. they are e_j-critical for all j. □

Let $Z_t^{x,h}$ denote the minimal (càdlàg) Markov chain in $x + h\mathbb{Z}_+^d \subset \mathbb{R}^d$ generated by Λ_K^h.

Proposition 6.9.3. *Suppose Λ_K of form (6.30), (6.31) is L-subcritical with some L having full support.*

(i) *There exists a solution to the martingale problem for Λ_K^h with sample paths $D_{\bar{\mathbb{R}}_+^d}[0,\infty)$ for any $x \in \mathbb{R}_+^d$.*

(ii) *The family of processes $Z_t^{hN,h}$, $h \in (0,1]$, $N = x/h$, with any given $x \in \mathbb{R}_+^d$ is tight and it contains a subsequence that converges (in the sense of distributions) as $h \to 0$ to a solution of the martingale problem for Λ_K.*

Proof. Part (i) is a consequence of (ii). Let us prove part (ii).

Step 1. Let us show that the family of processes $Z_t^{Nh,h}$, $h \in (0,1]$, $Nh = x$ is tight.

First we observe that the compact containment condition holds, i.e. for every $\epsilon > 0$ and every $T > 0$ there exists a compact set $\Gamma_{\epsilon,T} \subset \mathbb{R}_+^d$ such that

$$\inf_h P(Z_t^{Nh,h} \in \Gamma_{\epsilon,T}\ \forall t \in [0,T]) \geq 1 - \epsilon$$

uniformly for all starting points x from any compact subset of $\mathbb{R}_+^d$. In fact, the compact containment condition for $(L, Z_t^{Nh,h})$ follows directly from maximal inequalities for positive supermartingales and Proposition 6.9.1. It implies the compact containment condition for X_t^x, because L is assumed to have full support. The tightness can now be deduced by standard methods, that is using Theorem 4.9.2 and Step 2 below.

Step 2. Let us show that the operators Λ_K^h approximate Λ_K on the space $C^3(\mathbb{R}_+^d) \cap C_c(\mathbb{R}^d)$, i.e. for an arbitrary function f in this space

$$\|(\Lambda_K^h - \Lambda_K)f\| = o(1) \sup_x (1 + |x|^{|K|}) \max_{|y| \geq |x| - h} (|f'(y)| + |f''(y)| + |f'''(y)|), \tag{6.36}$$

with $o(1)$ as $h \to 0$ not depending on f (but only on the family of measures μ_Ψ, ν_Ψ (see below for a precise dependence of $o(1)$ on h).

Estimate (6.34) shows that the diffusion part of Λ_K is approximated by finite sums of the form

$$\sum C_\Psi^h(x) \Delta_h(\Psi, G)$$

in the required sense. It is obvious that the drift part of λ_K is approximated by the sum in (6.35) depending on β and γ. Let us prove that the integral part of $-p_\Psi(-i\nabla)$ depending on ν_Ψ is approximated by the corresponding sum from (6.35) (a similar fact for the integral part depending on μ is simpler and is omitted).

Since

$$(f(x-he_j)-f(x)) = -hf'(x)+\frac{1}{2}h^2 f''(x-\theta e_j), \quad \theta\in[0,h],$$

and

$$\sum_{M:M_j\ge h^{-\omega}}\sum_{j=1}^d M_j h^2\nu(M,h) \le 2\sum_{M:M_j\ge h^{-\omega}}\sum_{j=1}^d M_j h^2\nu(Bh+Mh)$$
$$\le 2h\sum_{j=1}^d\int_{y:y_j\ge h^{1-\omega}\,\forall j} y_j\nu(dy) \le 2h^\omega\int |y|^2\nu(dy),$$

the sum in (6.35) depending on ν can be written in the form

$$\sum_{M:M_j\ge h^{-\omega}}(f(x+Mh)-f(x)-h(f'(x),M))\nu(M,h)$$
$$+\,O(h^\omega)\sup_{|y|\ge|x|-h}|f''(y)|\int|y|^2\nu(dy),$$

and hence the difference between this sum and the corresponding integral from (6.31) has the form

$$\begin{aligned}&\sum_{M:M_j\ge h^{-\omega}}(f(x+Mh)-f(x)-h(f'(x),M))\nu(M,h)\\&-\int_{y:y_j\ge h^{-\omega}\,\forall j}(f(x+y)-f(x)-f'(x)y)\nu(dy)\\&+\sup_{|y|\ge|x|-h}|f''(y)|\left(O(1)\int_0^{h^{1-\omega}}|y|^2\nu(dy)+O(h^\omega)\int|y|^2\nu(dy)\right).\end{aligned}\tag{6.37}$$

To estimate the difference between the sum and the integral here, we shall use the following simple estimate:

$$\left|\sum_M g(Mh)\tilde\nu(Mh+B_h)-\int g(x)\tilde\nu(dx)\right|\le h\|g'(x)\|\int\tilde\nu(dx)\tag{6.38}$$

(valid for any continuously differentiable function g in the cube $\bar B_1$) with

$$g(y)=|y|^{-2}(f(x+y)-f(x)-f'(x)y)$$

(that clearly satisfies the estimate $\|g'\| \leq \sup_{|y|\geq|x|} |f'''(y)|$), yielding for this difference the estimate

$$\sup_{|y|\geq|x|} |f'''(y)| O(h) \int |y|^2 \nu(dy). \tag{6.39}$$

Clearly (6.36) follows from (6.37), (6.39) and the observation that $C_\Psi^h(x) = O(1 + |x|^{|K|})$ for $\Psi \leq K$.

Step 3. End of the proof. As the coefficients of Λ_K grow at most polynomially as $x \to \infty$, similarly to (6.36) one shows that the operators Λ_K^h, $h > 0$, approximate Λ_K on the Schwartz space $S(\mathbb{R}^d)$, i.e. for an arbitrary $f \in S(\mathbb{R}^d)$ the estimate $\|(\Lambda_K^h - \Lambda_K)f\| = o(1)$ as $h \to 0$ holds uniformly for all f from the ball $\sup_x (1 + |x|)^{|K|+4} |f'''(x)| < R$ with any R. Again using Theorem 4.9.2 we conclude that the distribution of the limit of a converging subsequence of the family $Z_t^{Nh,h}$ solves the martingale problem for Λ_K. □

We shall prove now a uniqueness result for solutions to the martingale problem discussed above, using results from Section 6.5.

First we shall need some assumptions on the measures μ and ν. Let

$$p_0(\xi) = \sum_{\Psi\leq K} p_\Psi(\xi).$$

We shall suppose that there exists $c > 0$ and constants $\alpha_\Psi > 0$, $\beta_\Psi < \alpha_\Psi$ such that for each Ψ

(A1) $|\mathrm{Im}\, p_\Psi^\mu(\xi) + Im\, p_\Psi^\nu(\xi)| \leq c|p_0(\xi)|$,

(A2) $\mathrm{Re}\, p_\Psi^\nu(\xi) \geq c^{-1} |\mathrm{pr}_{\nu_\Psi}(\xi)|^{\alpha_\Psi}$ and $|(p_\Psi^\nu)'(\xi)| \leq c|\mathrm{pr}_{\nu_\Psi}(\xi)|^{\beta_\Psi}$, where pr_{ν_Ψ} is the orthogonal projection on the minimal subspace containing the support of the measure ν_Ψ.

Let us say that a type j of particles is *immortal* if for any solution of the martingale problem for Λ_K, the j-th coordinate of the process X_t^x is positive for all times almost surely whenever the j-th coordinate of x was positive. In other words this means that the boundary $\bar{U}_j = \{x \in \bar{\mathbb{R}}^d : x_j = 0\}$ is inaccessible.

Theorem 6.9.1. (i) *Let the conditions of Proposition* 6.9.3 *together with* (A1), (A2) *be satisfied. If, in addition, all types of particles are immortal, then the martingale problem of Λ_K is well-posed and has sample paths in $D_{\mathbb{R}_+^d}[0,\infty)$; i.e. the boundary is almost surely inaccessible. Hence this solution defines a strong Markov process in $\mathbb{R}_+^d$, which is a limit (in the sense of distributions) of the Markov chains $Z_t^{Nh,h}$, as $h \to 0$ with Nh tending to a constant.*

(ii) *If, in addition to the hypotheses in* (i), $\psi_j \geq 2$ *whenever either* $G_{jj}(\Psi) \neq 0$ *or* $\int (x_j)^2 \nu_\Psi(dx) \neq 0$, *and* $\psi_j \geq 1$ *whenever either* $\beta_j(\Psi) \neq 0$ *or* $\int x_j \mu_\Psi(dx) \neq 0$, *the semigroup of the corresponding Markov process preserves the space of bounded continuous functions on* $\bar{\mathbb{R}}^d_+$ *vanishing on the boundary. If, moreover,* $|K| \leq 2$ *(i.e. only binary interactions are allowed) and for* $|K| = 2$ *the drift term and the integral term depending on* μ_Ψ *vanish, then the corresponding semigroup is Feller, i.e. it preserves the space of continuous functions on* $\mathbb{R}^d$ *that tend to zero when the argument approaches either the boundary or infinity.*

Proof. This is a consequence of a more general Theorem 6.5.1. □

Our second result on uniqueness will be more general. Let us say that a type j of particles is *not revivable* if $\beta_j(\Psi) = 0$ whenever j is not contained in the support of Ψ, and supp μ_Ψ belongs to the subspace spanned by the vectors e_j with $j \in$ supp Ψ, that is, the boundary hyperspace $\bar{U}_j = \{x \in \bar{\mathbb{R}}^d : x_j = 0\}$ is gluing. The meaning of the term revivable is revealed in the following result.

Theorem 6.9.2. *Let the conditions of Proposition* 6.9.3 *together with* (A1), (A2) *be satisfied. Suppose additionally that all particles are either immortal or are not revivable. Then for any* $x \in \mathbb{R}^d_+$ *there exists a unique solution to the martingale problem for* Λ_K *under the additional assumption that, for any* j, *if at some (random) time* τ *the* j*-th coordinate of* X_t *vanishes, then it remains zero for all future times almost surely (i.e. once dead, the particles of type* j *are never revived). Moreover, the family of Markov chains* $Z_t^{Nh,h}$ *converges in distribution to this solution to the martingale problem.*

Proof. The uniqueness follow from more general Theorem 6.6.1. Since the family of processes $Z_t^{Nh,h}$ converges in distributional sense to the martingale solution X_t, Proposition 6.9.3 applies to the effect that X_t inherits the non-revivability property, because each process $Z_t^{Nh,h}$ is non-revivable. □

Let us discuss some examples of k-ary interactions from statistical mechanics and population biology.

1. *Branching processes and finite-dimensional superprocesses.* Branching without interaction in our model corresponds clearly to the cases with $K = 1$ and hence represents the simplest possible example. In this case the limiting processes in $\mathbb{R}^d$ have pseudo-differential generators with symbols $p(x, \xi)$ depending linearly on the position x. The corresponding processes are called (finite-dimensional) superprocesses and are well studied, see e.g. Etheridge [109].

2. *Coagulation-fragmentation and general mass preserving interactions.* These are natural models for the applications of our results in statistical mechanics. For these

models, the function L from Proposition 6.9.3 usually represents the mass of a particle, see Kolokoltsov [188] and [196] for a detailed discussion. Notice only that in the present finite-dimensional situation we always get an inaccessible boundary so that Theorem 6.9.1 applies.

3. *Local interactions (birth and death processes).* Generalizing the notion of local branching widely used in the theory of superprocesses, let us say that the interaction of particles of d types is *local* if a group of particles specified by a profile Ψ can produce particles only of type $j \in \text{supp}\ \Psi$. Processes subject to this restriction include a variety of the so-called birth-and-death processes from the theory of multidimensional population processes (see, e.g. Anderson [18] and references therein), such as competition processes, predator-prey processes, general stochastic epidemics and their natural generalizations (seemingly not much studied yet) that take into account the possibility of birth from groups of not only two (male, female) but also of more large number of species (say, for animals, living in groups containing a male and several females, such as gorillas). Excluded by the assumption of locality are clearly migration processes. In the framework of our general model, the assumption of locality gives the following additional restrictions to the generators (6.31): $\beta_j(\Psi) = 0$ whenever j is not contained in the support of Ψ, and supp μ_Ψ belongs to the subspace spanned by the vectors e_j with $j \in \text{supp}\ \Psi$. This clearly implies that the whole boundary of the corresponding process in $\mathbb{R}^d_+$ is gluing and Theorem 6.9.2 is valid giving uniqueness and convergence.

6.10 Application to evolutionary games

A popular way of modeling the evolution of behavioral patterns in populations is given by the replicator dynamics (see Kolokoltsov and Malafeyev [199] or any other text on evolutionary game theory), which is usually deduced by the following arguments. Suppose a population consists of individuals with d different types of behavior specified by their strategies in a symmetric two-player game given by the matrix A whose elements A_{ij} designate the payoffs to a player with strategy i whenever his opponent plays j. Suppose the number of individuals playing strategy i at time t is $x_i = x_i(t)$, with the whole size of the population being $\mu(x) = \sum_{j=1}^d x_j$. If the payoff represents an individual's fitness measured as the number of offsprings per time unit, the average fitness $A_{ij}x_j/\mu(x)$ of a player with strategy i coincides with the payoff of the pure strategy i playing against the mixed strategy $x/\mu(x) = \{x_1/\mu(x), \dots, x_d/\mu(x)\}$. Assuming additionally that the background fitness and death rates of individuals (independent of outcomes in the game) are given by some constants B and C yields the following dynamics:

$$\dot{x}_i = \left(B - C + \sum_{j=1}^d A_{ij} \frac{x_j}{\mu(x)} \right) x_i, \tag{6.40}$$

called the standard *replicator dynamics* (usually written in terms of the normalized vector $x/\mu(x)$); see Boylan [67] or Kolokoltsov and Malafeyev [199]. Of course, it follows as a trivial consequence of Theorem 6.9.2 that this dynamics describes the law of large numbers limit for the corresponding Markov process of interaction.

Having in mind the recent increase in interest in stochastic versions of replicator dynamics (see Corradi and Sarin [89] and references therein), let us consider now a general model of this kind and analyze the possible stochastic processes that may arise as continuous state (or measure-valued) limits. Denoting by N_j the number of individuals playing strategy j and by $N = \sum_{j=1}^d N_j$ the whole size of the population, assuming that the outcome of a game between players with strategies i and j is a probability distribution $A_{ij} = \{A_{ij}^m\}$ of the number of offsprings $m \geq -1$ of the players ($\sum_{m=-1}^\infty A_{ij}^m = 1$) and the intensity a_{ij} of the reproduction per time unit ($m = -1$ means the death of the individual) yields the Markov chain on $\mathbb{Z}_+^d$ with generator

$$Gf(N) = \sum_{j=1}^d N_j \sum_{m=-1}^\infty \left(B_j^m + \sum_{k=1}^d a_{jk} A_{jk}^m \frac{N_k}{|N|} \right) (f(N + me_j) - f(N)) \quad (6.41)$$

(where B_j^m describe the background reproduction process), which is similar to the generator of binary interaction G_2 of form (6.24), but has an additional multiplier $1/|N|$ on the intensity of binary interaction that implies that in the corresponding scaled version of type (6.26) one has to put a simple common multiplier h instead of $h^{|\Psi|}$. Apart from this modification, the same procedure as for (6.24)–(6.26) applies, leading to the limiting process on $\mathbb{R}_+^d$ with generator of type

$$\Lambda_{\mathrm{EG}} = \sum_{j=1}^d x_j \left(\phi_j + \sum_{k=1}^d \frac{x_k}{\mu(x)} \phi_{jk} \right), \quad (6.42)$$

where all ϕ_j and ϕ_{jk} are the generators of one-dimensional Lévy processes, more precisely

$$\begin{aligned} \phi_{jk} f(x) &= g_{jk} \frac{\partial^2 f}{\partial x_j^2}(x) + \beta_{jk} \frac{\partial f}{\partial x_j}(x) \\ &\quad + \int \left(f(x + ye_j) - f(x) - \mathbf{1}_{y \leq 1}(y) \frac{\partial f}{\partial x_j}(x) y_j \right) \nu_{jk}(dy), \quad (6.43) \\ \phi_j f(x) &= g_j \frac{\partial^2 f}{\partial x_j^2}(x) + \beta_j \frac{\partial f}{\partial x_j}(x) \\ &\quad + \int \left(f(x + ye_j) - f(x) - \mathbf{1}_{y \leq 1}(y) \frac{\partial f}{\partial x_j}(x) y \right) \nu_j(dy), \quad (6.44) \end{aligned}$$

where $\mathbf{1}_M$ denotes as usual the indicator function of the set M and all ν_{jk}, ν_j are Borel measures on $(0, \infty)$ such that the function $\min(y, y^2)$ is integrable with respect to these measures, g_j and g_{jk} are non-negative. Let $\tilde{\nu}_{jk}(dy) = y^2\nu_{jk}(dy)$, $\tilde{\nu}_j(dy) = y^2\nu_j(dy)$ and $\upsilon_{jk} = (hl)^{-2}\tilde{\nu}_{jk}([lh, lh+1))$, $\upsilon_j = (hl)^{-2}\tilde{\nu}_j([lh, lh+1))$. Then the corresponding approximation to (6.42) of type (6.41) can be written after scaling in the form

$$\Lambda_{\mathrm{EG}}^h f(Nh) = h\sum_{j=1}^d N_j\left(\phi_j^h + \sum_{k=1}^d \frac{N_k}{\mu(N)}\phi_{jk}^h\right) f(Nh),$$

where $\mu(N) = \sum_{j=1}^d N_j$ and $\phi_{jk}^h f(Nh)$ equals

$$\begin{aligned}
&\frac{1}{h^2} g_{jk}(f(Nh+he_j) + f(Nh-he_j) - 2f(Nh)) \\
&\quad + \frac{1}{h}|\beta_{jk}|(f(Nh+he_j \ \mathrm{sgn}\,(\beta_{jk})) - f(Nh)) \\
&\quad + \sum_{l\geq h^{-\omega}} [f(Nh+lhe_j) - f(Nh) + l(f(Nh-he_j) - f(Nh))]\upsilon_{jk}(l, h) \\
&\quad + \sum_{l=1}^{\infty}[f(Nh+(1+lh)e_j) - f(Nh)]\nu_{jk}([1+lh, 1+lh+h)),
\end{aligned} \tag{6.45}$$

and similarly ϕ_j^h are defined. The terms in (6.45) that approximate diffusion, drift and integral part of (6.43) have different scaling and have different interpretation in terms of population dynamics. Clearly the first term (approximation for diffusion) stands for a game that can be called "death or birth" game, which describes some sort of fighting for reproduction, whose outcome is that an individual either dies or produces offspring. The second term (approximating drift) describes games for death or for life depending on the sign of β_{jk}. Other terms describe games for a large number of offsprings and are analogues of usual branching but with game-theoretic interaction. The same arguments as given for the proof of Proposition 6.9.3 and Theorem 6.9.2 yield the following result (observe only that no particles are revivable in this model, and no additional assumption of subcriticallity is required, since the coefficients grow at most linearly):

Proposition 6.10.1. *Suppose conditions* (A1), (A2) *hold for all measures* ν_{jk}, ν_j. *Then for any* $x \in \mathbb{R}_+^d$ *there exists a unique solution to the martingale problem for* Λ_{EG} *under the additional assumption that, for any* j, *if at some (random) time* τ *the* j*-th coordinate of* X_t *vanishes, then it remains zero for all future times almost surely. Moreover, the family of Markov chains* $Z_t^{Nh,h}$ *defined by* Λ_{EG}^h *converges in distribution to this solution to the martingale problem as* $h \to 0$ *and* $Nh \to x$.

If the limiting operator is chosen to be deterministic (i.e. the diffusion and non-local term vanish and only a drift term is left), we get the standard replicator dynamics (6.40).

Similarly one obtains the corresponding generalization to the case of non-binary (k-nary) evolutionary games (see again Corradi and Sarin [89] for biological and social science examples of such games), the corresponding limiting generator having the form $\sum_{j=1}^{d} x_j \Phi_j$, where Φ_j are polynomials of the frequencies $y_j = x_j / \sum_{i=1}^{d} x_i$ with coefficients being again generators of one-dimensional Lévy processes.

6.11 Application to finances: barrier options, credit derivatives, etc.

Here we describe briefly the relevance of the boundary-value problems to some basic questions in financial mathematics.

In the classical insurance models of *Cramér–Lundberg* and *Sparre–Andersen*, see e.g. Ramasubramanian [269], the evolution of the capital of an insurance firm, called also *risk*, or *renewal risk process*, or *surplus process*, is given by the formula

$$R_t = u + ct - \sum_{j=1}^{N_t} X_j \tag{6.46}$$

until the time t_R when R_t becomes negative (the *ruin* time). Here $c > 0$ is a *premium rate*, $u > 0$ is the initial capital of the company, and the compound Poisson process $\sum_{j=1}^{N_t} X_j$ with positive i.i.d. random variables X_j denotes the flow of claims of insured customers.

Clearly R_t is a stochastically monotone Markov process on $\mathbb{R}_+$. Under general assumptions on X_j the transmission property does not hold and the process eventually leaves $\mathbb{R}_+$ with some probability $p_R \in (0, 1)$ called the *ruin probability* (in infinite time). One is also interested in the calculation of the *ruin probability* p_R^t in finite time t, which is the probability of exiting $\mathbb{R}_+$ before a given time t. Though from the point of view of general Markov processes, the process (6.46) is very simple, a rich theory was developed to describe its behavior, in particular to calculate the ruin probabilities (see Ramasubramanian [269] or Korolev, Bening and Shorgin [208] and references therein).

In reality, the capital of an insurance firm usually does not follow the simple law (6.46). In particular, this is due to the possibility of investing free (not reclaimed) capital aiming at additional gains. Amazingly enough, the models of *insurance with investment* appeared in the mathematical finance literature quite recently, see Hipp and Plum [130] and references therein, where the evolution of the invested capital is described by a diffusion process. The resulting Markov process is still stochastically monotone. Using other models for the dynamics of this capital leads to the analysis of general Markov process on $\mathbb{R}_+$. For any practical calculations, sensitivity analysis with respect to the initial data and the natural parameters of the model is crucial, for which the results of this and the previous chapters provide a natural starting point. Note that one-dimensional processes represent a reasonable setting for these models.

Let us turn to option pricing. The value of a barrier option based on a stock price dynamics described by a Markov process S_t in $\mathbb{R}_+$ is given by

$$\mathbf{E}_x(e^{-rT} g(S_T)\mathbf{1}_{\tau_A>T}) + \mathbf{E}_x(e^{-r\tau_A} h(S_{\tau_A})\mathbf{1}_{\tau_A\le T}),$$

where $r > 0$ is the risk free rat, T is the maturity time, τ_A is the random time of entering the knock-out set

$$A = [0, l] \cup [u, \infty), \quad 0 \le l < u \le \infty,$$

and g, f are given continuous functions, see detailed discussion and bibliography in Mijatovic and Pistorius [245], so that we again in the setting of Section 6.2 (up to some clear modifications). More practical issues are addressed in the collection edited by Lipton and Rennie [221].

Unlike the previous model from insurance mathematics, here a multi-dimensional extension is very natural, as investors are usually dealing with some baskets of options. Moreover, a similar multi-dimensional setting appear in the analysis of portfolios of CDOs (credit derivative obligations), which can be described by a Markov process in $\mathbb{R}^d_+$. Reaching the boundary of dimension $d - m$ means that k out of d bonds underlying the portfolio of CDOs have defaulted. The boundary in this model is of course gluing (defaulted firms are not revivable, as described in Section 6.6). It is worth stressing the necessity of considering general Markov processes in this setting, as simple Lévy processes are spatially homogeneous and do not feel an approach to the boundary, which cannot be the case in reality (a firm approaching a default should be reflected in the decision making of the governing body of the firm).

Finally, of special interest for finance are the processes living in $\mathbb{R}^m \times \mathbb{R}^d_+$, where the positive part may represent stochastic volatilities. Only quite particular classes of these models, namely the *affine processes*, are studied in detail, see Duffie, Filipovic and Schachermayer [104].

6.12 Comments

The exposition of this chapter is based essentially on Kolokoltsov [187], [186], though many important improvements were introduced. The classification of boundary points that we introduced are more general than in usual texts on diffusions (see e.g. Freidlin [117] and Taira [307], but at the same time more concrete than in general potential theory (see Bliedtner and Hansen [60]). In particular, we generalize the notion of entrance boundary from the theory of one-dimensional diffusions (see e.g. Mandl [228]) to the case of processes with general pseudo-differential generators. For the case of one-dimensional problems special tools are available, see [183].

Some bibliographical comments on the Dirichlet problem for the generators of Markov processes seem to be in order. For degenerate diffusions the essential progress was begun with the papers of Keldysh [157] and Fichera [112]. In particular, in [112] the Fichera function was introduced giving the partition of a smooth boundary into subsets $\Sigma_0, \Sigma_1, \Sigma_3, \Sigma_4$, which in one-dimensional case correspond to natural boundary, entrance boundary, exit boundary and regular boundary respectively studied by Feller (see e.g. Mandl [228] for one-dimensional theory). Hard analytic work was done afterwards on degenerate diffusions (see e.g. Kohn and Nirenberg [168], [169], or Oleinik and Radkevich [261] or more recent development in Taira [307], [308],[306]. The strongest results obtained by analytic methods require very strong assumptions on the boundary, namely that it is smooth and the four basic parts $\Sigma_0, \Sigma_1, \Sigma_3, \Sigma_4$ are disjoint smooth manifolds.

Probability theory suggests very natural notions of generalized solutions to the Dirichlet problem that can be defined and proved to exist in rather general situations (see Stroock and Varadhan [303] for a definition based on the martingale problem approach, Bliedtner and Hansen [60] for the approach based on the general balayage-space technique, Jacob [141] for comparison of different approaches and the generalized Dirichlet space approach). A natural language for the abstract approach (both probabilistic and analytic) is that of a Martin space. Roughly speaking a *Martin space* for a Markov process with a state space S is defined as a completion S^* of S with a certain metric in such a way that all harmonic functions are in one-to-one correspondence with Borel measures on the set $S^* \setminus S$, called the *Martin boundary* of S. This theory is mostly developed for Markov chains, that is, when either the state space is discrete, see e.g. [86], or the time is discrete, see [277], and for diffusions, see e.g. [265] or [251] and references therein.

The interpretation of the abstract results of potential theory in terms of a given concrete integro-differential generator is a non-trivial problem. Usually it is supposed, in particular, that the process can be extended beyond the boundary. For degenerate diffusions, a well developed theory of regularity of solutions can be found e.g. in Freidlin [117] and Taira [307]. But for non-local generators of Feller processes with jumps, the results obtained so far seem to deal only with the situations when the boundary is infinitely smooth and there is a dominating non-degenerate diffusion term in the generator, see e.g. Taira [306], Taira, Favini and Romanelli [309].

From Proposition 6.3.1, one can deduce some criteria for transience and recurrence for processes with pseudo-differential generators, see Kolokoltsov, Schilling and Tyukov [202]. The method of Lyapunov function is also traditionally used for analyzing stability and control, see e.g. [213].

Our discussion of reflected processes in Section 6.8 is only meant to give a very brief introduction. For related development, we refer to Richou [279], Ghosh and Kumar [122], Menaldi and Robin [240], [241], Garroni and Menaldi [119], Bass and Burdzy [29] and references therein.

Chapter 7
Heat kernels for stable-like processes

This chapter deals with the transition probability densities of stable-like processes (in analytic language, with the Green function of the corresponding integro-differential evolutionary equations), i.e. of the processes with generators of stable type, but with coefficients depending on the position of the process, as well as their extensions by pure jumps. More precisely, we shall analyze semigroups generated by operators of the form

$$Lu(x) = -a(x)|\nabla|^{\alpha(x)}u(x)$$

and their time non-homogeneous versions (propagators). Roughly speaking, the main result states that if the coefficient functions α, a are bounded and h-Hölder continuous with an arbitrary $h \in (0, 1]$, then this L generates a conservative Feller semigroup that has a continuous transition density, which for finite times is bounded by, and for small times and near diagonal is close to, the stable density, and is of the same order of smoothness as the coefficients of the operator L.

After a warm-up Section 7.1 describing the well-known asymptotic expansions of one-dimensional stable laws, the asymptotic expansions of finite-dimensional stable laws are studied for small and large distances in Section 7.2. The section ends with curious identities expressing the multidimensional stable laws in terms of certain derivatives of one-dimensional ones.

Next, in Section 7.3, we obtain global estimates for the derivatives of stable densities with respect to basic parameters in terms of the stable densities themselves, using the asymptotics obtained above and the unimodality of stable laws.

Section 7.4 is devoted to some auxiliary estimates on the convolutions of stable laws. In Sections 7.5, 7.6 the main results of this chapter are given, yielding the existence and asymptotics for the Green function of stable-like processes and consequently a construction of the corresponding Feller semigroup.

In Section 7.7, illustrating possible applications of our analytic results, some sample-path properties of stable-like processes are proved, which extend to variable coefficients some standard facts about the stable paths. Finally, Section 7.8 explains how the heat kernel bounds obtained can be used to solve the related problems of optimal control.

7.1 One-dimensional stable laws: asymptotic expansions

This section is devoted to a beautiful classical result on the asymptotic expansions of stable laws (see Feller [111] for historical comments). We give it here anticipating more general expansion of finite-dimensional laws given in Section 7.2.

The characteristic function of the general (up to a shift) one-dimensional stable law with the index of stability $\alpha \in (0, 2), \alpha \neq 1$, is

$$\exp\{-\sigma |y|^\alpha e^{i\frac{\pi}{2}\gamma \operatorname{sgn} y}\} \tag{7.1}$$

(see Section 1.4), where the parameter γ (which measures the skewness of the distribution) satisfies the conditions $|\gamma| \leq \alpha$, if $0 < \alpha < 1$, and $|\gamma| \leq 2 - \alpha$, if $1 < \alpha < 2$. The parameter $\sigma > 0$ is called the scale. For $\alpha = 1$ only in symmetric case, i.e. for $\gamma = 0$, the characteristic function can be written in form (7.1). In order to have unified formulas we exclude the non-symmetric stable laws with the index of stability $\alpha = 1$ from our exposition and will always consider $\gamma = 0$ whenever $\alpha = 1$. The probability density corresponding to characteristic function (7.1) is

$$S(x; \alpha, \gamma, \sigma) = \frac{1}{2\pi} \int_{-\infty}^{+\infty} \exp\{-ixy - \sigma |y|^\alpha e^{i\frac{\pi}{2}\gamma \operatorname{sgn} y}\}\, dy. \tag{7.2}$$

Due to the evident relations

$$S(-x; \alpha, \gamma, \sigma) = S(x; \alpha, -\gamma, \sigma), \tag{7.3}$$

$$S(x; \alpha, \gamma, \sigma) = \sigma^{-1/\alpha} S(x\sigma^{-1/\alpha}; \alpha, \gamma, 1), \tag{7.4}$$

it is enough to investigate the properties of the normalized density $S(x; \alpha, \gamma, 1)$ for positive values of x. Clearly for these x

$$S(x; \alpha, \gamma, 1) = \frac{1}{\pi} \operatorname{Re} \int_0^\infty \exp\{-ixy - y^\alpha e^{i\frac{\pi}{2}\gamma}\}\, dy. \tag{7.5}$$

It follows that all S are infinitely differentiable with bounded derivatives. Using a linear change of the variable in (7.5) yields for $x > 0$

$$S(x; \alpha, \gamma, 1) = \frac{1}{\pi x} \operatorname{Re} \int_0^\infty \exp\left\{-\frac{y^\alpha}{x^\alpha} e^{i\frac{\pi}{2}\gamma}\right\} e^{-iy}\, dy. \tag{7.6}$$

Proposition 7.1.1. *For small $x > 0$ and any $\alpha \in (0, 2)$, the function $S(x; \alpha, \gamma, 1)$ has the following asymptotic expansion:*

$$S(x; \alpha, \gamma, 1) \sim \frac{1}{\pi x} \sum_{k=1}^{\infty} \frac{\Gamma(1 + k/\alpha)}{k!} \sin \frac{k\pi(\gamma - \alpha)}{2\alpha} (-x)^k. \tag{7.7}$$

Moreover, for $\alpha \in (1, 2)$ (resp. for $\alpha = 1$), the series on the r.h.s. of (7.7) is absolutely convergent for all x (resp. for x from a neighborhood of the origin) and its sum is equal to $S(x; \alpha, \gamma, 1)$. The asymptotic expansion can be differentiated infinitely many times.

Proof. Expanding the function e^{-ixy} in (7.5) in power series, yields for $S(x;\alpha,\gamma,1)$ the expression

$$\frac{1}{\pi}\,\mathrm{Re}\int_0^\infty \exp\{-y^\alpha e^{i\pi\gamma/2}\}\left(1-ixy+\cdots+\frac{(-ixy)^k}{k!}+\theta\frac{(xy)^{k+1}}{(k+1)!}\right)dy$$

with $|\theta|\le 1$. Since

$$\int_0^\infty y^{\beta-1}\exp\{-\lambda y^\alpha\}\,dy=\alpha^{-1}\lambda^{-\beta/\alpha}\Gamma(\beta/\alpha),\quad \mathrm{Re}\,\lambda>0 \tag{7.8}$$

(and these integrals are absolutely convergent for $\mathrm{Re}\,\lambda>0$), one obtains

$$S(x;\alpha,\gamma,1)=\frac{1}{\pi\alpha}\,\mathrm{Re}\sum_{m=0}^{k}\exp\left\{-i\frac{\pi\gamma(m+1)}{2\alpha}\right\}\frac{(-ix)^m}{m!}\Gamma\left(\frac{m+1}{\alpha}\right)+R_{k+1}$$

with

$$|R_{k+1}|\le\frac{1}{\pi\alpha}\Gamma\left(\frac{k+2}{\alpha}\right)\frac{|x|^{k+1}}{(k+1)!}.$$

Therefore, we have an asymptotic expansion for S. It is convenient to rewrite this expansion in the form

$$S(x;\alpha,\gamma,1)\sim\frac{1}{\pi x\alpha}\,\mathrm{Re}-\sum_{k=1}^{\infty}(-x)^k\frac{\Gamma(k/\alpha)}{(k-1)!}\exp\left\{-i\frac{\pi}{2}\left(\frac{\gamma}{\alpha}k-k+1\right)\right\}.$$

Using the formula $\Gamma(k/\alpha)=\Gamma(1+k/\alpha)\alpha/k$ and taking the real part yields (7.7). The statement about convergence follows from the asymptotics of the Gamma function,

$$\Gamma(x+1)=\sqrt{2\pi x}\,x^x e^{-x}(1+o(1)),\quad x\to\infty$$

(*Stirling's formula*), which implies that the radius of convergence of series (7.7) is equal to infinity, is finite, or is zero, respectively if $\alpha\in(1,2)$, $\alpha=1$, or $\alpha\in(0,1)$. □

Proposition 7.1.2. *For any $\alpha\in(0,2)$ and $x\to\infty$, the function $S(x;\alpha,\gamma,1)$ has the following asymptotic expansion:*

$$S(x;\alpha,\gamma,1)\sim\frac{1}{\pi x}\sum_{k=1}^{\infty}\frac{\Gamma(1+k\alpha)}{k!}\sin\frac{k\pi(\gamma-\alpha)}{2}(-x^{-\alpha})^k. \tag{7.9}$$

For $\alpha\in(0,1)$ (resp. $\alpha=1$, $\gamma=0$), the series on the r.h.s. of (7.9) is absolutely convergent for all finite $x^{-\alpha}$ (resp. for $x^{-\alpha}$ in a neighbourhood of the origin) and its sum is equal to $S(x;\alpha,\gamma)$. The asymptotic expansion (7.9) can be differentiated infinitely many times.

Moreover, if $x>0$ and either $\alpha\in(\frac{1}{2},1)$ or $\alpha\in(1,2)$, Zolotarev's identity hold:

$$S(x;\alpha,\gamma,1)=x^{-(1+\alpha)}S\left(x^{-\alpha};\frac{1}{\alpha},\frac{1}{\alpha}(\gamma+1)-1,1\right). \tag{7.10}$$

Proof. First let $\alpha \in (0, 1]$. Due to the Cauchy theorem, one can change the path of integration in (7.6) to the negative imaginary axes, i.e.

$$S(x; \alpha, \gamma, 1) = \frac{1}{\pi x} \operatorname{Re} \int_0^{-i\infty} \exp\left\{-\frac{y^\alpha}{x^\alpha} e^{i\frac{\pi}{2}\gamma}\right\} e^{-iy}\, dy \tag{7.11}$$

because the magnitude of the integral along the arch $l = \{y = re^{-i\phi}, \phi \in [0, \frac{\pi}{2}]\}$ does not exceed

$$\int_0^{\pi/2} r \exp\left\{-r \sin\phi - \frac{r^\alpha}{x^\alpha} \cos\left(\alpha\phi - \frac{\pi}{2}\gamma\right)\right\} d\phi,$$

and tends to zero as $r \to \infty$, due to the assumptions on α and γ. Changing now the variable of integration y to $z = ye^{i\pi/2} = iy$ in (7.11) yields

$$S(x; \alpha, \gamma, 1) = \operatorname{Re} -\frac{i}{\pi x} \int_0^\infty \exp\left\{-z - \frac{z^\alpha}{x^\alpha} e^{i\frac{\pi}{2}(\gamma-\alpha)}\right\} dz.$$

By expanding $\exp\{-\frac{z^\alpha}{x^\alpha} e^{i\frac{\pi}{2}(\gamma-\alpha)}\}$ in power series, this rewrites as

$$\operatorname{Re} -\frac{i}{\pi x} \int_0^\infty e^{-z} \sum_{k=0}^\infty \frac{1}{k!} \left(-\frac{z^\alpha}{x^\alpha} e^{i\frac{\pi}{2}(\gamma-\alpha)}\right)^k dz.$$

Evaluating the standard integrals yields

$$S(x; \alpha, \gamma, 1) = \operatorname{Re} -\frac{i}{\pi x} \sum_{k=1}^\infty \frac{\Gamma(1 + k\alpha)}{k!} (-x^{-\alpha})^k \exp\left\{i\frac{k\pi}{2}(\gamma - \alpha)\right\},$$

which implies (7.9). As in the proof of Proposition 7.1.1, one sees from the asymptotic formula for the Gamma function that the radius of convergence of series (7.9) is equal to infinity for $\alpha \in (0, 1)$ and is finite non-vanishing for $\alpha = 1$. Therefore, we have proved (7.9) for $\alpha \in (0, 1]$. Comparing formulas (7.9) for $\alpha \in (1/2, 1)$ and (7.7) for $\alpha \in (1, 2)$ one gets Zolotarev's identity (7.10). Using this identity and asymptotic expansion (7.7) for $\alpha \in (\frac{1}{2}, 1)$ one obtains asymptotic formula (7.9) for $\alpha \in (1, 2)$. □

For the analysis of processes with a variable index α, the following more general integrals appear (that we consider only for the case of vanishing γ):

$$\phi_b(x, \alpha) = \frac{1}{\pi} \operatorname{Re} \int_0^\infty y^b e^{-ixy - y^\alpha}\, dy.$$

This function can be analyzed like the function S above. For instance, the following holds

$$\phi_b(x, \alpha) = \frac{1}{\pi\alpha} \sum_{m=0}^\infty (-1)^m \frac{x^{2m}}{(2m)!} \Gamma\left(\frac{2m + 1 + b}{\alpha}\right) + R_{m+1}, \tag{7.12}$$

where

$$R_{m+1} \leq \Gamma\left(\frac{2m+3+b}{\alpha}\right)\frac{x^{2m+2}}{(2m+2)!}.$$

Moreover, the corresponding power series converges for $\alpha \in (1,2)$ and yields an asymptotic expansion for all $\alpha \in (0,2)$.

7.2 Stable laws: asymptotic expansions and identities

This section is devoted to the large- and small-distance asymptotics of finite-dimensional stable laws that generalize the classical one-dimensional asymptotic expansions given in Section 7.1. For simplicity, we reduce our attention to uniform spectral measures, in other words, to the *stable densities* in $\mathbb{R}^d$ of the form

$$S(x;\alpha,\sigma) = \frac{1}{(2\pi)^d}\int_{\mathbb{R}^d} \exp\{-\sigma|p|^\alpha\}e^{-ipx}\,dp. \tag{7.13}$$

Remark 50. Many results given below have a natural extension to non-uniform spectral measures (with $\int_{S^{d-1}} |(s,p)|^\alpha \mu(ds)$ instead of $\sigma|p|^\alpha$), see [178].

As in the one-dimensional cases there is a straightforward reduction to the case of unit σ, i.e. the obvious change of the variable of integration yields

$$S(x;\alpha,\sigma) = \sigma^{-d/\alpha}S(x\sigma^{-1/\alpha},\alpha,1). \tag{7.14}$$

In particular, this implies the simple global estimate

$$S(x;\alpha,\sigma) \leq c\sigma^{-d/\alpha}, \tag{7.15}$$

which holds with a constant c that can be chosen uniformly for all $\sigma > 0$, $x \in \mathbb{R}^d$ and α from any compact subinterval of $(0,2)$.

We shall denote by $\bar{p}$ the unit vector in the direction of p, i.e. $\bar{p} = p/|p|$. Using for $\bar{p}$ spherical coordinates (θ,ϕ), $\theta \in [0,\pi]$, $\phi \in S^{d-2}$ with the main axis directed along x, so that

$$dp = |p|^{d-1}d|p|\sin\theta^{d-2}d\theta d\phi$$

(θ being the angle between x and p), integrating over ϕ, and then changing the variable θ to $t = \cos\theta$ yields

$$S(x;\alpha,\sigma) = \frac{|S^{d-2}|}{(2\pi)^d}\int_0^\infty d|p|\int_{-1}^1 dt\exp\{-\sigma|p|^\alpha\}\cos(|p||x|t)|p|^{d-1}(1-t^2)^{(d-3)/2}, \tag{7.16}$$

where $|S^0| = 2$ and

$$|S^{d-2}| = 2\frac{\pi^{(d-1)/2}}{\Gamma(\frac{d-1}{2})} \tag{7.17}$$

for $d > 2$ is the area of the unit sphere S^{d-2} in $\mathbb{R}^{d-1}$. Changing the variable of integration $|p|$ to $y = |p||x|$ implies

$$S(x;\alpha,\sigma) = \frac{|S^{d-2}|}{(2\pi|x|)^d}\int_0^\infty dy \int_{-1}^1 dt \exp\left\{-\sigma\frac{y^\alpha}{|x|^\alpha}\right\}\cos(yt)y^{d-1}(1-t^2)^{(d-3)/2}. \tag{7.18}$$

Theorem 7.2.1. *For small $|x|/\sigma^{1/\alpha}$ the density S has the asymptotic expansion*

$$S(x;\alpha,\sigma) \sim \frac{|S^{d-2}|}{(2\pi\sigma^{1/\alpha})^d}\sum_{k=0}^\infty \frac{(-1)^k}{(2k)!}a_k\left(\frac{|x|}{\sigma^{1/\alpha}}\right)^{2k} \tag{7.19}$$

with

$$a_k = \alpha^{-1}\Gamma\left(\frac{2k+d}{\alpha}\right)B\left(k+\frac{1}{2},\frac{d-1}{2}\right), \tag{7.20}$$

where

$$B(q,p) = \int_0^1 x^{p-1}(1-x)^{q-1}\,dx = \frac{\Gamma(p)\Gamma(q)}{\Gamma(p+q)}$$

is the standard Beta function. The modulus of each term in expansion (7.19) *serves also as an estimate for the remainder in this asymptotic representation, i.e. for each m, $S(x;\alpha,\sigma)$ equals*

$$\frac{|S^{d-2}|}{(2\pi\sigma^{1/\alpha})^d}\left(\sum_{k=0}^m \frac{(-1)^k}{(2k)!}a_k\left(\frac{|x|}{\sigma^{1/\alpha}}\right)^{2k} + \theta\frac{a_{m+1}}{(2m+1)!}\left(\frac{|x|}{\sigma^{1/\alpha}}\right)^{2m+1}\right) \tag{7.21}$$

with $|\theta| \le 1$. Finally, if $\alpha > 1$ (resp. $\alpha = 1$), the series on the r.h.s. of (7.19) *is absolutely convergent for all $|x|$ (resp. in a neighbourhood of the origin) and its sum equals $S(x;\alpha,\sigma)$.*

Proof. This is the same as the case $d = 1$ discussed in Section 7.1. Expanding the function $\cos(|p||x|)$ in (7.16) in power series yields

$$S(x;\alpha,\sigma) = \frac{|S^{d-2}|}{(2\pi)^d}\int_0^\infty d|p|\int_{-1}^1 dt \exp\{-\sigma|p|^\alpha\}|p|^{d-1}(1-t^2)^{(d-3)/2}$$
$$\times\left(\sum_{m=0}^k (-1)^m\frac{(|p||x|t)^{2m}}{(2m)!} + \theta\frac{(|p||x|t)^{2m+2}}{(2m+2)!}\right)$$

with $|\theta| \le 1$. Due to (7.8) and the equation

$$\int_{-1}^1 t^{2m}(1-t^2)^{(d-3)/2}\,dt = B\left(m+\frac{1}{2},\frac{d-1}{2}\right),$$

we can integrate in $|p|$ and t to obtain for $S(x;\alpha,\sigma)$ the expression (7.21). The statement about the convergence of the series for $\alpha \ge 1$ follows from the Stirling formula for the Gamma function. □

Next, we construct the asymptotic expansion of S in a more involved case, namely for large distances. To this end, let us recall first some facts on the Bessel and Whittaker functions (see e.g. Whittaker and Watson [318] for details). For any complex z that is not a negative real, and any real $n > 1/2$ the *Bessel function* $J_n(z)$ and the *Whittaker function* $W_{0,n}(z)$ can be defined by the integral formulae

$$J_n(z) = \frac{(z/2)^n}{\Gamma(n+1/2)\sqrt{\pi}} \int_{-1}^{1} (1-t^2)^{n-1/2} \cos(zt)\, dt,$$

$$W_{0,n}(z) = \frac{e^{-z/2}}{\Gamma(n+1/2)} \int_0^\infty [t(1+t/z)]^{n-1/2} e^{-t}\, dt,$$

where the principle value of $\arg z$ is chosen: $|\arg z| < \pi$. These functions are connected by the formula

$$J_n(z) = \frac{1}{\sqrt{2\pi z}} \Big[\exp\Big\{\frac{1}{2}\Big(n+\frac{1}{2}\Big)\pi i\Big\} W_{0,n}(2iz) + \exp\Big\{-\frac{1}{2}\Big(n+\frac{1}{2}\Big)\pi i\Big\} W_{0,n}(-2iz)\Big],$$

which for real positive z implies

$$J_n(z) = 2\,\mathrm{Re}\Big[\frac{1}{\sqrt{2\pi z}} \exp\Big\{\frac{1}{2}\Big(n+\frac{1}{2}\Big)\pi i\Big\} W_{0,n}(2iz)\Big]. \tag{7.22}$$

If $n = m + 1/2$ with nonnegative integer m, $W_{0,n}$ can be expressed in the closed form

$$W_{0,n}(z) = e^{-z/2}\Big(1 + \frac{n^2-(1/2)^2}{z} + \frac{(n^2-(1/2)^2)(n^2-(3/2)^2)}{2z^2} + \cdots + \frac{(n^2-(1/2)^2)\cdots(n^2-(m-1/2)^2)}{m!z^m}\Big). \tag{7.23}$$

In particular, $W_{0,1/2}(z) = e^{-z/2}$. More generally, for any $n > 1/2$ one has the following asymptotic expansion as $z \to \infty$, $|\arg z| \le \pi - \epsilon$ with some $\epsilon > 0$:

$$W_{0,n}(z) \sim e^{-z/2}\Big[1 + \frac{n^2-(1/2)^2}{z} + \frac{(n^2-(1/2)^2)(n^2-(3/2)^2)}{2z^2} + \cdots\Big]. \tag{7.24}$$

The following is the main result of this section.

Theorem 7.2.2. *For $|x|/\sigma^{1/\alpha} \to \infty$, the asymptotic expansion*

$$S(x;\alpha,\sigma) \sim (2\pi)^{-(d+1)/2} \frac{2}{|x|^d} \sum_{k=1}^{\infty} \frac{a_k}{k!} (\sigma|x|^{-\alpha})^k \tag{7.25}$$

holds, where

$$a_k = (-1)^{k+1} \sin \frac{k\pi\alpha}{2} \int_0^\infty \xi^{\alpha k+(d-1)/2} W_{0,\frac{d}{2}-1}(2\xi)\, d\xi. \tag{7.26}$$

In particular, a_1 is positive for all d, and for odd dimensions $d = 2m+3$, $m \geq 0$,

$$\begin{aligned} a_k = (-1)^{k+1} \sin \frac{k\pi\alpha}{2} \Bigg[&\Gamma(m+2+\alpha k) \\ &+ \frac{1}{2}\Gamma(m+1+\alpha k)\left(\left(m+\frac{1}{2}\right)^2 - \left(\frac{1}{2}\right)^2\right) \\ &+ \frac{1}{2^2 2!}\Gamma(m+\alpha k)\left(\left(m+\frac{1}{2}\right)^2 - \left(\frac{1}{2}\right)^2\right)\left(\left(m+\frac{1}{2}\right)^2 - \left(\frac{3}{2}\right)^2\right) + \cdots \\ &+ \frac{1}{2^m m!}\Gamma(2+\alpha k)\left(\left(m+\frac{1}{2}\right)^2 - \left(\frac{1}{2}\right)^2\right)\cdots\left(\left(m+\frac{1}{2}\right)^2 - \left(m-\frac{1}{2}\right)^2\right)\Bigg]. \end{aligned} \tag{7.27}$$

Moreover, for $\alpha \in (0,1)$ (resp. $\alpha = 1$) this series is convergent for all $|x|^{-1}$ (resp. in a neighborhood of the origin) and its sum equals $S(x;\alpha,\sigma)$. Furthermore, as in the case of the expansion of Theorem 7.2.1, each term in (7.18) serves also as an estimate for the remainder, in the sense that the difference between S and the sum of the $(k-1)$ terms of the expansion does not exceed in magnitude the magnitude of the kth term.

Proof. Due to (7.18) and the definition of the Bessel functions,

$$\begin{aligned} S(x;\alpha,\sigma) = \frac{|S^{d-2}|}{(2\pi|x|)^d} \int_0^\infty 2^{(d-2)/2}\Gamma\left(\frac{d-1}{2}\right)\sqrt{\pi} J_{\frac{d}{2}-1}(y) y^{d/2} \\ \times \exp\left\{-\sigma \frac{y^\alpha}{|x|^\alpha}\right\} dy. \end{aligned} \tag{7.28}$$

The key idea of the proof is to use (7.22) and to rewrite the last expression in the form

$$\begin{aligned} S(x;\alpha,\sigma) = \frac{|S^{d-2}|}{(2\pi|x|)^d} \operatorname{Re} \int_0^\infty \Gamma\left(\frac{d-1}{2}\right) W_{0,\frac{d}{2}-1}(2iy)(2ye^{i\pi/2})^{(d-1)/2} \\ \times \exp\left\{-\sigma \frac{y^\alpha}{|x|^\alpha}\right\} dy. \end{aligned} \tag{7.29}$$

First let $\alpha \in (0,1]$. By (7.24) we can justify the change of the variable of the path of integration in (7.29) to the negative imaginary half-line. Taking this new path of integration and then changing the variable of integration $y = -i\xi$ yields

$$\begin{aligned} S(x;\alpha,\sigma) = \frac{|S^{d-2}|}{(2\pi|x|)^d} \operatorname{Re} -i \int_0^\infty \Gamma\left(\frac{d-1}{2}\right) W_{0,\frac{d}{2}-1}(2\xi)(2\xi)^{(d-1)/2} \\ \times \exp\left\{-\sigma \frac{\xi^\alpha}{|x|^\alpha} e^{-i\alpha\pi/2}\right\} d\xi. \end{aligned} \tag{7.30}$$

Expanding the exponent under this integral in power series yields

$$S(x;\alpha,\sigma) = \frac{|S^{d-2}|}{(2\pi|x|)^d} \operatorname{Re} -i \int_0^\infty \Gamma\left(\frac{d-1}{2}\right) W_{0,\frac{d}{2}-1}(2\xi)(2\xi)^{(d-1)/2}$$
$$\times \sum_{k=0}^{\infty} (-1)^k \left(\frac{\sigma}{|x|^\alpha}\right)^k \xi^{\alpha k} \frac{1}{k!} \left(\cos\frac{k\pi\alpha}{2} - i \sin\frac{k\pi\alpha}{2}\right) d\xi. \quad (7.31)$$

Taking the real part yields

$$S(x;\alpha,\sigma) \sim \frac{|S^{d-2}|}{(2\pi|x|)^d} \Gamma\left(\frac{d-1}{2}\right) 2^{(d-1)/2} \sum_{k=1}^{\infty} \frac{a_k}{k!} (\sigma|x|^{-\alpha})^k, \quad (7.32)$$

with a_k given by (7.26). By (7.17) it rewrites as (7.25). Estimating coefficients (7.26) using the asymptotic formula (7.24) and the fact that $z^{m-1/2}W_{0,m}(z)$ is continuous for $z \geq 0$ (which follows from the definition of $W_{0,n}$ given above) one gets the convergence of series (7.25) and the estimate $a_1 > 0$. In the case of odd dimension one calculates coefficients (7.26) explicitly using (7.23).

Let $\alpha \in (1,2)$. In this case one cannot rotate the contour of integration in (7.29) through the whole angle $\pi/2$, but one can rotate it through the angle $\pi/(2\alpha)$. This amounts to the possibility of making the change of the variable in (7.29) $y = ze^{-i\pi/2\alpha}$ and then considering z to be again real, which gives

$$S(x;\alpha,\sigma) = \frac{|S^{d-2}|}{(2\pi|x|)^d} \operatorname{Re}\left[\int_0^\infty \Gamma\left(\frac{d-1}{2}\right) W_{0,\frac{d}{2}-1}\left(2z \exp\left\{\frac{i\pi(\alpha-1)}{2\alpha}\right\}\right)\right.$$
$$\left.\times (2z)^{(d-1)/2} \exp\left\{i\sigma\frac{z^\alpha}{|x|^\alpha} + \frac{i}{4}(d-1)\pi - i\frac{\pi(d+1)}{4\alpha}\right\} dz\right].$$

Using the Taylor formula for $\exp\{i\sigma\frac{z^\alpha}{|x|^\alpha}\}$ yields

$$S(x;\alpha,\sigma) = \frac{|S^{d-2}|}{(2\pi|x|)^d} \operatorname{Re}\exp\left\{\frac{\pi i}{4\alpha}(\alpha(d-1)-(d+1))\right\}\Gamma\left(\frac{d-1}{2}\right)$$
$$\times \int_0^\infty \left[1 + \sum_{k=1}^{m} \frac{(i\sigma z^\alpha)^k}{x^{\alpha k} k!} + \frac{\theta}{(m+1)!}\frac{(\sigma z^\alpha)^{m+1}}{x^{\alpha(m+1)}}\right]$$
$$\times (2z)^{(d-1)/2} W_{0,\frac{d}{2}-1}\left[2z \exp\left\{\frac{i\pi(\alpha-1)}{2\alpha}\right\}\right] dz$$

with $|\theta| \leq 1$. It implies the asymptotic expansion (7.25) with

$$a_k = |S^{d-2}| \operatorname{Re}\exp\left\{\frac{\pi i}{4\alpha}(\alpha(d-1)-(d+1))\right\}\Gamma\left(\frac{d-1}{2}\right) 2^{(d-1)/2} i^k$$
$$\times \int_0^\infty z^{\alpha k+(d-1)/2} W_{0,\frac{d}{2}-1}\left(2z\exp\left\{\frac{i\pi(\alpha-1)}{2\alpha}\right\}\right) dz.$$

To simplify this expression, one makes here a new rotation of the path of integration, which amounts to changing the variable of integration z to $\xi = 2z\exp\{\frac{i\pi(\alpha-1)}{2\alpha}\}$ and then considering ξ to be real. After simple manipulations one obtains the same formula (7.26) as for the case $\alpha \in (0,1)$. □

As a corollary of the formulas of this section, let us deduce some curious *identities for odd-dimensional stable densities*, allowing to express them in terms of one-dimensional densities.

Theorem 7.2.3. *Let*

$$S_d(|x|;\alpha) = \frac{1}{(2\pi)^d}\int_{\mathbb{R}^d} \exp\{-i(p,x) - |p|^{-\alpha}\}\,dp$$

be the density of the d-dimensional stable law of the index α with the uniform spectral measure, which obviously depends on $|x|$ only. Let $S_d^{(k)}(|x|;\alpha)$ denote its k-th derivative with respect to $|x|$. Then

$$S_3(|x|;\alpha) = \frac{-1}{2\pi|x|}S_1'(|x|;\alpha),$$

$$S_5(|x|;\alpha) = \frac{A_3}{8\pi^4|x|^2}\left(S_1''(|x|;\alpha) - \frac{1}{2|x|}S_1'(|x|;\alpha)\right),$$

and in general for each positive integer m

$$\begin{aligned} S_{2m+3} = -\frac{\Gamma(m+1)A_{2m+1}}{(2\pi)^{2m+2}}\Bigg[&\frac{(-2)^m}{|x|^{m+1}}S_1^{(m+1)} \\ &+ \frac{(-2)^{(m-1)}}{|x|^{m+2}}\left(\left(m+\frac{1}{2}\right)^2 - \left(\frac{1}{2}\right)^2\right)S_1^{(m)} \\ &+ \frac{(-2)^{(m-2)}}{2!|x|^{m+3}}\left(\left(m+\frac{1}{2}\right)^2 - \left(\frac{1}{2}\right)^2\right)\left(\left(m+\frac{1}{2}\right)^2 - \left(\frac{3}{2}\right)^2\right)S_1^{(m-1)} + \cdots \\ &+ \frac{1}{m!|x|^{2m+1}}\left(\left(m+\frac{1}{2}\right)^2 - \left(\frac{1}{2}\right)^2\right)\cdots\left(\left(m+\frac{1}{2}\right)^2 - \left(m-\frac{1}{2}\right)^2\right)S_1'\Bigg]. \end{aligned}$$

Proof. By (7.27) one can represent S_{2m+3} as the sum of $m+1$ series. Each of these series is obtained by differentiating series (7.10) with $\gamma = 0$ respectively $1, 2, \ldots, m+1$ times (up to a multiplier). □

7.3 Stable laws: bounds

In this section we use the asymptotics obtained in the previous section to obtain effective global bounds for stable densities and their derivatives with respect to the parameters of stable laws. These bounds are of independent interest. Moreover, they form the basis for the analysis of the case of variable coefficients, dealt with below.

Theorems 7.2.1, 7.2.2 describe the asymptotic behavior of the densities S for small and large $|x|/\sigma^{1/\alpha}$. The next result supplies an important two-sided estimate for the intermediate regime.

Proposition 7.3.1. *For any $K > 1$ there exists $c = c(K) > 1$ such that $c^{-1}|x|^{-d} \le S(x;\alpha,\sigma) \le c|x|^{-d}$ whenever $K^{-1} \le |x|/\sigma^{1/\alpha} \le K$, uniformly for α from any compact subinterval of the interval* $(0,2)$.

Proof. Due to the small-distance and large-distance asymptotics and the property of unimodality of symmetric stable laws (see Section 1.4), it follows that the stable densities are always (strictly) positive. On the other hand, it follows from (7.18) that $|x|^d S(x,\alpha,\sigma)$ is a continuous function of $|x|/\sigma^{1/\alpha}$ and $\alpha \in (0,2)$. Since on any compact set it achieves its minimal and maximal values, which are both positive, the statement of the proposition readily follows. □

The next result supplies the basic two-sided estimate for stable laws.

Theorem 7.3.1. *For any $K > 0$ there exists a constant $c = c(K)$ such that*

$$\begin{aligned}
\frac{1}{c}\sigma^{-d/\alpha} &< S(x;\alpha,\sigma) \le c\sigma^{-d/\alpha} \quad \text{if } \frac{|x|}{\sigma^{1/\alpha}} \le K \\
\frac{\sigma}{c|x|^{\alpha+d}} &< S(x;\alpha,\sigma) \le \frac{c\sigma}{|x|^{\alpha+d}} \quad \text{if } \frac{|x|}{\sigma^{1/\alpha}} \ge K
\end{aligned} \tag{7.33}$$

The constant $c(K)$ can be chosen uniformly for α from any compact interval in $(0,2)$.

Proof. This follows by combining the bounds of Proposition 7.3.1 and the asymptotic expansions (only the first terms of these expansions are needed) of Theorems 7.2.1 and 7.2.2. □

For the analysis of stable-like processes one needs also estimates for the derivatives of stable densities with respect to their parameters.

Theorem 7.3.2. *There exists a constant C such that the following global estimates hold uniformly for α from any compact subinterval of the interval* $(0,2)$*:*

$$\left|\frac{\partial S}{\partial \sigma}(x;\alpha,\sigma)\right| \le \frac{C}{\sigma} S(x;\alpha,\sigma), \tag{7.34}$$

$$\left|\frac{\partial^l S}{\partial x^l}(x;\alpha,\sigma)\right| \le C\min(\sigma^{-l/\alpha}, |x|^{-l}) S(x;\alpha,\sigma), \quad l = 1,2,\dots \tag{7.35}$$

and

$$\left|\frac{\partial S}{\partial \alpha}(x;\alpha,\sigma)\right| \le C(1+|\log\sigma|+|\log|x||)S(x;\alpha,\sigma). \qquad (7.36)$$

Moreover, one has the asymptotic relation

$$\frac{\partial S}{\partial \sigma}(x;\alpha,\sigma) = \frac{1}{\sigma}S(x;\alpha,\sigma)(1+O(\sigma|x|^{-\alpha})). \qquad (7.37)$$

Proof. Clearly $\frac{\partial S}{\partial \sigma}$ has the expression of the r.h.s. of (7.16) but with additional multiplier $-|p|^\alpha$, leading to the expansion (7.19), (7.20), but with $\Gamma(\frac{2k+d+\alpha}{\alpha})/\sigma$ instead of $\Gamma(\frac{2k+d}{\alpha})$ in the coefficients, which gives (7.34) for small $|x|/\sigma^{1/\alpha}$. On the other hand, $\frac{\partial S}{\partial \sigma}$ has the expression (7.28), (7.29), but with additional multiplier $-y^\alpha/|x|^\alpha$, and hence (7.30) with the multiplier $-\xi^\alpha \exp\{-i\pi\alpha/2\}/|x|^\alpha$. Hence instead of (7.31) one obtains

$$\frac{\partial S}{\partial \sigma}(x;\alpha,\sigma) = \frac{|S^{d-2}|}{(2\pi|x|)^d}\,\mathrm{Re}\, i\int_0^\infty \Gamma\left(\frac{d-1}{2}\right)W_{0,\frac{d}{2}-1}(2\xi)(2\xi)^{(d-1)/2}$$
$$\times \frac{\xi^\alpha}{|x|^\alpha}\sum_{k=0}^\infty (-1)^k\left(\frac{\sigma}{|x|^\alpha}\right)^k \xi^{\alpha k}\frac{1}{k!}\left(\cos\frac{(k+1)\pi\alpha}{2} - i\sin\frac{(k+1)\pi\alpha}{2}\right)d\xi.$$

Unlike (7.31) the term with $k=0$ does not vanish here, but equals

$$\frac{1}{\sigma}\frac{|S^{d-2}|}{(2\pi|x|)^d}\Gamma\left(\frac{d-1}{2}\right)\frac{\sigma}{|x|^\alpha}\int_0^\infty W_{0,\frac{d}{2}-1}(2\xi)(2\xi)^{(d-1)/2}\xi^\alpha \sin\frac{\pi\alpha}{2}\,d\xi,$$

which coincides up to a multiplier σ^{-1} with the main term with $k=1$ of (7.25). This implies (7.37) and (7.34) for large $|x|/\sigma^{1/\alpha}$. Consequently (7.34) is obtained as in Proposition 7.3.1, because, by (7.18), $\sigma|x|^d\frac{\partial S}{\partial \sigma}$ is a continuous function of $|x|/\sigma^{1/\alpha}$.

Next let us analyze the first derivative with respect to x (higher derivatives are considered analogously). As we noted, our asymptotic expansions for S can be differentiated with respect to x. Consequently, for small $|x|/\sigma^{1/\alpha}$

$$\frac{\partial S}{\partial x}(x;\alpha,\sigma) \sim \frac{|S^{d-2}|}{(2\pi\sigma^{1/\alpha})^d}\sum_{k=1}^\infty \frac{(-1)^k}{(2k)!}a_k(2k)\frac{|x|^{2k-2}}{\sigma^{2k/\alpha}}x.$$

The first term here is bounded in magnitude by $|x|/\sigma^{-2/\alpha}$ (up to a constant), leading to the estimate

$$\left|\frac{\partial S}{\partial x}(x;\alpha,\sigma)\right| \le C\sigma^{-1/\alpha}S(x;\alpha,\sigma)$$

for small $|x|/\sigma^{1/\alpha}$, so that

$$\left|\frac{\partial S}{\partial x}(x;\alpha,\sigma)\right| \le C\sigma^{-1/\alpha}S(x;\alpha,\sigma) \le C\epsilon|x|^{-1}S(x;\alpha,\sigma), \quad \frac{|x|}{\sigma^{1/\alpha}} < \epsilon, \qquad (7.38)$$

for small enough ϵ. For large $|x|/\sigma^{1/\alpha}$ one obtains the expansion

$$\frac{\partial S}{\partial x}(x;\alpha,\sigma) \sim \frac{|S^{d-2}|}{(2\pi|x|)^d}\Gamma\left(\frac{d-1}{2}\right)\sum_{k=1}^{\infty}\frac{a_k}{k!}\sigma^k(\alpha k)|x|^{-\alpha k-2}x,$$

leading to the estimate

$$\left|\frac{\partial S}{\partial x}(x;\alpha,\sigma)\right| \leq C|x|^{-1}S(x;\alpha,\sigma)$$

for large $|x|/\sigma^{1/\alpha}$. Moreover, as

$$\frac{\partial S}{\partial x}(x;\alpha,\sigma)|x|^d|x|$$

equals a continuous function of $|x|/\sigma^{1/\alpha}$ multiplied by the bounded function $x/|x|$, it follows that for $|x|/\sigma^{1/\alpha}$ from any bounded interval $|\frac{\partial S}{\partial x}(x)|$ is bounded by a multiple of S, implying that the previous bound holds not only for large $|x|/\sigma^{1/\alpha}$, but actually for all $|x|/\sigma^{1/\alpha}$ bounded from below, i.e.

$$\left|\frac{\partial S}{\partial x}(x;\alpha,\sigma)\right| \leq C|x|^{-1}S(x;\alpha,\sigma) \leq C\epsilon^{-1}\sigma^{-1/\alpha}S(x;\alpha,\sigma), \quad \frac{|x|}{\sigma^{1/\alpha}} > \epsilon. \quad (7.39)$$

Clearly (7.35) follows from (7.38) and (7.39).

It remains to estimate the derivative of S with respect to α. It has the form of the r.h.s. of (7.16) with the additional multiplier $-\sigma|p|^\alpha \log|p|$. Taking into account that

$$\int_0^\infty x^{\beta-1}\log x\exp\{-\sigma x^\alpha\}\,dx = \alpha^{-2}\sigma^{-\beta/\alpha}[\Gamma'(\beta/\alpha) - \Gamma(\beta/\alpha)\log\sigma]$$

(which follows from $\int_0^\infty x^{\beta-1}\log x e^{-x}\,dx = \Gamma'(\beta)$), one concludes that

$$\left|\frac{\partial S}{\partial \alpha}(x;\alpha,\sigma)\right| \leq C(1+|\log\sigma|)S(x;\alpha,\sigma)$$

for small $|x|/\sigma^{1/\alpha}$. For large $|x|/\sigma^{1/\alpha}$, this derivative is given by the r.h.s. of (7.16) with the additional multiplier

$$-\sigma\frac{y^\alpha}{|x|^\alpha}\log\frac{y}{|x|},$$

and hence by the r.h.s. of (7.31) with the multiplier

$$-\sigma\frac{y^\alpha}{|x|^\alpha}e^{-i\pi\alpha/2}(\log\xi - i\frac{\pi}{2} - \log|x|),$$

implying the estimate

$$\left|\frac{\partial S}{\partial \alpha}(x;\alpha,\sigma)\right| \le C(1+|\log|x||)S(x;\alpha,\sigma)$$

for large $|x|/\sigma^{1/\alpha}$. As above the latter estimate can be expanded to all $|x|/\sigma^{1/\alpha}$ bounded from below, which finally implies (7.36). □

By the same argument as in Theorem 7.3.2, we can get the following slight extension of its results.

Proposition 7.3.2. *Let*

$$\phi(x,\alpha,\beta,\sigma) = \frac{1}{(2\pi)^d}\int_{\mathbb{R}^d} |p|^\beta \exp\{-i(p,x) - \sigma|p|^{-\alpha}\}\,dp,$$

so that

$$\frac{\partial \phi}{\partial \beta}(x,\alpha,\beta,\sigma) = \frac{1}{(2\pi)^d}\int_{\mathbb{R}^d} |p|^\beta \log|p| \exp\{-i(p,x) - \sigma|p|^{-\alpha}\}\,dp.$$

Then

$$|\phi(x,\alpha,\beta,\sigma)| \le \begin{cases} c\sigma^{-\beta/\alpha} S(x,\alpha,\sigma), & \frac{|x|}{\sigma^{1/\alpha}} \le K, \\ c\sigma^{-1}|x|^{\alpha-\beta} S(x,\alpha,\sigma), & \frac{|x|}{\sigma^{1/\alpha}} > K, \end{cases} \tag{7.40}$$

and

$$\left|\frac{\partial \phi}{\partial \beta}(x,\alpha,\beta,\sigma)\right| \le \begin{cases} c(1+|\log\sigma|)\sigma^{-\beta/\alpha} S(x,\alpha,\sigma), & \frac{|x|}{\sigma^{1/\alpha}} \le K, \\ c(1+|\log\sigma|+|\log|x||)\sigma^{-1}|x|^{\alpha-\beta} S(x,\alpha,\sigma), & \frac{|x|}{\sigma^{1/\alpha}} > K. \end{cases} \tag{7.41}$$

7.4 Stable laws: auxiliary convolution estimates

In this section we obtain auxiliary estimates for the convolution of stable laws that are needed for the proof of the main result of the next section.

Lemma 7.4.1. (i) *If α, a are measurable functions on $\mathbb{R}^d$ ranging in compact subsets of $(0,2)$ and $(0,\infty)$ respectively, then uniformly for all x and bounded t, one has*

$$\int S(x-\eta,\alpha(\eta),a(\eta)t)\,d\eta = O(1)t^{-d(\alpha_u-\alpha_d)/(\alpha_u\alpha_d)}. \tag{7.42}$$

(ii) *If additionally the function $\alpha(\eta)$ is Hölder continuous, i.e. $|\alpha(x)-\alpha(y)| = O(|x-y|^\beta)$ with a $\beta \in (0,1]$, then for all x and bounded t*

$$\int S(x-\eta,\alpha(\eta),a(\eta)t)\,d\eta = O(1). \tag{7.43}$$

Proof. Let $|x-\eta| \geq 1$. Then $|x-\eta| \geq ct^{1/\alpha(\eta)}$ with a constant c depending on the bound for t. Consequently, by (7.33),

$$S(x-\eta, \alpha(\eta), a(\eta)t) = \frac{O(t)}{|x-\eta|^{d+\alpha(\eta)}} = \frac{O(t)}{|x-\eta|^{d+\alpha_d}}.$$

Since this function is integrable outside any neighbourhood of the origin, it follows that the integral from the l.h.s. of (7.42) over the set $|x-\eta| \geq 1$ is bounded. Next, let $t^{1/\alpha_u} \leq |x-\eta| \leq 1$. Again $|x-\eta| \geq t^{1/\alpha(\eta)}$ in this set, and therefore

$$S(x-\eta, \alpha(\eta), a(\eta)t) = \frac{O(t)}{|x-\eta|^{d+\alpha(\eta)}} = \frac{O(t)}{|x-\eta|^{d+\alpha_u}}.$$

Consequently, the integral over this set can be estimated in magnitude by the expression

$$O(t)\int_{t^{1/\alpha_u}}^{\infty} \frac{dy}{y^{1+\alpha_u}} = O(t)t^{-1} = O(1).$$

Lastly, by (7.15),

$$\begin{aligned}\int_{\{|x-\eta|\leq t^{1/\alpha_u}\}} S(x-\eta, \alpha(\eta), a(\eta)t)\, d\eta &= O(1)t^{-d/\alpha_d}\int_{\{|x-\eta|\leq t^{1/\alpha_u}\}} d\eta \\ &= O(1)t^{-d/\alpha_d}t^{-d/\alpha_u},\end{aligned}$$

which has the form of the r.h.s. of (7.42), and therefore (7.42) is proven.

To obtain (7.43) under assumption (ii) we need now to consider only the integral over the set $\{|x-\eta| \leq t^{1/\alpha_u}\}$. For a positive c, denote

$$\alpha_u(x,c) = \max\{\alpha(y) : |y-x| \leq c\}, \quad \alpha_d(x,c) = \min\{\alpha(y) : |y-x| \leq c\}. \tag{7.44}$$

Let $b = b(t) = t^{1/\alpha_u}$ and $M(x) = \{\eta : t^{1/\alpha_u(x,b)} \leq |\eta - x| \leq b\}$. Since $\alpha_u(x,b) \geq \alpha(\eta)$ for $\eta \in M(x)$, $|\eta - x| > t^{1/\alpha(\eta)}$ in $M(x)$. Hence the integral on the l.h.s. of (7.43) over the set $M(x)$ can be estimated by

$$\begin{aligned}O(t)\int_{M(x)} \frac{d\eta}{|x-\eta|^{d+\alpha(\eta)}} &= O(t)\int_{t^{1/\alpha_u(x,b)}}^{t^{1/\alpha_u}} \frac{dy}{y^{1+\alpha_u(x,b)}} \\ &= O(t)[t^{1/\alpha_u(x,b)}]^{-\alpha_u(x,b)} = O(1).\end{aligned}$$

By (7.15), the integral on the l.h.s. of (7.43) over the set $\{|\eta-x| \leq t^{1/\alpha_u(x,b)}\}$ can be estimated by

$$\begin{aligned}\int_{|\eta-x|\leq t^{1/\alpha_u(x,b)}} t^{-d/\alpha(\eta)}\, d\eta &= O(t^{-d/\alpha_d(x,b)})t^{d/\alpha_u(x,b)} \\ &= O(1)\exp\left\{-\frac{d(\alpha_u(x,b)-\alpha_d(x,b))}{\alpha_u(x,b)\alpha_d(x,b)}\log t\right\} \\ &= O(1)\exp\{O(t^{\beta/\alpha_u}\log t\} = O(1),\end{aligned}$$

which completes the proof of Lemma 7.4.1. □

Let for a $\beta > 0$ the functions f_β^d be defined on $\mathbb{R}^d$ by the formulas

$$f_\beta^d(x) = (1 + |x|^{\beta+d})^{-1}. \tag{7.45}$$

Lemma 7.4.2. *Under the assumptions of Lemma* 7.4.1 (ii) *one has*

$$\int S(x-\eta, \alpha(\eta), a(\eta)(t-\tau))S(\eta - x_0, \alpha(x_0), a(x_0)\tau)\, d\eta$$
$$= O(1)S(x-x_0, \alpha(x_0), a(x_0)t) + O(t) f_{\alpha_d}^d(x-x_0). \tag{7.46}$$

Proof. We shall use the notations of the previous proof, in particular (7.44) and $b = b(t) = t^{1/\alpha_u}$.

Step 1. Assume $|x - x_0| \geq t^{1/\alpha_u(x_0,b)}$. In this case $|x - x_0| \geq t^{1/\alpha(x_0)}$ and $S(x - x_0, \alpha(x_0), a(x_0)t)$ is of order $t/|x - x_0|^{d+\alpha(x_0)}$.

In the domain $\{\eta : |\eta - x_0| > |x - x_0|/2\}$,

$$\begin{aligned} S(\eta - x_0, \alpha(x_0), a(x_0)\tau) &= \frac{O(\tau)}{|\eta - x_0|^{d+\alpha(x_0)}} \\ &= \frac{O(t)}{|x - x_0|^{d+\alpha(x_0)}} = O(1)S(x-x_0, \alpha(x_0), a(x_0)t), \end{aligned}$$

and therefore, due to Lemma 7.4.1, the corresponding part of the integral on the l.h.s. of (7.46) is $O(1)S(x - x_0, \alpha(x_0), a(x_0)t)$. Consequently we can reduce our attention to the region

$$D = \{\eta : |\eta - x_0| \leq |x - x_0|/2 \iff |x - \eta| > |x - x_0|/2\}.$$

If $\eta \in D$,

$$S(x - \eta, \alpha(\eta), a(\eta)(t-\tau)) = \frac{O(t-\tau)}{|x-\eta|^{d+\alpha(\eta)}} = \frac{O(t)}{|x - x_0|^{d+\alpha(\eta)}}. \tag{7.47}$$

Let us consider three cases, keeping in mind throughout that $\eta \in D$.

If $|x - x_0| \geq 1$,

$$S(x - \eta, \alpha(\eta), a(\eta)(t-\tau)) = \frac{O(t)}{|x - x_0|^{d+\alpha_d}},$$

so that the corresponding part of the integral on the l.h.s. of (7.46) is of order $O(t) \times f_{\alpha_d}^d(x - x_0)$.

If $t^{1/\alpha_u} \le |x - x_0| \le 1$,

$$\begin{aligned} S(x-\eta, \alpha(\eta), a(\eta)(t-\tau)) &= \frac{O(t)}{|x-x_0|^{d+\alpha_u(x_0,|x-x_0|)}} \\ &= \frac{O(t)}{|x-x_0|^{d+\alpha(x_0)}} |x-x_0|^{\alpha(x_0)-\alpha_u(x_0,|x-x_0|)} \\ &= O(1) S(x-x_0, \alpha(x_0), a(x_0)t) \exp\{O(|x-x_0|^{\beta}) \log|x-x_0|\} \\ &= O(1) S(x-x_0, \alpha(x_0), a(x_0)t), \end{aligned}$$

so that the corresponding part integral on the l.h.s. of (7.46) is $O(1)S(x - x_0, \alpha(x_0), a(x_0)t)$.

Finally, if $t^{1/\alpha_u(x_0,b)} \le |x - x_0| \le t^{1/\alpha_u}$,

$$\begin{aligned} S(x-\eta, \alpha(\eta), a(\eta)(t-\tau)) &= \frac{O(t)}{|x-x_0|^{d+\alpha_u(x_0,b)}} \\ &= \frac{O(t)}{|x-x_0|^{d+\alpha(x_0)}} \exp\{O(t^{\beta/\alpha_u}) \log t\} \\ &= O(1) S(x-x_0, \alpha(x_0), a(x_0)t), \end{aligned}$$

so that again the corresponding integral on the l.h.s. of (7.46) over D is $O(1)S(x-x_0, \alpha(x_0), a(x_0)t)$.

Step 2. Let us show that if $|x - x_0| \le t^{1/\alpha_u(x_0,b)}$, then $S(x - x_0, \alpha(x_0), a(x_0)t)$ is of order $t^{-d/\alpha(x_0)}$. If $|x - x_0| \le t^{1/\alpha(x_0)}$, this holds directly by (7.33). Assume

$$t^{1/\alpha(x_0)} < |x - x_0| \le t^{1/\alpha_u(x_0,b)}.$$

Then

$$S(x - x_0, \alpha(x_0), a(x_0)t)/t^{-d/\alpha(x)}$$

is of order

$$t^{d/\alpha(x_0)} \frac{t}{|x-x_0|^{d+\alpha(x_0)}} = \exp\left\{-\left[\frac{d}{\alpha_u(x_0,b)} - \frac{d}{\alpha(x_0)} - 1 + \frac{\alpha(x_0)}{\alpha_u(x_0,b)}\right] \log t\right\},$$

which is bounded from above and from below, due to the Hölder continuity of the function $\alpha(x)$.

Step 3. Assume now that $|x - x_0| \le t^{1/\alpha_u(x_0,b)}$. By Step 2, it is sufficient to prove that the integral on the l.h.s. of (7.46) can be estimated by $O(1)t^{-d/\alpha(x_0)}$. Decompose our integral into the sum $I_1 + I_2$ decomposing the domain of integration into the union $D_1 \cup D_2$ with $D_1 = \{\eta : |\eta - x_0| \ge t^{1/\alpha_u(x_0,b)}\}$ and D_2 being its complement. If

$\eta \in D_1$,

$$S(\eta - x_0, \alpha(x_0), a(x_0)\tau) = \frac{O(\tau)}{|\eta - x_0|^{d+\alpha(x_0)}} = O(t)t^{-(d+\alpha(x_0))/\alpha_u(x_0,b)}$$
$$= O(t^{-d/\alpha(x_0)}) \exp\left\{\left[\frac{d}{\alpha(x_0)} - \frac{d}{\alpha_u(x_0,b)} + 1 - \frac{\alpha(x_0)}{\alpha_u(x_0,b)}\right] \log t\right\},$$

which is of order $O(1)t^{-d/\alpha(x_0)}$. Consequently, one obtains for I_1 the required estimate by Lemma 7.4.1. Turning to I_2 we distinguish two cases. If $\tau \geq t/2$ we estimate the second multiplier under the integral on the l.h.s. of (7.46) by

$$O(1)\tau^{-d/\alpha(x_0)} = O(1)t^{-d/\alpha(x_0)},$$

so that Lemma 7.4.1 again does the job. Finally, if $t - \tau \geq t/2$ we estimate the first multiplier under the integral by

$$O(1)(t-\tau)^{-d/\alpha(\eta)} = O(1)t^{-d/\alpha(\eta)},$$

which by the Hölder continuity of the function $\alpha(x)$ again reduces to $O(1)t^{-d/\alpha(x_0)}$, implying the required estimate also in this case. □

It is often useful to know that the solution to a Cauchy problem preserves certain rates of decay at infinity. We present now a result of this kind for stable semigroups.

Lemma 7.4.3. *There exists a constant c such that*

$$\int_{\mathbb{R}^d} S(x-\eta, \alpha, t) f_\beta^d(\eta - x_0)\, d\eta \leq c f^d_{\min(\alpha,\beta)}(x - x_0) \tag{7.48}$$

uniformly for bounded t, β from any compact subset of $\mathbb{R}_+$ *and α from a compact subinterval of* (0, 2).

Proof. It is enough to prove the statement for $x_0 = 0$. For brevity let us consider only the case $\beta = \alpha$.

For $|\eta| > \frac{1}{2}|x|$, one can estimate

$$f_\alpha^d(\eta) \leq (1 + (|x|/2)^{\alpha+d})^{-1} \leq 2^{\alpha+d} f_\alpha^d(x),$$

and therefore, for $x_0 = 0$, the integral on the l.h.s. of (7.48) over the set $\{|\eta| > \frac{1}{2}|x|\}$ does not exceed

$$\int_{\mathbb{R}^d} G_0(t,x,\eta) 2^{\alpha+d} f_\alpha^d(x)\, d\eta \leq 2^{\alpha+d} f_\alpha^d(x).$$

Hence, it remains to show that

$$\int_{\{|\eta| \leq |x|/2\}} S(x-\eta; \alpha, t) f_\alpha^d(\eta)\, d\eta \leq c f_\alpha^d(x).$$

To this end, consider separately three domains $|x| \le t^{1/\alpha}$, $t^{1/\alpha} \le |x| \le 1$ and $|x| > 1$. In the first case

$$I = O(t^{-d/\alpha}) \int_{\{|\eta| \le t^{1/\alpha}\}} d\eta = O(1) = O(1) f_\alpha^d(x).$$

In the second case

$$I = \int_{\{|\eta| \le |x|/2\}} \frac{O(t)}{|x|^{d+\alpha}} f_\alpha^d(\eta)\, d\eta = \frac{O(t)}{|x|^\alpha} = O(1) = O(1) f_\alpha^d(x).$$

In the third case

$$I = \frac{O(t)}{|x|^{d+\alpha}} \int_{\{|\eta| \le |x|/2\}} f_\alpha^d(\eta)\, d\eta = \frac{O(t)}{|x|^{\alpha+d}} = O(t) f_\alpha^d(x),$$

which completes the proof. □

Exercise 7.4.1. Check that

$$\int f_\alpha^d(x-\eta) f_\alpha^d(\eta)\, d\eta = O(1) f_\alpha^d(x) \tag{7.49}$$

Hint: decompose the integral into two parts with $|\eta| \le |x-\eta|$ and $|\eta| \le |x-\eta|$.

7.5 Stable-like processes: heat kernel estimates

This section contains the key results of this chapter. Here we analyze the heat kernels (or Green functions) of the stable-like evolution equation

$$\frac{\partial u}{\partial t} = -a(x)|\nabla|^{\alpha(x)} u, \quad x \in \mathbb{R}^d, \quad t \ge 0. \tag{7.50}$$

If $a(x)$ and $\alpha(x)$ are constants, the Green function for the Cauchy problem of this equation is given by the stable density (see Section 4.1)

$$S(x_0 - x; \alpha, at) = (2\pi)^{-d} \int_{\mathbb{R}^d} \exp\{-at|p|^\alpha + ip(x-x_0)\}\, dp. \tag{7.51}$$

In the theory of pseudo-differential operators, equation (7.50) is written in the pseudo-differential form as

$$\frac{\partial u}{\partial t} = \Phi(x, -i\nabla) u(x) \tag{7.52}$$

with the symbol

$$\Phi(x, p) = -a(x)|p|^{\alpha(x)}. \tag{7.53}$$

As follows from direct computations (see equation (1.72)), an equivalent form of equation (7.50) is the following integro-differential form of Lévy–Khintchine type:

$$\frac{\partial u}{\partial t} = -a(x)c(\alpha)\int_0^\infty (u(x+y) - u(x) - (y, \nabla u))\frac{d|y|}{|y|^{1+\alpha}} \tag{7.54}$$

with a certain constant $c(\alpha)$. We shall not need this form much, but it will be important for us to have in mind that the operator on the r.h.s. of (7.50) satisfies PMP, which is clear from the representation given in (7.54), but is not so obvious from (7.50).

Recall that the functions f_β^d is defined by (7.45).

Naturally, one expects that for small times the Green function of equation (7.50) with varying coefficients can be approximated by the Green function of the corresponding problem with constant coefficients, i.e. by the function

$$G_0(t, x, x_0) = S(x - x_0, \alpha(x_0), a(x_0)t). \tag{7.55}$$

This is in fact true, as the following main result of this chapter shows.

Theorem 7.5.1. *Let $\beta \in (0, 1]$ be arbitrary and let $\alpha \in [\alpha_d, \alpha_u]$, $a \in [a_d, a_u]$ be β-Hölder continuous functions on $\mathbb{R}^d$ with values in compact subsets of $(0, 2)$ and $(0, \infty)$ respectively. Then there exists a function $G(t, x, x_0)$, $t > 0, x, x_0 \in \mathbb{R}^d$, the Green function of the Cauchy problem for equation (7.50), such that*

(i) *G is continuous and continuously differentiable with respect to the time variable t for $t > 0$,*

(ii) *it has Dirac initial data, in other words $\lim_{t\to 0} G(t, x, x_0) = \delta(x - x_0)$, and*

(iii) *it solves equation (7.50) in distribution, i.e.*

$$\int \frac{\partial G}{\partial t}(t, x, x_0)\phi(x)\,dx = \int G(t, x, x_0)L'\phi(x)\,dx \tag{7.56}$$

for any $\phi \in C^2(\mathbb{R}^d)$ with compact support, where L' is the dual operator to the operator L on the r.h.s. of equation (7.50). Moreover,

$$\begin{aligned} G(t, x, x_0) = S(x - x_0; \alpha(x_0), ta(x_0))&(1 + O(t^{\beta/\alpha_u})(1 + |\log t|)) \\ &+ O(t) f_{\alpha_d}^d(x - x_0) \end{aligned} \tag{7.57}$$

and

$$\frac{\partial G}{\partial t} = O(t^{-1})S(x - x_0; \alpha(x_0), ta(x_0)) + O(1) f_{\alpha_d}^d(x - x_0). \tag{7.58}$$

Furthermore, if the functions α, a are of class $C^2(\mathbb{R}^2)$, then $G(t, x, x_0)$ solves the equation (7.50) classically for $t > 0$, is twice continuously differentiable in x and

$$\frac{\partial^l G}{\partial x^l}(t, x, x_0) = O(t^{-l/\alpha})[S(x - x_0; \alpha(x_0), ta(x_0)) + O(t) f_{\alpha_d}^d(x - x_0)] \tag{7.59}$$

for $l = 1, 2$.

Proof. The function (7.55) clearly satisfies the required initial conditions

$$G_0(0, x, x_0) = \delta(x - x_0)$$

and the equation

$$\frac{\partial u}{\partial t}(t, x) = -a(x_0)|\nabla|^{\alpha(x_0)} u(t, x),$$

which can be equivalently rewritten as

$$\frac{\partial u}{\partial t}(t, x) = -a(x)|-i\nabla|^{\alpha(x)} u(t, x) - F(t, x, x_0) - \tilde{F}(t, x, x_0), \tag{7.60}$$

with

$$\begin{aligned} F(t, x, x_0) &= (a(x_0) - a(x))|\nabla|^{\alpha(x_0)} G_0(t, x, x_0) \\ &= \frac{a(x_0) - a(x)}{(2\pi)^d} \int_{\mathbb{R}^d} |p|^{\alpha(x_0)} \exp\{-a(x_0)|p|^{\alpha(x_0)} t + ip(x - x_0)\}\, dp, \end{aligned}$$

and

$$\tilde{F} = \frac{a(x)}{(2\pi)^d} \int (|p|^{\alpha(x_0)} - |p|^{\alpha(x)}) \exp\{-a(x_0)|p|^{\alpha(x_0)} t + i(p, x - x_0)\}\, dp. \tag{7.61}$$

where, in order to evaluate $\tilde{F}$, we used the fact (see Section 1.8) that applying $|\nabla|^\alpha$ to a function is equivalent to multiplication of its Fourier transform on $|p|^\alpha$.

Therefore, due to the Duhamel principle (see formula (4.4)), if G is the Green function for equation (7.50), then

$$G_0(t, x, x_0) = (G - \mathcal{F} G)(t, x, x_0), \tag{7.62}$$

where the integral operator $\mathcal{F}$ is defined by the formula

$$(\mathcal{F}\phi)(t, x, \xi) = \int_0^t \int_{\mathbb{R}^d} \phi(t - \tau, x, \eta)(F + \tilde{F})(\tau, \eta, \xi)\, d\eta\, d\tau. \tag{7.63}$$

Introducing a special notation for the integral on the r.h.s. of (7.63), i.e. denoting

$$\phi \otimes \psi(t, x, \xi) = \int_0^t \int_{\mathbb{R}^d} \phi(t - \tau, x, \eta)\psi(\tau, \eta, \xi)\, d\eta\, d\tau,$$

one can represent a solution to equation (7.62) by the series

$$G = (1 - \mathcal{F})^{-1} G_0 = (1 + \mathcal{F} + \mathcal{F}^2 + \cdots) G_0, \quad \mathcal{F}^k G_0 = G_0 \otimes (F + \tilde{F})^{\otimes k}. \tag{7.64}$$

Let us prove the convergence of this series and the required estimate for its sum.

Due to the mean-value theorem and Hölder continuity of $\alpha(x)$,

$$\tilde{F} = O(1)\min(1, |x - x_0|^\beta) \times \max_b \left| \frac{\partial}{\partial b} \int |p|^b \exp\{-a(x_0)|p|^{\alpha(x_0)}t + i(p, x - x_0)\}\, dp \right|,$$

where the max is taken over $b \in [\min(\alpha(x), \alpha(x_0)), \max(\alpha(x), \alpha(x_0))]$. If $|x - x_0| \le t^{1/\alpha(x_0)}$, one finds using (7.41) that

$$\begin{aligned}\tilde{F} &= O(|x - x_0|^\beta)t^{-1}(1 + |\log t|)\max_b t^{-|b-\alpha(x_0)|/\alpha(x_0)} S(x - x_0, \alpha(x_0), a(x_0)t)\\ &= O(|x - x_0|^\beta)t^{-1}(1 + |\log t|)S(x - x_0, \alpha(x_0), a(x_0)t),\end{aligned}$$

since

$$t^{-|b-\alpha(x_0)|/\alpha(x_0)} = \exp\{O(|x - x_0|^\beta \log t\} = \exp\{O(t^{\beta/\alpha(x_0)}) \log t\} = O(1).$$

If $t^{1/\alpha(x_0)} \le |x - x_0| \le t^{1/\alpha_u}$, one finds using (7.41) that

$$\begin{aligned}\tilde{F} &= O(|x - x_0|^\beta)t^{-1}(1 + |\log t|)|x - x_0|^{-|\alpha(x_0)-\alpha(x)|} S(x - x_0, \alpha(x_0), a(x_0)t)\\ &= O(|x - x_0|^\beta)t^{-1}(1 + |\log t|)S(x - x_0, \alpha(x_0), a(x_0)t).\end{aligned}$$

In case $|x - x_0| \le t^{1/\alpha_u}$, one gets similar estimates for F.

If $|x - x_0| > t^{1/\alpha_u}$, we estimate each term in the expression

$$(F + \tilde{F})(t, x, x_0) = (a(x)|\nabla|^{\alpha(x)} - a(x_0)|\nabla|^{\alpha(x_0)})S(x - x_0, \alpha(x_0), a(x_0)t)$$

separately, using (7.40) to obtain

$$|(F + \tilde{F})(t, x, x_0)| = O(1) f^d_{\alpha_d}(x - x_0).$$

Therefore

$$|(F + \tilde{F})(t, x, x_0)| = O(t^{-1})(1 + |\log t|)t^{\beta/\alpha_u} G_0(t, x, x_0) + O(1) f^d_{\alpha_d}(x - x_0). \tag{7.65}$$

Consequently, by Lemmas 7.4.2, 7.4.3,

$$|G_0 \otimes (F + \tilde{F}))(t, x, x_0)| = O(t^{\beta/\alpha_u})(1 + |\log t|)G_0(t, x, x_0) + O(t) f^d_{\alpha_d}(x - x_0). \tag{7.66}$$

Using Lemmas 7.4.2, 7.4.3, estimate (7.49) and induction yields for any k

$$\begin{aligned}|G_0 \otimes (F + \tilde{F}))(t, x, x_0)^{\otimes k}| &\le c^k t^{(k-1)\beta/\alpha_u})(1 + |\log t|)^{k-1}\\ &\quad \times [t^{\beta/\alpha_u}(1 + |\log t|)G_0(t, x, x_0) + t f^d_{\alpha_d}(x - x_0)]\end{aligned} \tag{7.67}$$

with a constant $c > 0$. This implies the convergence of series (7.64), the continuity of its limit and estimate (7.57).

Let us turn to the derivatives of G. We want to show that term-by-term differentiation of series (7.64) yield a convergent series. If one differentiates directly the terms of these series and uses the estimates of Theorem 7.3.1, one obtains the expressions which are not defined (because τ^{-1} is not an integrable function for small τ). To avoid this difficulty, one needs to rearrange the variables of integration in (7.63) appropriately before using the estimates for the derivatives.

First, by (7.40),

$$\frac{\partial(F+\tilde{F})}{\partial t}(t,x,x_0) = O(t^{-2})(1+|\log t|)t^{\beta/\alpha_u}G_0(t,x,x_0) + O(t^{-1})f_{\alpha_d}^d(x-x_0).$$

The convolution (7.63) after the change of the variable $\tau = st$ can be presented in the equivalent form

$$(\phi \otimes (F+\tilde{F}))(t,x,\xi) = t\int_0^1 \int \phi(t(1-s),x,\eta)(F+\tilde{F})(ts,\eta,\xi)\,d\eta ds.$$

One can now estimate the derivative of the second term in (7.64) as

$$\begin{aligned}\frac{\partial}{\partial t}(G_0 \otimes (F+\tilde{F}))(t,x,x_0) &= \int_0^1 \int G_0(t(1-s),x,\eta)(F+\tilde{F})(ts,\eta,x_0)\,d\eta ds \\ &+ t\int_0^1 \int \Big[(1-s)\frac{\partial G_0}{\partial t}(t(1-s),x,\eta)(F+\tilde{F})(ts,\eta,x_0) \\ &+ sG_0(t(1-s),x,\eta)\frac{\partial(F+\tilde{F})}{\partial t}(ts,\eta,x_0)\Big]d\eta ds,\end{aligned}$$

and all three terms of this expression are of the order $O(t^{-1})(G_0 \otimes (F+\tilde{F})(t,x,x_0)$. Estimating other terms similarly, one obtains (7.58) for the sum of the series obtained by term-by-term differentiation of series (7.64) with respect to time.

Suppose now that α, a are smooth functions. We restrict ourselves to the estimate of the first derivative only, second derivatives being estimated similarly.

Remark 51. The consideration of the case when $\alpha_d > 1$ is trivial, because in that case $\tau^{-1/\alpha}$ is an integrable function for small τ, and consequently, differentiating expansion (7.64) term by term and using estimate (7.35) yields the required result straightforwardly.

In the general case, let observe first that by (7.40) (and due to the assumption of smoothness of a ad α),

$$\begin{aligned}\frac{\partial(F+\tilde{F})}{\partial x}(t,x,x_0) &= O(t^{-1-1/\alpha_d})(1+|\log t|)t^{\beta/\alpha_u}G_0(t,x,x_0) \\ &+ O(t^{-1/\alpha_d})f_{\alpha_d}^d(x-x_0).\end{aligned} \tag{7.68}$$

To shorten the lengthy formulas we shall write below F instead of $F + \tilde{F}$. The idea is that in the convolution-type integrals under consideration we have to rearrange differentiation to act on the term with the time τ or $t - \tau$ being of order t. Thus, to estimate the derivative of the second term in (7.64) let us rewrite it in the following form:

$$\begin{aligned}\mathcal{F} G_0(t,x,x_0) &= (G_0 \otimes F)(t,x,x_0)\\ &= \int_0^{t/2} \int G_0(t-\tau,x,\eta) F(\tau,\eta,x_0)\, d\eta\, d\tau\\ &\quad + \int_{t/2}^{t} \int G_0(t-\tau,x,x-\eta) F(\tau,x-\eta,x_0)\, d\eta\, d\tau.\end{aligned}$$

Differentiating with respect to x, using the equation $G_0(t-\tau,x,x-\eta) = S(\eta,\alpha(x-\eta),ta(x-\eta))$ and estimates (7.35) and (7.68), yields for the magnitude of the derivative of $\mathcal{F} G_0$ the same estimate as for $\mathcal{F} G_0$ itself but with an additional multiplier of the order $O(t^{-1/\alpha_d})$. We now estimate similarly the derivative of the term $\mathcal{F}^k G_0(t,x,x_0)$ in (7.64), which equals

$$\int_{\sigma_t} d\tau_1 \cdots d\tau_k \int_{\mathbb{R}^{kd}} d\eta_1 \cdots d\eta_k \\ \times G_0(t-\tau_1-\cdots-\tau_k,x,\eta_1) F(\tau_1,\eta_1,\eta_2) \cdots F(\tau_k,\eta_k,x_0),$$

where we denoted by σ_t the simplex

$$\sigma_t = \{\tau_1 \ge 0, \ldots, \tau_k \ge 0 : \tau_1 + \cdots + \tau_k \le t\}.$$

To this end, we partition this simplex as the union of the $k+1$ domains D_l, $l = 0,\ldots,k$, (which clearly have disjoint interiors) with $D_0 = \sigma_{t/2}$ and

$$D_l = \{(\tau_1,\ldots,\tau_k) \in \sigma_t \setminus \sigma_{t/2} : \tau_l = \max\{\tau_j, j = 1,\ldots,k\}\}, \quad l = 1,\ldots,k.$$

and then change variables η to obtain $\mathcal{F}^k G_0 = \mathcal{F}_0^k + \cdots + \mathcal{F}_k^k$ with $\mathcal{F}_0^k(t,x,x_0)$ being equal to

$$\int_{\sigma_{t/2}} d\tau_1 \cdots d\tau_k \int_{\mathbb{R}^{kd}} d\eta_1 \cdots d\eta_k \\ \times G_0(t-\tau_1-\cdots-\tau_k,x,\eta_1) F(\tau_1,\eta_1,\eta_2) \cdots F(\tau_k,\eta_k,x_0)$$

and with $\mathcal{F}_l^k(t,x,x_0)$ being equal to

$$\int_{D_l} d\tau_1 \cdots d\tau_k \int_{\mathbb{R}^{kd}} d\eta_1 \cdots d\eta_k G_0(t-\tau_1-\cdots-\tau_k,x,x-y_1) \\ \times F(\tau_1,x-y_1,x-y_2) \cdots F(\tau_{l-1},x-y_{l-1},x-y_l) \\ \times F(\tau_l,x-y_l,\eta_{l+1}) \cdots F(\tau_k,\eta_k,x_0)$$

for $l = 1,\dots,k$. Now, differentiating $\mathcal{F}_0^k$ with respect to x we use estimate (7.35), and differentiating $\mathcal{F}_l^k$, $l = 1,\dots,k$, we use (7.68) for the derivative of $F(\tau_l, x - y_l, \eta_{l+1})$ and the estimate

$$\frac{\partial}{\partial x}F(t, x-\eta, x-\xi) = O(t^{-1})(1+|\log t|)t^{\beta/\alpha_u}G_0(t,x,x_0) + O(1)f_{\alpha_d}^d(x-x_0), \tag{7.69}$$

i.e. the same estimate as for F itself. In this way one obtains (noticing also that $\tau_l \geq t/(2k)$ in D_l) for the derivative of the term $\mathcal{F}^k u$ in expansion (7.64) the same estimate as for $\mathcal{F}^k u$ itself, but with an additional multiplier of order

$$= O(1)t^{-1/\alpha}k^{1+1/\alpha}.$$

Multiplying the terms of a power series by k^q with any fixed positive q does not change the radius of convergence, which implies the required estimate for the derivative of G.

Now observe that though in previous arguments we presupposed the existence of the Green function G, one verifies directly that the sum (7.64) satisfies equation (7.50) whenever it converges together with its derivatives.

If α, a are not smooth, they can be approximated by smooth functions. As our estimates for G do not depend on smoothness and the coefficients of series (7.64) depend continuously on α, a, one can pass to the limit in equation (7.56). □

Remark 52. If $\alpha(x)$ is a constant, the term $(1 + |\log t|)$ can be omitted from formula (7.57). In fact it appears only when estimating $\tilde{F}$, which vanishes in case of a constant α.

As a direct consequence, we get the following global (in space) upper bound for the transition densities.

Theorem 7.5.2. *Under the assumptions of Theorem* 7.5.1

$$G(t,x,x_0) \leq K[S(x-x_0,\alpha(x_0),t) + S(x-x_0,\alpha_d,t)] \tag{7.70}$$

for all $t \leq T$ with an arbitrary T and all x, x_0, where $K = K(T)$ is a constant.

Exercise 7.5.1. Show that the theorem holds also for the equation

$$\frac{\partial u}{\partial t} = (A(x), \nabla u(x)) - a(x)|-i\nabla|^{\alpha(x)}u, \quad x \in \mathbb{R}^d,\ t \geq 0, \tag{7.71}$$

if A is a bounded continuous function, α takes value in a compact subset of $(1,2)$ and other conditions on α, a are the same as above. Hint: see [178] if necessary.

Finally, let us give a two-sided estimate for the heat kernel, which is valid at least for the case of a constant stability index.

Theorem 7.5.3. *Under the assumptions of Theorem 7.5.1 assume that the index $\alpha(x)$ does not depend on x. Then for any $T > 0$ there exists a constant K such that for all $t \le T$ and all x, x_0*

$$K^{-1} S(x - x_0, \alpha, t) \le G(t, x, x_0) \le K S(x - x_0, \alpha, t).$$

Proof. The upper bound was obtained above. The details concerning the lower bound can be found in [178]. □

7.6 Stable-like processes: Feller property

The existence of a Feller process generated by a stable-like generators was obtained under rather general assumptions in Proposition 4.6.2 and further extended in Proposition 4.6.2. Here we show how the same result can be obtained from the existence of the Green function, but for non-degenerate stable-like processes only. However, in this non-degenerate setting, we can essentially weaken the smoothness requirements. Moreover, having the Green function and its asymptotics yields much more than just a Feller process, which we shall demonstrate in the next two sections.

For an arbitrary $f \in C(\mathbb{R}^d)$ and $t > 0$, set

$$(R_t f)(x) = \int_{\mathbb{R}^d} G(t, x, \xi) f(\xi)\, d\xi, \tag{7.72}$$

where the Green function G was constructed in the previous section.

Theorem 7.6.1. *Suppose the assumptions of Theorem 7.5.1 hold. Let a and α are of class $C^2(\mathbb{R}^d)$. Then the following statements hold.*

(i) *$(R_t f)(x)$ tends to $f(x)$ as $t \to 0$ for each x and any $f \in C(\mathbb{R}^d)$; moreover, if $f \in C_\infty(\mathbb{R}^d)$, then $R_t f$ tends to f uniformly, as $t \to 0$.*

(ii) *R_t is a continuous operator $C(\mathbb{R}^d) \mapsto C^2(\mathbb{R}^d)$ with the norm of the order $O(t^{-2/\alpha_d})$.*

(iii) *If $f \in C(\mathbb{R}^d)$, the function $R_t f(x)$ satisfies equation (7.50) for $t > 0$.*

(iv) *R_t, $t \in [0, T]$, is a uniformly bounded family of operators $C^l(\mathbb{R}^d) \mapsto C^l(\mathbb{R}^d)$ and $C^l_\infty(\mathbb{R}^d) \mapsto C^l_\infty(\mathbb{R}^d)$ for $l = 1, 2$.*

(v) *The Cauchy problem for equation (7.50) can have at most one solution in the class of continuous functions belonging to $C_\infty(\mathbb{R}^d)$ for each t; this solution is necessarily non-negative whenever the initial function is non-negative.*

(vi) *The Green function G is everywhere non-negative, satisfies the semigroup identity (the Chapman–Kolmogorov equation) and the conservativity condition $\int G(t, x, \eta)\, d\eta = 1$ holds for all $t > 0$, $x \in \mathbb{R}^d$. In particular, equation (7.72) specifies a conservative Feller semigroup, which therefore corresponds to a certain Feller process.*

(vii) *The space $C^2_\infty(\mathbb{R}^d)$ is an invariant core for the Feller semigroup of (vi).*

Proof. Statements (i)–(iii) follows directly from (7.72) and the properties of G obtained in Theorem 7.5.1. To prove (iv), we rewrite the expression for R_t in the following equivalent form:

$$R_t f(x) = \int_{\mathbb{R}^d} G(t, x, x - y) f(x - y)\, dy,$$

and use the estimates

$$\frac{\partial^l}{\partial x^l} G(t, x, x - y) = O(1) G_0(t, y, 0), \quad l = 1, 2,$$

which follow again from the series representation of G, formula

$$G_0(t, x, x - y) = \frac{1}{(2\pi)^d} \int_{\mathbb{R}^d} \exp\{-a(x - y)|p|^{\alpha(x-y)} t + ipy\}\, dp$$

and the estimates of Theorem 7.3.2. Statements (v), (vi) follow from Theorems 4.1.2, 4.1.3. □

Approximating Hölder continuous functions a, α by smooth ones, one deduces the following corollary.

Theorem 7.6.2. *Under the assumptions of Theorem* 7.5.1 *(not assuming additionally the smoothness of a, α) the statements* (i) *and* (vi) *of Theorem* 7.6.1 *still hold.*

7.7 Application to sample-path properties

Let $X(t)$ (resp. $X(t, x_0)$) be the Feller process starting at the origin (resp. at x_0) and corresponding to the Feller semigroup defined by Theorem 7.6.1.

In this section we show how the analytic results of the previous sections can be used in studying the sample-path properties of this process. We start with estimates for the distribution of maximal magnitude of stable-like processes, generalizing partially the corresponding well-known estimates for stable processes (see e.g. Bingham [57]). Then we obtain the *principle of approximate scaling* and the *principle of approximate independence of increments*, and finally apply these results to the study of the lim sup behaviour of $|X(t)|$ as $t \to 0$. For simplicity we reduce the discussion to the case of a fixed index of stability. In other words, we shall suppose throughout this section that the assumptions of Theorem 7.5.1 are satisfied and that $\alpha(x) = \alpha$ is a constant. From the general theory of Feller processes it follows that we can choose a càdlàg modification of $X(t, x_0)$ with the natural filtration of σ-algebras corresponding to $X(t)$ satisfying usual conditions.

The Green function $G(t, x, x_0)$ constructed in this theorem defines the transition probability density (from x to x_0 in time t) for the process X. The connection between the analytic language and the probabilistic language that we use in this section can be expressed essentially by the formula

$$\int_A \mathbf{P}(X(t,y) \in d\xi) f(\xi) = \int_A G(t,y,\xi) f(\xi)\, d\xi$$

for $A \subset \mathbb{R}^d$. Let

$$X^\star(t,x_0) = \sup\{|X(s,x_0) - x_0| : s \in [0,t]\}, \quad X^\star(t) = X^\star(t,0).$$

Theorem 7.7.1. *For any $T > 0$ there exist positive constants C, K such that for all $t \le T$, x_0, and $\lambda > Kt^{1/\alpha}$*

$$C^{-1} t\lambda^{-\alpha} \le \mathbf{P}(X^\star(t,x_0) > \lambda) \le Ct\lambda^{-\alpha}. \tag{7.73}$$

Proof. Since we shall use the uniform estimates from Theorem 7.5.3, it will be enough to consider only $x_0 = 0$ in (7.73). Plainly $\mathbf{P}(X^\star(t) > \lambda) \ge \mathbf{P}(|X(t)| > \lambda)$, and thus the left hand side inequality in (7.73) follows directly from Theorem 7.5.3. Turning to the proof of the r.h.s. inequality, denote by T_a the first time when the process $X(t)$ leaves the ball $B(a)$, i.e. $T_a = \inf\{t \ge 0 : |X(t)| > a\}$. Notice now that due to Theorem 7.5.3,

$$\mathbf{P}(|X(s) - x_0| \ge \lambda/2) = O(s)\lambda^{-\alpha} = O(K^{-\alpha})$$

uniformly for all x_0 and $s \le t$. Therefore, due to the homogeneity and the strong Markov property, one has that

$$\begin{aligned}\mathbf{P}(|X(t)| > \lambda/2) &\ge \mathbf{P}((X^\star(t) > \lambda) \cap (|X(t)| > \lambda/2)) \\ &\ge \int_0^t \mathbf{P}(T_\lambda \in ds)\mathbf{P}(|X(t - T_\lambda, X(T_\lambda)) - X(T_\lambda)| \le \lambda/2) \\ &\ge (1 - O(K^{-\alpha})) \int_0^t \mathbf{P}(T_\lambda \in ds) \ge (1 - O(K^{-\alpha})\mathbf{P}((X^\star(t) > \lambda).\end{aligned}$$

It follows that

$$\mathbf{P}((X^\star(t) > \lambda) \le (1 + O(K^{-\alpha}))\mathbf{P}(|X(t)| > \lambda/2),$$

which implies the r.h.s. inequality in (7.73) again by Theorem 7.5.3. $\square$

Let us formulate now explicitly the main tools in the investigation of the sample-path properties of stable diffusions that can be used as substitutes to the scaling property and the independence of increments, which constitute the main tools in studying stable Lévy motions.

Theorem 7.7.2 (Local principle of approximate scaling). *There exists C such that for $t \leq 1$ and all x, x_0*

$$C^{-1}G(1, xt^{1/\alpha}, x_0t^{1/\alpha}) \leq G(t, x, x_0) \leq CG(1, xt^{1/\alpha}, x_0t^{1/\alpha}) \tag{7.74}$$

and

$$C^{-1}S((x - x_0)t^{1/\alpha}, \alpha, 1) \leq G(t, x, x_0) \leq CS((x - x_0)t^{1/\alpha}, \alpha, 1). \tag{7.75}$$

Proof. This follows directly from Theorem 7.5.3 and the scaling properties of the stable densities themselves. □

Theorem 7.7.3 (Local principle of approximate independence of increments). *For any t_0 there exists a constant C such that if $0 \leq s_1 < t_1 \leq s_2 < t_2 \leq t_0$, M_1, M_2 are any measurable sets in $\mathbb{R}^d$ and x_0 is any point in $\mathbb{R}^d$, then*

$$\begin{aligned} &C^{-1}\mathbf{P}(X(t_1, x_0) - X(s_1, x_0) \in M_1)\mathbf{P}(X(t_2, x_0) - X(s_2, x_0) \in M_2) \\ &\quad \leq \mathbf{P}((X(t_1, x_0) - X(s_1, x_0) \in M_1) \cap (X(t_2, x_0) - X(s_2, x_0) \in M_2)) \\ &\quad \leq C\mathbf{P}(X(t_1, x_0) - X(s_1, x_0) \in M_1)\mathbf{P}(X(t_2, x_0) - X(s_2, x_0) \in M_2). \end{aligned} \tag{7.76}$$

Proof. Consider for brevity only the case $s_2 = t_1$, the case $s_2 > t_1$ being similar. Also by homogeneity one can set $s_1 = 0$ without loss of generality. Then, by Theorem 7.5.3 and the Markov property, one has

$$\begin{aligned} &\int_{M_1 + x_0} \mathbf{P}(X(t_1, x_0) \in dy)\mathbf{P}(X(t_2, x_0) - X(t_1, x_0) \in M_2) \\ &\qquad\qquad \leq C\mathbf{P}(X(t_2 - t_1) \in M_2) \int_{M_1 + x_0} \mathbf{P}(X(t_1, x_0) \in dy), \end{aligned}$$

which implies the r.h.s. inequality in (7.76) again by Theorem 7.5.3. Similarly one obtains the l.h.s. inequality in (7.76). □

As an application of these facts, let us prove now that the integral test on the limsup behaviour of the stable processes (discovered first in Khintchine [158]) is valid also for stable-like processes.

Theorem 7.7.4. *Let $f : (0, \infty) \to (0, \infty)$ be an increasing function. Then* $\limsup_{t \to 0}$ *of the function $(|X(t)|/f(t))$ is equal to 0 or ∞ almost surely according to whether the integral $\int_0^1 f(t)^{-\alpha}\, dt$ converges or diverges.*

Proof. The proof generalizes the arguments given in Bertoin [49] for the proof of the corresponding result for one-dimensional stable Lévy motions. Suppose first that the integral converges. Then

$$tf^{-\alpha}(t) \leq \int_0^t f^{-\alpha}(s)\, ds = o(1), \quad t \to 0.$$

Consequently there exists $t_0 > 0$ such that $f(t/2)t^{-1/\alpha} > K$ for $t \le t_0$ with small enough t_0, where K is the constant from Theorem 7.7.1. Consequently, for every positive integer n,

$$\mathbf{P}(X^\star(2^{-n}) > f(2^{-n-1})) \le C2^{-n}(f(2^{-n-1}))^{-\alpha}.$$

Since the series $\sum 2^{-n}(f(2^{-n-1}))^{-\alpha}$ converges, the Borel–Cantelli lemma yields $X^\star(2^{-n}) \le f(2^{-n-1})$ for all n large enough, almost surely. It follows that $X^\star(t) \le f(t)$ for all $t > 0$ small enough, almost surely (because, if $n = n(t)$ denotes the maximal natural number such that $t \le 2^{-n}$, then $X^\star(t) \le X^\star(2^{-n}) \le f(2^{-n-1}) \le f(t)$). Using the function ϵf instead of f in the above arguments yields $X^\star(t)/f(t) \le \epsilon$ for any ϵ and all small enough t almost surely. Consequently, $\lim_{t\to 0}(X^\star(t)/f(t)) = 0$, almost surely.

Now let the integral diverge. Let $a > 0$. For any integer $n > 0$ consider the event $A_n = A'_n \cap B_n$ with

$$B_n = \{|X(2^{-n})| \le a2^{-n/\alpha}\},$$
$$A'_n = \{|X(2^{-n-1}) - X(2^{-n})| \ge f(2^{-n+1}) + a2^{-n/\alpha}\}.$$

By the local principle of approximate scaling,

$$C^{-1}\mathbf{P}(|X(1)| \le a) \le \mathbf{P}(B_n) \le C\mathbf{P}(|X(1)| \le a)$$

for all n uniformly. Consequently, by the Markov property, homogeneity and again approximate scaling, one has

$$\begin{aligned}\mathbf{P}(A_n) &\ge \mathbf{P}(B_n) \min_{|\xi|\le a2^{-n}} \mathbf{P}(|X(2^{-n},\xi) - \xi| \ge f(2^{-n+1}) + a2^{-n/\alpha}) \\ &\ge C^{-2}\mathbf{P}(|X(1)| \le a)\mathbf{P}(|X(1)| \ge 2^{n/\alpha} f(2^{-n+1}) + a) \\ &\ge \tilde{C}\mathbf{P}(|X(1)| \le a)(2^{n/\alpha} f(2^{-n+1}) + a)^{-\alpha}.\end{aligned}$$

for some positive $\tilde{C}$. The sum $\sum 2^{-n}(f(2^{-n+1}) + a2^{-n/\alpha})^{-\alpha}$ diverges, because it is a Riemann sum for the integral of the function

$$(f(t) + a(t/2)^{1/\alpha})^{-\alpha} \ge 2^{-\alpha}\max(a^{-\alpha}(2t)^{-1}, f(t)^{-\alpha})$$

and consequently the sum $\sum \mathbf{P}(A_n)$ is divergent. Notice now that though the events A_n are not independent, the events A'_n and A'_m are "approximately independent" in the sense of Theorem 7.7.3, if $n \ne m$. The same remark concerns the events B_n and A_n. Therefore, if $n \ne m$,

$$\begin{aligned}\mathbf{P}(A_n \cap A_m) \le \mathbf{P}(A'_n \cap A'_m) &= O(1)\mathbf{P}(A'_n)\mathbf{P}(A'_m) \\ &= O(1)\mathbf{P}(A_n)\mathbf{P}(A_m)\mathbf{P}(B_n)^{-1}\mathbf{P}(B_m)^{-1} \\ &= O(1)\mathbf{P}(A_n)\mathbf{P}(A_m)\mathbf{P}(|X(1)| \le a)^{-2},\end{aligned}$$

and one can use the generalization of the second Borel–Cantelli lemma for dependent events (see e.g. Spitzer [298]) to conclude that

$$\limsup_{t\to 0}(X(t)/f(t)) \geq 1$$

with positive probability, and therefore almost surely, due to the Blumenthal zero-one law (Theorem 3.11.3). Repeating the same arguments for the function $\epsilon^{-1} f$ instead of f one gets

$$\limsup_{t\to 0}(X(t)/f(t)) = \infty$$

a.s., which completes the proof of the theorem. □

7.8 Application to stochastic control

This section deviates from the mainstream of our exposition. The idea is to demonstrate that the solution to the main optimization equation for Markov processes with controlled drift can be easily constructed, once these Markov processes have regular enough transition probabilities (or Green functions). This section can be well placed at the end of Chapter 5, but now we have at our disposal an appropriate nontrivial class of examples of processes (stable-like ones) with regular transition probability.

Suppose a space $D \subset C^1_\infty(\mathbb{R}^d)$ is a core for an operator L in $C_\infty(\mathbb{R}^d)$ that generates a Feller semigroup Φ_t in $C_\infty(\mathbb{R}^d)$. Suppose the control parameters α and β belong to certain metric compact spaces, an integral payoff is specified by a function $g(x,\alpha,\beta)$, and the control of the process can be carried out by controlling the additional drift $f(x,\alpha,\beta)$. This leads (see any textbook on stochastic control) to the Cauchy problem for the stochastic *Hamilton–Jacobi–Bellman* (*HJB*) equation

$$\frac{\partial S}{\partial t} = LS + H\left(x, \frac{\partial S}{\partial x}\right), \tag{7.77}$$

where

$$H(x,p) = \max_\alpha \min_\beta \left(f(x,\alpha,\beta)\frac{\partial S}{\partial x} + g(x,\alpha,\beta)\right).$$

If the process generated by L has a heat kernel (Green function) $G(t,x,\xi)$, it is often easier to work with the *mild form* of the HJB equation

$$S_t(x) = \int G(t,x,\xi)S_0(\xi)d\xi + \int_0^t ds \int G(t-s,x,\xi)H\left(\xi, \frac{\partial S_s(\xi)}{\partial \xi}\right)d\xi, \tag{7.78}$$

which, according to Duhamel principle (see Theorem 4.1.3), is formally equivalent (i.e. when both solutions are sufficiently smooth) to solving equation (7.78) with the initial data S_0.

Theorem 7.8.1. *Suppose*

(i) *$H(x,p)$ is Lipschitz in p uniformly in x with a Lipschitz constant κ, and $|H(x,0)| \le h$ for a constant h and all x;*

(ii) *the space $C^1_\infty(\mathbb{R}^d)$ is invariant under Φ_t and the operators Φ_t form a strongly continuous semigroup in the Banach space $C^1_\infty(\mathbb{R}^d)$;*

(iii) *the operators Φ_t are integral with an integral kernel (heat kernel or Green function) $G(t,x,\xi)$, of class $C^1_\infty(\mathbb{R}^d)$ with respect to x and such that*

$$\sup_x \int_0^t ds \int \left(\left| \frac{\partial}{\partial x} G(s,x,\xi) \right| + |G(s,x,\xi)| \right) d\xi = o(t) \tag{7.79}$$

for $t > 0$.

Then for any $S_0 \in C^1_\infty(\mathbb{R}^d)$ there exists a unique solution $S_t(x)$ of equation (7.78), *which is of class $C^1_\infty(\mathbb{R}^d)$ for all t.*

Proof. Recall that $C([0,T], C^1_\infty(\mathbb{R}^d))$ denotes the space of continuous functions $t \mapsto \phi_t \in C^1_\infty(R^d)$ with the norm

$$\|\phi.\| = \sup_{t\in[0,T]} \|\phi_t\|_{C^1(\mathbb{R}^d)}.$$

Let $B^T_{S_0}$ denote the closed convex subset of $C([0,T], C^1_\infty(\mathbb{R}^d))$ consisting of functions with a given $\phi_0 = S_0$. Let Ψ be a mapping in $B^T_{S_0}$ given by the formula

$$\Psi_t(\phi) = \int G(t,x,\xi) S_0(\xi) d\xi + \int_0^t ds \int G(t-s,x,\xi) H\left(\xi, \frac{\partial \phi_s}{\partial \xi}\right) d\xi.$$

In fact, it follows from assumptions (i)–(iii) that Φ maps $B^T_{S_0}$ into itself, because these assumptions imply the inequality

$$\|\Psi_t(\phi)\|_{C^1(\mathbb{R}^d)} \le \|T_t S_0\|_{C^1(\mathbb{R}^d)} + o(t)(h + \kappa \sup_{s\le t} \|\phi_s\|_{C^1(\mathbb{R}^d)}).$$

Fixed points of Ψ are obviously solutions of equation (7.78). By (7.79), for $\phi^1, \phi^2 \in B^T_{S_0}$,

$$\|\Psi_t(\phi^1) - \Psi_t(\phi^2)\|_{C^1(\mathbb{R}^d)} \le o(t)\kappa \sup_{s\le t} \|\nabla\phi^1_s - \nabla\phi^2_s\|,$$

where ∇ denotes the derivative with respect to the spatial variable x. Hence Φ is a contraction in $B^T_{S_0}$ for small enough T, and consequently it has a unique fixed point. For finite t, the solutions are constructed by iteration, as usual in such cases. □

It is not difficult to see that a classical solution to the HJB equation, which belongs to D for all times, solves the mild equation (7.78). Thus Theorem 7.8.1 provides bypassing the uniqueness result for the solutions of the HJB equation. However, in order to conclude that a solution to (7.78) actually solves the HJB equation (7.77) one has to prove that it actually belongs to D, and in order to be able to do this, additional regularity assumptions on the semigroup Φ_t are needed. Next theorem gives an example of such assumptions.

Recall that we denote by $C_{\mathrm{Lip}}(\mathbb{R}^d)$ the Banach space of Lipschitz continuous functions on $\mathbb{R}^d$ equipped with the norm

$$\|f\|_{\mathrm{Lip}} = \sup_x |f(x)| + \sup_{x \neq y} |f(x) - f(y)|/|x - y|.$$

Theorem 7.8.2. *Under the assumptions of Theorem* 7.8.1 *suppose additionally that*

(i)
$$|H(x_1, p) - H(x_2, p)| \leq \tilde{\kappa}|x_1 - x_2|(1 + |p|) \tag{7.80}$$
with a certain constant $\tilde{\kappa}$*;*

(ii) $D = C^2_\infty(\mathbb{R}^d)$ *and the operators* Φ_t *are uniformly bounded for finite times as the operators in the Banach space* $C^2_\infty(\mathbb{R}^d)$*;*

(iii) *for* $t > 0$, G *is continuously differentiable in* t, Φ_t *maps the space* $C_{\mathrm{Lip}}(\mathbb{R}^d)$ *to* $C^2_\infty(\mathbb{R}^d)$, *and*
$$\left\| \int G(t, x, \xi)\phi(\xi)\, d\xi \right\|_{C^2(\mathbb{R}^d)} \leq \omega(t)\|\phi\|_{\mathrm{Lip}} \tag{7.81}$$
for all $\phi \in C_{\mathrm{Lip}}(\mathbb{R}^d)$, *where* $\omega(t)$ *is a positive integrable function of* $t > 0$.

Then, if $S_0 \in C^2_\infty(\mathbb{R}^d)$, *the unique solution* $S_t(x)$ *of equation* (7.78) *constructed in Theorem* 7.8.1 *belongs to* $C^2_\infty(\mathbb{R}^d)$ *for all* $t > 0$ *and represents a classical solution of the HJB equation* (7.77).

Proof. Let $B^{T,R,2}_{S_0}$ denote a subset of $B^T_{S_0}$ consisting of functions, which are twice continuously differentiable in x with

$$\sup_{t \leq T} \|\phi_t\|_{C^2_\infty(\mathbb{R}^d)} \leq R.$$

From (7.81) it follows that

$$\|\Psi_t(\phi)\|_{C^2(\mathbb{R}^d)} \leq \|T_t S_0\|_{C^2(\mathbb{R}^d)} + \int_0^t \omega(s)\, ds(\tilde{\kappa} + \kappa) \sup_{s \leq t} \|\nabla\phi_s\|_{\mathrm{Lip}}. \tag{7.82}$$

Hence Ψ maps $B^{T,R,2}_{S_0}$ to itself whenever

$$\sup_{t \leq T} \|\Phi_T S_0\|_{C^2(\mathbb{R}^d)} + \int_0^T \omega(s)\, ds(\tilde{\kappa} + \kappa)R \leq R$$

or if

$$R \geq \sup_{t \leq T} \|\Phi_t S_0\|_{C^2(\mathbb{R}^d)} \left(1 - \int_0^T \omega(s)\, ds (\tilde{\kappa} + \kappa)\right)^{-1}, \tag{7.83}$$

where T is reduced, if necessary, to make the dominator on the r.h.s. positive. By Theorem 7.8.1, for any $S_0 \in C^2_\infty(\mathbb{R}^d)$, the iterations $\Psi^k(S_0)$ (of course, S_0 here is considered as embedded in $B^T_{S_0}$ as a constant, in t, function) converge, as $n \to \infty$, to the unique solution $S_t(x)$ of equation (7.78). Moreover, by the above argument, if $\|S_0\|_{C^2_\infty(\mathbb{R}^d)} \leq R$ with R satisfying (7.83), then all iterations also belong to $B^{T,R,2}_{S_0}$. Consequently, the limit $S_t(x) \in B^T_{S_0}$ has an additional property that its derivatives in x are Lipschitz continuous with the Lipschitz constant bounded by R. Hence it follows from (7.82) and equation (7.78) that $S_t(x)$ is twice continuously differentiable in x and hence represents a classical solution to the HJB equation (7.77). □

By Theorem 7.5.1, the 'stable-like' operator

$$L = -a(x)|\nabla|^{\alpha(x)} u + (b(x), \nabla f(x)) + \int_{\mathbb{R}^d \setminus \{0\}} (f(x+y) - f(x)) \nu(x, dy) \tag{7.84}$$

where (i) $\inf \alpha(x) > 1$ and a, α satisfy the requirements of Theorem 7.5.1, (ii) b, ν satisfy the requirements of Theorem 5.1.1, fits into the conditions of Theorem 7.8.1. As it follows from Section 7.6 the additional regularity conditions needed for the validity of Theorem 7.8.2 follow from the appropriate additional regularity assumptions on the coefficients of the stable-like operator L.

Remark 53. Showing that a classical (smooth) solution of the HJB equation yields a solution to the corresponding optimization problem is a standard procedure, called the *verification theorem*, see Fleming and Soner [114]. Moreover, by Theorem 7.8.1, the solutions to the mild form of HJB equation form a semigroup, and hence, according to a general result of control theory, see Theorem 5.1 in Chapter 2 of Fleming and Soner [114], represent *viscosity solutions*, closely related to the corresponding optimization problem.

The fixed-point argument used above for proving well-posedness of HJB equation have a straightforward extension to time non-homogeneous situations. Notice however, that in applications to optimal control, the Cauchy problem for HJB equation is usually given in inverse time. In time-homogeneous situations, one can easily reformulate it as a usual Cauchy problem by time reversion (which is often done in practice). But in time non-homogeneous cases, one has to stick to the inverse time formulation, as we are going to do now.

Namely, consider the *time non-homogeneous stochastic HJB equation*

$$\frac{\partial S}{\partial t} + L_t S + H_t\left(x, \frac{\partial S}{\partial x}\right) = 0, \quad t < T, \tag{7.85}$$

with a given $S|_{t=T} = S_T$, where H_t is a time-dependent family of functions on $\mathbb{R}^{2d}$ and L_t a time-dependent family of conditionally positive operators in $C_\infty(\mathbb{R}^d)$. If the family L_t generate a backward propagator $\{U^{t,r}\}$ on some common invariant domain $D \subset C^1_\infty(\mathbb{R}^d)$ (see Section 1.9 for definitions related to propagators), one often introduces the following *mild form* of HJB equation (7.85):

$$S_t = U^{t,T} S_T + \int_t^T U^{t,s} H_s(., \nabla S_s(.))\, ds. \tag{7.86}$$

If $U^{t,s}$ are integral operators with kernels $G(t,s,x,y)$, then equation (7.86) can be written in the form

$$S_t(x) = \int G(t,T,x,\xi) S_0(\xi) d\xi + \int_t^T ds \int G(t,s,x,\xi) H\Big(\xi, \frac{\partial S_s(\xi)}{\partial \xi}\Big)\, d\xi. \tag{7.87}$$

The proof of the following two results is almost literally the same as the proof of Theorems 7.8.1, 7.8.2 and is omitted.

Theorem 7.8.3. *Suppose*

(i) *$H_t(x,p)$ is continuous in t and Lipschitz continuous in p uniformly in x and t with a Lipschitz constant κ, and $|H_t(x,0)| \le h$ for a constant h and all x, t;*

(ii) *the operators $U^{t,s}$ form a strongly continuous backward propagator in the Banach space $C^1_\infty(\mathbb{R}^d)$;*

(iii) *the operators $U^{t,s}$ are smoothing, in the sense that they map $C_\infty(\mathbb{R}^d)$ to $C^1_\infty(\mathbb{R}^d)$ for $t < s$, and*

$$\|U^{t,r}\phi\|_{C^1_\infty(\mathbb{R}^d)} \le \omega(r-t)\|\phi\|_{C_\infty(\mathbb{R}^d)} \tag{7.88}$$

for $t < r < T$, with an integrable positive function ω on $[0,T]$.

Then for any $S_0 \in C^1_\infty(\mathbb{R}^d)$ there exists a unique solution $S_t(x)$ of equation (7.86), *which is of class $C^1_\infty(\mathbb{R}^d)$ for all t.*

Theorem 7.8.4. *Under the assumptions of Theorem* 7.8.3 *suppose additionally that*

(i) (7.80) *hold for all H_t uniformly in t;*

(ii) *$D = C^2_\infty(\mathbb{R}^d)$ and the operators $U^{t,s}$ are uniformly bounded for finite times as the operators in the Banach space $C^2_\infty(\mathbb{R}^d)$;*

(iii) *L_t are bounded operators $D \to C_\infty(\mathbb{R}^d)$ that depend continuously on t, $U^{t,s}$ maps the space $C_{\mathrm{Lip}}(\mathbb{R}^d)$ to $C^2_\infty(\mathbb{R}^d)$ for $t < s$, and*

$$\|U^{t,r}\phi\|_{C^2_\infty(\mathbb{R}^d)} \le \omega(r-t)\|\phi\|_{\mathrm{Lip}} \tag{7.89}$$

for all $\phi \in C_{\mathrm{Lip}}(\mathbb{R}^d)$.

Then, if $S_0 \in C^2_\infty(\mathbb{R}^d)$*, the unique solution* $S_t(x)$ *of equation* (7.86) *constructed in Theorem* 7.8.3 *belongs to* $C^2_\infty(\mathbb{R}^d)$ *for all* $t > 0$ *and represents a classical solution of the HJB equation* (7.77).

To apply these results to time nonhomogeneous stable-like process one has to use the corresponding time non-homogeneous extensions of the theory of Section 7.6. This extension is more or less straightforward and is discussed in Kolokoltsov [192] and [196]. Of course, the conditions of Theorems 7.8.1–7.8.3 are satisfied for L_t generating non-degenerate diffusion processes with sufficiently smooth coefficients.

7.9 Application to Langevin equations driven by a stable noise

As we already mentioned, the system of SDE $dx = v\,dt$, $dv = dB_t$, describes a behavior of a Newtonian particle with the white noise force and velocity being a Brownian motion. The process (x, v) is often called a *physical Brownian motion* or *Kolmogorov's diffusion.* It is a Gaussian process with transition probability densities that can be written in a simple closed form. Its natural extension to the stable case is described by a process solving the corresponding *stable noise driven Langevin equation* $dx = v\,dt$, $dv = dW_t^\alpha$, where W^α is an α-stable Lévy motion. Like with the stable motion itself, the corresponding transition probabilities cannot be written in closed form. However, one can obtain their asymptotic expansions. We shall not develop here this story fully, but discuss only finite distance asymptotic expansions in symmetric case.

Thus, suppose W_t^α is the simplest d-dimensional α-stable process, so that

$$\mathbf{E}e^{iyW_t} = e^{-\sigma|y|^\alpha}$$

with $\sigma > 0$, $\alpha \in (0, 2)$.

The solution (X_t, V_t) of the corresponding Langevin equation with the initial condition (x_0, v_0) is

$$\begin{cases} X_t = x_0 + tv_0 + \int_0^t W_s^\alpha\,ds, \\ V_t = v_0 + W_t^\alpha. \end{cases}$$

Like with the stable laws, the starting point for the analysis of the densities is the explicit expression for the characteristic function of the process (X_t, V_t), which is given by

$$\begin{aligned} \phi_{X,V}(p, q, t) &= \mathbf{E}\exp\{ipX_t + iqV_t\} \\ &= \exp\left\{ip(x_0 + tv_0) + iqv_0 - \sigma t \int_0^1 |q + stp|^\alpha\,ds\right\}. \end{aligned} \tag{7.90}$$

This expression can be easily obtained from the methods of stochastic integration with respect to stable random measure developed in Samorodnitski and Taqqu [288] (details are given e.g. in Kolokoltsov and Tyukov [204] and will not be reproduced here). Hence the transition density of the process (X_t, V_t) is given by the Fourier transform

$$S_t(x, v; x_0, v_0) = \frac{1}{(2\pi)^{2d}} \int dp\, dq$$
$$\times \exp\left\{-ip(x - x_0 - tv_0) - iq(v - v_0) - \sigma t \int_0^1 |q + stp|^\alpha\, ds\right\}. \quad (7.91)$$

Proposition 7.9.1. (a) *For any $t > 0$, the density* (7.91) *is well defined and is an infinitely smooth function of x, v such that*

$$S_t(x, v; x_0, v_0) \le \mathbb{C}(\alpha, d) t^{-d} (t\sigma)^{-2d/\alpha} \quad (7.92)$$

with $C(\alpha)$ being a constant depending only on α.

(b) *For $\alpha \in (1, 2)$ the density* (7.91) *is a holomorphic function of x, v.*

(c) *Finally, the density* (7.91) *can be represented via an explicit asymptotic power series (given below) which is absolutely convergent for $\alpha \in (1, 2)$.*

Proof. Let us start with the case $d = 1$. Changing p to $w = tp/q$ in (7.90) yields for $S_t(x, v; x_0, v_0)$ the expression

$$\frac{1}{(2\pi)^2 t} \iint dq dw |q| \exp\left\{-iq\left[\frac{w}{t}(x - x_0 - tv_0) + (v - v_0)\right] - \sigma t |q|^\alpha I(w)\right\}, \quad (7.93)$$

where $I(0) = 1$ and

$$I(w) = \int_0^1 |1 + sw|^\alpha\, ds = \frac{1}{w(\alpha + 1)} (|1 + w|^{1+\alpha} \operatorname{sgn}(1 + w) - 1), \quad w \neq 0.$$

Notice that $I(w)$ is a bounded below (by a strictly positive constant) function that behaves like $|z|^\alpha/(1 + \alpha)$ for $z \to \infty$.

Next, changing the variable of integration q to $q[t\sigma I(w)]^{1/\alpha}$ in (7.93) yields

$$S_t(x, v; x_0, v_0) = \frac{1}{2\pi t} \int (t\sigma I(w))^{-2/\alpha}\, dw \quad (7.94)$$
$$\times \phi_1\left((t\sigma I(w))^{-1/\alpha}\left[\frac{w}{t}(x - x_0 - tv_0) + (v - v_0)\right], \alpha\right),$$

where the function ϕ_b introduced at the end of Section 7.1 is given by

$$\phi_b(x, \alpha) = \frac{1}{2\pi} \int_{-\infty}^{\infty} |y|^b e^{-ixy - |y|^\alpha}\, dy = \frac{1}{\pi} \int_0^\infty y^b \cos(xy) e^{-y^\alpha}\, dy.$$

Since the function ϕ_1 is clearly uniformly bounded, it follows that $S_t(x, v; x_0, v_0)$ is well defined and enjoys the estimate (7.92). Moreover, from expansion (7.12) it follows that S_t is given by the asymptotic expansion

$$\frac{1}{2t\pi^2\alpha}\sum_{m=0}^{\infty}\frac{(-1)^m}{(2m)!}\Gamma\left(\frac{2m+2}{\alpha}\right) \times \int (t\sigma I(w))^{-2(m+1)/\alpha}\left[\frac{w}{t}(x-x_0-tv_0)+(v-v_0)\right]^{2m} dw,$$

which can be written as

$$\frac{1}{2t\pi^2\alpha}\sum_{m=0}^{\infty}\sum_{l=0}^{m}\frac{(-1)^m}{(2m)!}\Gamma\left(\frac{2m+2}{\alpha}\right)C_{2m}^l \times (t\sigma)^{-(2m+1)/\alpha}(v-v_0)^{2m-l}\left[\frac{x-x_0-tv_0}{t}\right]^l \Omega_{lm}, \quad (7.95)$$

where C_{2m}^l are binomial coefficients and

$$\Omega_{lm} = \int w^l (I(w))^{-2(m+1)/\alpha} dw.$$

As these integrals are convergent and bounded by ω^m with some ω uniformly in l and m, all claims follow.

Let now d be arbitrary. Introducing spherical coordinate $|p|, \bar{p} = p/|p|, |q|, \bar{q} = q/|q|$, and using

$$|q + tsp|^\alpha = (q^2 + 2ts(q,p) + t^2s^2p^2)^{\alpha/2},$$

allows one to rewrite (7.91) as

$$S_t(x, v; x_0, v_0) = \frac{1}{(2\pi)^{2d}}\int_0^\infty d|q| \int_0^\infty d|p| \int_{S^{d-1}} d\bar{q} \int_{S^{d-1}} d\bar{p}\, |q|^{d-1}|p|^{d-1} \times \cos(|p|(\bar{p}, x-x_0-tv_0) + |q|(\bar{q}, v-v_0)) \times \exp\left\{-\sigma t \int_0^1 (|q|^2 + 2st|q|\,|p|\tau + t^2s^2|p|^2)^{\alpha/2} ds\right\},$$

where τ denotes the cosine of the angle between q and p. Next, as in one-dimensional cases, change $|p|$ to $w = t|p|/|q|$ yielding

$$S_t(x, v; x_0, v_0) = \frac{1}{(2\pi)^{2d}t^d}\int_0^\infty d|q| \int_0^\infty dw \int_{S^{d-1}} d\bar{q} \int_{S^{d-1}} d\bar{p}\, |q|^{2d-1}w^{d-1} \times \cos\left(|q|\left[\frac{w}{t}(\bar{p}, x-x_0-tv_0) + (\bar{q}, v-v_0)\right]\right)\exp\{-\sigma t|q|^\alpha I(\tau, w)\},$$

where

$$I(\tau, w) = \int_0^1 (1 + 2sw\tau + s^2w^2)^{\alpha/2}\, ds.$$

Finally, changing $|q|$ to $r = |q|(\sigma t I(\tau, w))^{1/\alpha}$ yields

$$S_t(x, v; x_0, v_0) = \frac{\pi}{(2\pi)^{2d} t^d} \int_0^\infty dw \int_{S^{d-1}} d\bar{q} \int_{S^{d-1}} d\bar{p}\, w^{d-1} (t\sigma I(\tau, w))^{-2d/\alpha}$$
$$\times \phi_{2d-1}\left((t\sigma I(\tau, w))^{-1/\alpha} \left[\frac{w}{t}(\bar{p}, x - x_0 - tv_0) + (\bar{q}, v - v_0) \right], \alpha \right). \quad (7.96)$$

The proof is now complete again by the properties of the function ϕ_{2d-1}. □

7.10 Comments

The extension of one-dimensional asymptotic expansions of stable laws to the finite-dimensional case was obtained in Kolokoltsov [178], while the main term of the large-distance asymptotic was previously calculated in Blumenthal, Getoor [61] and Bendikov [43] and in particular cases by S. Chandrasekhar [75]. Turning to transition probability densities (Green functions) for stable processes with varying coefficients, in the case of a fixed index $\alpha > 1$, let us first mention that the existence of the transition probability density for such a process was first proved in a more general framework by Kochubei [165], based on the theory of hyper-singular integrals from Samko, Kilbas and Marichev [286]. For variable index α the existence of a measurable density was proved by Negoro [253] and Kikuchi, Negoro, [163] for uniform spectral measure and infinitely smooth coefficients.

Our exposition follows essentially Kolokoltsov [178], [179], where the existence of a smooth density for the case of variable α was proved under mild regularity assumption on the coefficients (Hölder continuity), and global (in space) bounds and local multiplicative asymptotics for the density were provided. In the latter paper also the extension to nonuniform symmetric spectral measures was developed, i.e. for generators of the form $\int |(s, \nabla)^{\alpha(x)}| S(x, s)\, ds$ (with a continuous symmetric in $s \in S^{d-1}$ function $S(x, s)$) instead of $|\nabla|^{\alpha(x)}$. The extensions to the time-dependent case and to stable-like generators perturbed by integral generators, i.e. having form

$$\frac{\partial u}{\partial t} = (A(x), \nabla u(x)) - a(x)|\nabla|^{\alpha(x)} u(x) + \int (u(x + y) - u(x)) f(x, \xi)\, d\xi,$$

are developed in Kolokoltsov [192] and [178]. It is worth stressing that the basic idea of reconstructing the exact heat kernel of stable processes from its asymptotic approximation is the same as for diffusions. For non-degenerate diffusions it was seemingly first introduced by P. Lévy. With some modifications it works also for degenerate and complex diffusions, see Sections 9.5, as well as the book [179] and bibliography therein.

In Chapter 6 of [179] one can also find the analogs of the *small diffusion* or *quasi-classical* asymptotics for *truncated stable-like processes*, namely the asymptotics as $h \to 0$ to the Green function of Markov processes with generator

$$Lu(x) = \left(A(x), \frac{\partial u}{\partial x}\right) + \frac{1}{h}\int_{\mathbb{R}^d}\left[u(x+h\xi) - u(x) - \frac{h}{1+\xi^2}\left(\xi, \frac{\partial u}{\partial x}\right)\right]\nu(x, d\xi),$$

where

$$\nu(x, d\xi) = G(x)\mathbf{1}_{|\xi|\le a(x)}(|\xi|)|\xi|^{-(d+\alpha)}\, d\xi.$$

An approach to obtaining bounds for stable-like kernels and certain their extensions based on the Poincaré inequality was developed by Chen, Kim and Kumagai [82]. The analysis of stable-like processes by means of the martingale problem approach was carried out in Bass [28] and Bass and Tang [33], see also Abels and Kassmann [1]. For an approach based on Dirichlet forms and the related notion of the carré du champ operator we refer to Barlow et al [27] and Uemura [313] and references therein, for numerical analysis to Zhuang et al. [326]. A study of the related Harnack inequality was conducted in Song and Vondracek [297].

Extensions to related classes of nonlinear equations are numerous, see e.g. Brandolese and Karch [68], where asymptotic expansions are developed for solutions to the equation

$$\partial_t u + (-\Delta)^{\alpha/2} u + \nabla(f(u)) = 0,$$

with $1 < \alpha < 2$, or Truman and Wu [310], devoted to the equation

$$(\partial_t - A)u + \partial_x q(u) = f(u) + g(u)F_{t,x},$$

where A generates a strong Feller semigroup and F is a Lévy space-time white noise, or Kolokoltsov [192], devoted to rather general interacting stable-like processes.

For infinite-dimensional extensions we refer e.g. to Chen and Kumagai [85]. Estimates for heat kernels of stable-like processes on *d-sets* (which are defined as closed subsets F of R^n, $0 < d \le n$, equipped with a Borel measure μ such that, for some constants $C_2 > C_1 > 0$, the μ-measures of balls centered at any point from F and with any radius r belong to the interval $[C_1 r^d, C_2 r^d]$), are obtained by Chen and Kumagai [84]. These d-sets arise naturally in the theory of function spaces and in fractal geometry. An extension into a different direction can be performed by applying the Feynman–Kac formula, see Wang [315] and references therein.

Applications of the stable-like heat kernel estimates to the analysis of the scaling limit behavior of periodic stable-like processes can be found in Franke [116], and to optimization and finance in Bennett and Wu [47].

On the intersection of the themes developed in this and the preceding chapters lies an interesting topic of censored stable and stable-like processes, that represent stable-like processes forced to stay inside a given domain, see Bogdan, Burdzy and Chen

[62], Chen, Kim and Song [81], [83]. These *censored stable-like process*, in an open domain D of $\mathbb{R}^d$, can be defined via the Dirichlet forms

$$E(u,v) = \frac{1}{2}\int_D\int_D (u(x)-u(y))(v(x)-v(y))\frac{C(x,y)}{|x-y|^{d+\alpha}}dxdy,$$

where $C(x,y)$ is a bounded function on $D \times D$.

Stochastic control is of course widely developed. However, the mainstream of research on Markov controls is still devoted mostly to either Markov chains, see e.g. Yin and Zhang [321], Guo and Hernández-Lerma [124], Hernández-Lerma and Lasserre [129] and references therein, or diffusion processes, see Bensoussan [48] or Yong and Zhou [322] and references therein. More difficult analysis of general controlled Markov process is slowly developing, stimulated mostly by the needs of financial mathematics, where Lévy processes are firmly establishing themselves.

Our Theorem 7.8.1 is an extension to stable-like processes of a result from Droniou and Imbert [103] devoted to nonlinear equations arising from controlled stable processes. On the other hand, Theorems 7.8.3 and 7.8.4 extend to general Feller processes with a smoothing propagator (including stable-like processes) some known results from the control of linear and nonlinear diffusions, see e.g. Huang, Caines and Malhamé [135], [136], where such result was obtained as an intermediate step in the analysis of controlled interacting diffusions. Our approach is different, as it is direct and does not rely on the control theory interpretation and related tools (like the verification theorem). Similar results are obtained in Jacob, Potrykus and Wu [144], where the authors study the solutions of the related nonlinear equations in spaces of integrable functions.

As in the analysis of Markov processes themselves, in analyzing general controlled processes, a crucial simplification is achieved by assuming the presence of a non-degenerate diffusion term that induces a regularizing effect on the dynamics. Under this assumption an existence and uniqueness result for the classical solutions of the HJB equation can be found e.g. in Mikami [246] or Garroni and Menaldi [119].

Of course, in many situations the classical solutions to general nonlinear integro-differential HJB equations do not exist, and one needs an appropriate notion of a generalized solutions. These solutions can be introduced via vanishing viscosity approach, see e.g. Fleming and Soner [114], [296] or Ishikawa [139] and references therein, or via the so called *idempotent* or *max-plus* algebras, see e.g. Fleming and McEneaney [113], Kolokoltsov and Maslov [200], [201], Litvinov [224], Akian, David and Gaubert [2], or Akian, Gaubert and Walsh [3].

Chapter 8
CTRW and fractional dynamics

Suppose $(X_1, T_1), (X_2, T_2), \dots$ is a sequence of i.i.d. pairs of random variables such that $X_i \in \mathbb{R}^d$, $T_i \in \mathbb{R}_+$ (jump sizes and waiting times between the jumps), the distribution of each (X_i, T_i) being given by a probability measure $\psi(dx\,dt)$ on $\mathbb{R}^d \times \mathbb{R}_+$. Let

$$N_t = \max\left\{n : \sum_{i=1}^{n} T_i \le t\right\}.$$

The process

$$S_{N_t} = X_1 + X_2 + \cdots + X_{N_t} \tag{8.1}$$

is called the *continuous time random walk* (CTRW) arising from ψ.

Of particular interest are the situations where T_i belong to the domain of attraction of a $\beta \in (0,1)$-stable law and X_i belong to the domain of attraction of a $\alpha \in (0,2)$-stable law. Here we extend the theory much further to include possible dependence of (T_n, X_n) on the current position, i.e. to spatially non-homogeneous situations.

As a basis for our limit theorems, we develop the general theory of subordination of Markov processes by the hitting-time process, showing that this procedure leads naturally to (generalized) fractional evolutions.

After two introductory sections, we demonstrate our approach to the limits of CTRW in Section 8.3 by obtaining simple limit theorems for position-dependent random walks with jump sizes from the domain of attraction of stable laws. Section 8.4 is devoted to the theory of subordination by hitting times. Finally in Section 8.5 we combine the two bits of the theory from Sections 8.3 and 8.4 giving our main results on CTRW.

By a continuous family of transition probabilities (CFTP) in X we mean as usual a family $p(x; dy)$ of probability measures on X depending continuously on $x \in X$, where probability measures are considered in their weak topology ($\mu_n \to \mu$ as $n \to \infty$ means that $(f, \mu_n) \to (f, \mu)$ as $n \to \infty$ for any $f \in C(X)$).

8.1 Convergence of Markov semigroups and processes

Let Ω be a locally compact space. For any $h > 0$, let $\Omega_h \subset \Omega$ be a closed subset. In basic examples $\Omega = \mathbb{R}^d$ and Ω_h either coincides with Ω or is a lattice in $\mathbb{R}^d$ with steps in each direction depending on h. By π_h we shall denote here the projection $C_\infty(\Omega) \mapsto C_\infty(\Omega_h)$ obtained by restriction. We shall mostly need the case with

$\Omega_h = \Omega$, but for numerical calculations and simulations the case of Ω_h a lattice is important.

Theorem 8.1.1 (Convergence of semigroups with rates). *Let T_t be a Feller semigroup in $C_\infty(\Omega)$ with generator L and B be a dense invariant (under T_t) subspace of the domain of L such that B is itself a normed space under a norm $\|.\|_B$ and*

$$\|T_t f\|_B \le c(t)\|f\|_B \tag{8.2}$$

for all f, for some non-decreasing continuous function c on $\mathbb{R}_+$.

(i) *Let T_t^h be a Feller semigroup in $C_\infty(\Omega_h)$ with generator L_h having domain containing $\pi_h B$ such that*

$$\|(L_h\pi_h - \pi_h L)f\| \le \omega(h)\|f\|_B \tag{8.3}$$

with $\omega(h) = o(1)$ as $h \to 0$ (i.e. $\omega(h) \to 0$ as $h \to 0$). Then for any $t > 0$ and $f \in C_\infty(\Omega)$ the sequence $T_s^h f$ converges to $T_s f$ as $h \to 0$ uniformly for $s \in [0, t]$, in the sense that

$$\sup_{s\le t} \|T_s^h \pi_h f - \pi_h T_s f\| \to 0, \quad h \to 0 \tag{8.4}$$

(in case $\Omega_h = \Omega$ this is the usual convergence in $C_\infty(\Omega)$). Specifically, if $f \in B$, then

$$\sup_{s\le t} \|T_s^h \pi_h f - \pi_h T_s f\| \le \omega(h) \int_0^t c(s)\, ds \|f\|_B. \tag{8.5}$$

(ii) *Let a function $h(\tau)$, $\tau > 0$, be specified (that defines the scale of approximation in the space, which is appropriate to a given time step τ), together with a family U^τ of contractions in $C_\infty(\Omega_{h(\tau)})$ such that*

$$\left\| \left(\frac{U^\tau - 1}{\tau} \pi_{h(\tau)} - \pi_{h(\tau)} L \right) f \right\| \le \omega(\tau)\|f\|_B, \tag{8.6}$$

where $\lim_{t\to 0} \omega(t) = 0$. Assume also that the limit defining L is uniform on bounded sets of B, more precisely that

$$\left\| \frac{T_t f - 1}{t} f - Lf \right\| \le \kappa(t)\|f\|_B \tag{8.7}$$

with $\lim_{t\to 0} \kappa(t) = 0$. Then for all $f \in C_\infty(\Omega)$

$$\|U^{\tau k} \pi_{h(\tau)} f - \pi_{h(\tau)} T_{k\tau} f\| \le (\kappa(\tau) + \omega(\tau)) \int_0^{k\tau} c(s)\, ds \|f\|_B, \tag{8.8}$$

and

$$\lim_{h\to 0} \sup_{s\le t} \|(U^\tau)^{[s/\tau]} \pi_{h(\tau)} f - \pi_{h(\tau)} T_s f\| = 0. \tag{8.9}$$

Proof. In case $\Omega_h = \Omega$ the required statement follows from a more general, time non-homogeneous, Theorem 1.9.5. In our new setting of different Banach spaces for approximating semigroups, we shall still follow the same reasoning.

(i) For $f \in B$

$$T_t^h \pi_h f - \pi_h T_t f = -T_{t-s}^h \pi_h T_s f|_0^t = -\int_0^t \frac{d}{ds}(T_{t-s}^h \pi_h T_s) f \, ds$$
$$= \int_0^t T_{t-s}^h (L_h \pi_h - \pi_h L) T_s f \, ds.$$

Consequently, by (8.3)

$$\|T_t^h \pi_h f - \pi_h T_t f\| \le \int_0^t \|(L_h \pi_h - \pi_h L) T_s f\| \, ds \le \int_0^t \omega(h) \|T_s f\|_B \, ds,$$

implying (8.5) by (8.2). Convergence for all $f \in C_\infty(\Omega)$ follows now as usual by approximation of f by the elements of B.

(ii) Notice that (8.7) is equivalent to

$$\left\| \frac{1}{t} \int_0^t (T_s - 1) L f \right\| ds \le \kappa(t) \|f\|_B,$$

because

$$T_t f - f = tLf + \int_0^t (T_s - 1) L f \, ds.$$

Consequently, by (8.6) and the equations

$$U^\tau \pi_h - \pi_h T_\tau = (U^\tau - 1)\pi_h - \pi_h \int_0^\tau T_s L \, ds$$
$$= (U^\tau - 1)\pi_h - \tau \pi_h L - \pi_h \int_0^\tau (T_s - 1) L \, ds,$$

one concludes that

$$\|U^\tau \pi_{h(\tau)} f - \pi_{h(\tau)} T_\tau f\| \le \tau(\kappa(\tau) + \omega(\tau)) \|f\|_B. \tag{8.10}$$

Consequently, since

$$U^{\tau k} \pi_{h(\tau)} - \pi_{h(\tau)} T_{k\tau}$$
$$= U^{\tau k} \pi_{h(\tau)} - U^{\tau(k-1)} \pi_{h(\tau)} T_\tau + U^{\tau(k-1)} \pi_{h(\tau)} T_\tau - U^{\tau(k-2)} \pi_{h(\tau)} T_{2\tau}$$
$$+ \cdots + U^\tau \pi_{h(\tau)} T_{(k-1)\tau} - \pi_{h(\tau)} T_{k\tau}$$
$$= U^{\tau(k-1)} (U^\tau \pi_{h(\tau)} - \pi_{h(\tau)} T_\tau) + U^{\tau(k-2)} (U^\tau \pi_{h(\tau)} - \pi_{h(\tau)} T_\tau) T_\tau$$
$$+ \cdots + (U^\tau \pi_{h(\tau)} - \pi_{h(\tau)} T_\tau) T_{(k-1)\tau},$$

it follows from (8.10) that

$$\|U^{\tau k}\pi_{h(\tau)}f - \pi_{h(\tau)}T_{k\tau}f\| \\ \leq \tau(\kappa(\tau)+\omega(\tau))\|f\|_B(1+c(\tau)+c(2\tau)+\cdots+c((k-1)\tau)),$$

implying (8.8) by the assumed monotonicity of $c(t)$. The last statement follows again by approximation. □

8.2 Diffusive approximations for random walks and CLT

Let $p(x,dy)$ be a continuous family of transition probabilities in $\mathbb{R}^d$ such that

$$\int yp(x,dy) = 0, \quad \sup_x \int |y|^3 p(x,dy) = C < \infty. \tag{8.11}$$

Consider the family of jump-type Markov process $Z_x^h(t)$, $h > 0$, generated by

$$(L_h f)(x) = \frac{1}{h^2}\int (f(x+hy) - f(x))p(x;dy). \tag{8.12}$$

For each h the operator L_h is bounded in $C_\infty(\mathbb{R}^d)$ (the second assumptions in (8.11) ensures that $C_\infty(\mathbb{R}^d)$ is invariant under L_h) and hence specifies a Feller semigroup there (see Theorem 3.7.3 for its properties). If p does not depend on x,

$$Z_x^h(t) = x + h(Y_1 + \cdots + Y_{N_t})$$

is a normalized random walk with i.i.d. Y_j being distributed according to p and the number of jumps N_t being a Poisson process with parameter h^{-2}, so that $\mathbf{E}N_1 = h^{-2}$.

We are also interested in the discrete Markov chain $S_x^\tau(k)$, $k \in \mathbb{N}$, with the transition operator

$$U^\tau f(x) = \int f(x+\sqrt{\tau}y)p(x,dy).$$

If p does not depend on x, this Markov chain is represented by the sum (normalized random walk)

$$S_x^\tau(k) = x + \sqrt{\tau}(Y_1 + \cdots + Y_k).$$

On the other hand, let

$$Lf = \frac{1}{2}\int \operatorname{tr}\left(a(x)\frac{\partial^2 f}{\partial x^2}\right), \quad a_{ij}(x) = \int y_i y_j p(x,dy). \tag{8.13}$$

Assume that the matrix square root $\sigma(x) = \sqrt{a}(x)$ is well defined with elements belonging to $C^3(\mathbb{R}^d)$. By Section 4.6, L generates a semigroup T_t specifying a diffusion process in $\mathbb{R}^d$ such that $C_\infty^2 \cap C_{\mathrm{Lip}}^2(\mathbb{R}^d)$ is an invariant core, and

$$\kappa(T_t f) \leq c(t)\kappa(f), \quad \|T_f\|_{C_{\mathrm{Lip}}^2} \leq c(t)\|f\|_{C_{\mathrm{Lip}}^2} \tag{8.14}$$

for some continuous $c(t)$, where $\kappa(f)$ for $f \in C^2_{\mathrm{Lip}}$ denotes the Lipschitz constant of its second derivative.

Theorem 8.2.1 (CLT for position-dependent random walks). *Assume that the matrix square root $\sigma(x) = \sqrt{a}(x)$ is well defined with elements belonging to $C^3(\mathbb{R}^d)$. Then the semigroup T^h_t generated by L_h and the discrete semigroup $(U^{\tau k})^{[t/\tau]}$ both converge to the semigroup T_t generated by L. And more specifically, for $f \in C^2_\infty \cap C^2_{\mathrm{Lip}}(\mathbb{R}^d)$*

$$\sup_{s\le t} \|T^h_s \pi_h f - \pi_h T_s f\| \le \omega(h) \int_0^t c(s)\, ds \|f\|_B, \tag{8.15}$$

and

$$\|U^{\tau k} \pi_{h(\tau)} f - \pi_{h(\tau)} T_{k\tau} f\| \le (\kappa(\tau) + \omega(\tau)) \int_0^{k\tau} c(s)\, ds \|f\|_B. \tag{8.16}$$

Proof. From the Taylor series estimate

$$\left| g(h) - g(0) - g'(0)h - \frac{1}{2} g''(0) h^2 \right| \le \frac{1}{6} \kappa(g) h^3,$$

where κ denotes the Lipschitz constant of g'', one gets from (8.11), (8.12), (8.13) that

$$|L_h f - Lf| \le \frac{h}{6} C \kappa(f),$$

with $\kappa(f)$ the Lipschitz constant of f''. The rest is a consequence of (8.14) and Theorem 8.1.1. □

Corollary 18 (Functional CLT with convergence rates). *If $a(x)$ is the unit matrix for all x, then $c(t) = 1$ in* (8.14), *implying the estimate*

$$\|U^{\tau k} \pi_{h(\tau)} f - \pi_{h(\tau)} T_{k\tau} f\| \le (\kappa(\tau) + \omega(\tau)) \int_0^{k\tau} c(s)\, ds \|f\|_B. \tag{8.17}$$

According to Section 4.8 the convergence of semigroups can be recast in terms of the convergence of distributions on the space of càdlàg trajectories, showing that the distributions of the processes $Z^h(t)$ and $S^\tau([t/\tau])$ converge to the distribution of the diffusion process generated by L.

8.3 Stable-like limits for position-dependent random walks

For a measure $\mu(dy)$ in $\mathbb{R}^d$ and a positive number h, we denote by $\mu(dy/h)$ the scaled measure defined via its action

$$\int g(z)\mu(dz/h) = \int g(hy)\mu(dy)$$

on functions $g \in C(\mathbb{R}^d)$.

For a vector $y \in \mathbb{R}^d$ we shall always denote by $\bar{y}$ its normalization $\bar{y} = y/|y|$, where $|y|$ means the usual Euclidean norm.

Fix an arbitrary $\alpha \in (0,2)$. Let $S : \mathbb{R}^d \times S^{d-1} \to \mathbb{R}_+$ be a continuous positive function that is symmetric with respect to the second variable, i.e. $S(x,y) = S(x,-y)$. It defines a family of α-stable d-dimensional symmetric random vectors (depending on $x \in \mathbb{R}^d$) specified by its characteristic function ϕ_x with

$$\ln \phi_x(p) = \int_0^\infty \int_{S^{d-1}} \left(e^{i(p,\xi)} - 1 - \frac{i(p,\xi)}{1+\xi^2} \right) \frac{d|\xi|}{|\xi|^{1+\alpha}} S(x,\bar{\xi})\, d_S\bar{\xi}, \tag{8.18}$$

where d_S denotes Lebesgue measure on the sphere S^{d-1}. It can be also rewritten in the form

$$\ln \phi_x(p) = C_\alpha \int_{S^{d-1}} |(p,\bar{\xi})|^\alpha S(x,\bar{\xi})\, d_S\bar{\xi}$$

for some constant C_α (see Section 1.4), but for the present discussion the above integral representation will be the most convenient one.

By Section 4.6 (or Section 7.6 in case of non-degenerate spectral measure S), assuming that $S(x,s)$ has bounded derivatives with respect to x up to and inclusive order $q \geq 3$ (if $\alpha < 1$, the assumption $q \geq 2$ is sufficient), the pseudo-differential operator

$$Lf(x) = \ln \phi_x\left(\frac{1}{i}\frac{\partial}{\partial x}\right) f(x) = \int_0^\infty \int_{S^{d-1}} (f(x+y) - f(x)) \frac{d|y|}{|y|^{1+\alpha}} S(x,\bar{y})\, d_S\bar{y} \tag{8.19}$$

generates a Feller semigroup T_t in $C_\infty(\mathbb{R}^d)$ with the space $C^2_{\text{Lip}}(\mathbb{R}^d) \cap C^2_\infty(\mathbb{R}^d)$ as its invariant core.

Denote by $Z_x(t)$ the Feller process corresponding to the semigroup T_t. We are interested here in discrete approximations to T_t and $Z_x(t)$.

We shall start with the following technical result.

Proposition 8.3.1. *Assume that $p(x;dy)$ is a CFTP in $\mathbb{R}^d$ from the normal domain of attraction of the stable law specified by* (8.18). *More precisely, assume that for an arbitrary open $\Omega \subset S^{d-1}$ with a boundary of Lebesgue measure zero*

$$\int_{|y|>n} \int_{\bar{y}\in\Omega} p(x;dy) \sim \frac{1}{\alpha n^\alpha} \int_\Omega S(x,s)\, d_S s, \quad n \to \infty, \tag{8.20}$$

(i.e. the ratio of the two sides of this formula tends to one as $n \to \infty$) uniformly in x. Assume also that $p(x,\{0\}) = 0$ for all x. Then

$$\min(1,|y|^2) p(x, dy/h) h^{-\alpha} \to \min(1,|y|^2) \frac{d|y|}{|y|^{\alpha+1}} S(x,\bar{y}) d_S\bar{y}, \quad h \to 0, \tag{8.21}$$

where both sides are finite measures on $\mathbb{R}^d \setminus \{0\}$ and the convergence is in the weak sense and is uniform in $x \in \mathbb{R}^d$. If $\alpha < 1$, then also

$$\min(1,|y|)p(x,dy/h)h^{-\alpha} \to \min(1,|y|)\frac{d|y|}{|y|^{\alpha+1}}\int_\Omega S(x,\bar{y})d_S\bar{y}, \quad h \to 0,$$

holds in the same sense.

Remark 54. As the limiting measure has a density with respect to Lebesgue measure, the uniform weak convergence means simply that the measures of any open set with boundaries of Lebesgue measure zero converge uniformly in x.

Proof. By (8.20)

$$\int_{|z|>A}\int_{\bar{z}\in\Omega} p(x;dz/h)h^{-\alpha} = \int_{|y|>A/h}\int_{\bar{y}\in\Omega} p(x;dy)h^{-\alpha} \sim \frac{1}{\alpha A^\alpha}\int_\Omega S(x,s)d_S s$$

as $h \to 0$. Hence

$$\int_{A<|z|<B}\int_{\bar{z}\in\Omega} p(x;dz/h)h^{-\alpha} \to \int_A^B \frac{d|z|}{|z|^{\alpha+1}}\int_\Omega S(x,s)d_S s.$$

Hence $p(x;dz/h)h^{-\alpha}$ converges weakly to $|z|^{-(\alpha+1)}d|z|S(x,z/|z|)d_S(z/|z|)$ on any set separated from the origin. Consequently, (8.21) follows from the uniform bound

$$\int_{|y|<\epsilon} \min(1,|y|^2)p(x,dy/h)h^{-\alpha} \le C\epsilon^{2-\alpha} \tag{8.22}$$

for some constant C. In order to prove (8.22), observe that

$$\int_{|y|>n} p(x,dy) \le Cn^{-\alpha}$$

for some constant C, uniformly for all x and $n > 0$ (in fact it holds for large enough n by 8.20 and is extended to all n, because all $p(x,dy)$ are probability measures). Hence for an arbitrary $\epsilon < 1$ one has

$$\int_{|y|<\epsilon} \min(1,|y|^2)p(x,dy/h)h^{-\alpha} = \int_{|z|<\epsilon/h} h^2|z|^2 p(x,dy/h)h^{-\alpha}.$$

Representing this integral as the countable sum of the integrals over the regions

$$\epsilon/(2^{k+1}h) < y \le \epsilon/(2^k h),$$

it can be estimated by

$$\sum_{k=0}^\infty h^2\left(\frac{\epsilon}{2^k h}\right)^2 h^{-\alpha}Ch^\alpha 2^{\alpha(k+1)}\epsilon^{-\alpha} = \sum_{k=0}^\infty C\epsilon^{2-\alpha}2^\alpha 2^{-(2-\alpha)k}.$$

This yields (8.22), since the sum on the r.h.s. converges.

The improvement concerning the case $\alpha < 1$ is obtained similarly. □

Consider the jump-type Markov process $Z_x^h(t)$ generated by

$$(L_h f)(x) = \frac{1}{h^\alpha} \int (f(x+hy) - f(x)) p(x; dy). \tag{8.23}$$

For each h the operator L_h is bounded in $C_\infty(\mathbb{R}^d)$, and hence specifies a Feller semigroup T_t^h there. In case when p does not depend on x,

$$Z_x^h(t) = x + h(Y_1 + \cdots + Y_{N_t})$$

is a normalized random walk with the number of jumps N_t being a Poisson process with parameter $h^{-\alpha}$, so that $\mathbf{E}N_t = th^{-\alpha}$. In particular, the number of jumps $n = N_t \sim th^{-\alpha}$ for small h, so that $Z^h(1) \sim n^{-1/\alpha}(Y_1 + \cdots + Y_n)$.

On the other hand, approximations with a non-random number of jumps are specified by the process $S_x^\tau(t) = S_x^\tau([t])$ (by the square bracket the integer part of a real number was denoted) defined by

$$S_x^\tau(0) = x, \quad S_x^\tau(1) = x + \tau^{1/\alpha} Y_1, \quad \ldots, \quad S_x^\tau(j) = S_x^\tau(j-1) + \tau^{1/\alpha} Y_j, \quad \ldots,$$

where each Y_j is distributed according to $p(S_{j-1}, dy)$. The corresponding transition operator

$$U^\tau f(x) = \int f(x + \tau^{1/\alpha} y) p(x, dy)$$

specifies the discrete-time semigroup $U^{k\tau} f(x) = \mathbf{E} f(S_x^\tau(k))$. If $p(x; dy)$ does not depend on x, then

$$S_x^\tau(n) = x + \tau^{1/\alpha}(Y_1 + \cdots + Y_n)$$

is just a standard random walk.

Theorem 8.3.1. *Under the assumptions of Proposition* 8.3.1, *the semigroups* T_t^h *and* $(U^\tau)^{[t/\tau]}$ *converge to the semigroup* T_t *generated by* L. *In particular, the corresponding processes converge in distribution.*

Proof. By (8.23)

$$(L_h f)(x) = \frac{1}{h^\alpha} \int (f(x+z) - f(x)) p(x; dz/h),$$

and by Proposition 8.3.1, these operators converge to $Lf(x)$ as $h \to 0$ uniformly in x for $f \in C_\infty(\mathbb{R}^d) \cap C^2(\mathbb{R}^d)$. Hence our claim follows from Theorem 8.1.1. □

In case of p not depending on x, Theorem 8.3.1 reduces to the known fact on the convergence of random walks with the distribution of jumps from the domain of normal attraction of a stable law to the corresponding stable Lévy motion.

For applications to CTRW a generalization of the above results to multi-scaled walks is of importance. To this end, we shall discuss the process in $\mathbb{R}^d \times \mathbb{R}_+$ specified by the generator

$$\mathcal{L}f(x,u) = \int_0^\infty \int_{S^{d-1}} (f(x+y,u) - f(x,u)) \frac{d|y|}{|y|^{1+\alpha}} S(x,u,\bar{y})\, d_S \bar{y} + \int_0^\infty (f(x,u+v) - f(x,u)) \frac{1}{v^{1+\beta}} w(x,u) dv. \tag{8.24}$$

Again the results of Section 4.6 imply that if $S(x,s)$ and $w(x,u)$ have bounded derivatives with respect to x and u up to and inclusive order $q \geq 3$, then the pseudo-differential operator (8.24) generates a Feller semigroup $\mathcal{T}_t$ in $C_\infty(\mathbb{R}^d \times \mathbb{R}_+)$ (continuous functions up to the boundary) with the space $C_\infty^{q-1}(\mathbb{R}^d \times \mathbb{R}_+)$ as its invariant core and hence a Feller process $(Y, V)(t)$ in $\mathbb{R}^d \times \mathbb{R}_+$.

We shall obtain now the corresponding extension of Theorem 8.3.1.

Theorem 8.3.2. *Suppose $S(x,s)$ and $w(x,u)$ have bounded derivatives with respect to x and u up to and inclusive order $q \geq 3$. Let $p(x,u;dydv)$ be a CFTP in $\mathbb{R}^d \times \mathbb{R}_+$, which is symmetric with respect to the reflection $y \mapsto -y$ and for which*

$$p(x,u;\{0\} \times \mathbb{R}_+) + p(x,u;\mathbb{R}^d \times \{0\}) = 0.$$

Assume also that the projections belong to the domain of normal attraction of stable laws; more precisely, that uniformly in (x,u)

$$\int_{|y|>n} \int_{\bar{y}\in\Omega} p(x,u;dydv) \sim \frac{1}{\alpha n^\alpha} \int_\Omega S(x,u,s)\, d_S s, \quad n \to \infty, \tag{8.25}$$

and

$$\int_{v>n} \int_{|y|>A} p(x,u;dydv) \sim \frac{1}{\beta n^\beta} w(x,u,A), \quad n \to \infty, \tag{8.26}$$

for any $A \geq 0$ with a measurable function w of three arguments such that

$$w(x,u,0) = w(x,u), \quad \lim_{A\to\infty} w(x,u,A) = 0 \tag{8.27}$$

(so that $w(x,u,A)$ is a measure on $\mathbb{R}_+$ for any x,u).

Consider the jump-type processes generated by

$$(\mathcal{L}_\tau f)(x,u) - \frac{1}{\tau} \int (f(x+\tau^{1/\alpha} y, u + \tau^{1/\beta} v)) - f(x,u)) p(x,u;dydv). \tag{8.28}$$

Then the Feller semigroups $\mathcal{T}_t^h$ in $C_\infty(\mathbb{R}^d \times \mathbb{R}_+)$ of these processes (which are Feller, because $\mathcal{L}_h$ is bounded in $C_\infty(\mathbb{R}^d \times \mathbb{R}_+)$ for any h) converge to the semigroup $\mathcal{T}_t$.

Proof. As in Proposition 8.3.1 one deduces from (8.25), (8.26) that uniformly in x, u

$$\min(1, |y|^2) \int_0^\infty p(x, u; dy/h\, dv) h^{-\alpha} \to \min(1, |y|^2) \frac{d|y|}{|y|^{\alpha+1}} S(x, \bar{y}) d_S \bar{y}, \quad h \to 0, \tag{8.29}$$

and

$$\min(1, v) \int_{|y|>A} p(x, u; dy dv/h) h^{-\beta} \to \min(1, v) w(x, u, A) \frac{dv}{v^{\beta+1}}, \quad h \to 0. \tag{8.30}$$

Next, assuming $f \in C^2_\infty(\mathbb{R}^d \times \mathbb{R}_+)$ and writing

$$\mathcal{L}_\tau f(x, u) = \mathrm{I} + \mathrm{II}$$

with

$$\begin{aligned} \mathrm{I} = {} & \frac{1}{\tau} \int (f(x + \tau^{1/\alpha} y, u) - f(x, u)) p(x, u; dy dv) \\ & + \frac{1}{\tau} \int (f(x, u + \tau^{1/\beta} v) - f(x, u)) p(x, u; dy dv) \end{aligned}$$

and

$$\begin{aligned} \mathrm{II} = {} & \frac{1}{\tau} \int [f(x + \tau^{1/\alpha} y, u + \tau^{1/\beta} v) - f(x + \tau^{1/\alpha} y, u)] p(x, u; dy dv) \\ & - \frac{1}{\tau} \int [f(x, u + \tau^{1/\beta} v) - f(x, u)] p(x, u; dy dv), \end{aligned}$$

one observes that, as in the proof of Theorem 8.3.1, (8.29) and (8.30) (the latter with $A = 0$) imply that I converges to $\mathcal{L} f(x, u)$ uniformly in x, u. Thus in order to complete our proof we have to show that the function II converges to zero, as $\tau \to 0$. We have

$$\mathrm{II} = \int (g(x + \tau^{1/\alpha} y, u, v) - g(x, u, v)) p(x, u; dy dv/\tau^{1/\beta}) \frac{1}{\tau}$$

with

$$g(x, u, v) = f(x, u + v) - f(x, u).$$

By our assumptions on f,

$$|g(x, u, v)| \le C \min(1, v) \left(\max \left| \frac{\partial f}{\partial u} \right| + \max |f| \right) \le \tilde{C} \min(1, v),$$

and

$$\left| \frac{\partial g}{\partial x}(x, u, v) \right| \le C \min(1, v) \left(\max \left| \frac{\partial^2 f}{\partial u \partial x} \right| + \max \left| \frac{\partial f}{\partial x} \right| \right) \le \tilde{C} \min(1, v)$$

for some constants C and $\tilde{C}$. Hence by (8.30) and (8.27), for an arbitrary $\epsilon > 0$ there exists a A such that

$$\int_{|y|>A} (g(x + \tau^{1/\alpha} y, u, v) - g(x, u, v)) p(x, u; dy dv/\tau^{1/\beta}) \frac{1}{\tau} < \epsilon;$$

and on the other hand, for arbitrary A

$$\int_{|y|<A} (g(x + \tau^{1/\alpha} y, u, v) - g(x, u, v)) p(x, u; dy dv/\tau^{1/\beta}) \frac{1}{\tau} \leq \tau^{1/\alpha} A\kappa$$

for some constant κ, so that II can be made arbitrary small by first choosing large enough A and then choosing small enough τ. □

Define now the process $(Y, V)^{\tau}_{x,u}(t/\tau) = (Y, V)^{\tau}_{x,u}([t/\tau])$, where

$$(Y, V)^{\tau}_{x,u}(0) = (x, u), \quad (Y, V)^{\tau}_{x,u}(1) = (x + \tau^{1/\alpha} Y_1, u + \tau^{1/\beta} V_1), \quad \dots,$$
$$(Y, V)^{\tau}_{x,u}(j) = (Y, V)^{\tau}_{x,u}(j-1) + (\tau^{1/\alpha} Y_j, \tau^{1/\beta} V_j), \quad \dots$$

and each pair (Y_j, V_j) is distributed according to $p((Y, V)^{\tau}_{x,u}(j-1); dy dv)$. If $p(x, u; dy dv)$ does not depend on x, u, then

$$(Y, V)^{\tau}_{x,u}(n) = (x, u) + (\tau^{1/\alpha}(Y_1 + \dots + Y_n), \tau^{1/\beta}(V_1 + \dots + V_n)).$$

The following result is proved similarly to Theorem 8.3.2.

Theorem 8.3.3. *Under the assumptions of Theorem* 8.3.2, *the linear contractions* $Ef((Y, V)^{\tau}_{x,u}(t/\tau))$ *in* $C_{\infty}(\mathbb{R}^d \times \mathbb{R}_+)$ *converge to the semigroup* $\mathcal{T}_t f(x, u)$ *of the process* $(Y, V)(t)$ *uniformly on* $t \in [0, t_0]$, *as* $\tau \to 0$.

8.4 Subordination by hitting times and generalized fractional evolutions

Let $X(u)$, $u \geq 0$ be a Lévy subordinator, i.e. an increasing i.i.d. càdlàg Feller process (adapted to a filtration on a suitable probability space) with generator

$$Af(x) = \int_0^{\infty} (f(x + y) - f(x)) \nu(dy) + a \frac{\partial f}{\partial x}, \tag{8.31}$$

where $a \geq 0$ and ν is a Borel measure on $\{y > 0\}$ such that

$$\int_0^{\infty} \min(1, y) \nu(dy) < \infty.$$

We are interested in the inverse-function process or the first-hitting time process $Z(t)$ defined as

$$Z_X(t) = Z(t) = \inf\{u : X(u) > t\} = \sup\{u : X(u) \leq t\}, \tag{8.32}$$

which is of course also an increasing càdlàd process. To make our further analysis more transparent (avoiding heavy technicalities of the most general case) we shall assume that there exist $\epsilon > 0$ and $\beta \in (0, 1)$ such that

$$\nu(dy) \geq y^{1+\beta}\, dy, \quad 0 < y < \epsilon. \tag{8.33}$$

For convenient reference we collect in the next statement (without proofs) the elementary (well known) properties of $X(u)$.

Proposition 8.4.1. *Under condition* (8.33),

(i) *the process $X(u)$ is a.s. increasing at each point, i.e. it is not a constant on any finite time interval;*

(ii) *the distribution of $X(u)$ for $u > 0$ has a density $G(u, y)$ vanishing for $y < 0$, which is infinitely differentiable in both variable and satisfies the equation*

$$\frac{\partial G}{\partial u} = A^\star G, \tag{8.34}$$

where $A^\star$ is the dual operator to A given by

$$A^\star f(x) = \int_0^\infty (f(x-y) - f(x))\nu(dy) - a\frac{\partial f}{\partial x},$$

(iii) *if extended by zero to the half-space $\{u < 0\}$ the locally integrable function $G(u, y)$ on $\mathbb{R}^2$ specifies a generalized function (which is in fact infinitely smooth outside* $(0, 0)$*) satisfying (in the sense of distribution) the equation*

$$\frac{\partial G}{\partial u} = A^\star G + \delta(u)\delta(y). \tag{8.35}$$

Corollary 19. *Under condition* (8.33),

(i) *the process $Z(t)$ is a.s. continuous and $Z(0) = 0$;*

(ii) *the distribution of $Z(t)$ has a continuously differentiable probability density function $Q(t, u)$ for $u > 0$ given by*

$$Q(t, u) = -\frac{\partial}{\partial u}\int_{-\infty}^t G(u, y)\, dy. \tag{8.36}$$

Proof. (i) follows from Proposition 8.4.1 (i) and for (ii) one observes that

$$\mathbf{P}(Z(t) \le u) = \mathbf{P}(X(u) \ge t) = \int_t^\infty G(u, y)\, dy = 1 - \int_0^t G(u, y)\, dy,$$

which implies (8.36) by the differentiability of G. Let us stress for clarity that (8.36) defines $Q(t, u)$ as a smooth function for all t as long as $u > 0$ and $Q(t, u) = 0$ for $t \le 0$ and $u > 0$. □

Theorem 8.4.1. *Under condition* (8.33), *the density Q satisfies the equation*

$$A^\star Q = \frac{\partial Q}{\partial u} \tag{8.37}$$

for $u > 0$, where $A^\star$ acts on the variable t, and the boundary condition

$$\lim_{u \to 0_+} Q(t, u) = -A^\star \theta(t), \tag{8.38}$$

where $\theta(t)$ is the indicator function equal one (respectively 0*) for positive (respectively negative) t. If Q is extended by zero to the half-space $\{u < 0\}$, it satisfies the equation*

$$A^\star Q = \frac{\partial Q}{\partial u} + \delta(u) A^\star \theta(t), \tag{8.39}$$

in the sense of distribution (generalized functions).

Moreover the (point-wise) derivative $\frac{\partial Q}{\partial t}$ also satisfies equation (8.37) *for $u > 0$ and satisfies the equation*

$$A^\star \frac{\partial Q}{\partial t} = \frac{\partial}{\partial u} \frac{\partial Q}{\partial t} + \delta(u) \frac{d}{dt} A^\star \theta(t) \tag{8.40}$$

in the sense of distributions.

Remark 55. In the case of a β-stable subordinator $X(u)$ with the generator

$$Af(x) = -\frac{1}{\Gamma(-\beta)} \int_0^\infty (f(x+y) - f(x)) y^{-1-\beta}\, dy, \tag{8.41}$$

one has

$$A = -\frac{d^\beta}{d(-x)^\beta}, \quad A^\star = -\frac{d^\beta}{dx^\beta} \tag{8.42}$$

(see Section 1.8), in which case equation (8.39) takes the form

$$\frac{d^\beta Q}{dt^\beta} + \frac{\partial Q}{\partial u} = \delta(u) \frac{t^{-\beta}}{\Gamma(1-\beta)} \tag{8.43}$$

coinciding with equation (B14) from Saichev and Zaslavsky [285]. This remark gives rise to the idea of calling the operator (8.31) a generalized fractional derivative.

Proof. Notice that by (8.36), (8.34) and by the commutativity of the integration and $A^\star$, one has

$$\begin{aligned} Q(t,u) &= -\int_{-\infty}^t \frac{\partial}{\partial u} G(u,y)\,dy \\ &= -\int_{-\infty}^t (A^\star G(u,.))(y)\,dy = -A^\star \int_{-\infty}^t G(u,y)\,dy. \end{aligned}$$

This implies (8.37) (by differentiating with respect to u and again using (8.36)) and (8.38), because $G(0,y) = \delta(y)$.

Assume now that Q is extended by zero to $\{u < 0\}$. Let ϕ be a test function (infinitely differentiable with a compact support) in $\mathbb{R}^2$. Then in the sense of distributions,

$$\begin{aligned} &\left(\left(\frac{\partial}{\partial u} - A^\star\right)Q, \phi\right) = \left(Q, \left(-\frac{\partial}{\partial u} - A\right)\phi\right) \\ &= \lim_{\epsilon \to 0} \int_\epsilon^\infty du \int_{\mathbb{R}} dt\, Q(t,u)\left(-\frac{\partial}{\partial u} - A\right)\phi(t,u) \\ &= \lim_{\epsilon \to 0} \left[\int_\epsilon^\infty du \int_{\mathbb{R}} dt\, \phi(t,u)\left(\frac{\partial}{\partial u} - A^\star\right)Q(t,u) + \int_{\mathbb{R}} \phi(t,\epsilon)Q(t,\epsilon)\,dt\right]. \end{aligned}$$

The first term here vanishes by (8.37). Hence by (8.38)

$$\left(\left(\frac{\partial}{\partial u} - A^\star\right)Q, \phi\right) = -\int_{\mathbb{R}} \phi(t,0)A^\star\theta(t)\,dt,$$

which clearly implies (8.39). The required properties of $\frac{\partial Q}{\partial t}$ follows similarly from the representation

$$\frac{\partial Q}{\partial t}(t,u) = -\frac{\partial G}{\partial u}(u,t).$$

For instance for $u > 0$

$$A^\star \frac{\partial Q}{\partial t}(t,u) = -\frac{\partial}{\partial u} A^\star G(u,t) = -\frac{\partial}{\partial u}\frac{\partial}{\partial u} G(u,t) = \frac{\partial}{\partial u}\frac{\partial Q}{\partial t}.$$ □

Remark 56. Let us stress that the generalized function Q coincides with an infinitely differentiable function outside the ray $\{t \geq 0, u = 0\}$, vanishes on the set $\{t < 0, u < 0\}$ and satisfies the limiting condition $\lim_{t \to 0_+} Q(t,u) = \delta(u)$. The latter holds, because for $t > 0$ and a smooth test function ϕ

$$\begin{aligned} \int_{-\infty}^\infty Q(t,u)\phi(u)\,du &= \int_0^\infty du \frac{\partial}{\partial u} \int_t^\infty G(u,y)\,dy\phi(u) \\ &= -\int_0^\infty du\phi'(u) \int_t^\infty G(u,y)\,dy \to -\int_0^\infty \phi'(u)du = \phi(0), \end{aligned}$$

as $t \to 0$.

We are interested now in the random time-change of Markov processes specified by the process $Z(t)$.

Theorem 8.4.2. *Under the conditions of Theorem* 8.4.1 *let* $Y(t)$ *be a Feller process in* $\mathbb{R}^d$*, independent of* $Z(t)$*, and with the domain of the generator* L *containing* $C^2_\infty(\mathbb{R}^d)$*. Denote the transition probabilities of* $Y(t)$ *by*

$$T(t, x, dy) = \mathbf{P}(Y_x(t) \in dy) = \mathbf{P}_x(Y(t) \in dy).$$

Then the distributions of the (time changed or subordinated) process $Y(Z(t))$ *for* $t > 0$ *are given by*

$$\mathbf{P}_x(Y(Z(t)) \in dy) = \int_0^\infty T(u, x, dy) Q(t, u)\, du, \tag{8.44}$$

the averages $f(t, x) = \mathbf{E} f(Y_x(Z(t)))$ *of* $f \in C^2_\infty(\mathbb{R}^d)$ *satisfy the (generalized) fractional evolution equation*

$$A^\star_t f(t, x) = -L_x f(t, x) + f(x) A^\star \theta(t)$$

(where the subscripts indicate the variables, on which the operators act), and their time derivatives $h = \partial f/\partial t$ *satisfy for* $t > 0$ *the equation*

$$A^\star_t h = -L_x h + f(x) \frac{d}{dt} A^\star \theta(t).$$

Moreover, if $Y(t)$ *has a smooth transition probability density so that* $T(t, x, dy) = T(t, x, y)dy$ *and the forward and backward equations*

$$\frac{\partial T}{\partial t}(t, x, y) = L_x T(t, x, y) = L^\star_y T(t, x, y) \tag{8.45}$$

hold, then the distributions of $Y(Z(t))$ *have smooth density*

$$g(t, x, y) = \int_0^\infty T(u, x, y) Q(t, u)\, du \tag{8.46}$$

satisfying the forward (generalized) fractional evolution equation

$$A^\star_t g = -L^\star_y g + \delta(x - y) A^\star \theta(t) \tag{8.47}$$

and the backward (generalized) fractional evolution equation

$$A^\star_t g = -L_x g + \delta(x - y) A^\star \theta(t) \tag{8.48}$$

(when g *is continued by zero to the domain* $\{t < 0\}$*) with the time derivative* $h = \partial g/\partial t$ *satisfying for the equation*

$$A^\star_t h = -L^\star_y h + \delta(x - y) \frac{d}{dt} A^\star \theta(t) \tag{8.49}$$

Remark 57. In the case of a β-stable Lévy subordinator $X(u)$ with the generator (8.41), where (8.42) hold, the l.h.s. of the above equations become fractional derivatives per se. In particular, if $Y(t)$ is a symmetric α-stable Lévy motion, equation (8.47) takes the form

$$\frac{\partial^\beta}{\partial t^\beta} g(t, y-x) = \frac{\partial^\alpha}{\partial |y|^\alpha} g(t, y-x) + \delta(y-x)\frac{t^{-\beta}}{\Gamma(1-\beta)}, \tag{8.50}$$

deduced in Saichev and Zaslavsky [285] and Uchaikin [311]. The corresponding particular case of (8.46) also appeared in Meerschaert and Scheffler [239] as well as in [285], where it is called a formula of separation of variables. Our general approach makes it clear that this separation of variables comes from the independence of $Y(t)$ and the subordinator $X(u)$ (see Theorem 8.4.3 for a more general situation).

Proof. For a continuous bounded function f, one has for $t > 0$ that

$$\begin{aligned}\mathbf{E}f(Y_x(Z(t))) &= \int_0^\infty \mathbf{E}(f(Y_x(Z(t)))|Z(t) = u)Q(t,u)\,du \\ &= \int_0^\infty \mathbf{E}f(Y_x(u))Q(t,u)\,du,\end{aligned}$$

by independence of Z and Y. This implies (8.44) and (8.46).

From Theorem 8.4.1 it follows that for $t > 0$

$$\begin{aligned}A_t^\star g &= \lim_{\epsilon\to 0}\int_\epsilon^\infty T(u,x,y)A_t^\star Q(t,u)\,du = \lim_{\epsilon\to 0}\int_\epsilon^\infty T(u,x,y)\frac{\partial}{\partial u}Q(t,u)\,du \\ &= -\int_0^\infty \frac{\partial}{\partial u}T(u,x,y)Q(t,u)\,du + \delta(x-y)A^\star\theta(t),\end{aligned}$$

where by (8.45) the first term equals $-L_y^\star g = L_x g$, implying (8.47) and (8.48). Of course for $t < 0$ both sides of (8.47) and (8.48) vanish. Other equations are proved analogously. □

Now we generalize this theory to the case of Lévy type subordinators $X(u)$ specified by the generators of the form

$$Af(x) = \int_0^\infty (f(x+y) - f(x))\nu(x,dy) + a(x)\frac{\partial f}{\partial x} \tag{8.51}$$

with position-dependent Lévy measure and drift. We need some regularity assumptions in order to have a smooth transition probability density as in case of Lévy motions.

Proposition 8.4.2. *Assume that*

(i) ν *has a density* $\nu(x,y)$ *with respect to Lebesgue measure such that*

$$C_1 \min(y^{-1-\beta_1}, y^{-1-\beta_2}) \leq \nu(x,y) \leq C_2 \max(y^{-1-\beta_1}, y^{-1-\beta_2}) \qquad (8.52)$$

with some constants $C_1, C_2 > 0$ *and* $0 < \beta_1 < \beta_2 < 1$

(ii) ν *is thrice continuously differentiable with respect to* x *with the derivatives satisfying the right estimate* (8.52),

(iii) $a(x)$ *is non-negative with bounded derivatives up to the order three.*

Then the generator (8.51) *specifies an increasing Feller process having for* $u > 0$ *a transition probability density* $G(u,y) = \mathbf{P}(X(u) \in dy)$ *(we assume that* $X(u)$ *starts at the origin) that is twice continuously differentiable in* u.

Remark 58. Condition (8.52) holds for popular stable-like processes with a position-dependent stability index.

Proof. The existence of the Feller process follows again from Proposition 4.6.2. A proof of the existence of a smooth transition probability density can be carried out in the same way as for symmetric multidimensional stable-like processes, see Section 7.5. □

One can see now that the hitting-time process defined by (8.32) with $X(u)$ from the previous Proposition is again continuous and has a continuously differentiable density $Q(t,u)$ for $t > 0$ given by (8.36). However (8.37) does not hold, because the operators A and integration do not commute. On the other hand, equation (8.40) remains true (as easily seen from the proof). This leads directly to the following partial generalization of Theorem 8.4.2.

Proposition 8.4.3. *Let* $Y(t)$ *be the same Feller process in* $\mathbb{R}^d$ *as in Theorem* 8.4.3, *but independent hitting-time process* $Z(t)$ *be constructed from* $X(u)$ *under the assumptions of Proposition* 8.4.2.

Then the distributions of the (time changed or subordinated) process $Y(Z(t))$ *for* $t > 0$ *are given by* (8.44) *and the time derivatives* $h = \partial f / \partial t$ *of the averages* $f(t,x) = Ef(Y_x(Z(t)))$ *of continuous bounded functions* f *satisfy* (8.49).

Lastly, we extend this to the case of dependent hitting times.

Theorem 8.4.3. *Let* $(Y,V)(t)$ *be a random process in* $\mathbb{R}^d \times \mathbb{R}_+$ *such that*

(i) *the components* $Y(t)$, $V(s)$ *at different times have a joint probability density*

$$\phi(s,u;y_0,v_0;y,v) = P_{(y_0,v_0)}(Y(s) \in dy, V(u) \in dv) \qquad (8.53)$$

that is continuously differentiable in u *for* $u, s > 0$, *and*

(ii) *the component $V(t)$ is increasing and is a.s. not constant on any finite interval. For instance, assuming that*

$$C_1 \leq \int S^{d-1} |(\bar{p}, s)|^{\alpha} S(x, u, s) d_S s \leq C_2, \quad C_1 \leq w(x, u) \leq C_2,$$

the process generated by (8.24) *enjoys these properties (by a straightforward extension of Theorem* 7.5.1*).*

Then

(i) *the hitting-time process $Z(t) = Z_V(t)$ (defined by* (8.32) *with V instead of X) is a.s. continuous;*

(ii) *there exists a continuous joint probability density of $Y(s), Z(t)$ given by*

$$g_{Y(s),Z(t)}(y_0, 0; y, u) = \frac{\partial}{\partial u} \int_t^{\infty} \phi(s, u; y, v) \, dv; \tag{8.54}$$

(iii) *the distribution of the composition $Y(Z(t))$ has the probability density*

$$\begin{aligned} \Phi_{Y(Z(t))}(y) &= \int_0^{\infty} g_{Y(s),Z(t)}(y_0, 0; y, s) \, ds \\ &= \int_0^{\infty} \left(\frac{\partial}{\partial u} \int_t^{\infty} \phi(s, u; y_0, 0; y, v) \, dv \right) \bigg|_{u=s} ds; \end{aligned} \tag{8.55}$$

(iv) *in particular, if $(Y, V)(t)$ is a Feller process with a transition probability density $G_{YV}(u, y_0, v_0, y, v)$ and a generator of the form $(L + A) f(y, v)$, where L acts on the variable y and does not depend on v (intuition: jumps do not depend on waiting times) and A acts on v (but may depend on y), then for $s \geq u$*

$$\phi(s, u; y_0, v_0; y, v) = \int G_Y(s - u, z, y) G_{YV}(u, y_0, v_0; z, v) \, dz, \tag{8.56}$$

where G_Y denotes of course the transition probability density of the component Y, and

$$\frac{\partial}{\partial t} \Phi_{Y(Z(t))}(y) = \int_0^{\infty} A^{\star} G_{YV}(u, y_0, 0; y, t) \, du, \tag{8.57}$$

i.e. the time-derivative of the density of the subordinated process equals the generalized fractional derivative of the 'time component V' of the integrated joint density of the process (Y, V). This derivative $h = \frac{\partial}{\partial t} \Phi$ satisfies instead of (8.49) *the more complicated equation*

$$(A^{\star} + L^{\star}) h = \delta(y - y_0) A^{\star} \delta(v) + [L^{\star}, A^{\star}] \int_0^{\infty} G_{YV}(u, y_0, 0; y, v) \, du. \tag{8.58}$$

Proof. (i) and (ii) are straightforward extensions of the corollary to Proposition 8.4.1. Statement (iii) follows from conditioning and the definition of the joint distribution. To prove (iv) we write for $s \geq u$ by conditioning at time u

$$\begin{aligned}\mathbf{E} f(Y_{y_0}(s), V_{(y_0,v_0)}(u)) &= \mathbf{E} \int G_Y(s-u, Y_{y_0}(u); y) f(y, V_{(y_0,v_0)})\, du \\ &= \iint G_Y(s-u, z; y) f(y, v) G_{YV}(u, y_0, v_0; z, v)\, dy\, dz\, dv,\end{aligned}$$

implying (8.56). Consequently

$$\begin{aligned}\frac{\partial}{\partial t}\Phi_{Y(Z(t))}(y) &= -\int_0^\infty \frac{\partial}{\partial u} \int G_Y(s-u, z, y) G_{YV}(u, y_0, 0; z, t)\, dz|_{u=s}\, ds \\ &= \int_0^\infty \int -(L_z G_Y(s-u, z, y)) G_{YV}(u, y_0, 0; z, t)\, dz|_{u=s}\, ds \\ &\quad + \int_0^\infty \int G_Y(s-u, z, y))(A^\star + L^\star) G_{YV}(u, y_0, 0; z, t)\, dz|_{u=s}\, ds,\end{aligned}$$

which yields (8.57), as L cancels due to the assumptions on the form of the generator. Finally (8.57) implies (8.58) by straightforward inspection. □

8.5 Limit theorems for position dependent CTRW

Now everything is ready for the main result of this chapter.

Theorem 8.5.1. *Under the assumptions of Theorem* 8.3.2 *let* $Z^\tau(t)$, $Z(t)$ *be the hitting-time processes for* $V^\tau(t/\tau)$ *and* $V(t)$ *respectively (defined by the corresponding formula* (8.32)*). Then the subordinated processes* $Y^\tau(Z^\tau(t)/\tau)$ *converge to the subordinated process* $Y(Z(t))$ *in the sense of marginal distributions, i.e.*

$$\mathbf{E}_{x,0} f(Y^\tau(Z^\tau(t)/\tau)) \to \mathbf{E}_{x,0}(Y(Z(t))), \quad \tau \to 0, \tag{8.59}$$

for arbitrary $x \in \mathbb{R}^d$, $f \in C_\infty(\mathbb{R}^d \times \mathbb{R}_+)$, *uniformly for t from any compact interval.*

Remark 59. (i) The distribution of the limiting process is described in Theorem 8.4.3.

(ii) We show the convergence in the weakest possible sense. One can extend it to the convergence in the Skorohod space of trajectories using standard compactness tools from Section 4.8.

(iii) Similar result holds for the continuous-time approximation from Theorem 8.3.2.

Proof. Since the time is effectively discrete in $V^\tau(t/\tau)$, it follows that

$$Z^\tau(t) = \max\{u : X(u) \le t\},$$

and that the events $(Z^\tau(t) \le u)$ and $(V^\tau(u/\tau) \ge t)$ coincide, which implies that the convergence of finite-dimensional distributions of $(Y^\tau(s/\tau), V^\tau(u/\tau))$ to $(Y(s), V(u))$ (proved in Theorem 8.3.3) is equivalent to the corresponding convergence of the distributions of $(Y^\tau(s/\tau), Z^\tau(t))$ to $(Y(s), Z(t))$.

Next, since $V(0) = 0$, $V(u)$ is continuous and $V(u) \to \infty$ as $u \to \infty$ and because the limiting distribution is absolutely continuous, to show (8.59) it is sufficient to show that

$$\mathbf{P}_{x,0}[Y^\tau(Z_K^\tau(t)/\tau) \in A] \to \mathbf{P}_{x,0}[Y(Z_K(t)) \in A], \quad \tau \to 0, \tag{8.60}$$

for large enough $K > 0$ and any compact set A, whose boundary has Lebesgue measure zero, where

$$Z_K^\tau(t) = Z^\tau(t), \quad K^{-1} \le Z^\tau(t) \le K,$$

and vanishes otherwise, and similarly $Z_K(t)$ is defined.

Now

$$\mathbf{P}[Y^\tau(Z_K^\tau(t)/\tau) \in A] = \sum_{k=1/K\tau}^{K/\tau} \mathbf{P}[V^\tau(k) \in A \text{ and } Z^\tau(t) \in [k\tau, (k+1)\tau)] \tag{8.61}$$

and

$$\mathbf{P}[Y(Z_K(t)) \in A] = \sum_{k=1/K\tau}^{K/\tau} \int_A dy \int_{\tau k}^{\tau(k+1)} g_{Y(s),Z(t)}(y,s)\, ds, \tag{8.62}$$

which can be rewritten as

$$\sum_{k=1/K\tau}^{K/\tau} \int_A dy \int_{\tau k}^{\tau(k+1)} g_{Y(\tau k),Z(t)}(y,s)\, ds \\ + \sum_{k=1/K\tau}^{K/\tau} \int_A dy \int_{\tau k}^{\tau(k+1)} (g_{Y(s),Z(t)} - g_{Y(\tau k),Z(t)})(y,s)\, ds. \tag{8.63}$$

The second term here tends to zero as $\tau \to 0$ due to the continuity of the function (8.54), and the difference between the first term and (8.61) tends to zero, because the distributions of $(Y^\tau(s/\tau), Z^\tau(t)$ converge to the distribution of $(Y(s), Z(t))$. Hence (8.60) follows. □

In the case when S does not depend on u and w does not depend on x in (8.24), the limiting process $(Y, V)(t)$ has independent components, so that the averages of the limiting subordinated process satisfy the generalized fractional evolution equation from Proposition 8.4.3, and if moreover w is a constant, they satisfy the fractional equations from Theorem 8.4.2. In particular, if $p(x, u, dydv)$ does not depend on (x, u) and decomposes into a product $p(dy)q(dv)$, and the limit $V(t)$ is stable, we recover the main result from Meerschaert and Scheffler [239] (in a slightly less general setting, since we worked with symmetric stable laws and not with operator-stable motions as in [239]), as well as the corresponding results from Kotulski [209] or Kolokoltsov, Korolev and Uchaikin [198] (put $t = 1$ in (8.59)) on the long-time behavior of the normalized subordinated sums (8.1).

8.6 Comments

CTRW were introduced in Montroll and Weiss [248] and found numerous applications in physics and economics, see e.g. Zaslavsky [324], Meerschaert, Nane and Vellaisamy [236], Bening et al [46], Korolev [207], Metzler and Klafter [243], Uchaikin [311] and references therein. The limit distributions of appropriately normalized sums S_{N_t} were first studied in Kotulski [209] in the case of independent T_i and X_i (see also Kolokoltsov, Korolev and Uchaikin [198]). In Bening, Korolev and Kolokoltsov [44] the rate of convergence in double-array schemes was analyzed and in Meerschaert and Scheffler [239] the corresponding functional limit was obtained, which was shown to be specified by a fractional differential equation. The basic cases of dependent T_i and X_i were developed in Becker-Kern, Meerschaert and Scheffler [34] in the framework of the theory of the operator stable processes (see monograph [238] for the latter).

An important observation that the fractional evolution can arise from the subordination of Levy processes by the hitting times of stable Levy subordinators was made in Meerschaert and Scheffler [239]. Implicitly this idea was present already in Saichev and Zaslavsky [285].

Our extension to (T_n, X_n) depending on a current position, i.e. to spatially non-homogeneous situations, is based on [193]. The approach is quite different from those used in Kotulski [209] or Kolokoltsov, Korolev and Uchaikin [198], Meerschaert and Scheffler [238] and is based on the general philosophy of the present book: to study processes from their generators. Analytically, this means using finite-difference approximations to continuous-time operator semigroups.

A detailed analytic study of the fractional evolution was recently carried out in Kochubei [166], [167].

In Econophysics, the CTRW model describes the evolution of log-prices, see e.g. Meerschaert and Scalas [237], Mainardi et al. [227], Raberto et al. [267], Sabatelli et al. [283]. Namely, let $J_1, J_2, \dots$ denote the waiting times between trades and $Y_1, Y_2, \dots$ denote the log-returns. The sum $T_n = J_1 + \dots + J_n$ represents the time of the nth trade. The log-returns are linked with the prices $\mathbf{P}(T_n)$ via the relation $Y_n = \log[\mathbf{P}(T_n)/\mathbf{P}(T_{n-1})]$. The log-price at time t is

$$\log \mathbf{P}(t) = S_{N_t} = Y_1 + \cdots + Y_{N_t},$$

where $N_t = \max\{n : T_n \leq t\}$. The asymptotic theory of CTRW models describes the behavior of the long-time limit. Empirical study shows the dependence of the variables J_n, Y_n. They can also depend on S_n, the position of the process at time n.

Chapter 9
Complex Markov chains and Feynman integral

There exist several approaches to the rigorous mathematical construction of the Feynman path integral. However, most of these methods cover only a very restrictive class of potentials, which is not sufficient for physical applications, where path integration is widely used without rigorous justification. On the other hand, most of the approaches define the path integral not as a genuine integral (in the sense of Lebesgue or Riemann), but as a certain generalized functional. In this chapter we give a rigorous construction of the path integral which, on the one hand, covers a wide class of potentials and can be applied in a uniform way to the Schrödinger, heat and complex diffusion equations, and on the other hand, it is defined as a genuine integral over a bona fide σ- additive (or even finite) measure on path space. Such a measure turns out to be connected with a jump-type Markov process.

Following our strategy of giving several points of view on the key objects, in the next two sections we develop two approaches for such a construction in the simplest setting. Afterwards we concentrate on the extension of the second approach fertilized by the important idea of regularization, whose various versions are discussed. The applications to the Schrödinger equation are then presented in various cases including singular potentials, polynomially growing potentials and curvilinear state spaces.

Let us note finally that the representation of Feynman integrals in terms of the infinite sum of finite-dimensional integrals given in Theorems 9.3.1–9.3.3 corresponds to the general methodology of quantum physics of representing quantum evolutions (propagators) in terms of the sums over the contributions of the Feynman diagrams. Each diagram describes a situation with several interactions (specified by the operator V in Theorems 9.3.1–9.3.3) that occur at certain random times, between which the system evolves freely (according to the evolution e^{tA} in the abstract formulation of Theorems 9.3.1–9.3.3).

9.1 Infinitely-divisible complex distributions and complex Markov chains

We present here a general approach to the construction of the measures on the path space, which is based on the application of the Riesz–Markov theorem to the space of trajectories in the (compact) Tikhonov topology. It can be used for the path-integral representation of various evolution equations.

Let us recall from Section 1.1 that a *complex* σ-finite measure μ on $\mathbb{R}^d$ is a set function of the form

$$\mu(dy) = f(y)M(dy) \tag{9.1}$$

with some positive measure M (which can be chosen to be finite whenever μ is finite) and some bounded complex-valued function f (a possible choice of M is $|\mathrm{Re}\,\mu| + |\mathrm{Im}\,\mu|$). Representation (9.1) can be also considered as a definition of a complex measure. Though it is not unique, one way to specify the measure M uniquely is by imposing the additional assumption that $|f(y)| = 1$ for all y. If this condition is fulfilled, the positive measure M is called the *total variation measure* of the complex measure μ and is denoted by $|\mu|$. If a complex measure μ is presented in form (9.1) with some positive measure M, then $\|\mu\| = \int |f(y)|M(dy)$.

We say that a map ν from $\mathbb{R}^d \times \mathcal{B}(\mathbb{R}^d)$ into $\mathbb{C}$ is a *complex transition kernel* if for every x, the map $A \mapsto \nu(x, A)$ is a finite complex measure on $\mathbb{R}^d$, and for every $A \in \mathcal{B}(\mathbb{R}^d)$, the map $x \mapsto \nu(x, A)$ is $\mathcal{B}$-measurable. A (time homogeneous) *complex transition function* (abbreviated CTF) on $\mathbb{R}^d$ is a family $\nu_t, t \geq 0$, of complex transition kernels such that $\nu_0(x, dy) = \delta(y - x)$ for all x, where $\delta_x(y) = \delta(y - x)$ is the Dirac measure in x, and for every non-negative s, t, the Chapman–Kolmogorov equation

$$\int \nu_s(x, dy)\nu_t(y, A) = \nu_{s+t}(x, A)$$

holds. For simplicity, we consider here only time homogeneous CTF (the generalization to non-homogeneous case is straightforward).

A CTF is said to be *spatially homogeneous* if $\nu_t(x, A)$ depends on x, A only through the difference $A - x$. If a CTF is spatially homogeneous it is natural to denote $\nu_t(0, A)$ by $\nu_t(A)$ and to write the Chapman–Kolmogorov equation in the form

$$\int \nu_t(dy)\nu_s(A - y) = \nu_{t+s}(A).$$

A CTF will be called *regular* if there exists a positive constant K such that for all x and $t > 0$, the norm $\|\nu_t(x, .)\|$ of the measure $A \mapsto \nu_t(x, A)$ does not exceed $\exp\{Kt\}$.

CTF appear naturally in the theory of evolution equations: if T_t is a strongly continuous semigroup of bounded linear operators in $C_\infty(\mathbb{R}^d)$, then there exists a time-homogeneous CTF ν such that

$$T_t f(x) = \int \nu_t(x, dy) f(y). \tag{9.2}$$

In fact, the existence of a measure $\nu_t(x, .)$ such that (9.2) is satisfied follows from the Riesz–Markov theorem, and the semigroup identity $T_s T_t = T_{s+t}$ is equivalent to the Chapman–Kolmogorov equation. Since $\int \nu_t(x, dy) f(y)$ is continuous for all $f \in C_\infty(\mathbb{R}^d)$, it follows by the monotone convergence theorem (and the fact that each complex measure is a linear combination of four positive measures) that $\nu_t(x, A)$ is a Borel function of x.

Let us say that the semigroup T_t is *regular* if the corresponding CTF is regular. This is equivalent to the assumption that $\|T_t\| \le e^{Kt}$ for all $t > 0$ and some constant K.

Now we construct a measure on the path space corresponding to each regular CTF, introducing first some (rather standard) notations. Let $\dot{\mathbb{R}}_d$ denote the one-point compactification of the Euclidean space $\mathbb{R}^d$ (i.e. $\dot{\mathbb{R}}_d = \mathbb{R}^d \cup \{\infty\}$ and is homeomorphic to the sphere S^d). Let $\dot{\mathbb{R}}_d^{[s,t]}$ denote the infinite product of $[s,t]$ copies of $\dot{\mathbb{R}}_d$, i.e. it is the set of all functions from $[s,t]$ to $\dot{\mathbb{R}}_d$, the path space. As usual, we equip this set with the product topology, in which it is a compact space (due to the celebrated Tikhonov theorem). Let $\mathrm{Cyl}_{[s,t]}^k$ denote the set of functions on $\dot{\mathbb{R}}_d^{[s,t]}$ having the form

$$\phi^f_{t_0,t_1,\dots,t_{k+1}}(y(.)) = f(y(t_0),\dots,y(t_{k+1}))$$

for some bounded complex Borel function f on $(\dot{\mathbb{R}}^d)^{k+2}$ and some points t_j, $j = 0,\dots,k+1$, such that $s = t_0 < t_1 < t_2 < \cdots < t_k < t_{k+1} = t$. The union $\mathrm{Cyl}_{[s,t]} = \bigcup_{k\in\mathbb{N}} \mathrm{Cyl}_{[s,t]}^k$ is called the set of cylindrical functions (or functionals) on $\dot{\mathbb{R}}_d^{[s,t]}$. It follows from the Stone–Weierstrass theorem that the linear span of all continuous cylindrical functions is dense in the space $C(\dot{\mathbb{R}}_d^{[s,t]})$ of all complex continuous functions on $\dot{\mathbb{R}}_d^{[s,t]}$. Any CTF ν defines a family of linear functionals $\nu_{s,t}^x$, $x \in \mathbb{R}^d$, on $\mathrm{Cyl}_{[s,t]}$ by the formula

$$\nu^x_{s,t}(\phi^f_{t_0,\dots,t_{k+1}}) \tag{9.3}$$
$$= \int f(x, y_1,\dots,y_{k+1})\nu_{t_1-t_0}(x, dy_1)\nu_{t_2-t_1}(y_1, dy_2)\cdots\nu_{t_{k+1}-t_k}(y_k, dy_{k+1}).$$

Due to the Chapman–Kolmogorov equation, this definition is correct, i.e. if one considers an element from $\mathrm{Cyl}_{[s,t]}^k$ as an element from $\mathrm{Cyl}_{[s,t]}^{k+1}$ (any function of l variables $y_1,\dots,y_l$ can be considered as a function of $l+1$ variables $y_1,\dots,y_{l+1}$, which does not depend on y_{l+1}), then the two corresponding formulae (9.3) will be consistent.

Proposition 9.1.1. *If a semigroup T_t in $C_\infty(\mathbb{R}^d)$ is regular and ν is its corresponding CTF, then the functional* (9.3) *is bounded. Hence, it can be extended by continuity to a unique bounded linear functional ν^x on $C(\dot{\mathbb{R}}_d^{[s,t]})$, and consequently there exists a complex Borel measure $D_x^{s,t}$ on the path space $\dot{\mathbb{R}}_d^{[s,t]}$ such that*

$$\nu^x_{s,t}(F) = \int F(y(.))D_x^{s,t}y(.) \tag{9.4}$$

for all $F \in C(\dot{\mathbb{R}}_d^{[s,t]})$. In particular,

$$(T_t f)(x) = \int f(y(t))D_x^{s,t}y(.).$$

Proof. This is a direct consequence of the Riesz–Markov theorem, because the regularity of CTF implies that the norm of the functional $\nu_{s,t}^x$ does not exceed $\exp\{K(t-s)\}$. □

Formula (9.3) defines the family of finite-dimensional complex distributions on the path space. It gives rise to a finite complex measure on this path space and to a linear evolution of marginal measures, and can be called a *complex Markov process*. Unlike the case of the standard Markov processes, the generator, say A, of the corresponding semigroup T_t is not conditionally positive, and the corresponding bilinear Dirichlet form (Av, v) is complex.

The following simple fact can be used for proving the regularity of a semigroup.

Proposition 9.1.2. *Let B and A be linear operators in $C_\infty(\mathbb{R}^d)$ such that A is bounded and B is the generator of a strongly continuous regular semigroup T_t. Then $A + B$ is also the generator of a regular semigroup, which we denote by $\tilde{T}_t$.*

Proof. This follows directly from the fact that $\tilde{T}_t$ can be presented as the convergent (in the sense of the norm) series of standard perturbation theory

$$\tilde{T}_t = T_t + \int_0^t T_{t-s} A T_s \, ds + \int_0^t ds \int_0^s d\tau \, T_{t-s} A T_{s-\tau} A T_\tau + \cdots . \qquad \square$$

Of major importance for our purposes are the spatially homogeneous CTF. Let us discuss them in greater detail, in particular, their connection with infinitely divisible characteristic functions. Let $\mathcal{F}(\mathbb{R}^d)$ denote the Banach space of Fourier transforms of the elements of the space $\mathcal{M}_{\mathbb{C}}(\mathbb{R}^d)$ of finite complex Borel measures on $\mathbb{R}^d$, i.e. the space of (automatically continuous) functions on $\mathbb{R}^d$ of form

$$V(x) = V_\mu(x) = \int_{\mathbb{R}^d} e^{-ipx} \, \mu(dp) \tag{9.5}$$

for some $\mu \in \mathcal{M}_{\mathbb{C}}(\mathbb{R}^d)$, with the induced norm $\|V_\mu\| = \|\mu\|$. Since $\mathcal{M}_{\mathbb{C}}(\mathbb{R}^d)$ is a Banach algebra with convolution as the multiplication, it follows that $\mathcal{F}(\mathbb{R}^d)$ is also a Banach algebra with respect to the standard (pointwise) multiplication. We say that an element $f \in \mathcal{F}(\mathbb{R}^d)$ is *infinitely divisible* if there exists a family $(f_t, t \geq 0,)$ of elements of $\mathcal{F}(\mathbb{R}^d)$ depending continuously on t such that $f_0 = 1$, $f_1 = f$, and $f_{t+s} = f_t f_s$ for all positive s, t. Clearly if f is infinitely divisible, then it has no zeros and a continuous function $g = \log f$ is well defined (and is unique up to an imaginary shift). Moreover, the family f_t has the form $f_t = \exp\{tg\}$ and is defined uniquely up to a multiplier of the form $e^{2\pi ikt}$, $k \in \mathbb{N}$. Let us say that a continuous function g on $\mathbb{R}^d$ is a *complex characteristic exponent* (abbreviated CCE) if e^g is an infinitely divisible element of $\mathcal{F}(\mathbb{R}^d)$, or equivalently, if e^{tg} belongs to $\mathcal{F}(\mathbb{R}^d)$ for all $t > 0$.

Remark 60. The problem of the explicit characterization of the whole class of infinite divisible functions (or of the corresponding complex CCEs) seems to be quite non-trivial. When dealing with this problem, it is reasonable to describe first some natural subclasses. For example, it is easy to show that if $f_1 \in \mathcal{F}(\mathbb{R})$ is infinitely divisible and such that the measures corresponding to all functions f_t, $t > 0$, are concentrated on the half line $\mathbb{R}_+$ (complex generalization of subordinators) and have densities from $L^2(\mathbb{R}_+)$, then f_1 belongs to the Hardy space H_2 of analytic functions on the upper half plane, which have no Blaschke product in their canonical decomposition.

It follows from the definitions that the set of spatially homogeneous CTF $\nu_t(dx)$ is in one-to-one correspondence with CCE g, in such a way that for any positive t the function e^{tg} is the Fourier transform of the transition measure $\nu_t(dx)$.

Proposition 9.1.3. *If V is a CCE, then the solution to the Cauchy problem*

$$\frac{\partial u}{\partial t} = V\left(i\frac{\partial}{\partial y}\right)u \tag{9.6}$$

defines a strongly continuous and spatially homogeneous semigroup T_t of bounded linear operators in $C_\infty(\mathbb{R}^d)$ (i.e. $(T_t u_0)(y)$ is the solution to equation (9.6) *with the initial function u_0). Conversely, each such semigroup is the solution to the Cauchy problem of an equation of type* (9.6) *with some CCE g.*

Proof. This is straightforward. Since (9.6) is a pseudo-differential equation, it follows that the inverse Fourier transform $\psi(t, x)$ of the function $u(t, y)$ satisfies the ordinary differential equation

$$\frac{\partial \psi}{\partial t}(t, x) = V(x)\psi(t, x),$$

whose solution is $\psi_0(x)\exp\{tV(x)\}$. Since e^{tV} is the Fourier transform of a complex measure $\nu_t(dy)$, it follows that the solution to the Cauchy problem of equation (9.6) is given by the formula $(T_t u_0)(y) = \int u_0(z)\nu_t(dz - y)$, which is as required. □

We say that a CCE is *regular* if equation (9.6) defines a regular semigroup.

It would be very interesting to describe explicitly all regular CCE. We only give here two classes of examples. First of all, if a CCE is given by the Lévy–Khintchine formula (i.e. it defines a transition function consisting of probability measures), then this CCE is regular, because all CTF consisting of probability measures are regular. Another class is given by the following result.

Proposition 9.1.4. *Let $V \in \mathcal{F}(\mathbb{R}^d)$, i.e. it is given by* (9.5) *with $\mu \in \mathcal{M}(\mathbb{R}^d)$. Then V is a regular CCE. Moreover, if the positive measure M in the representation* (9.1) *for μ has no atom at the origin, i.e. $M(\{0\}) = 0$, then the corresponding measure $D_x^{0,t}$ on the path space from Proposition* 9.1.1 *is concentrated on the set of piecewise-constant paths in $\dot{\mathbb{R}}_d^{[0,t]}$ with a finite number of jumps. In other words, $D_x^{0,t}$ is the measure of a jump-process.*

Proof. Let $W = W_M$ be defined by the formula

$$W(x) = \int_{\mathbb{R}^d} e^{-ipx} M(dp). \tag{9.7}$$

The function $\exp\{tV\}$ is the Fourier transform of the measure $\delta_0 + t\mu + \frac{t^2}{2}\mu \star \mu + \cdots$ which can be denoted by $\exp^\star(t\mu)$ (it is equal to the sum of the standard exponential series, but with the convolution of measures instead of the standard multiplication). Clearly $\|\exp^\star(t\mu)\| \le \|\exp^\star(t\bar{f}M)\|$, where we denoted by $\bar{f}$ the supremum of the function f, and both these series are convergent series in the Banach algebra $\mathcal{M}(\mathbb{R}^d)$. Therefore $\|e^{Vt}\| \le \|e^{Wt}\| \le \exp\{t\bar{f}\|\mu\|\}$, and consequently V is a regular CCE. Moreover, the same estimate shows that the measure on the path space corresponding to the CCE V is absolutely continuous with respect to the measure on the path space corresponding to the CCE W. But the latter coincides up to a positive constant multiplier with the probability measure of the compound Poisson process with the Lévy measure M defined by the equation

$$\frac{\partial u}{\partial t} = [W(i\nabla) - \lambda_M]u, \tag{9.8}$$

where $\lambda_M = M(\mathbb{R}^d)$, or equivalently

$$\frac{\partial u}{\partial t} = \int (u(y+\xi) - u(y))\, M(d\xi), \tag{9.9}$$

because the condition $M(\{0\}) = 0$ ensures that M is actually a measure on $\mathbb{R}^d \setminus \{0\}$, i.e. it is a finite Lévy measure. It remains to note that the measures of compound Poisson processes are concentrated on piecewise-constant paths (see Theorem 3.7.3). □

Therefore, we have two different classes (essentially different, because they obviously are not disjoint) of regular CCE: those given by the Lévy–Khintchine formula, and those given by Proposition 9.1.4. It is easy to prove that one can combine these regular CCEs, more precisely that the class of regular CCE is a convex cone.

Let us apply the results obtained so far to the case of the pseudo-differential equation of Schrödinger type

$$\frac{\partial \psi}{\partial t} = \Phi(-i\nabla)\psi - iV(x)\psi, \tag{9.10}$$

where $\Phi(-i\nabla)$ is a ΨDO with a continuous symbol $\Phi(p)$ having bounded real part on real arguments:

$$\operatorname{Re}\Phi(p) \le c, \quad p \in \mathbb{R}^d, \tag{9.11}$$

for some constant c, and where V is a complex-valued function of form (9.5). As the main example we have in mind the equation

$$\frac{\partial \psi}{\partial t} = -G(-\Delta)^{\alpha}\psi + \left(A, \frac{\partial}{\partial x}\right)\psi - iV(x)\tilde{u}, \tag{9.12}$$

where G is a complex constant with a non-negative real part, α is any positive constant, A is a real-valued vector. In this case $\Phi(p) = -G|p|^{2\alpha} + i(A, p)$. The standard Schrödinger equation corresponds to the case $\alpha = 1$, $G = i$, $A = 0$ and V being purely imaginary. Our general equation includes the Schrödinger equation, the heat equation with drifts and sources, and their stable (when $\alpha \in (0, 1)$) and complex generalizations in one formula. This general consideration also shows directly how the functional integral corresponding to the Schrödinger equation can be obtained by analytic continuation from the functional integral corresponding to the heat equation, which gives a connection with other approaches to the path integration. The equation on the Fourier transform

$$u(y) = \int_{\mathbb{R}^d} e^{-iyx}\psi(x)\, dx$$

of ψ (or equation (9.12) in momentum representation) clearly has the form

$$\frac{\partial u}{\partial t} = \Phi(y)u - iV(i\nabla)u, \tag{9.13}$$

which in case of equation (9.12) takes the form

$$\frac{\partial u}{\partial t} = -G(y^2)^{\alpha}u + i(A, y)u - iV\left(i\frac{\partial}{\partial y}\right)u. \tag{9.14}$$

One easily sees that already in the trivial case $V = 0$, $A = 0$, $\alpha = 1$, equation (9.12) defines a regular semigroup only in the case of real positive G, i.e. only in the case of the heat equation. It turns out however that for equation (9.14) the situation is completely different.

Proposition 9.1.5. *The solution to the Cauchy problem of equation* (9.13) *can be written in the form of the complex Feynman–Kac formula*

$$u(t, y) = \int \exp\left\{\int_0^t \Phi(q(\tau))\, d\tau\right\} u_0(q(t)) D_y^{0,t} q(.), \tag{9.15}$$

or more explicitly

$$u(t, y) = \int \exp\left\{-\int_0^t [G(q(\tau)^2)^{\alpha} - i(A, q(\tau))]\, d\tau\right\} u_0(q(t)) D_y^{0,t} q(.) \tag{9.16}$$

in the case of equation (9.14), *where D_y is the measure of the jump process corresponding to the equation*

$$\frac{\partial u}{\partial t} = -iV\left(i\frac{\partial}{\partial y}\right)u. \tag{9.17}$$

Proof. Let T_t be the regular semigroup corresponding to equation (9.17). By the Trotter formula, the solution to the Cauchy problem of equation (9.14) can be written in the form

$$\begin{aligned} u(t,y) &= \lim_{n\to\infty} \left(\left(\exp\left\{ \frac{t}{n} \Phi(y) \right\} T_{t/n} \right)^n u_0 \right)(y) \\ &= \lim_{n\to\infty} \int \exp\left\{ \frac{t}{n} \sum_{k=1}^{n} \Phi(q_k) \right\} u_0(q_n) \prod_{k=1}^{n} \nu_{t/n}(q_{k-1}, dq_k), \end{aligned}$$

where in the last product $q_0 = y$. Using (9.3), we can rewrite this as

$$u(t,y) = \lim_{n\to\infty} \nu_{0,t}^{y}(F_n) = \lim_{n\to\infty} \int F_n(q(.)) D_y^{0,t} q(.),$$

where F_n is the cylindrical function

$$F_n(q(.)) = \exp\left\{ \frac{t}{n} [\Phi(q(t/n)) + \Phi(q(2t/n)) + \cdots + \Phi(q(t))] \right\} u_0(q(t)).$$

By the dominated convergence theorem this implies (9.16). □

The statement of the proposition can be generalized easily to the following situation, which includes all Schrödinger equations, namely to the case of the equation

$$\frac{\partial \phi}{\partial t} = i(A - B)\phi,$$

where A is a selfadjoint operator, for which therefore there exists (according to spectral theory, see any textbook on functional analysis) a unitary transformation U such that UAU^{-1} is the multiplication operator in some $L^2(X, d\mu)$, where X is locally compact, and B is such that UBU^{-1} is a bounded operator in $C_\infty(X)$. We shall expand on this observation in Section 9.3.

It is of course more convenient to write a path integral as an integral over a positive, and not a complex, measure. This surely can be done, because any complex measure has a density with respect to its total variation measure. Explicit calculation of the density for the complex measure in (9.16) can be found e.g. in [179]. We shall deduce the corresponding representation with respect to a positive measure in the next section by using a different, more straightforward, approach.

9.2 Path integral and perturbation theory

In this section we introduce an alternative approach to the construction of path integrals specially suited to the case of underlying pure-jump processes. We begin with a simple proof of a version of formula (9.15), where the complex measure on paths is expressed in terms of a density with respect to a positive measure. This proof clearly indicates the route for the generalizations that are the subject of this chapter.

We shall work with equation (9.10), where Φ satisfies (9.11) and V is given by (9.5), that is

$$V(x) = V_\mu(x) = \int_{\mathbb{R}^d} e^{-ipx} \mu(dp)$$

with a finite complex Borel measure μ on $\mathbb{R}^d$ that is presented in the form

$$\mu(dy) = f(y)M(dy) \tag{9.18}$$

with a finite positive measure M. The equation (9.10) in momentum representation (i.e. after the Fourier transform) has the form (9.13), i.e.

$$\frac{\partial u}{\partial t} = \Phi(y)u - iV\left(i\frac{\partial}{\partial y}\right)u. \tag{9.19}$$

In order to represent Feynman's integral probabilistically, it is convenient to assume that M has no atom at the origin, i.e. $M(\{0\}) = 0$. This assumption is by no means restrictive, because one can ensure its validity by shifting V by an appropriate constant. Under this assumption, if

$$W(x) = \int_{\mathbb{R}^d} e^{-ipx} M(dp), \tag{9.20}$$

the equation

$$\frac{\partial u}{\partial t} = \left(W\left(i\frac{\partial}{\partial y}\right) - \lambda_M\right)u, \tag{9.21}$$

where $\lambda_M = M(\mathbb{R}^d)$, or equivalently

$$\frac{\partial u}{\partial t} = \int (u(y+\xi) - u(y))\, M(d\xi), \tag{9.22}$$

defines a Feller semigroup, which is the semigroup associated with the compound Poisson process having Lévy measure M. Such a process has a.s. piecewise constant paths. More precisely, a sample path Y of this process on the time interval $[0,t]$ starting at a point y is defined by a finite number, say n, of jump-times $0 < s_1 < \cdots < s_n \le t$, which are distributed according to the Poisson process N with intensity $\lambda_M = M(\mathbb{R}^d)$, and by independent jumps $\delta_1, \dots, \delta_n$ at these times, each of which is a random variable with values in $\mathbb{R}^d \setminus \{0\}$ and with distribution defined by the probability measure M/λ_M. This path has the form

$$Y_y(s) = y + Y^{s_1\dots s_n}_{\delta_1\dots\delta_n}(s) = \begin{cases} Y_0 = y, & s < s_1, \\ Y_1 = y + \delta_1, & s_1 \le s < s_2, \\ \vdots & \\ Y_n = y + \delta_1 + \delta_2 + \cdots + \delta_n, & s_n \le s \le t. \end{cases} \tag{9.23}$$

We shall denote by $\mathbf{E}_y^{[0,t]}$ the expectation with respect to this process. To visualize such an expectation, let us introduce some notations. Let $\mathrm{PC}_y(s,t)$ (abbreviated to $\mathrm{PC}_y(t)$, if $s = 0$) denote the set of all right-continuous and piecewise-constant paths $[s,t] \mapsto \mathbb{R}^d$ starting from the point y, and let $\mathrm{PC}_y^n(s,t)$ denote the subset of paths with exactly n discontinuities. Topologically, PC_y^0 is a point and $\mathrm{PC}_y^n = \mathrm{Sim}_t^n \times (\mathbb{R}^d \setminus \{0\})^n$, $n = 1, 2, \dots$, where

$$\mathrm{Sim}_t^n = \{s_1, \dots, s_n : 0 < s_1 < s_2 < \cdots < s_n \le t\} \tag{9.24}$$

denotes the standard n-dimensional simplex. In fact, the numbers s_j are the jump-times, and the n copies of $\mathbb{R}^d \setminus \{0\}$ represent the magnitudes of these jumps. To each σ-finite measure M on $\mathbb{R}^d \setminus \{0\}$ (or on $\mathbb{R}^d$, but without an atom at the origin), there corresponds a σ-finite measure $M^{\mathrm{PC}} = M^{\mathrm{PC}}(t,p)$ on $\mathrm{PC}_p(t)$, which is defined as the sum of measures M_n^{PC}, $n = 0, 1, \dots$, where each M_n^{PC} is the product-measure on $\mathrm{PC}_p^n(t)$ of the Lebesgue measure on Sim_t^n and of n copies of the measure M on $\mathbb{R}^d$. Thus if Y is parametrized as in (9.23), then

$$M_n^{\mathrm{PC}}(dY(.)) = ds_1 \cdots ds_n M(d\delta_1) \cdots M(d\delta_n).$$

From properties of the Poisson process (or more generally pure-jump Markov processes, see Section 3.7), it follows that, for $u_0 \in C_b(\mathbb{R}^d)$, the function

$$u(t,y) = \exp\{t\lambda_M\} \mathbf{E}_y^{[0,t]} \left[F(Y(.)) u_0(Y(t))\right] \tag{9.25}$$

can be rewritten in the form

$$u(t,y) = \int_{\mathrm{PC}_y(t)} M^{\mathrm{PC}}(dY(.)) F(Y(.)) u_0(Y(t)), \tag{9.26}$$

or, equivalently, as the sum

$$u(t,y) = \sum_{n=0}^{\infty} u_n(t,y) = \sum_{n=0}^{\infty} \int_{\mathrm{PC}_y^n(t)} M_n^{\mathrm{PC}}(dY(.)) F(Y(.)) u_0(Y(t)). \tag{9.27}$$

The integrals in this series can be written more explicitly as

$$\begin{aligned} u_n(t,y) &= \int_{\mathrm{PC}_y^n(t)} M_n^{\mathrm{PC}}(dY(.)) F(Y(.)) u_0(Y(t)) \\ &= \int_{\mathrm{Sim}_t^n} ds_1 \cdots ds_n \int_{\mathbb{R}^d} \cdots \int_{\mathbb{R}^d} M(d\delta_1) \cdots M(d\delta_n) \\ &\qquad \times F(y + Y_{\delta_1 \dots \delta_n}^{s_1 \dots s_n}) u_0(y + \delta_1 + \cdots + \delta_n). \end{aligned} \tag{9.28}$$

Notice that the multiplier $\exp\{t\lambda_M\}$ in (9.25) arises because this integral is not over the measure M^{PC}, but over a probability measure obtained from M^{PC} by an appropriate normalization, where the jumps are distributed according to the probability law M/λ_M. The following path-integral representation for solution to equation (9.19) can now be obtained.

Proposition 9.2.1. *Let V be given by* (9.5) *and u_0 be a bounded continuous function. Then the solution to the Cauchy problem of equation* (9.19) *with initial function u_0 has the form* (9.25) *or equivalently* (9.26), *where Y is given by* (9.23) *and*

$$\begin{aligned} F(Y(.)) &= \exp\left\{\sum_{j=0}^{n} \Phi(Y_j)(s_{j+1}-s_j)\right\} \prod_{j=1}^{n}(-if(\delta_j)) \\ &= \exp\left\{\int_0^t \Phi(Y(s))\,ds + \sum_{j=1}^{n} \ln(-if(\delta_j))\right\} \end{aligned} \tag{9.29}$$

(here $s_{n+1} = t$, $s_0 = 0$, and the function f is as from (9.18). *In particular, choosing u_0 to be the exponential function e^{iyx_0}, one obtains a path-integral representation for the Green function of equation* (9.10) *in momentum representation.*

Proof. Let us rewrite equation (9.19) in the mild integral form using Duhamel principle (see Theorem 1.9.2) considering the second operator as a perturbation to the free motion described by $\Phi(y)$:

$$u(t,y) = e^{t\Phi(y)}u_0(y) - i\int_0^t e^{(t-s)\Phi(y)}\left(V\left(i\frac{\partial}{\partial y}\right)u(s,.)\right)(y)\,ds. \tag{9.30}$$

Since the operator $V(-i(\partial/\partial y))$ acts as the convolution with the measure μ, the corresponding perturbation series representation for the solution of this equation is precisely series (9.28). It remains to observe that this series is absolutely convergent in the Banach spaces $C_\infty(\mathbb{R}^d)$ or $C_b(\mathbb{R}^d)$ (and all its terms are absolutely convergent integrals). □

Proposition 9.1.5 is a direct consequence of Proposition 9.2.1. In fact, formula (9.25) with F from (9.29) rewrites as

$$u(t,y) = \int_{\mathrm{PC}_y(t)} \exp\left\{\int_0^t \Phi(Y(s))\,ds\right\} u_0(Y(t))(-i\mu)^{\mathrm{PC}}(dY(.)), \tag{9.31}$$

showing that the measure $D_y^{0,t}q(.)$ in (9.15) is $(-i\mu)^{\mathrm{PC}} = (-ifM)^{\mathrm{PC}}$ in the notation of this section.

Proposition 9.2.2. *Under the assumptions and notations of Proposition 9.2.1, suppose X_s is any continuous curve. Then*

$$\exp\left\{-i\int_0^t V(X(s))\,ds\right\} = \int_{\mathrm{PC}_0(t)} \exp\left\{-i\int_0^t X_s\,dY_s\right\}(-i\mu)^{\mathrm{PC}}(dY(.)). \tag{9.32}$$

Proof. Expanding the l.h.s. of (9.32) in the exponential power series and ordering the time arguments in each term, we get

$$\exp\left\{-i\int_0^t V(X(s))\,ds\right\} = 1+\sum_{n=1}^{\infty}(-i)^n \int_{\mathrm{Sim}_t^n} V(X_{s_n})\cdots V(X_{s_1})ds_1\cdots ds_n,$$

which by the definition of V rewrites as

$$1+\sum_{n=1}^{\infty}(-i)^n \int_{\mathrm{Sim}_t^n} \exp\left\{\sum_{k=1}^{n} -iX_k\delta_k\right\}\mu(d\delta_1)\cdots\mu(d\delta_n)\,ds_1\cdots ds_n,$$

and hence coincides with the r.h.s. of (9.32). □

The r.h.s. of (9.32) is sometimes called the *Fourier–Feynman transform* of the complex measure $(-i\mu)^{\mathrm{PC}}$.

Remark 61. We introduced the Fourier–Feynman transform in order to be able to make the link with the theory of infinite-dimensional oscillatory integrals of Albeverio and Hoegh-Krohn. This approach for the construction of the path-integral representation of Schrödinger equations also works for potentials being Fourier transforms of a finite measure. This method is based on the possibility of representing the function $\exp\{-i\int_0^t V(X(s))\,ds\}$ as the Fourier transform of a finite measure $\mathcal{M}_V$ on the Cameron–Martin space of curves with square-integrable derivatives. Formula (9.32) yields a precise description of this measure. Namely, as on the classical paths of the free dynamics the position and velocity are connected via the trivial ODE $\dot{x} = p$, the set PC of piecewise-constant paths in the velocity space corresponds to the set CPL of continuous piecewise-linear paths in the position space. The path $X_x(s)$ starting at x_0 and such that $\dot{X}_{x_0}(s) = Y_y(s)$ with Y_y of form (9.23) is

$$X_{x_0}(s) = \begin{cases} x_0 + sY_0, & s < s_1, \\ x_0 + s_1Y_0 + (s-s_1)Y_1, & s_1 \le s < s_2, \\ \vdots & \\ x_0 + s_1Y_0 + (s_2-s_1)Y_1 + \cdots + (t-s_n)Y_n, & s_n \le s \le t. \end{cases} \tag{9.33}$$

One can thus transform a measure on the set PC to the measure on CPL. It is not difficult to deduce from (9.32) that the function $\exp\{-i\int_0^t V(X(s))\,ds\}$ can be represented as the Fourier transform of the measure on CPL (which is a subspace of the Cameron–Martin space) that is obtained by transforming the measure $(-i\mu)^{\mathrm{PC}}$ from PC to PC via the transformation (9.33).

9.3 Extensions

Even if V is not the Fourier transform of a finite measure, one can hope to prove the convergence of series (9.28), and hence a path-integral representation for the solutions of the corresponding equation, if the functional (9.29) is decreasing strongly enough for large Y_j. We shall start with the following extension of Proposition 9.2.1.

Proposition 9.3.1. *Assume that*

(i) *the continuous function* Φ *is such that*

$$\int e^{t \operatorname{Re} \Phi(y)}\, dy \le ct^{-\beta} \tag{9.34}$$

with $\beta > 0$ *with a constant* c *uniformly for finite positive times* t *(for instance, if* $\Phi(y) = -|y|^\alpha$*, then* $\beta = d/\alpha$*),*

(ii) *the measure* M *in* (9.1) *is Lebesgue measure, i.e.*

$$V(x) = \int_{\mathbb{R}^d} e^{-ipx} f(p)\, dp \tag{9.35}$$

and finally

(iii) *that* $f \in L_q(\mathbb{R}^d)$ *with arbitrary* $q > 1$ *in case* $\beta \le 1$ *and* $q < \beta/(\beta - 1)$ *otherwise.*

Then the statement of Proposition 9.2.1, *more precisely formula* (9.26) *(not necessarily* (9.25)*), still holds.*

Remark 62. The integral in (9.35) may not be defined in the Lebesgue sense, but it is defined as the Fourier transform of a distribution.

Proof. We want to show that series (9.28) converges in $C_b(\mathbb{R}^d)$. Hence we need to estimate the integrals

$$\int_{\mathrm{Sim}_t^n} ds_1 \cdots ds_n \int_{\mathbb{R}^{nd}} \exp\Big\{\sum_{j=0}^{n} \Phi(Y_j)(s_{j+1} - s_j)\Big\} \prod_{j=1}^{n} (-if(\delta_j))\, d\delta_1 \cdots d\delta_n. \tag{9.36}$$

Let us estimate the last integral (over $d\delta_n$), namely

$$\int_{\mathbb{R}^d} \exp\{(t - s_n) \operatorname{Re} \Phi(Y_{n-1} + \delta_n)\} |f(\delta_n)| d\delta_n.$$

By Hölder's inequality, this does not exceed

$$\|f\|_{L_q} \left(\int_{\mathbb{R}^d} \exp\{p(t - s_n) \operatorname{Re} \Phi(Y_{n-1} + \delta_n)\} d\delta_n\right)^{1/p},$$

where $p = q/(q-1)$, and hence, by (9.34), it is bounded by

$$\|f\|_{L_q} c[p(t-s_n)]^{-\beta/p}.$$

Thus by induction, one gets for integral (9.36) the estimate

$$(\|f\|_{L_q} c p^{-\beta/p})^n \int_{\mathrm{Sim}_t^n} e^{s_1 \Phi(y)} \prod_{j=1}^{n} (s_{j+1} - s_j)^{-\beta/p} ds_1 \cdots ds_n,$$

implying the convergence of (9.28) for $\beta < p$. □

We shall give now simple abstract formulations that will form the basis for the path-integral representation of regularized Schrödinger equations in momentum, position, or energy representations.

Let us consider the linear equation

$$\dot{\psi} = A\psi + V\psi, \tag{9.37}$$

where A and V are operators in the complex Hilbert space $\mathcal{H}$. As was already exploited above, if A generates a strongly continuous semigroup e^{tA} in $\mathcal{H}$, solutions to the Cauchy problem for this equation are often constructed via the integral equation (mild form of (9.37)):

$$\psi_t = e^{tA}\psi_0 + \int_0^t e^{(t-s)A} V \psi_s \, ds. \tag{9.38}$$

Plugging the r.h.s. of this equation repeatedly into its integral part leads to the following perturbation series representation of the solution (if necessary, see Theorem 1.9.2 for detail):

$$\begin{aligned} \psi_t = e^{tA}\psi_0 &+ \int_0^t e^{(t-s)A} V e^{sA} \psi_0 \, ds + \cdots \\ &+ \int_{0 \le s_1 \le \cdots \le s_n \le t} e^{(t-s_n)A} V e^{(s_n - s_{n-1})A} V + \cdots \\ &+ V e^{s_1 A} \psi_0 \, ds_1 \cdots ds_n + \cdots . \end{aligned} \tag{9.39}$$

Theorem 9.3.1. *Let $(\Omega, \mathcal{F}, M)$ be a Borel measure space such that in $L_2(\Omega)$ the operator A is represented as the multiplication operator on the function $A(x)$, and V be an integral operator of the form $Vf(x) = \int f(y)V(x,dy)$ with certain (possibly unbounded) transition kernel $V(x,dy)$. Suppose* Re $A(x) \le c$ *and*

$$\|Ve^{tA}\|_{C_b(\Omega)} \le ct^{-\beta}, \tag{9.40}$$

with $c > 0$ and $\beta < 1$. Then series (9.39) converges in $C_b(\Omega)$ for any $\psi_0 \in C_b(\Omega)$ and all finite $t > 0$. Its sum ψ_t solves equation (9.38) and is represented as a path integral similar to (9.26), that is

$$\psi_t(y) = \int_{\mathrm{PC}_y(t)} \mathcal{V}^{\mathrm{PC}}(dY(.))F(Y(.))\psi_0(Y(t)),$$
$$= \sum_{n=0}^{\infty} \int_{\mathrm{PC}_y^n(t)} \mathcal{V}_n^{\mathrm{PC}}(dY(.))F(Y(.))\psi_0(Y(t)), \tag{9.41}$$

where

$$\int_{\mathrm{PC}_y^n(t)} \mathcal{V}_n^{\mathrm{PC}}(dY(.))F(Y(.))\psi_0(Y(t)) = \int_{\mathrm{Sim}_t^n} ds_1 \cdots ds_n \int_{\Omega^n} V(y, dy_n) \tag{9.42}$$
$$\times \prod_{j=1}^{n-1} V(y_{j+1}, dy_j) \exp\{(t-s_n)A(y) + (s_n - s_{n-1})A(y_n) + \cdots + s_1 A(y_1)\}\psi_0(y_1).$$

Proof. Condition (9.40) ensures that the terms of series (9.41) are estimated by

$$\int_{0 \le s_1 \le \cdots \le s_n \le t} e^{c(t-s_n)} c^n (s_n - s_{n-1})^{-\beta} s_1^{-\beta},$$

which implies the required convergence. □

Clearly Proposition 9.3.1 is a concrete version of Theorem 9.3.1, where the kernels V are spatially homogeneous, i.e.

$$\int V(x, dy)u(y) = -i \int f(y)u(x+y)\,dy.$$

The momentum representation for wave functions is known to be usually convenient for the study of interacting quantum fields. In quantum mechanics one usually deals with the Schrödinger equation in position representation. Since in p-representation our measure is concentrated on the space PC of piecewise constant paths, and since classically trajectories $x(t)$ and momenta $p(t)$ are connected by the equation $\dot{x} = p$, one can expect that in position representation the corresponding measure is concentrated on the set of continuous piecewise-linear paths. To anticipate this application to the Schrödinger equation, our next result will be devoted to measures on path spaces that are concentrated on continuous piecewise-linear paths so that their velocities are piecewise constant. Denote this set by CPL. Let $\mathrm{CPL}^{x,y}(0,t)$ denote the class of paths $q : [0,t] \mapsto \mathbb{R}^d$ from CPL joining x and y in time t, i.e. such that $q(0) = y$, $q(t) = x$. By $\mathrm{CPL}_n^{x,y}(0,t)$ we denote the subclass consisting of all paths from $\mathrm{CPL}^{x,y}(0,t)$ that have exactly n jumps of their derivative. Obviously,

$$\mathrm{CPL}^{x,y}(0,t) = \bigcup_{n=0}^{\infty} \mathrm{CPL}_n^{x,y}(0,t).$$

Notice also that the set $\mathrm{CPL}^{x,y}(0,t)$ belongs to the Cameron–Martin space of curves that have derivatives in $L^2([0,t])$.

To any σ-finite measure M on $\mathbb{R}^d$ there corresponds a unique σ-finite measure M^{CPL} on $\mathrm{CPL}^{x,y}(0,t)$, which is the sum of the measures M_n^{CPL} on $\mathrm{CPL}_n^{x,y}(0,t)$, where M_0^{CPL} is just the unit measure on the one-point set $\mathrm{CPL}_0^{x,y}(0,t)$ and each M_n^{CPL}, $n > 0$, is the direct product of the Lebesgue measure on the simplex (9.24) of the jump times $0 < s_1 < \cdots < s_n < t$ of the derivatives of the paths $q(.)$ and of n copies of the measure M on the values $q(s_j)$ of the paths at these times. In other words, if

$$q(s) = q_{\eta_1\ldots\eta_n}^{s_1\ldots s_n}(s) = \eta_j + (s - s_j)\frac{\eta_{j+1} - \eta_j}{s_{j+1} - s_j}, \quad s \in [s_j, s_{j+1}] \tag{9.43}$$

(where $s_0 = 0, s_{n+1} = t, \eta_0 = y, \eta_{n+1} = x$) is a typical path in $\mathrm{CPL}_n^{x,y}(0,t)$ and Φ is a functional on $\mathrm{CPL}^{x,y}(0,t)$, then

$$\int_{\mathrm{CPL}^{x,y}(0,t)} \Phi(q(.))M^{\mathrm{CPL}}(dq(.)) = \sum_{n=0}^{\infty} \int_{\mathrm{CPL}_n^{x,y}(0,t)} \Phi(q(.))M_n^{\mathrm{CPL}}(dq(.))$$

$$= \sum_{n=0}^{\infty} \int_{\mathrm{Sim}_t^n} ds_1 \cdots ds_n \int_{\mathbb{R}^d} \cdots \int_{\mathbb{R}^d} M(d\eta_1)\cdots M(d\eta_n)\Phi(q(.)). \tag{9.44}$$

Theorem 9.3.2. *Let the operator V in* (9.37) *be the multiplication operator by a function V on $\mathbb{R}^d$ or, more generally, by a measure V on $\mathbb{R}^d$, and the operator e^{tA} in $L_2(\mathbb{R}^d)$ be an integral operator of the form $e^{tA}f(x) = \int A_t(x,y)f(y)dy$ with certain measurable function $A_t(x,y)$. Suppose $\|e^{tA}\|_{C_b(\Omega)} \le c$ and*

$$\|e^{tA}V\|_{C_b(\Omega)} \le ct^{-\beta} \tag{9.45}$$

with $c > 0$ and $\beta < 1$. Then series (9.39) *converges in $C_b(\Omega)$ for any $\psi_0 \in C_b(\Omega)$ and all finite $t > 0$. Its sum ψ_t solves equation* (9.38) *and is represented as a path integral on the space* CPL*:*

$$\psi_t(x) = \int_{\mathrm{CPL}^{x,y}(0,t)} \int_{\mathbb{R}^d} \psi_0(y)\Phi^A(q(.))V^{\mathrm{CPL}}(dq(.)), \tag{9.46}$$

where

$$\Phi^A(q(.)) = A_{t-s_n}(x,\eta_n)A_{s_n-s_{n-1}}(\eta_n,\eta_{n-1})\cdots A_{s_1}(\eta_1,y). \tag{9.47}$$

Proof. By the definition of the measure V^{CPL} the integral (9.46) is nothing else but series (9.39). Condition (9.45) ensures that the terms of this series are estimated as

$$\int_{0\le s_1\le\cdots\le s_n\le t} c^{n+1}(t-s_n)^{-\beta}(s_n - s_{n-1})^{-\beta}s_1^{-\beta},$$

which implies the required convergence. □

For applications to the Schödinger equation with magnetic fields one needs to handle the case when V is the composition of a multiplication and the derivation operator, which is the subject of the next result.

Theorem 9.3.3. *As in Theorem* 9.3.3, *assume that e^{tA} is an integral operator of the form $e^{tA}f(x) = \int A_t(x,y)f(y)dy$ with certain measurable function $A_t(x,y)$ such that $\|e^{tA}\|_{C_b(\Omega)} \leq c$. But suppose now that $V = (\nabla, F)$ with a bounded measurable vector-function F on $\mathbb{R}^d$, i.e.*

$$V(f) = \sum_{j=1}^{d} \nabla^j (F_j f)(x).$$

Assume that $A_t(x,y)$ is differentiable with respect to the second variable for $t > 0$ and $A_t(x,y) \to 0$ as $y \to \infty$ for any x. Denote by $\nabla_2 e^{tA}$ the integral operator with the kernel being the derivative of A with respect to the second variable, i.e.

$$[(\nabla_2 e^{tA}) f(x)]^j = \int \left[\frac{\partial}{\partial y_j} A_t(x,y)\right] f(y)\, dy.$$

Finally, assume

$$\|(\nabla_2 e^{tA}, F)\|_{C_b(\Omega)} \leq c t^{-\beta} \tag{9.48}$$

with $c > 0$ and $\beta < 1$. Then series (9.39) *converges in $C_b(\Omega)$ for any $\psi_0 \in C_b(\Omega)$ and all finite $t > 0$. Its sum ψ_t solves equation* (9.38) *and is represented as a path integral on the space* CPL*:*

$$\psi_t(x) = \int_{\mathrm{CPL}^{x,y}(0,t)} \int_{\mathbb{R}^d} \psi_0(y) \tilde{\Phi}^A(q(.)) F^{\mathrm{CPL}}(dq(.)),$$

where

$$\tilde{\Phi}^A(q(.)) = \frac{\partial}{\partial \eta_n} A_{t-s_n}(x,\eta_n) \cdots \frac{\partial}{\partial \eta_1} A_{s_2-s_1}(\eta_2,\eta_1) A_{s_1}(\eta_1,y).$$

Proof. This is a consequence of Theorem 9.3.3, if one notices that

$$e^{tA}V = e^{tA}(\nabla, F) = (\nabla_2 e^{tA}, F)$$

by integration by parts. □

Finally, when working with Schrödinger equation, the most natural convergence is mean-square. The following statement is a mean-square version of the above results. Its proof is straightforward.

Proposition 9.3.2. *Under the assumptions of Theorems* 9.3.1 *or* 9.3.2 *suppose instead of* (9.40) *and* (9.45) *one has the estimates*

$$\|Ve^{tA}\|_{L_2(\Omega)} \le ct^{-\beta} \tag{9.49}$$

or respectively

$$\|e^{tA}V\|_{L_2(\mathbb{R}^d)} \le ct^{-\beta}. \tag{9.50}$$

Then the statements of the theorems remain true, but for $\psi_0 \in L_2(\Omega)$, *and with the convergence of the series understood in the mean-square sense (meaning that the corresponding path integral should be considered as an improper Riemann integral).*

9.4 Regularization of the Schrödinger equation by complex time or mass, or continuous observation

To apply the path-integral construction (as well as their extensions) to the Schrödinger equations beyond the case of potentials representing Fourier transform of finite measures, one often needs certain regularization. For instance, one can use the same regularization as is used to define the finite-dimensional integral

$$(U_0 f)(x) = (2\pi t i)^{-d/2} \int_{\mathbb{R}^d} \exp\left\{-\frac{|x-\xi|^2}{2ti}\right\} f(\xi)\, d\xi \tag{9.51}$$

giving the free propagator $e^{it\Delta/2} f$. Namely, this integral is not well-defined for general $f \in L^2(\mathbb{R}^d)$. The most natural way to define it is based on the observation that, according to the spectral theorem (see e.g. Reed and Simon [273] or any other text on functional analysis), for all $t > 0$

$$e^{it\Delta/2} f = \lim_{\epsilon \to 0_+} e^{it(1-i\epsilon)\Delta/2} f \tag{9.52}$$

in $L^2(\mathbb{R}^d)$. Since

$$(e^{it(1-i\epsilon)\Delta/2} f)(x) = (2\pi t(i+\epsilon))^{-d/2} \int_{\mathbb{R}^d} \exp\left\{-\frac{|x-\xi|^2}{2t(i+\epsilon)}\right\} f(\xi)\, d\xi$$

(the argument of $\sqrt{i+\epsilon}$ is assumed to belong to $(0,\pi)$) and the integral on the r.h.s. of this equation is already absolutely convergent for all $f \in L^2(\mathbb{R}^d)$, one can define the integral (9.51) by the formula

$$(U_0 f)(x) = \lim_{\epsilon \to 0_+} (2\pi t(i+\epsilon))^{-d/2} \int_{\mathbb{R}^d} \exp\left\{-\frac{|x-\xi|^2}{2t(i+\epsilon)}\right\} f(\xi)\, d\xi, \tag{9.53}$$

where convergence holds in $L^2(\mathbb{R}^d)$.

More generally, if the operator $-\Delta/2 + V(x)$ is self-adjoint and bounded from below, again by the spectral theorem,

$$\exp\{it(\Delta/2 - V(x))\}f = \lim_{\epsilon \to 0_+} \exp\{it(1 - i\epsilon)(\Delta/2 - V(x))\}f. \tag{9.54}$$

In other words, solutions to the Schrödinger equation

$$\frac{\partial \psi_t(x)}{\partial t} = \frac{i}{2}\Delta\psi_t(x) - iV(x)\psi_t(x) \tag{9.55}$$

can be approximated by the solutions to the equation

$$\frac{\partial \psi_t(x)}{\partial t} = \frac{1}{2}(i + \epsilon)\Delta\psi_t(x) - (i + \epsilon)V(x)\psi_t(x), \tag{9.56}$$

which describes the Schrödinger evolution in complex time. The corresponding equation on the Fourier transform u is

$$\frac{\partial u}{\partial t} = -\frac{1}{2}(i + \epsilon)y^2 u - (i + \epsilon)V\left(i\frac{\partial}{\partial y}\right)u. \tag{9.57}$$

As we shall see, the results of the previous section are often applicable to regularized equations (9.56) with arbitrary $\epsilon > 0$, so that (9.54) yields an improper Riemann integral representation for $\epsilon = 0$, i.e. to the Schrödinger equation per se. Thus, unlike the usual method of analytical continuation often used for defining Feynman integrals, where the rigorous integral is defined only for purely imaginary Planck's constant h, and for real h the integral is defined as the analytical continuation by rotating h through a right angle, in our approach, the measure is defined rigorously and is the same for all complex h with non-negative real part. Only on the boundary $\operatorname{Im} h = 0$ does the corresponding integral usually become an improper Riemann integral.

Equation (9.56) is certainly only one of many different ways to regularize the Feynman integral. However, this method is one of the simplest methods, because the limit (9.54) follows directly from the spectral theorem, and other methods may require additional work to obtain the corresponding convergence result. For instance, one can use regularization by introducing complex mass, which boils down to solving the equation

$$\frac{\partial \psi}{\partial t} = \frac{1}{2}(i + \epsilon)\Delta\psi - iV(x)\psi.$$

A more physically motivated regularization can be obtained from the quantum theory of continuous measurement describing spontaneous collapse of quantum states, which regularizes the divergences of Feynman's integral for large position or momentum. The work with this regularization is technically more difficult than with the above regularization by complex times, but it is a matter of independent interest. We only sketch it, referring for detail to original papers.

The standard Schrödinger equation describes an isolated quantum system. In the quantum theory of open systems one considers a quantum system under observation in a quantum environment (reservoir). This leads to a generalization of the Schrödinger equation, which is called *stochastic Schrödinger equation (SSE)*, or *quantum state diffusion model*, or *Belavkin's quantum filtering equation* (see Section 9.7 for references). In the case of a *non-demolition measurement of diffusion type*, the SSE has the form

$$du + \left(iH + \frac{1}{2}\lambda^2 R^\star R\right)u\,dt = \lambda Ru\,dW, \tag{9.58}$$

where u is the unknown a posterior (non-normalised) wave function of the given continuously observed quantum system in a Hilbert space $\mathcal{H}$, the self-adjoint operator $H = H^\star$ in $\mathcal{H}$ is the Hamiltonian of a free (unobserved) quantum system, the vector-valued operator $R = (R^1, \dots, R^d)$ in $\mathcal{H}$ represents the observed physical values, W is the standard d-dimensional Brownian motion, and the positive constant λ represents the precision of measurement. The simplest natural examples of (9.58) concern the case when H is the standard quantum mechanical Hamiltonian and the observed physical value R is either the position or momentum of the particle. For the purposes of regularization, the second case is usually more handy, that is the equation

$$d\psi = \left(\frac{1}{2}\left(i + \frac{\lambda}{2}\right)\Delta\psi - iV(x)\psi\right)dt + \frac{1}{i}\sqrt{\frac{\lambda}{2}}\frac{\partial}{\partial x}\psi\,dW. \tag{9.59}$$

The equation on the Fourier transform $u(y)$ of ψ has the form

$$du = \left(-\frac{1}{2}\left(i + \frac{\lambda}{2}\right)y^2 u - iV\left(i\frac{\partial}{\partial y}\right)u\right)dt + \sqrt{\frac{\lambda}{2}}yu\,dW. \tag{9.60}$$

Solutions to this equation are quite peculiar. Say, if $V = 0$, one can write the solutions to the Cauchy problem with initial function u_0 explicitly:

$$[T^t_{\lambda,W}u_0](y) = \exp\left\{-\frac{1}{2}(i+\lambda)y^2 t + \sqrt{\frac{\lambda}{2}}yW(t)\right\}u_0(y) \tag{9.61}$$

(notice the change of the coefficient at λ which is due to the correction term in Itô formula). The operators $T^t_{\lambda,W}$ are bounded both in $C(\mathbb{R}^d)$ and $C_\infty(\mathbb{R}^d)$ for each $t \geq 0$, but their norms tend to ∞ a.s. as $t \to 0$.

As $\lambda \to 0$, equation (9.59) approaches the standard Schrödinger equation. We refer to Kolokoltsov [179] for the path-integral representation of the solutions to this equation and its use as regularization to the path integral of the Schrödinger equation. In what follows, we restrict our attention only to the simpler regularization by complex times.

9.5 Singular and growing potentials, magnetic fields and curvilinear state spaces

Here we use the ideas from the previous two sections to give a rigorous path-integral representation for solutions to the Schrödinger equation with various kinds of potentials and with a possibly curvilinear state space. Our first result is a direct corollary of Proposition 9.3.1.

Proposition 9.5.1. *Let* V *have form* (9.35) *(again in the sense of distributions) and* $f \in L^1 + L^q$*, i.e.* $f = f_1 + f_2$ *with* $f_1 \in L^1(\mathbb{R}^d)$*,* $f_2 \in L^q(\mathbb{R}^d)$*, with* q *in the interval* $(1, d/(d-2))$*,* $d > 2$*. Then for any* $\epsilon > 0$ *the regularized Schrödinger equation in momentum representation* (9.57) *satisfies the conditions of Proposition* 9.3.1 *yielding a representation for its solutions in terms of the path integral. Moreover, the operator* $-\Delta/2 + V$ *is self-adjoint, so that* (9.54) *yields an improper Riemann integral representation for* $\epsilon = 0$*, i.e. to the Schrödinger equation per se.*

Proof. Self-adjointness of the Schrödinger operators for this class of potentials is well known, see e.g. Section X.2 in Reed and Simon [274]. Condition (9.34) clearly holds with $\beta = d/2$ for any $\epsilon > 0$. Thus $q < d/(d-2)$ is equivalent to $q < \beta/(\beta - 1)$. Finally one observes that one can combine (take sums of) the potentials satisfying Propositions 9.5.1 and 9.2.1. □

Remark 63. The class of potentials from Proposition 9.5.1 includes the Coulomb case $V(x) = |x|^{-1}$ in $\mathbb{R}^3$, because for this case $f(y) = |y|^{-2}$ in representation (9.35).

We shall turn now to the Schrödinger equation in the position representation, aiming at the application of Theorem 9.3.2. Of course, if V is a bounded function, the conditions of this theorem for regularized Schrödinger equation (9.56) are trivially satisfied (with $\beta = 0$). Let us discuss singular potentials. The most important class of these potentials represent Radon measures supported by null sets such as discrete sets (point interaction), smooth surfaces (surface delta interactions), Brownian paths and more general fractals. Less exotic examples of potentials satisfying the assumptions of Proposition 9.5.3 below are given by measures with a density $V(x)$ having bounded support and such that $V \in L^p(\mathbb{R}^d)$ with $p > d/2$.

The one-dimensional situation turns out to be particularly simple in our approach, because in this case no regularization is needed to express the solutions to the corresponding Schrödinger equation and its propagator in terms of path integrals.

Proposition 9.5.2. *Let* V *be a bounded (complex) measure on* $\mathbb{R}$*. Then the solution* ψ_G *to equation* (9.56) *with* $\epsilon \geq 0$ *(i.e. including equation* (9.55)*) and the initial function* $\psi_0(x) = \delta(x - x_0)$ *(i.e. the propagator or the Green function of* (9.56)*)* *exists and is a continuous function of the form*

$$\psi_G(t,x) = (2\pi(i+\epsilon)t)^{-1/2} \exp\left\{-\frac{|x-x_0|^2}{2t(i+\epsilon)}\right\} + O(1),$$

uniformly for finite times. Moreover, one has the path-integral representation for ψ_G *of the form*

$$\psi_G(t,x) = \int_{\mathrm{CPL}^{x,y}(0,t)} \Phi(q(.)) V^{\mathrm{CPL}}(dq(.)),$$

where V^{CPL} *is related to* V *as* M^{CPL} *is to* M *in Theorem* 9.3.2, *and*

$$\Phi(q(.)) = \prod_{j=1}^{n+1} [2\pi(s_j - s_{j-1})(i+\epsilon)]^{-1/2} \exp\left\{-\frac{1}{2(i+\epsilon)} \int_0^t \dot{q}^2(s)\,ds\right\}.$$

Remark 64. The same holds for $\epsilon = 1 - i$, i.e. for the heat equation.

Proof. Clearly the condition of Theorem 9.3.2 are satisfied with $\beta = 1/2$ (one-dimensional effect). □

For the Schrödinger equation in the finite-dimensional case one needs a regularization, say (9.56) with $\epsilon > 0$ or (9.59) with $\lambda > 0$. As we mentioned, here we work only with the simpler regularization (9.56).

A number $\dim(V)$ is called the *dimensionality* of a measure V if it is the least upper bound of all $\alpha \geq 0$ such that there exists a constant $C = C(\alpha)$ such that

$$|V(B_r(x))| \leq C r^\alpha$$

for all $x \in \mathbb{R}^d$ and all $r > 0$.

Proposition 9.5.3. *Let* V *be a finite complex Borel measure on* $\mathbb{R}^d$ *with* $\dim(V) > d-2$. *Then*

(i) *for any* $\epsilon > 0$ *the regularized Schrödinger equation* (9.56) *satisfies the conditions of Theorem* 9.3.2 *with* $A = (i+\epsilon)\Delta/2$ *and* V *replaced by* $-(i+\epsilon)V$, *yielding a representation for its solutions in terms of the path integral:*

$$\psi_\epsilon(t,x) = \int_{\mathrm{CPL}^{x,y}(0,t)} \int_{\mathbb{R}^d} \psi_0(y) \Phi_\epsilon(q(.)) V^{\mathrm{CPL}}(dq(.))dy, \qquad (9.62)$$

where

$$\Phi_\epsilon = \prod_{j=1}^{n+1} \left[2\pi(s_j - s_{j-1})(i+\epsilon)\right]^{-d/2} (-\epsilon - i)^n \exp\left\{-\frac{1}{2(i+\epsilon)} \int_0^t \dot{q}^2(s)\,ds\right\}; \qquad (9.63)$$

(ii) *one can give rigorous meaning to the formal expression* $H = -\Delta/2 + V$ *as a bounded-below self-adjoint operator in such a way that the operators* $\exp\{-t(i+\epsilon)H\}$ *defined by means of functional operator calculus are given by the path integral of Theorem* 9.3.2;

(iii) *formula* (9.54) *yields an improper Riemann integral representation for* $\epsilon = 0$.

Proof. To check the conditions of Theorem 9.3.2 for the regularized Schrödinger equation (9.56) we need to show that for any $\epsilon > 0$

$$\int_{\mathbb{R}^d} [2\pi t(i+\epsilon)]^{-d/2} \exp\left\{-\frac{|x-\xi|^2}{2t(i+\epsilon)}\right\} V(d\xi) \le c(\epsilon) t^{-\beta}$$

with $\beta < 1$ uniformly for all x. To this end, let us decompose this integral into the sum of two integrals $I_1 + I_2$ by decomposing the domain of integration into two parts:

$$D_1 = \{\xi : |x-\xi| \ge t^{-\delta+1/2}\}, \quad D_2 = \{\xi : |x-\xi| < t^{-\delta+1/2}\}.$$

Then I_1 is exponentially small for small t and

$$I_2 \le [2\pi t\sqrt{1+\epsilon^2}]^{-d/2} \int_{D_2} V(d\xi) \le c(\alpha,\epsilon) t^{-d/2} (t^{-\delta+1/2})^{\alpha}$$

with $\alpha > d-2$. This expression is of order $t^{-\beta}$ with $\beta = \delta\alpha + (d-\alpha)/2 < 1$ whenever $\delta < (\alpha - d + 2)/(2\alpha)$. It remains to prove self-adjointness. This can be obtained from the properties of the corresponding semigroup. We refer for details to Kolokoltsov [185],[1] where this is done in a more general setting of the equation including magnetic fields discussed below. □

Remark 65. It is not difficult to see that expression (9.62) still converges for the Green function, that is for the the solution $G^{\epsilon}(t,x,y)$ with initial function $\psi_0(x) = \delta_y(x)$ with an arbitrary y, yielding

$$G^{\epsilon}(t,x,y) = \int_{\mathrm{CPL}^{x,y}(0,t)} \int_{\mathbb{R}^d} \Phi_{\epsilon}(q(.)) V^{\mathrm{CPL}}(dq(.)), \tag{9.64}$$

see again [185] for detail.

Let us extend this result to the case of a formal Schrödinger operator with magnetic fields in $L^2(\mathbb{R}^d)$ of the form

$$H = \frac{1}{2}\left(\frac{1}{i}\frac{\partial}{\partial x} + A(x)\right)^2 + V(x) \tag{9.65}$$

under the following conditions:

[1]This paper contains nasty typos messing up the coefficient $1/2$ and signs $\pm$ before Δ in several places.

(C1) the magnetic vector-potential A is a bounded measurable mapping $\mathbb{R}^d \to \mathbb{R}^d$,

(C2) the potential V and the divergence $\operatorname{div} A = \sum_{j=1}^d \frac{\partial A^j}{\partial x^j}$ (defined in hte sense of distributions) are both (signed) Borel measures,

(C3) if $d > 1$ there exist $\alpha > d-2$ and $C > 0$ such that for all $x \in \mathbb{R}^d$ and $r \in (0, 1]$

$$|\operatorname{div} A|(B_r(x)) \le Cr^\alpha, \quad |V|(B_r(x)) \le Cr^\alpha,$$

if $d = 1$ the same holds for $\alpha = 0$.

The corresponding regularized Schrödinger equation can be written in the form

$$\frac{\partial \psi}{\partial t} = -DH\psi,$$

where D is a complex number with $\operatorname{Re} D = \epsilon \ge 0$, $|D| > 0$, and the corresponding integral (mild) equation is

$$\psi_t = e^{Dt\Delta/2}\psi_0 - D\int_0^t e^{D(t-s)\Delta/2}(W + i(\nabla, A))\psi_s\, ds, \tag{9.66}$$

where

$$W(x) = V(x) + \frac{1}{2}|A(x)|^2 + \frac{i}{2}\operatorname{div} A(x).$$

More precisely, W is a measure, which is the sum of the measure $V(x) + i \operatorname{div} A(x)/2$ and the measure having the density $|A(x)|^2/2$ with respect to Lebesgue measure.

Theorem 9.5.1. *Suppose* (C1)–(C3) *hold for operator* (9.65). *Then all the statements of Proposition* 9.5.3 *are valid for the operator H (only the explicit expression* (9.63) *for Φ should be appropriately modified).*

Proof. This is the same as for Proposition 9.5.3 above, but one needs to use Theorem 9.3.3 instead of Theorem 9.3.2. □

In fact much more can be proved. One can get rather precise estimates for the *heat kernel* of e^{-tDH}. Namely, let $G^D(t, x, y)$ denotes the fundamental solution (Green function or heat kernel) of the Cauchy problem generated by DH, i.e. the solution of (9.66) with the Dirac initial condition $\psi_0(x) = \delta(x - y)$. It can be sought (as for the vanishing A above) via the perturbation series representation for equation (9.66).

Theorem 9.5.2. *Under the assumptions of Theorem* 9.5.1 *the following holds.*

(i) *If $\epsilon = \operatorname{Re} D > 0$, the perturbation series expressing G^D is absolutely convergent and its sum $G^D(t, x, y)$ is continuous in $x, y \in \mathbb{R}^d$, $t > 0$ and satisfies the estimate*

$$|G^D(t, x, y)| \le cG_{\text{free}}^{|D|^2/\epsilon}(t, x - y)\exp\{c|x - y|\} \tag{9.67}$$

uniformly for $t \le t_0$ *with any* t_0 *and a constant* $c = c(t_0)$, *where* G^D_{free} *denotes the kernel of the 'free' propagator:*

$$G^D_{\text{free}}(t, x - y) = (2\pi t D)^{-d/2} \exp\left\{-\frac{(x-y)^2}{2Dt}\right\}.$$

(ii) *The integral operators*

$$(U^D_t \psi_0)(t, x) = \int G^D(t, x, y)\psi_0(y)\, dy \tag{9.68}$$

defining the solutions to equation (9.66) *for* $t \in [0, t_0]$ *form a uniformly bounded family of operators* $L^2(\mathbb{R}^d) \to L^2(\mathbb{R}^d)$.

(iii) *If* D *is real, i.e.* $D = \epsilon > 0$, *then there exists a constant* $\omega > 0$ *such that the Green function* G^ϵ *has the asymptotic representation*

$$G^\epsilon(t, x, y) = G^\epsilon_{\text{free}}(t, x, y)[1 + O(t^\omega) + O(|x - y|)] \tag{9.69}$$

for small t *and* $x - y$. *In case of vanishing* A, *the multiplier* $\exp\{c|x - y|\}$ *in* (9.67) *can be dropped and formula* (9.69) *gives global (uniform for all* x, Y*) small time asymptotics for* G^ϵ.

Proof. See [185]. Notice only that the estimates for G^D can be obtained by estimating the first term of the corresponding perturbation series representation with $\psi_0(x) = \delta(x - y)$. Thus obtained estimates (9.69) allows us to deduce statement (ii), which in turn implies selfadjointness of H. □

Let us discuss now shortly the case of operators with discrete spectrum. Namely, let us consider the Schrödinger equation

$$\frac{\partial \psi_t(x)}{\partial t} = -iH\psi_t(x) - iV(x)\psi_t(x), \tag{9.70}$$

where V is the operator of multiplication by a function V and H is a selfadjoint operator in $L^2(\Omega)$ with discrete spectrum, where Ω is a Borel measurable space. Basic examples of interest are as follows:

(i) H is the Laplace operator (or more generally an elliptic operator) on a compact Riemann manifold (curvilinear state space),

(ii) $H = -\Delta + W(x)$ in $L^2(\mathbb{R}^d)$, where W is bounded below and $W(x) \to \infty$ for $x \to \infty$ (see e.g. [276] for a proof that such H has a discrete spectrum),

(iii) many-particle versions of the situation from (ii).

In this case the most natural representation for the Schrödinger equation is the energy representation. In other words, if $\lambda_1 \leq \lambda_2 \leq \cdots$ are eigenvalues of H and $\psi_1, \psi_2, \ldots$ are the corresponding normalized eigenfunctions, then any $\psi \in L^2(\Omega)$ can be represented by its Fourier coefficients $\{c_n\}$, where $\psi = \sum_{n=1}^{\infty} c_n \psi_n$ is the expansion of ψ with respect to the orthonormal basis $\{\psi_j\}$. In terms of $\{c_n\}$ the operator e^{-itH} acts as the multiplication $c_n \mapsto \exp\{-it\lambda_n\}c_n$, and V is represented by the infinite-dimensional symmetric matrix $V_{nm} = \int \psi_n(x)V(x)\psi_m(x)\,dx$ (i.e. it is a discrete integral operator). If V is a bounded function, condition (9.49) of Proposition 9.3.2 is trivially satisfied (with $\beta = 0$) yielding a path-integral representation for the solutions of equation (9.70) in the spectral representation of the operator H. It is not difficult to find examples when the conditions of Theorem 9.3.1 hold, but these examples do not seem to be generic.

9.6 Fock-space representation

Let us start with the simplest probabilistic interpretation of the solutions to the Cauchy problem of equation (9.56) in terms of an expectation with respect to a compound Poisson process. The following statement is a direct consequence of Proposition 9.5.3 and the standard properties of Poisson processes.

Proposition 9.6.1. *Suppose a measure V satisfies the assumptions of Proposition 9.5.3. Let $\lambda_V = V(\mathbb{R}^d)$. Let paths of* CPL *be parametrized by* (9.43) *and let* **E** *denote the expectation with respect to the process of jumps η_j which are identically independently distributed according to the probability measure V/λ_V and which occur at times s_j from $[0, t]$ that are distributed according to Poisson process of intensity λ_V. Then the solution to the Cauchy problem of* (9.56) *can be written in the form*

$$\psi_\epsilon(t, x) = e^{t\lambda_V} \int_{\mathcal{R}^d} \psi_0(y)\mathbf{E}(\Phi_\epsilon(q(.)))\,dy. \tag{9.71}$$

Next let us move to the representations in terms of Wiener measures, for which we shall use the Fock spaces.

The paths of the spaces PC and CPL used above are parametrised by finite sequences $(s_1, x_1), \ldots, (s_n, x_n)$ with $s_1 < \cdots < s_n$ and $x_j \in \mathbb{R}^d$, $j = 1, \ldots, d$. Denote by $\mathcal{P}^d$ the set of all these sequences and by $\mathcal{P}_n^d$ its subset consisting of sequences of the length n. Thus, functionals on the path spaces PC or CPL can be considered as functions on $\mathcal{P}^d$. To each measure ν on $\mathbb{R}^d$ there corresponds a measure $\nu_{\mathcal{P}}$ on $\mathcal{P}^d$ which is the sum of the measures ν^n on $\mathcal{P}_n^d$, where ν^n are the product measures $ds_1 \cdots ds_n d\nu(x_1) \cdots d\nu(x_n)$. The Hilbert space $L^2(\mathcal{P}^d, \nu_{\mathcal{P}})$ is isomorphic to the Fock space Γ_ν^d over the Hilbert space $L^2(\mathbb{R}_+ \times \mathbb{R}^d, dx \times \nu)$ (which is isomorphic to the space of square-integrable functions on $\mathbb{R}_+$ with values in $L^2(\mathbb{R}^d, \nu)$). Therefore, square-integrable functionals on CPL can be considered as vectors in the Fock space $\Gamma_{V(dx)}^d$.

Let us consider the Green function of equation (9.56), given by (9.64). First let us rewrite it as the integral of an element of the Fock space $\Gamma^0 = L^2(\mathrm{Sim}_t)$ with $\mathrm{Sim}_t = \bigcup_{n=0}^{\infty} \mathrm{Sim}_t^n$ (which was denoted $\mathcal{P}^0$ above), where Sim_t^n is as usual the simplex (9.24):

$$\mathrm{Sim}_t^n = \{s_1, \dots, s_n : 0 < s_1 < s_2 < \cdots < s_n \leq t\}.$$

Let

$$g_0^V = g_0^V(t, x, y) = (2\pi(i + \epsilon 0))^{-d/2} \exp\left\{-\frac{(x-y)^2}{2t(i+\epsilon)}\right\}$$

and

$$g_n^V(s_1, \dots, s_n) = g_n^V(s_1, \dots, s_n; t, x, y) = \int_{\mathbb{R}^{nd}} \Phi_\epsilon(q_{\eta_1 \dots \eta_n}^{s_1 \dots s_n})\, d\eta_1 \cdots d\eta_n$$

for $n = 1, 2, \dots$, where Φ_ϵ and $q_{\eta_1 \dots \eta_n}^{s_1 \dots s_n}$ are given by (9.63) and (9.43). Considering the series of functions $\{g_n^V\}$ as a single function g^V on Sim_t, we shall rewrite the r.h.s. of (9.64) in the following concise notation:

$$\int_{\mathrm{Sim}_t} g^V(s)\, ds = \sum_{n=0}^{\infty} \int_{\mathrm{Sim}_t^n} g_n^V(s_1, \dots, s_n)\, ds_1 \cdots ds_n.$$

Now, the Wiener chaos decomposition theorem states (Section 2.9) that, if $dB_{s_1} \cdots dB_{s_n}$ denotes the n-dimensional stochastic Wiener differential, then to each $f = \{f_n\} \in L^2(\mathrm{Sim}_t)$ there corresponds an element $\phi_f \in L^2(\Omega_t)$, where Ω_t is the Wiener space of continuous real functions on $[0, t]$, given by the formula

$$\phi_f(B) = \sum_{n=0}^{\infty} \int_{\mathrm{Sim}_t^n} f_n(s_1, \dots, s_n)\, dB_{s_1} \cdots dB_{s_n},$$

or in concise notation

$$\phi_f(B) = \int_{\mathrm{Sim}_t} f(s)\, dB_s.$$

Moreover the mapping $f \mapsto \phi_f$ is an isometric isomorphism, i.e.

$$\mathbf{E}(\phi_f(B)\bar{\phi}_g(B)) = \int_{\mathrm{Sim}_t} f(s)\bar{g}(s)\, ds,$$

where $\mathbf{E}$ denotes the expectation with respect to the standard Wiener process.

Now, it is not difficult to show that

$$\int_{\mathrm{Sim}_t} dB_s = e^{B_t - t/2},$$

see e.g. Meyer [244] for detail.[2] Hence, if the function g^V belongs not only to $L^1(\mathrm{Sim}_t)$, but also to $L^2(\mathrm{Sim}_t)$ the Green function can be rewritten as

$$G_\epsilon(t, x, x_0) = \mathbf{E}(\phi_{g^V} \exp\{B_t - t/2\}). \tag{9.72}$$

This leads to the following result, whose full proof is given in Kolokoltsov [189].

Proposition 9.6.2. *Under the assumptions of Proposition* 9.5.3 *suppose additionally that* V *is finite and* $\dim(V) > d - 1$. *Then the Green function* (9.64) *can be written in the form* (9.72), *where* E_W *denotes the expectation with respect to the standard Wiener process.*

For general V from Propositions 9.5.3, when the corresponding function g^V belongs to $L^1(\mathrm{Sim}_t)$, but not to $L^2(\mathrm{Sim}_t)$, some regularized version of the above representation is available, see [179].

9.7 Comments

The idea of expressing the solutions to the Schrödinger equation in momentum representation in terms of the expectations with respect to a compound Poisson process belongs to Maslov and Chebotarev [232], [233], where Proposition 9.2.1 was proved for the Schrödinger equation, i.e. for $\Phi(y) = -|y|^2/2$, as well as Proposition 9.2.2. In Chebotarev, Konstantinov and Maslov [77] the extension to Pauli and Dirac equations was carried out. A review of the literature on various applications of these results can be found in [179]. The idea of going beyond the potentials that can be expressed as Fourier transforms of finite measures was developed systematically in Kolokoltsov [179], [182], which form the ideological basis of the exposition of this chapter. Section 9.1 follows essentially Kolokoltsov [177], [181]. A general construction of the measures on path space, given there, is an abstract version of the constructions of Nelson [254] and Ichinose [137], devoted respectively to the Wiener measure and to a measure corresponding to the hyperbolic systems of the first order.

Using equation in complex times for regularization of the Schrödinger equation was suggested by Gelfand and Yaglom [121]. However in this paper an erroneous attempt was made to use it for defining a version of the Wiener measure with a complex volatility. It was noted afterwards in Cameron [71] that there exists no direct extension of the Wiener measure that could be used to define Feynman integral for equation (9.56) for any real ϵ. As we have shown, the idea works fine for the jump-type processes.

The idea of using the theory of continuous observation for regularization of Feynman's integral was first discussed in physical literature on the heuristic level starting

[2] This formula follows from the observation that the process $Y_t = \exp\{B_t - t/2\}$ is the solution to the linear SDE $Y_t = 1 + \int_0^t Y_s dB_s$ and the required chaos expansion is obtained via standard Picard approximations.

with Feynman himself. It was revived by Menski [242], though this paper advocates an adaptation of the erroneous construction from the above cited paper [121]. The first mathematically rigorous construction of the path integral for the systems of continuous quantum measurement (i.e. for Belavkin's quantum filtering equation) was given by Albeverio, Kolokoltsov and Smolyanov [14], [15] via the method of infinite-dimensional oscillatory integrals and by Kolokoltsov [179] via the method of jump-type Markov processes.

Belavkin's quantum filtering equation was first deduced in full generality in Belavkin [36], see also [37]. Simple deductions of this equation can be found in Kolokoltsov [179] and Belavkin–Kolokoltsov [39] together with rather extensive literature on this subject. Let us also stress that Belavkin's equation can be written in two versions, linear and nonlinear, that correspond to the Kushner–Stratonovich equation and Duncan–Zakai equation in the classical filtering theory, see Bain and Crisan [25].

As our aim here was to illustrate the methods of jump-type Markov processes for the description of quantum systems, we do not review here other (mostly non-probabilistic) approaches to the rigorous definition of the Feynman path integral. A review of the rather extensive literature on this subject can be found in [179]. Let us only mention the recent monograph of S. Mazzucchi [234], which presents an up-to-date introduction to the infinite-dimensional oscillatory integral approach, which was first systematically developed by Albeverio and Hoegh-Krohn [12], and in a slightly modified version by Elworthy and Truman [108].

The definition of dimensionality used in Section 9.5 is taken from Albeverio et al. [6], but similar conditions were used by many authors. Paper Kolokoltsov [185] contains a review of the bibliography about the Schrödinger equation with singular potentials and magnetic fields. A partial extension of the results on the Schrödinger equation with singular magnetic fields to the case of the equations on manifold was given in Brüning, Geyler and Pankrashkin [69].

Section 9.6 is based on [189] and [179].

Bibliography

[1] H. Abels, M. Kassmann. The Cauchy problem and the martingale problem for integro-differential operators with non-smooth kernels. Osaka J. Math. **46:3** (2009), 661–683.

[2] M. Akian, B. David, St. Gaubert. Un théorème de représentation des solutions de viscosité d'une équation d'Hamilton-Jacobi-Bellman ergodique dégénérée sur le tore. (French) C. R. Math. Acad. Sci. Paris **346: 21–22** (2008), 1149–1154.

[3] M. Akian, St. Gaubert, C. Walsh. The max-plus Martin boundary. Doc. Math. **14** (2009), 195–240.

[4] S. Albeverio, Z. Brzezniak, L. Dabrowski. Fundamental solution of the heat and Schrödinger equations with point interaction. J. Funct. Anal. **130** (1995), 220–254.

[5] S. Albeverio, Ph. Combe, R. Hoegh-Krohn, G. Rideau, R. Stora (Eds.). Feynman path integrals. LNP 106, Springer 1979.

[6] S. Albeverio, J. E. Fenstad, R. Hoegh-Krohn, W. Karwowski, T. Lindstrom. Schrödinger operators with potentials supported by null sets. In: Ideas and Methods in Quantum and Statistical Physics. Vol. 2 in Memory of R. Hoegh-Krohn, ed. S Albeverio et al., Cambridge Univ. Press, 1992, 63–95.

[7] S. Albeverio, J. E. Fenstad, R. Hoegh-Krohn, T. Lindstrom. Nonstandard methods in stochastic analysis and mathematical physics. Academic Press, New York, 1986.

[8] S. Albeverio, A. Hilbert, V. N. Kolokoltsov. Transience for stochastically perturbed Newton systems. Stochastics and Stochastics Reports **60** (1997), 41–55.

[9] S. Albeverio, A. Hilbert, V. N. Kolokoltsov. Estimates uniform in time for the transition probability of diffusions with small drift and for stochastically perturbed Newton equations. J. Theor. Prob. **12:3** (1999), 293–300.

[10] S. Albeverio, A. Hilbert, V. N. Kolokoltsov. Sur le comportement asymptotique du noyau associé á une diffusion Dégénéré. C. R. Math. Rep. Acad. Sci. Canada. **22:4** (2000), 151–159.

[11] S. Albeverio, A. Hilbert, E. Zehnder. Hamiltonian systems with a stochastic force: nonlinear versus linear, and a Girsanov formula. Stochastics and Stochastics Reports **39** (1992), 159–188.

[12] S. Albeverio, R. J. Hoegh-Krohn. Mathematical Theory of Feynman Path Integrals. LNM **523**, Springer Verlag, Berlin, 1976.

[13] S. Albeverio, V. N. Kolokoltsov. The rate of escape for some Gaussian Processes and the scattering theory for their small perturbations. Stochastic Analysis and Appl. **67** (1997), 139–159.

[14] S. Albeverio, V. N. Kolokoltsov, O. G. Smolyanov. Représentation des solutions de l'équation de Belavkin pour la mesure quantique par une version rigoureuse de la formule d'intégration fonctionnelle de Menski. C. R. Acad. Sci.Paris, Sér. 1, **323** (1996), 661–664.

[15] S. Albeverio, V. N. Kolokoltsov, O. G. Smolyanov. Continuous Quantum Measurement: Local and Global Approaches. Rev. Math. Phys. **9:8** (1997), 907–920.

[16] D. J. Aldous. Deterministic and stochastic models for coalescence (aggregation and coagulation): a review of the mean-field theory for probabilists. Bernoulli **5:1** (1999), 3–48.

[17] U. T. Alparslan. Exceedance of power barriers for integrated continuous-time stationary ergodic stable processes. Adv. in Appl. Probab. **41:3** (2009), 874–892.

[18] W. J. Anderson. Continuous-Time Markov Chains. Probability and its Applications. Springer Series in Statistics. Springer 1991.

[19] D. Applebaum. Probability and Information. Cambridge University Press, 1996.

[20] D. Applebaum. Lévy Processes and Stochastic Calculus. Cambridge studies in advanced mathematics, v. 93. Cambridge Univ. Press, 2004.

[21] D. Applebaum. Some L^2 properties of semigroups of measures on Lie groups. Semigroup Forum **79:2** (2009), 217–228.

[22] D. Applebaum. Universal Malliavin calculus in Fock and Lévy–Itô spaces. Commun. Stoch. Anal. **3:1** (2009), 119–141.

[23] D. Applebaum. Infinitely Divisible Central Probability Measures on Compact Lie Groups – Regularity, Semigroups and Transition Kernels. To appear in Annals of Probablity 2010.

[24] S. Assing, W. M. Schmidt. Continuous strong Markov processes in dimension one. A stochastic calculus approach. Lecture Notes in Mathematics, v. 1688. Springer-Verlag, Berlin, 1998.

[25] A. Bain, D. Crisan. Fundamentals of stochastic filtering. Stochastic Modelling and Applied Probability, 60. Springer, New York, 2009.

[26] A. Barchielli, A. M. Paganoni. On the asymptotic behaviour of some stochastic differential equations for quantum states. Infin. Dimens. Anal. Quantum Probab. Relat. Top. **6:2** (2003), 223–243.

[27] M. T. Barlow, R. F. Bass, Z.-Q. Chen and M. Kassmann. Non-local Dirichlet forms and symmetric jump processes. Trans. Amer. Math. Soc. **361:4** (2009), 1963–1999.

[28] R. F. Bass. Uniqueness in law for pure jump type Markov processes. Probab. Theory Related Fields **79** (1988), 271–287.

[29] R. F. Bass, K. Burdzy. On pathwise uniqueness for reflecting Brownian motion in $C^{1+\gamma}$ domains. Ann. Probab. **36:6** (2008), 2311–2331.

[30] R. F. Bass, K. Burdzy, Z.-Q. Chen and M. Hairer. Stationary distributions for diffusions with inert drift. Probab. Theory Related Fields **146:1** (2010), 1–47.

[31] R. F. Bass, Z.-Q. Chen. Systems of equations driven by stable processes. Probab. theory Relat. Fields **134** (2006), 175–214.

[32] R. F. Bass, E. Perkins. A new technique for proving uniqueness for martingale problems. Astérisque No. 327 (2009), 47–53.

[33] R. F. Bass, H. Tang. The martingale problem for a class of stable-like processes. Stochastic Process. Appl. **119:4** (2009), 1144–1167.

[34] P. Becker-Kern, M. M. Meerschaert, H.-P. Scheffler. Limit Theorems for Coupled Continuous Time Random Walks. Annals of Probability **32:1B** (2004), 730–756.

[35] V. P. Belavkin. Quantum branching processes and nonlinear dynamics of multi-quantum systems Dokl. Acad. Nauk SSSR (in Russian) **301:6** (1988), 1348–1352.

[36] V. P. Belavkin. Nondemolition measurement, nonlinear filtering and dynamic programming of quantum stochastic processes. In: Modelling and Control of Systems. Proc. Bellman Continuous Workshop, Sophia-Antipolis, 1988. LNCIS **121** (1988), 245–265.

[37] V. P. Belavkin. A new wave equation for a continuous nondemolition measurement. Phys. Lett. A **140** (1989), 355–358.

[38] V. P. Belavkin, V. N. Kolokoltsov. Stochastic Evolutions As Boundary Value Problems. In: Infinite Dimensional Analysis and Quantum Probability, RIMS Kokyuroku **1227** (2001), 83–95.

[39] V. P. Belavkin, V. N. Kolokoltsov. Stochastic evolution as interaction representation of a boundary value problem for Dirac type equation. Infinite Dimensional Analysis, Quantum Probability and Related Fields **5:1** (2002), 61–92.

[40] V. P. Belavkin, V. Kolokoltsov. On general kinetic equation for many particle systems with interaction, fragmentation and coagulation. Proc. Royal Soc. Lond. A **459** (2003), 727–748.

[41] G. Ben Arous. Developpement asymptotique du noyau de la chaleur sur la diagonale. Ann. Inst. Fourier **39:1** (1989), 73–99.

[42] M. Benaïm and J.-Y. Le Boudec. A class of mean field interaction models for computer and communication systems. Performance Evaluation **65** (2008), 823–838.

[43] A. Bendikov. Asymptotic formulas for symmetric stable semigroups. Expo. Math. **12** (1994), 381–384.

[44] V. E. Bening, V. Yu. Korolev, V. N. Kolokoltsov. Limit theorems for continuous-time random walks in the double-array limit scheme. J. Math. Sci. (N. Y.) **138:1** (2006), 5348–5365.

[45] V. E. Bening, V. Yu. Korolev, V. N. Kolokoltsov, V. V. Saenko, V. V. Uchaikin, V. M. Zolotarev. Estimation of parameters of fractional stable distributions. J. Math. Sci. (N. Y.) **123:1** (2004), 3722–3732.

[46] V. Bening, V. Korolev, T. Suchorukova, G. Gusarov, V. Saenko, V. Kolokoltsov. Fractionally Stable Distributions. In: V. Korolev, N. Skvortsova (Eds.) "Stochastic Models of Plasma Turbulence", Moscow State University, Moscow, 2003, p. 291–360 (in Russian). Engl. transl. in V. Korolev, N. Skvortsova (Eds.) "Stochastic Models of Structural Plasma Turbulence", VSP, Boston 2006, p. 175–244.

[47] J. Bennett, J.-L. Wu. Stochastic control of SDEs associated with Lévy generators and application to financial optimization. Front. Math. China **5:1** (2010), 89–102.

[48] A. Bensoussan. Stochastic control by functional analysis methods. North-Holland, 1982.

[49] J. Bertoin. Lévy Processes. Cambridge Tracts in Mathematics, v. 121, Cambridge Univ. Press, 1996.

[50] Ph. Biane, Ph. Bougerol, N. O'Connell. Littelmann paths and Brownian paths. Duke Math. J. **130:1** (2005), 127–167.

[51] K. Bichteler. Stochastic Integration with Jumps. Encyclopedia of Mathematics and Applications. Cambridge Univ. Press 2002.

[52] K. Bichteler, J.-B. Gravereaux, J. Jacod. Malliavin Calculus for Processes with Jumps. Stochastic Monographs v. 2. Gordon and Breach 1987.

[53] P. Biler, M. Guedda, G. Karch. Asymptotic properties of solutions of the viscous Hamilton–Jacobi equation. J. Evol. Equ. **4:1** (2004), 75–97.

[54] P. Biler, G. Karch, W. Woyczyn'ski. Asymptotics for conservation laws involving Lévy diffusion generators. Studia Math. 148 (2001), no. 2, 171–192.

[55] P. Biler, G. Wu. Two-dimensional chemotaxis models with fractional diffusion. Math. Methods Appl. Sci. **32:1** (2009), 112–126.

[56] P. Billingsley. Convergence of Probability Measures. Wiley, NY, 1968.

[57] N. H. Bingham. Maxima of Sums of Random Variables and Suprema of Stable Processes. Z. Warscheinlichkeitstheorie verw. Geb. **26** (1973), 273–296.

[58] N. H. Bingham, R. Kiesel. Risk-neutral valuation. Pricing and hedging of financial derivatives. Second edition. Springer Finance. Springer-Verlag, London, 2004.

[59] J.-M. Bismut, G. Lebeau. The hypoelliptic Laplacian and Ray-Singer metrics. Annals of Mathematics Studies, 167. Princeton University Press, Princeton, NJ, 2008.

[60] H. Bliedtner, W. Hansen. Potential Theory – An Analytic Approach to Balayage. Universitext, Springer-Verlag, Berlin/Heidelberg/New York/Tokyo, 1986.

[61] R. M. Blumenthal, R. K. Getoor. Some Theorems on Stable Processes. Trans. Amer. Math. Soc. **95** (1960), 263–273.

[62] K. Bogdan, K. Burdzy, Z.-Q. Chen. Censored stable processes. Probab. Theory Relat. Fields 127, 89–152 (2003)

[63] J.-M. Bony, Ph. Courrège, P. Priouret. Semi-groupes de Feller sur une variété a bord compacte et problèmes aux limites intégro- différentiels du second ordre donnant lieu au principe du maximum. Ann. Inst. Fourier, Grenoble **18:2** (1968), 369–521.

[64] L. C. Botelho. Lecture notes in applied differential equations of mathematical physics. World Scientific, 2008.

[65] B. Böttcher, R. L. Schilling. Approximation of Feller processes by Markov chains with Lévy increments. Stoch. Dyn. **9:1** (2009), 71–80.

[66] G. Bouch, Ph. L. Bowers. Evaluating Feynman path integrals via the polynomial path family. Math. Phys. Electron. J. **13** (2007), Paper 1, 29 pp.

[67] R. Boylan. Laws of large numbers for dynamical systems with randomly matched individuals. Journal of Economic Theory **57** (1992), 473–504.

[68] L. Brandolese, G. Karch. Far field asymptotics of solutions to convection equation with anomalous diffusion. J. Evol. Equ. **8:2** (2008), 307–326.

[69] J. Brüning, V. Geyler, K. Pankrashkin. Continuity properties of integral kernels associated with Schrödinger operators on manifolds. Ann. Henri Poincaré **8:4** (2007), 781–816.

[70] Yu. D. Burago, V. A. Zalgaller. Geometric Inequalities. Springer, New York, 1988.

[71] R. H. Cameron. A family of integrals serving to connect the Wiener and Feynman integrals. J. Math. and Phys. **39:2** (1960), 126–140.

[72] R. Carbone, F. Fagnola. The Feller property of a class of quantum Markov semigroups. Stochastic models, II (Spanish) (Guanajuato, 2000), 143–158, Aportaciones Mat. Investig., 16, Soc. Mat. Mexicana, México, 2001.

[73] R. Carbone, F. Fagnola. The Feller property of a class of quantum Markov semigroups. II. Quantum probability and infinite dimensional analysis (Burg, 2001), 57–76, QP–PQ: Quantum Probab. White Noise Anal., 15, World Sci. Publ., River Edge, NJ, 2003.

[74] R. A. Carmona, D. Nualart. Nonlinear Stochastic Integrators, Equations and Flows. Stochastic Monographs, v. 6, Gordon and Breach, 1990.

[75] S. Chandrasekhar. Stochastic problems in physics and astronomy. Rev. Modern Phys. **15:1** (1943), 1–89.

[76] A. M. Chebotarev. What is a quantum stochastic equation from the point of view of functional analysis? Math. Notes **71:3–4** (2002), 408–427.

[77] A. M. Chebotarev, A. A. Konstantinov, V. P. Maslov. Probabilistic representation of solutions of the Cauchy problem for Schrödinger, Pauli and Dirac equations. Russian Math. Surv. **45 :6** (1990), 3–24.

[78] A. Chen, H. Zhang. Stochastic monotonicity and duality for continuous time Markov chains with general Q-matrix. Southeast Asian Bull. Math. **23:3** (1999), 383–408.

[79] M. F. Chen. From Markov Chains to Non-Equilibrium Particle Systems. World Scientific, Singapore, 2 nd Ed., 2004.

[80] M. F. Chen, F. Y. Wang. On order-preservation and positive correlations for multidimensional diffusion process. Prob. Th. Rel. Fields **95** (1993), 421–428.

[81] Z.-Q. Chen, P. Kim. Green function estimate for censored stable processes. Probab. Theory Relat. Fields 124, 595–610 (2002)

[82] Z.-Q. Chen, P. Kim, T. Kumagai. Weighted Poincaré inequality and heat kernel estimates for finite range jump processes. Math. Ann. **342:4** (2008), 833–883.

[83] Z.-Q. Chen, P. Kim, R. Song. Two-sided heat kernel estimates for censored stable-like processes. Probab. Theory Related Fields **146:3** (2010), 361–399.

[84] Z.-Q. Chen, T. Kumagai. Heat kernel estimates for stable-like processes on d-sets. Stochastic Process. Appl. **108:1** (2003), 27–62.

[85] Z.-Q. Chen, T. Kumagai. Heat kernel estimates for jump processes of mixed types on metric measure spaces. Probab. Theory Related Fields **140:1** (2008), 277–317.

[86] K. L. Chung. Markov Chains with Stationary Transition Probabilities. Die Grundlehren der mathematischen Wissenschaften in Einzeldarstellungen, Sec. Ed. v. 104. Springer 1967.

[87] K. L. Chung, J. B. Walsh. Markov processes, Brownian motion, and time symmetry. Second edition. Grundlehren der Mathematischen Wissenschaften, 249. Springer, New York, 2005.

[88] J. Conlisk. Monotone Mobility Matrices. Journal of Mathematical Sociology **15** (1990), 173–191.

[89] V. Corradi, R. Sarin. Continuous Approximations of Stochastic Evolutionary Game Dynamics. J. of Economic Theory **94** (2000), 163–191.

[90] Ph. Courrège. Sur la forme integro-différentiélle du générateur infinitésimal d'un semigroupe de Feller sur une variété. Sém. Théorie du Potentiel, 1965–66. Exposé 3.

[91] D. Crisan. Particle approximations for a class of stochastic partial differential equations. Appl. Math. Optim. **54:3** (2006), 293–314.

[92] V. Dardoni. Monotone Mobility Matrices and Income Distribution. Social Choice and Welfare **12** (1995), 181–192.

[93] R. W. R. Darling, J. R. Norris. Differential equation approximations for Markov chains. Probab. Surv. **5** (2008), 37–79.

[94] E. B. Davies. Heat Kernels and Spectral Theory. Cambridge Univ. Press., 1992.

[95] E. B. Davies. Linear operators and their spectra. Cambridge Studies in Advanced Mathematics, 106. Cambridge University Press, Cambridge, 2007.

[96] I. M. Davies, A. Truman, H. Zhao. Huaizhong Stochastic heat and Burgers equations and their singularities. II. Analytical properties and limiting distributions. J. Math. Phys. **46:4** (2005), 043515.

[97] D. Dawson. Measure-Valued Markov Processes. P. L. Hennequin (ed.) Ecole d'Eté de probabilités de Saint-Flour XXI-1991. Springer Lect. Notes Math. v. 1541 (1993), 1–260.

[98] C. Dellacherie, P.-A. Meyer. Probabilities and potential. B. Theory of martingales (translated from French). North-Holland Mathematics Studies, 151. North-Holland Publishing, Amsterdam, 1982.

[99] S. Dharmadhikari, K. Joag-Dev. Unimodality, Convexity, and Applications. Academic Press 1988.

[100] L. Diosi. Continuous quantum measurement and Itô formalism. Phys. Let. A **129** (1988), 419–423.

[101] L. Diosi. Localised solution of a simple nonlinear quantum Langevin equation. Phys. Let. A **132** (1988), 233–236.

[102] J. L. Doob. Stochastic processes. Reprint of the 1953 original. Wiley Classics Library. A Wiley-Interscience Publication. John Wiley and Sons, New York, 1990.

[103] J. Droniou, C. Imbert. Fractal First-Order Partial Differential Equations. Arch. Rational Mech. Anal. **182** (2006), 299–331.

[104] D. Duffie, D. Filipovic, W. Schachermayer. Affine processes and applications in finance. Ann. Appl. Probab. **13:3** (2003), 984–1053.

[105] A. Dvoretsky, P. Erdős. Some problems on random walk in space. Proc. Second Berkley Symposium on Math. Stat. and Probability. Univ. of California Press (1951), 353–367.

[106] E. B. Dynkin. Theory of Markov processes (translated from Russian). Reprint of the 1961 Dover Publications, Mineola, NY, 2006.

[107] K. D. Elworthy, Y. Le Jan, X.-M. Li. On the geometry of diffusion operators and stochastic flows. Lecture Notes in Mathematics, 1720. Springer-Verlag, Berlin, 1999.

[108] K. D. Elworthy, A. Truman. Feynman maps, Cameron–Martin formulae and anharmonic oscillators. Ann. Inst. Henri Poincaré, **41:2** (1984), 115–142.

[109] A. M. Etheridge. An Introduction to Superprocesses. University Lecture Series **20**, AMS, Providence, RI, 2000.

[110] S. N. Ethier, Th. G. Kurtz. Markov Processes – Characterization and Convergence. Wiley Series in Probability and Mathematical Statistics, New York Chicester: Wiley 1986.

[111] W. Feller. An Introduction to Probability. Theory and Applications. Sec. Edition, v. 2., John Wiley and Sons, 1971.

[112] G. Fichera. Sulla equazioni differenziali lineari ellíptico-paraboliche del secondo ordine. Atti Naz. Lincei Mem., Ser 8 **5** (1956), 1–30.

[113] W. H. Fleming, W. M. McEneaney. Max-plus approaches to continuous space control and dynamic programming. Idempotent mathematics and mathematical physics, 145–160, Contemp. Math., 377, Amer. Math. Soc., Providence, RI, 2005.

[114] W. H. Fleming, H. M. Soner. Controlled Markov processes and viscosity solutions. Sec. Ed. Stochastic Modelling and Applied Probability, 25. Springer, New York, 2006.

[115] T. D. Frank. Nonlinear Markov processes. Phys. Lett. A **372:25** (2008), 4553–4555.

[116] B. Franke. The scaling limit behavior of periodic stable-like processes. Bernoulli **21:3** (2006), 551–570.

[117] M. Freidlin. Functional Integration and Partial Differential Equations. Princeton Univ. Press, Princeton, NY 1985.

[118] M. Fukushima, Y. Oshima, M. Takeda. Dirichlet Forms and Symmetric Markov processes. de Gruyter, 1994.

[119] M. G. Garroni, J. L. Menaldi. Second order elliptic integro-differential problems. Research Notes in Mathematics, v. 430. Chapman and Hall, 2002.

[120] J. Gärtner. On the McKean–Vlasov Limit for Interacting Diffusions. Math. Nachrichten **137** (1988), 197–248.

[121] I. M. Gelfand, A. M. Yaglom. Integration in functional spaces and applications in quantum physics. Uspehki Mat. Nauk **11:1** (1956), 77–114. Engl. transl. in J. Math. Phys. **1:1** (1960), 48–69.

[122] M. K. Ghosh, K. S. Kumar. A stochastic differential game in the orthrant. J. Math. Anal. Appl. **265:1** (2002), 12–37.

[123] S. Graf, R. D. Mauldin. A classification of disintegrations of measures. In: Measures and measurable dynamics. Contemp. Math. v. 94, AMS, Providence 1989, 147–158.

[124] X. Guo, O. Hernández-Lerma. Continuous-time Markov decision processes. Theory and applications. Stochastic Modelling and Applied Probability, 62. Springer-Verlag, Berlin, 2009.

[125] M. Hairer, A. Ohashi. Ergodic theory for SDEs with extrinsic memory. Ann. Probab. **35:5** (2007), 1950–1977.

[126] B. Harlamov. Continuous semi-Markov processes. Applied Stochastic Methods Series. ISTE, London; John Wiley and Sons, Hoboken, NJ, 2008.

[127] B. Helffer, F. Nier. Quantitative analysis of metastability in reversible diffusion processes via a Witten complex approach: the case with boundary. Mém. Soc. Math. Fr. (N. S.) **105**, 2006.

[128] L. L. Helms. Introduction to Potential Theory. Huntington, New York, 1975.

[129] O. Hernández-Lerma, J. B. Lasserre. Further topics on discrete-time Markov control processes. Applications of Mathematics (New York), 42. Springer-Verlag, New York, 1999.

[130] C. Hipp, M. Plum. Optimal investment for investors with state dependent income, and for insurers. Finance and Stochastics **7** (2003), 299–321.

[131] J. Hofbauer, K. Sigmund. Evolutionary Games and Population Dynamics. Cambridge University Press, 1998.

[132] W. Hoh. The martingale problem for a class of pseudo differential operators. Mathematische Annalen **300** (1994), 121–147.

[133] W. Hoh, N. Jacob. On the Dirichlet Problem for Pseudodifferential Operators Generating Feller Semigroups. Journ. Funct. Anal. **137:1** (1996), 19–48.

[134] J. Holtzmark. Über die Verbreiterung von Spektrallinien. Annalen der Physik **58** (1939), 577–630.

[135] M. Huang, P. E. Caines, R. P. Malhamé. Large population stochastic dynamic games: closed-loop McKean–Vlasov systems and the Nash certainty-equivalence principle. Commun. Inf. Syst. **6:3** (2006), 221–251.

[136] M. Huang, P. E. Caines, R. P. Malhamé. Large-population cost-coupled LQG problems with nonuniform agents: individual-mass behavior and decentralized ϵ-Nash equilibria. IEEE Trans. Automat. Control **52:9** (2007), 1560–1571.

[137] T. Ichinose. Path integral for a hyperbolic system of the first order. Duke Math. Journ. **51:1** (1984), 1–36.

[138] P. Imkeller, I. Pavlyukevich. First exit times of SDEs driven by stable Lévy processes. Stochastic Process. Appl. **116:4** (2006), 611–642.

[139] Y. Ishikawa. Optimal Control Problem Associated with Jump Processes. Appl. Math. Optim. **50** (2004), 21–65.

[140] S. Itô. Diffusion equations. Translations of Mathematical Monographs, 114. American Mathematical Society, Providence, RI, 1992.

[141] N. Jacob. Further pseudo differential operators generating Feller semigroups and Dirichlet forms. Rev. Mat. Iberoam. **9** (1993), 373–407.

[142] N. Jacob. Pseudo-differential operators and Markov processes. Vol. I, II, III. London: Imperial College Press, 2001, 2002, 2005.

[143] N. Jacob et al. Non-local (semi-)Dirichlet forms generated by pseudo differential operators. In: Z. M. Ma et al. (Eds.). Dirichlet Forms and Stochastic Processes. Proc. Intern. Conf. Beijing 1993, de Gruyter, Berlin, 1995, 223–233.

[144] N. Jacob, A. Potrykus, J.-L. Wu. Solving a nonlinear pseudo-differential equation of Burgers type. Stoch. Dyn. **8:4** (2008), 613–624.

[145] N. Jacob, R. L. Schilling. Lévy-Type Processes and Pseudodifferential Operators. In: O. E. Barndorff-Nielsen et al. (Eds). Lévy Processes, Theory and Applications. Birkhäuser 2001, p. 139–168.

[146] J. Jacod, Ph. Protter. Probability Essentials. Springer 2004.

[147] J. Jacod, A. N. Shiryaev. Limit Theorems for Stochastic Processes. Springer 1987, Sec. Edition 2003.

[148] A. Jakubowski. On the Skorohod topology. Ann. Inst. H. Poincaré. B22 (1986), 263–285.

[149] M. Jeanblanc, M. Yor, M. Chesney. Mathematical Methods for Financial Markets. Springer 2009.

[150] A. Joffe, M. Métivier. Weak convergence of sequence of semimartingales with applications to multitype branching processes. Adv. Appl. Probab. **18** (1986), 20–65.

[151] B. Jourdain, S. Méléard, W. A. Woyczynski. Probabilistic approximation and inviscid limits for one-dimensional fractional conbservation laws. Ann. Appl. Probab. **12:1** (2002), 334–360.

[152] D. Juriev. Belavkin–Kolokoltsov watch-dog effect in interactively controlled stochastic computer-graphic dynamical systems. A mathematical study. E-print (LANL Electronic Archive on Chaos Dyn.): chao-dyn/9406013+9504008 (1994, 1995).

[153] D. Juriev. Belavkin–Kolokoltsov watch-dog effect in interactively controlled stochastic computer-graphic dynamical systems. Theor. Math. Phys. **106:2** (1996), 276–290.

[154] O. Kallenberg. Foundations of Modern Probability. Second ed., Springer 2002.

[155] I. Karatzas, S. Shreve. Brownnian Motion and Stochastic Calculus. Springer 1998.

[156] G. Karch, W. A. Woyczyn'ski. Fractal Hamilton–Jacobi-KPZ equations. Trans. Amer. Math. Soc. **360:5** (2008), 2423–2442.

[157] M. V. Keldys. On certain cases of degeneration of equations of elliptic type on the boundary of a domain. Dokl. Acad. Nauk SSSR **77** (1951), 181–183.

[158] A. Khintchine. Sur la crosissance locale des prosessus stochastiques homogènes à acroissements indépendants. Isvestia Akad. Nauk SSSR, Ser Math. (1939), 487–508.

[159] D. Khoshnevisan. Escape rates for Lévy processes. Studia Sci. Math. Hungar. **33:1–3** (1997), 177–183.

[160] D. Khoshnevisan. Probability. Graduate Studies in Mathematics, 80. American Mathematical Society, Providence, RI, 2007.

[161] D. Khoshnevisan, Zh. Shi. Chung's law for integrated Brownian motion. Trans. Amer. Math. Soc. **350:10** (1998), 4253–4264.

[162] A. Khrennikov. Interpretations of probability. Second edition. Walter de Gruyter, Berlin, 2009.

[163] K. Kikuchi, A. Negoro. On Markov processes generated by pseudodifferential operator of variable order. Osaka J. Math. **34** (1997), 319–335.

[164] A. A. Kilbas, H. M. Srivastava, J. J. Trujillo. Theory and Applications of Fractional Differential Equations. Elsevier, Amsterdam, 2006.

[165] A. N. Kochubei. Parabolic pseudo-differentiable equation, supersingular integrals and Markov processes. Izvestia Akad. Nauk, Ser. Matem. **52:5** (1988), 909–934 (Russian), Math.-USSR Izv. **33:2** (1989), 233–259 (English).

[166] A. N. Kochubei. Distributed order calculus: an operator-theoretic interpretation. Ukrain. Mat. Zh. **60:4** (2008), 478–486. Transl. in Ukrainian Math. J. **60:4** (2008), 551–562.

[167] A. N. Kochubei. Distributed order derivatives and relaxation patterns. J. Phys. A **42:31** (2009), 315203.

[168] J. J. Kohn, L. Nirenberg. Non-Coercive Boundary Value Problems. Comm. Pure Appl. Math. **18** (1965), 443–492.

[169] J. J. Kohn, L. Nirenberg. Degenerate elliptic-parabolic equations of second order. Comm. Pure Appl. Math. **20:3** (1967), 443–492.

[170] V. N. Kolokoltsov. Application of quasi-classical method to the investigation of the Belavkin quantum filtering equation. Mat. Zametki **50** (1993), 153–156. Engl. transl. in Math. Notes.

[171] V. N. Kolokoltsov. Scattering theory for the Belavkin equation describing a quantum particle with continuously observed coordinate. J. Math. Phys. **36:6** (1995) 2741–2760.

[172] V. N. Kolokoltsov. Long time behavior of the solutions of the Belavkin quantum filtering equation. In: V. P. Belavkin, R. Hudson, R. Hirota (Eds). Quantum Communications and Measurements. N. Y., Plenum Press, 1995, p. 429–439.

[173] V. N. Kolokoltsov. Stochastic Hamilton–Jacobi–Bellman equation and stochastic Hamiltonian systems. J. Dynamic Contr. Syst. **2:3** (1996), 299–319.

[174] V. N. Kolokoltsov. A note on the long time asymptotics of the Brownian motion with application to the theory of quantum measurement. Potential Analysis **7** (1997), 759–764.

[175] V. N. Kolokoltsov. Localization and analytic properties of the simplest quantum filtering equation. Rev Math. Phys. **10:6** (1998), 801–828.

[176] V. N. Kolokoltsov. Long time behavior of continuously observed and controlled quantum systems (a study of the Belavkin quantum filtering equation). In: Quantum Probability Communications, QP-PQ, V. 10, Ed. R. L. Hudson, J. M. Lindsay, World Scientific, Singapore (1998), 229–243.

[177] V. N. Kolokoltsov. Complex measures on path space: an introduction to the Feynman integral applied to the Schrödinger equation. Methodology and Computing in Applied Probability **1:3** (1999), 349–365.

[178] V. N. Kolokoltsov. Symmetric Stable Laws and Stable-Like Jump-Diffusions. Proc. London Math. Soc. **3:80** (2000), 725–768.

[179] V. N. Kolokoltsov. Semiclassical Analysis for Diffusions and Stochastic Processes. Springer Lecture Notes in Math. v. 1724, 2000.

[180] V. N. Kolokoltsov. Small diffusion and fast dying out asymptotics for superprocesses as non-Hamiltonian quasi-classics for evolution equations. Electronic Journal of Probability www.math.washington.edu/~ejpecp **6** (2001), paper 21.

[181] V. N. Kolokoltsov. Mathematics of the Feynman path integral. Proc. of the International Mathematical conference FILOMAT 2001, University of Nis, FILOMAT **15** (2001), 293–312.

[182] V. N. Kolokoltsov. A new path integral representation for the solutions of the Schrödinger and stochastic Schrödinger equation. Math. Proc. Cam. Phil.Soc. **132** (2002), 353–375.

[183] V. N. Kolokoltsov. Measure-valued limits of interacting particle systems with k-nary interactions I. Probab. Theory Relat. Fields **126** (2003), 364–394.

[184] V. N. Kolokoltsov. On Extension of Mollified Boltzmann and Smoluchovski Equations to Particle Systems with a k-nary Interaction. Russian Journal of Math. Phys. **10:3** (2003), 268–295.

[185] V. N. Kolokoltsov. On the singular Schrödinger equations with magnetic fields. Matem. Zbornik **194:6** (2003), 105–126.

[186] V. N. Kolokoltsov. Measure-valued limits of interacting particle systems with k-nary interactions II. Stochastics and Stochastics Reports **76:1** (2004), 45–58.

[187] V. N. Kolokoltsov. On Markov processes with decomposable pseudo-differential generators. Stochastics and Stochastics Reports **76:1** (2004), 1–44.

[188] V. N. Kolokoltsov. Hydrodynamic Limit of Coagulation-Fragmentation Type Models of k-nary Interacting Particles. Journal of Statistical Physics **115**, 5/6 (2004), 1621–1653.

[189] V. N. Kolokoltsov. Path integration: connecting pure jump and Wiener processes. In: E. C. Waymire, J. Duan (Eds.). "Probability and Partial Differential Equations in Modern Applied Mathematics", The IMA Volumes in Mathematics and its Applications, v. 140 (2005), p. 163–180.

[190] V. N. Kolokoltsov. Kinetic equations for the pure jump models of k-nary interacting particle systems. Markov Processes and Related Fields **12** (2006), 95–138.

[191] V. N. Kolokoltsov. On the regularity of solutions to the spatially homogeneous Boltzmann equation with polynomially growing collision kernel. Advanced Studies in Contemp. Math. **12** (2006), 9–38.

[192] V. N. Kolokoltsov. Nonlinear Markov Semigroups and Interacting Lévy Type Processes. Journ. Stat. Physics **126:3** (2007), 585–642.

[193] V. N. Kolokoltsov. Generalized Continuous-Time Random Walks (CTRW), Subordination by Hitting Times and Fractional Dynamics. arXiv:0706.1928v1[math.PR] 2007. Probab. Theory and Applications **53:4** (2009).

[194] V. N. Kolokoltsov. The central limit theorem for the Smoluchovski coagulation model. arXiv:0708.0329v1[math.PR] 2007. Prob. Theory Relat. Fields **146:1** (2010), 87–153.

[195] V. N. Kolokoltsov. The Lévy–Khintchine type operators with variable Lipschitz continuous coefficients generate linear or nonlinear Markov processes and semigroups. Probability Theory Related Fields. Online First DOI 10.1007/s00440–010–0293–8.

[196] V. N. Kolokoltsov. Nonlinear Markov processes and kinetic equations. Cambridge University Press, 2010.

[197] V. N. Kolokoltsov. Stochastic monotonicity and duality for one-dimensional Markov processes. arXiv:1002.4773v1[math.PR] 25 Feb 2010. To appear in Mathematical Notes.

[198] V. V. Kolokoltsov, V. Yu. Korolev, V. Uchaikin. Fractional Stable Distributions. J. Math. Sci. (N. Y.) **105:6** (2001), 2570–2577.

[199] V. N. Kolokoltsov, O. A. Malafeev. Understanding Game Theory. Introduction to the Analysis of Many Agent Systems of Competition and Cooperation. World Scientific, 2010. (Russian Edition St. Petersburg Univ. Press, 2008.)

[200] V. N. Kolokoltsov, V. P. Maslov. Idempotent Analysis and its Application to Optimal Control. Moscow, Nauka, 1994 (in Russian).

[201] V. N. Kolokoltsov, V. P. Maslov. Idempotent Analysis and its Applications. Kluwer Publishing House, 1997.

[202] V. N. Kolokoltsov, R. L. Schilling, A. E. Tyukov. Transience and non-explosion of certain stochastic newtonian systems. Electronic Journal of Probability **7** (2002), Paper no. 19.

[203] V. N. Kolokoltsov, R. L. Schilling, A. E. Tyukov. Estimates for Multiple Stochastic Integrals and stochastic Hamilton–Jacobi equations. Revista Matematika Iberoamerikana **20** (2004), 333–380.

[204] V. N. Kolokoltsov, A. Tyukov. The rate of escape of stable Ornstein–Uhlenbeck processes and the scattering theory for their perturbations. Markov Processes and Related Fields **7:4** (2001), 603–626.

[205] V. N. Kolokoltsov, A. E. Tyukov. Small-time and semiclassical asymptotics for stochastic heat equation driven by Lévy noise. Stochastics and Stochastics Reports **75** (1–2) (2003), 1–38.

[206] T. Komatsu. On the martingale problem for generators of stable processes with perturbations. Osaka J. Math. **21** (1984), 113–132.

[207] V. Korolev et al. Some methods of the analysis of time characteristics of catastrophes in non-homogeneous flows of extremal events. In: I. A. Sokolov (Ed.) Sistemi i sredstva informatiki. Matematicheskie modeli v informacionnich technologiach. Moscow, RAN, 2006 (In Russian), p. 5–23.

[208] V. Yu. Korolev, V. E. Bening, S. Ya. Shorgin. Mathematical foundations of the risk theory (in Russian). Moscow, Fismatlit, 2007.

[209] M. Kotulski. Asymptotic Distribution of Continuous-Time Random Walks: a Probabilistic Approach. J. Stat. Phys. **81:3/4** (1995), 777–792.

[210] A. M. Kulik. On a regularity of distribution for solution of SDE of a jump type with arbitrary Lévy measure of the noise. Ukraïn. Mat. Zh. **57:9** (2005), 1261–1283. Engl. transl. Ukrainian Math. J. **57:9** (2005), 1477–1501.

[211] H. Kunita. Stochastic Flows and Stochastic Differential Equations. Cambridge studies in advanced mathematics, v. 24. Cambridge Univ. Press, 1990.

[212] Th. G. Kurtz, Ph. E. Protter. Weak Convergence of Stochastic Integrals and Differential Equations. In: C. Graham et al (Eds). Probability Models for Nonlinear Partial Differential Equations. Springer LNM v. 1627, Springer 1996, 1–41.

[213] H. J. Kushner. Stochastic Stability and Control. Academic Press, New York, London, 1967.

[214] H. J. Kushner, P. Dupuis. Numerical methods for stochastic control problems in continuous time. Second edition. Applications of Mathematics (New York), 24. Stochastic Modelling and Applied Probability. Springer-Verlag, New York, 2001.

[215] A. E. Kyprianou. Introductory Lectures on Fluctuations of Lévy Processes with Applications. Universitext. Springer 2006.

[216] A. Lachal. Local asymptotic classes for the successive primitives of Brownian motion. Ann. Probab. **25:4** (1997), 1712–1734.

[217] V. Lakshmikantham, S. Leela, J. Vasundhara Devi. Theory of Fractional Dynamic Systems. Cambridge Academic Publishers, Cambridge, 2009.

[218] G. Q. Lang. Stochastic monotonicity and positive correlations of a type of particle systems on Polish spaces (Chinese). Acta Math. Sinica (Chin. Ser.) **52:2** (2009), 309–314.

[219] R. Léandre. Uniform upper bounds for hypoelliptic kernels with drift. J. Math. Kyoto Univ. **34:2** (1994), 263–271.

[220] M. A. Leontovich. Main equations of the kinetic theory from the point of view of random processes. Journal of Experimental and Theoretical Physics (in Russian) **5** (1935), 211–231.

[221] A. Lipton, A. Rennie (Editors). Credit correlation: life after copulas. World Scientific, 2008.

[222] R. Sh. Liptser, A. N. Shiryayev. Theory of martingales (translated from the Russian). Mathematics and its Applications (Soviet Series), 49. Kluwer Academic Publishers Group, Dordrecht, 1989.

[223] V. S. Lisitsa. Stark broadening of hydrogen lines in plasmas. Usp. Fiz. Nauk **122** (1977), 449–495.

[224] G. L. Litvinov. The Maslov dequantization, and idempotent and tropical mathematics: a brief introduction. (Russian) Zap. Nauchn. Sem. S.-Peterburg. Otdel. Mat. Inst. Steklov. (POMI) 326 (2005), Teor. Predst. Din. Sist. Komb. i Algoritm. Metody. 13, 145–182. Engl. transl. J. Math. Sci. (N. Y.) **140:3** (2007), 426–444.

[225] Z.-M. Ma, M. Röckner. Introduction to the Theory of Non-Symmetric Dirichlet forms, Springer-Verlag, 1992.

[226] E. Maasoumi. On Mobility. In: A. Ullah, D. E. A. Giles (Eds.) Handbook of Applied Economic Statistics. Marcel Dekker, New York, 1998, p. 119–175.

[227] F. Mainardi, M. Raberto, R. Gorenflo, E. Scalas. Fractional calculus and continuous-time finance II: the waiting time distribution. Physica A **287** (2000), 468–481.

[228] P. Mandl. Analytic Treatment of One-dimensional Markov Processes. Springer Verlag 1968.

[229] N. Martin, J. England. Mathematical Theory of Entropy. Addison-Wesley, 1981.

[230] R. H. Martin. Nonlinear operators and differential equations in Banach spaces. New York, 1976.

[231] V. P. Maslov. Perturbation Theory and Asymptotical Methods. Moscow State University Press, 1965 (in Russian). French Transl. Dunod, Paris, 1972.

[232] V. P. Maslov. Complex Markov Chains and Functional Feynman Integral. Moscow, Nauka, 1976 (in Russian).

[233] V. P. Maslov, A. M. Chebotarev. Processus à sauts et leur application dans la mécanique quantique, in [5], p. 58–72.

[234] S. Mazzucchi. Mathematical Feynman Path Integral and their Applications. World Scientific, 2009.

[235] H. P. McKean. A class of Markov processes associated with nonlinear parabolic equations. Proc. Nat. Acad. Sci. **56** (1966), 1907–1911.

[236] M. M. Meerschaert, E. Nane, P. Vellaisamy. Fractional Cauchy problems on bounded domains. The Annals of Probability **37:3** (2009), 979–1007.

[237] M. M. Meerschaert, E. Scala. Coupled continuous time random walks in finance. Physica A **370** (2006), 114–118.

[238] M. M. Meerschaert, H.-P. Scheffler. Limit Distributions for Sums of Independent Random Vectors. Wiley Series in Probability and Statistics. John Wiley and Son, 2001.

[239] M. M. Meerschaert, H.-P. Scheffler. Limit Theorems for Continuous-Time Random Walks with Infinite Mean Waiting Times. J. Appl. Prob. **41** (2004), 623–638.

[240] J.-L. Menaldi, M. Robin. Construction and control of reflected diffusion with jumps. Stochastic differential systems (Marseille-Luminy, 1984), 309–322, Lecture Notes in Control and Inform. Sci., 69, Springer, Berlin, 1985.

[241] J.-L. Menaldi, M. Robin. Reflected diffusion processes with jumps. Ann. Probab. 13 (1985), no. 2, 319–341.

[242] M. B. Menski. The difficulties in the mathematical definition of path integrals are overcome in the theory of continuous quantum measurements. Teoret. i Matem. Fizika **93:2** (1992), 264–272. Engl. transl. in Theor. Math. Phys.

[243] R. Metzler, J. Klafter. The Random Walk's Guide to Anomalous Diffusion: A Fractional Dynamic Approach. Phys. Rep. **339** (2000), 1–77.

[244] P. A. Meyer. Quantum Probability for Probabilists. Springer Lecture Notes Math. **1538**, Springer-Verlag 1991.

[245] A. Mijatovic, M. Pistorius. Continuously monitored barrier options under Markov processes. arXiv:0908.4028v1, August 2009.

[246] T. Mikami. A Uniqueness Result on Cauchy Problem Related to Jump-type Markov Processes with Unbounded Characteristics. Stochastic Analysis and Appl. **20:2** (2002), 389–413.

[247] K. S. Miller, B. Ross. An Introduction to the Fractional Differential Equations. Wiley, New York, 1993.

[248] E. W. Montroll, G. H. Weiss. Random Walks on Lattices, II. J. Math. Phys. **6** (1965), 167–181.

[249] C. Mora, R. Rebolledo. Basic properties of nonlinear stochastic Schrödinger equations driven by Brownian motions. Ann. Appl. Probab. **18:2** (2008), 591–619.

[250] H. Morimoto. Stochastic Control and Mathematic Modeling: Applications in Economics. Encyclopedia of Mathematics and its applications. Cambridge Univ. Press, 2010.

[251] M. Murata. Integral representations of nonnegative solutions for parabolic equations and elliptic Martin boundaries. Journ. Funct. Anal. **245** (2007), 177–212.

[252] A. D. Neate, A. Truman. A one dimensional analysis of turbulence and its intermittence for the d-dimensional stochastic Burgers equation. Markov Process. Related Fields **13:2** (2007), 213–238.

[253] A. Negoro. Stable-like processes: construction of the transition density and the behavior of sample paths near $t = 0$. Osaka J. Math. **31** (1994), 189–214.

[254] E. Nelson. Feynman integrals and the Schrödinger equation. J. Math. Phys. **5:3** (1964), 332–343.

[255] J. Norris. Cluster Coagulation. Comm. Math. Phys. **209** (2000), 407–435.

[256] J. Norris. Markov Chains. Cambridge Univ. Press, 1998.

[257] D. Nualart. The Malliavin Calculus and Related Topics. Probability and its Applications. Springer. Sec. Edition 2006.

[258] D. Nualart. Malliavin Calculus and its Applications. CBMS Regional Conference Series in Mathematics, v. 110, AMS 2009.

[259] B. K. Oksendal. Stochastic Differential Equations: An Introduction with Applications. Springer, Berlin, 2003.

[260] B. K. Oksendal, A. Sulem. Applied stochastic control of jump diffusions. Second edition. Universitext. Springer, Berlin, 2007.

[261] O. A. Oleinik, E. V. Radkevich. Second Order Equations with Nonnegative Characteristic Form. Itogi Nauki Moscow, 1971 (in Russian); English transl., AMS, Providence, RI and Plenum Press, New York 1973.

[262] M. Ondreját. Existence of global mild and strong solutions to stochastic hyperbolic evolution equations driven by a spatially homogeneous Wiener process. J. Evol. Equ. **4:2** (2004), 169–191.

[263] R. E. Paley, N. Wiener. Fourier transforms in the complex domain. Amer. Math. Soc. Coll. Publ., v. 19, 1934.

[264] K. R. Parthasarathy. An Introduction to Quantum Stochastic Calculus. Birkhäuser Verlag, Basel, 1992.

[265] R. G. Pinsky. Positive Harmonic Functions and Diffusion. Cambridge studies in advanced mathematics, v. 45. Cambridge Univ. Press, 1995.

[266] N. I. Portenko, S. I. Podolynny. On multidimensional stable processes with locally unbounded drift. Random Oper. and Stoch. Eq. **3:2** (1995), 113–124.

[267] M. Raberto, E. Scalas, F. Mainardi. Waiting times and returns in high-frequency financial data: an empirical study. Physica A **314** (2002), 749–755.

[268] S. Rachev, L. Rüschendorf. Mass Transportation Problems, v. I, II. Springer, 1998.

[269] S. Ramasubramanian. Lectures on Insurance Models. Hindustan Book Agency, 2009.

[270] L. Rass, J. Radcliffe. Spatial Deterministic Epidemics. Mathematical Surveys and Monographs, v. 102. AMS 2003.

[271] R. Rebolledo. La methode des martingales appliquée à l'etude de la convergence en loi de processus. (French) Bull. Soc. Math. France Mem. No. **62**, 1979.

[272] R. Rebolledo. Sur l'existence de solutions à certains problèmes de semimartingales. (French) C. R. Acad. Sci. Paris A-B **290** (1980), no. 18, A843–A846.

[273] M. Reed, B. Simon. Methods of Modern Mathematical Physics, v. 1, Functional Analysis. Academic Press, N. Y. 1972.

[274] M. Reed, B. Simon. Methods of Modern Mathematical Physics, v. 2, Harmonic Analysis. Academic Press, N. Y. 1975.

[275] M. Reed, B. Simon. Methods of Modern Mathematical Physics, v. 3, Scattering Theory. Academic Press, N. Y. 1978.

[276] M. Reed, B. Simon. Methods of Modern Mathematical Physics, v. 4, Analysis of Operators. Academic Press, N. Y. 1978.

[277] D. Revuz. Markov Chains. North-Holland Publishing Company, 1967.

[278] D. Revuz, M. Yor. Continuous Martingales and Brownian Motion. Springer 1999.

[279] A. Richou. Ergodic BSDEs and related PDEs with Neumann boundary conditions. Stochastic Process. Appl. **119:9** (2009), 2945–2969.

[280] L. C. Rogers, D. Williams. Diffusions, Markov processes, and martingales. Vol. 1. Foundations. Reprint of the second (1994) edition. Cambridge University Press, Cambridge, 2000.

[281] L. C. Rogers, D. Williams. Diffusions, Markov processes, and martingales. Vol. 2. Itô calculus. Reprint of the second (1994) edition. Cambridge University Press, Cambridge, 2000.

[282] R. Rudnicki, R. Wieczorek. Fragmentation-coagulation models of phytoplankton. Bull. Pol. Acad. Sci. Math. **54:2** (2006), 175–191.

[283] L. Sabatelli, S. Keating, J. Dudley, P. Richmond. Waiting time distributions in financial markets. Eur. Phys. J. B **27** (2002), 273–275.

[284] A. I. Saichev, W. A. Woyczynski. Distributions in the Physical and Engineering Sciences. Birkhäuser, Boston, 1997, v. 1.

[285] A. I. Saichev, G. M. Zaslavsky. Fractional kinetic equations: solutions and applications. Chaos **7:4** (1997), 753–764.

[286] S. G. Samko, A. A. Kilbas, O. A. Marichev. Fractional Integrals and Derivatives and Their Applications. Nauka i Teknika, Minsk, 1987. Engl. transl. by Harwood Academic.

[287] G. Samorodnitski, M. Taqqu. Stochastic monotonicity and slepian-type inequalities for infinitely divisible and stable random vectors. Ann. Probab **21:1** (1993), 143–160.

[288] G. Samorodnitski, M. S. Taqqu. Stable non-Gaussian Random Processes, Stochastic Models with Infinite Variance. Chapman and Hall, N. Y., 1994.

[289] K. Sato. Lévy processes and infinitely divisible distributions. Cambridge Studies in Advanced Mathematics, 68. Cambridge University Press, Cambridge, 1999.

[290] R. L. Schilling. Measures, integrals and martingales. Cambridge University Press, New York, 2005.

[291] R. Schneider. Convex Bodies: The Brunn–Minkowski Theory. Cambridge Univ. Press, 1993.

[292] M. Sharpe. General theory of Markov processes. Pure and Applied Mathematics, 133. Academic Press, Inc., Boston, MA, 1988.

[293] A. N. Shiryayev. Probability. Springer 1984.

[294] V. M. Shurenkov. Ergodic Markov processes (in Russian). Probability Theory and Mathematical Statistics, v. 41. Nauka, Moscow, 1989.

[295] A. V. Skorohod. Stochastic equations for complex systems. Transl. from Russian. Mathematics and its Applications (Soviet Series), 13. D. Reidel Publishing Co., Dordrecht, 1988.

[296] H. M. Soner. Optimal control of jump-Markov processes and viscosity solutions. In: Stochastic differential systems, stochastic control theory and applications (Minneapolis, Minn., 1986), 501–511, IMA Vol. Math. Appl., 10, Springer, New York, 1988.

[297] R. Song, Z. Vondracek. Harnack inequality for some classes of Markov processes. Math. Z. **246:1** (2004), 177–202.

[298] F. Spitzer. Principles of Random Walks. Van Nostrand, Princton, 1964.

[299] H. Spohn. Large Scaling Dynamics of Interacting Particles. Springer-Verlag 1991.

[300] D. W. Stroock. Diffusion Processes Associated with Lévy Generators. Z. Wahrscheinlichkeitstheorie verw. Gebiete **32** (1975), 209–244.

[301] D. W. Stroock. An introduction to the Analysis of Paths on a Riemannian Manifold. Mathematical Surveys and Monographs, v. 74. American Mathematical Society, RI, 2000.

[302] D. W. Stroock. Markov Processes from K. Itô's Perspective. Annals of Mathematics Studies. Princeton University Press 2003.

[303] D. Stroock, S. R. S. Varadhan. On Degenerate Elliptic-Parabolic Operators of Second Order and Their Associated Diffusions. Comm. Pure Appl. Math. **XXV** (1972), 651–713.

[304] D. W. Stroock, S. R. S. Varadhan. Multidimensional Diffusion Processes. Springer, Berlin, 1979.

[305] D. Talay, L. Tubaro (Eds.) Probabilistic Models for Nonlinear Partial Differential Equations. Montecatini Terme, 1995. Springer LNM v. 1627, 1996.

[306] K. Taira. On the Existence of Feller Semigroups with Boundary Conditions. Memoirs of the American Mathematical Society, v. 99, Number 475, 1992.

[307] K. Taira. On the existence of Feller semigroups with Dirichlet conditions. Tsukuba J. Math. **17** (1993), 377–427.

[308] K. Taira. Boundary value problems for elliptic pseudo-differential operators II. Proc. Royal Soc. Edinburgh **127 A** (1997), 395–405.

[309] K. Taira, A. Favini, S. Romanelli. Feller semigroups and degenerate elliptic operators with Wentzell boundary conditions. Studia Mathematica **145:1** (2001), 17–53.

[310] A. Truman, J. L. Wu. On a stochastic nonlinear equation arising from 1D integro-differential scalar conservation laws. J. Funct. Anal. **238:2** (2006), 612–635.

[311] V. V. Uchaikin. Montroll–Weisse Problem, Fractional Equations and Stable Distributions. Intern. Journ. Theor. Phys. **39:8** (2000), 2087–2105.

[312] T. Uemura. On symmetric stable-like processes: some path properties and generators. J. Theoret. Probab. **17** (2004), 541–555.

[313] T. Uemura. A remark on non-local operators with variable order. Osaka J. Math. **46:2** (2009), 503–514.

[314] C. Villani. Topics in Optimal Transportation. Graduate Studies in Mathematics 58, AMS 2003.

[315] C. Wang. On estimates of the density of Feynman–Kac semigroups of α-stable-like processes. J. Math. Anal. Appl. **348:2** (2008), 938–970.

[316] J.-M. Wang. Stochastic Comparison and Preservation of Positive Correlations for Lévy-type processes. Acta Math. Sinica, English Series, 25(2009), 741–758.

[317] W. Whitt. Stochastic-Processes Limits. Springer 2002.

[318] E. T. Whittaker, G. N. Watson. Modern Analysis. Third edition. Cambridge Univ. Press 1920.

[319] L. Wu. Large and moderate deviations and exponential convergence for stochastic damping Hamiltonian systems. Stochastic Process. Appl. **91:2** (2001), 205–238.

[320] M. Yamazato. Unimodality of infinitely divisible distribution functions of class L. Ann. Prob. **6** (1978), 523–531.

[321] G. G. Yin, Q. Zhang. Discrete-time Markov chains. Two-time-scale methods and applications. Applications of Mathematics (New York), 55. Stochastic Modelling and Applied Probability. Springer-Verlag, New York, 2005.

[322] J. Yong, X. Y. Zhou. Stochastic Controls. Hamilton Systems and HJB equations. Springer 1999.

[323] K. Yosida. Functional Analysis. Springer 1980.

[324] G. M. Zaslavsky. Fractional kinetic equation for Hamiltonian chaos. Physica D **76** (1994), 110–122.

[325] Y. H. Zhang. Sufficient and necessary conditions for stochastic comparability of jump processes. Acta Mathematica Sinica, English Series, 2000, **16:1** (2000), 99–102.

[326] P. Zhuang, F. Liu, V. Anh, I. Turner. Numerical methods for the variable-order fractional advection-diffusion equation with a nonlinear source term. SIAM J. Numer. Anal. **47:3** (2009), 1760–1781.

[327] V. M. Zolotarev. One-dimensional Stable Distributions. Moscow, Nauka, 1983 (in Russian). Engl. transl. in vol. 65 of Translations of Mathematical Monographs AMS, Providence, Rhode Island, 1986.

Index

Bei Fragen zur Produktsicherheit wenden Sie sich bitte an:
If you have any questions regarding product safety,
please contact:

Walter de Gruyter GmbH
Genthiner Straße 13
10785 Berlin
productsafety@degruyterbrill.com